STUDIES OF HIGH TEMPERATURE SUPERCONDUCTORS

VOLUME 4

STUDIES OF HIGH TEMPERATURE SUPERCONDUCTORS

Advances in Research and Applications

Edited by Anant Narlikar
National Physical Laboratory
New Delhi

VOLUME 4

Nova Science Publishers
New York•Budapest

Nova Science Publishers, Inc.
283 Commack Road
Suite 300
Commack, New York 11725

Library of Congress Cataloging-in-Publication Data available upon request

ISBN 0-941743-57-8

Graphic Design by Elenor Kallberg and Peggy Harvey

Printed in the United States of America

Contributors to Volume 4

A. Balzarotti, Universita di Roma, "Tor Vergeta", Roma, Italy

P. Boolchand,University of Cincinnati, Cincinnati, Ohio, U.S.A.

A.K. Ganguli, Indian Institute of Science, Bangalore, India

A. Jezowski, Polish Academy of Sciences, Wroclaw, Poland

J. Keinonen, University of Helsinki, Helsinki, Finland

J. Klamut, Polish Academy of Sciences, Wroclaw, Poland

V. Kovachev, Bulgarian Academy of Sciences, Sofia, Bulgaria

K. Kuroda, NTT LSI Laboratories, Kanagawa, Japan

J. Labbe, Group de Physique des Solides de Ecole Normale Superieure, Paris, France

V. Manivannan, Indian Institute of Science, Bangalore, India

A. Marcelli, INFN-Laboratori Nazionali de Frascati, Frascati, Italy

D. McDaniel, University of Cincinnati, Cincinnati, Ohio, U.S.A.

S. Miyazawa, NTT LSI Laboratories, Kanagawa, Japan

M. Mukaida, NTT LSI Laboratories, Kanagawa, Japan

A. Nakamura, Atomic Energy Research Institute, Naka-gun, Ibaraki, Japan

K. Nakamura, National Research Institute for Metals, Tsukuba, Ibaraki, Japan

M. Nevriva, Czechoslovak Academy of Sciences, Prague, Czechoslovakia

K. Ogawa, National Research Institute for Metals, Tsukuba, Ibaraki, Japan

J. Raisanen, University of Helsinki, Helsinki, Finland

C.N.R. Rao, Indian Institute of Science, Bangalore, India

E. Rauhala, University of Helsinki, Helsinki, Finland

D. Schmeltzer, Bar-Ilan University, Israel

M. Takano, Kyoto University, Kyoto, Japan

Y. Tazoh, NTT LSI Laboratories, Kanagawa, Japan

N.Y. Vasanthacharya, Indian Institute of Science, Bangalore, India

R. Vijayaraghavan, Indian Institute of Science, Bangalore, India

E. Vlakhov, Institute of Solid State Physics, Bulgarian Academy of Sciences, Sofia, Bulgaria

ARTICLES PLANNED FOR FUTURE VOLUMES

Tunneling Spectroscopy of High Tc Superconductors *(J. Akimitsu and T. Ekino)*

Polaron Theory of Strong Coupling High Tc Superconductors *(A.S. Alexandrov)*

NMR Studies of the Magnetic Properties of 1-2-3 Compounds *(H. Alloul)*

Electron Microscopic Studies of High Tc Compounds *(S. Amelinckx)*

Relation between Superconductivity and Magnetism with Special Attention to the Problem of Phase Diagram and Magnetic Structures *(G. Collin and D. Petigrand)*

Flux Flow and Resistive Transition in Bulk Ceramic Superconductors *(J.E. Evetts)*

Sputter Deposition of Thin Film Ceramic Superconductors *(J.E. Evetts, R.E. Somekh and B.A. Glowacki)*

Diffusive Phenomena and Microstructure in HTSCs *(J.E. Evetts, R.E. Somekh and B.A. Glowacki)*

Relaxation Properties and Critical Current on Bi-Sr-Ca-Cu-O Single Crystals *(D. Fiorani)*

Processing and Structural Effects in High Tc Materials *(J.M. Gonzalez-Calbet, M. Vallet-Regi and X. Obradors)*

Experimental Tests of High-Tc Mechanisms *(J.B. Goodenough and A. Manthiram)*

Studies of Tl-based Cuprates *(M. Greenblatt)*

The Materials Design for Very High Tc Superconductors *(H. Ihara)*

Lower Crictical Field of Polycrystalline High Tc Copper Oxides *(T. Ishii)*

Formation Processes of Polycrystalline High Tc Copper Oxides *(Takao Itoh and H. Uchikawa)*

Novel Processing Techniques for High Tc Superconductors *(M. Jergel)*

Organic Superconductors-Prospects for High Tc *(D. Jérome)*

Isotope Effect in High Tc Oxide Superconductors *(H. Katayama Yoshida and T. Suzuki)*

Flux Pinning and Thermally Activated Depinning in Single Crystals of HTSC *(P.H. Kes and J. van den Berg)*

Effect of Magnetic Rare-Earth Dopings on YBCO and Related Systems *(J. Klamut and A.J. Zaleski)*

Electric Field Effect on Epitaxial YBCO Thin Films *(T. Kobayashi and collaborators)*

Studies of Thin Films of Thallium Based High Tc Cuprates *(J.F. Kwak)*

Flux Phases and Superconductivity *(P. Lederer)*

Preparation and Properties of Superconducting Ceramics and Films in Unsubstituted and Partial Bi-Substituted Bi-Sr-Ca-Cu-O Systems *(G. Marbach, M. Klee, H. Passing, S. Stotz and J.W.C. de Vries)*

High Resolution TEM Study of Modulated Structures in Bi-based Superconductors *(Y. Matsui, H. Maeda Y. Tanaka and S. Horiuchi)*

Studies of HTSCs using EELS *(N. Nucker)*

Critical Current Problem of High Tc Superconductors *(V.M. Pan)*

Oxygen in 1-2-3 Compounds *(V.M. Pan, V.S. Melnikov and A.V. Zhalko-Titarenko)*

Response Function Characteristics of Pairing Correlation of New High Tc Superconductors *(A.K. Rajagopal, W. Jin and S.D. Mahanti)*

High Tc Superconductors *(T.V. Ramakrishnan)*

Microstructure-Property Relationships in Ceramic Superconductors *(R. Ramesh, S.M. Green and G. Thomas)*

Field Ion Microscopic Studies of High Temperature Superconductors *(S. Ranganathan)*

Strong Coupling Interaction in HTSCs *(D.K. Ray)*

Electromagnetic Behaviour of High Temperature Superconductors *(G.J. Russell, K.N.R. Taylor, D.N. Mathews and G. Alvarez)*

Film-Substrate Interaction in High Tc Superconducting Thin Films *(M. Sergent and A. Perrin)*

ARTICLES PLANNED FOR FUTURE VOLUMES
(Continued)

Fluorination of High Tc Superconductors *(M. Sergent and C. Perrin)*

Vortex Pinning, Lower Critical Fields and Microwave Properties of YBa2Cu3O7 Thin Films *(C. Schlenker, P.L. Reydet, C. Escribe-Filippini, J. Marcus, R. Budor, J. Dumas, Ch.J. Liu, and S. Revenaz)*

Magnetic Properteis of R2CuO4 Materials (R: rare earth) *(S. Schultz and collaborators)*

Some Model Mechanisms for High Temperature Superconductors *(K.P. Sinha)*

Neutron Scattering Studies of Magnetism in High Tc Materials *(S.K. Sinha)*

Electronic and Magnetic Structure of High Tc Cuprates and Nickelates *(V.H. Smith, Jr. and S. Szpunar)*

Electronic Phase Transitions and Exchange Induced Pairing in Correlated Systems: Fermi Liquids Metal-Insulator Transitions and High Tc Superconductivity *(J. Spalek and J.M. Honig)*

Studies of Superconducting Energy Gap in High Tc Superconductors *(R. Srinivasan)*

Carrier Dependent Superconductivity in Cuprates *(R. Suryanarayanan)*

Field Penetration into High Tc Superconductors *(S. Takacs, F. Gomory and P. Lobotky)*

High Tc Superconductors-Crystal Growth, Characterization and Properties *(K.N.R. Taylor)*

Anisotropic Properties of High Tc Single Crystals *(J.L. Tholence and collaborators)*

Contents of Volume 1

1. Oxygen Hole Mechanism of Superconductivity in Cuprates and other Metal Oxides — C.N.R. Rao
2. Electronically Driven Instability in High Tc Superconductors and Related Materials — K.Machida
3. The BCS Pairing and Possible Mechanisms for High Tc Superconductivity — S.S. Jha
4. Boson Exchange Mechanisms, Bounds and Asymptotic Limits — J.P. Carbotte and F.Marsiglio
5. A Strong Coupling Approach to High Temperature Superconductors — J. Rammer
6. Ion Beam Modification and Analysis of Thin YBaCuO Films — O. Meyer
7. Superconducting Properties Associated with Short Coherence Length - Fluctuation Effect and Flux Creep Phenomenon in HTSC — Y. Iye
8. Basic Thin Film Processing for High Tc Superconductors — K.Wasa, H. Adachi, Y.K. Setsune and K. Hirochi
9. Three Terminal High Tc Superconducting Devices with Large Current Gains — T. Kobayashi and U. Kabasawa
10. Twins in High Tc YBaCuO Superconductors — C.J. Jou and J. Washburn
11. The Role of Mott-insulation, Nonstoichiometry and Altered Valence in Oxide Superconductors — G.J. Hyland
12. Sound Velocity and Elastic Constants in Oxide Superconductors — R. Srinivasan
13. X-Ray Photoelectron Spectroscopic Studies of High Tc Oxide Superconductors — B.D. Padalia
14. Synthesis of High Tc oxide Superconductors in Y-, Bi- and Tl- Systems: The Role of Chemistry — R.M. Iyer and J.V. Yakhmi
15. Substitutional Studies of High Temperature Superconductors — A.V. Narlikar, C.V. Narasimha and S.K. Agarwal

Contents of Volume 2

1. Structural Chemistry of High Tc Superconductors — B. Raveau, C. Michel, and M. Hervieu
2. High Pressure Studies — R.J. Wijngaarden and R. Griessen
3. Nuclear Resonance Studies in YBaCuO and TlBaCaCuO — R. Vijayaraghavan, A.K. Rajarajan and L.C. Gupta
4. Thermal Properties of High Temperature Superconductors — E. Gmelin
5. Phonons in High Tc Superconductors — S. Mase
6. A Review of the Preparation and Structure of Bi-Sr-Ca-Cu-O Superconductors and Pb-Substituted Phases — P. Bordet, J.J. Capponi, C. Chaillout, J. Chenavas, A.W. Hewat, J.L. Hodeau and M. Marezio
7. Anisotropic Superconducting and Normal State Transport Properties of HTSC Single Crystals — Y. Iye

8. Multiple Heteroepitaxial Growth of High Tc Superconductors — T. Kobayashi
9. Coprecipitation Route to Ceramic Superconductors — P. Pramanik, B.K. Raul, S. Chakrabarti, S. Biswas, D. Bhattacharya and K.L. Chopra
10. Microwave Absorption Studies in High Tc Superconducting Oxides — M.D. Sastry
11. X-Ray Spectroscopic Study of High Temperature Superconductors — K.B. Garg
12. Normal and Superconducting Properties of the New High Tc Oxides and the Phonon-Plasmon Mechanism of Superconductivity — V.Z. Kresin and H. Morawitz
13. Magnetic Pairing and Superconductivity — K. Huang
14. Low Dimensionality Aspects of High Temperature Superconductors — S.N. Behera
15. Theory of High Temperature Superconductivity — C.M. Srivastava

Contents of Volume 3

1. Charge-Fluctuation Excitaion as Mechanisms for High Tc — J. Ashkenazi and C.G. Kuper
2. High Resolution Electron Microscopy Study of High Tc Superconductors — M.Hiervieu, C. Michel and B. Raveau
3. Synthesis of High Temperature Superconducting Films-Processing Science and Issues — C. Deshpandey and R.F. Bunshah
4. Effect of Oxygen Ordering on the Superconductivity of YBCO — Y. Hariharan, C.S. Sundar, J. Janaki, A.K. Sood, M.P. Janawadkar, A. Bharathi, V. Sankar Sastry, R. Baskaran and T.S. Radhakrishnan
5. On the Architecture of High Tc Superconductors — A.P.B. Sinha
6. Thermally Activated Magnetic Relaxation in High Tc Superconductors — C.W. Hagen and R. Griessen
7. Magnetization of Single Crystalline and Grain Aligned High Tc Superconductors — H.W. Weber
8. Synthesis of Bulk Sintered Samples Processing and Crystal Growth of YBCO — U.V. Vardaraju and G.V. Subba Rao
9. Superconducting Weak Links in High Tc Superconductors — M. Akinaga and L. Rinderer
10. Ginzburg Landau Theory for High Tc Superconductors — M.P. Das
11. BCS-like Approach to HTSC with Emphasis on Structural Transformation — R. Sridhar
12. Vibrational Spectra and Transport Properties of High Temperature Superconductors — H. Kuzmany, E. Faulques, M. Matus and S. Pakker
13. Twins in High Tc Superconductors — H.A. Hoff and C.S. Pande
14. Structure, Microstructure and Superconductivity on the Thallium-based Oxide Superconductors — Z. Iqbal, J.C. Barry, and B.L. Ramakrishna
15. Microstructural Characteristics and Structure Property Correlations in High Tc Superconductors — O.N. Srivastava

PREFACE

With the discovery of high T_c ceramic compounds, superconductivity has evolved into a rich and highly competitive field of research of not just condensed matter physics and chemistry but also of diverse engineering disciplines like ceramics, metallurgy and microelectronics. Clearly, significant advances in research and applications of HTSCs are expected only through consolidated efforts requiring the best possible awareness on the part of researchers in these widely differing disciplines. The "Studies of High Temperature Superconductors" looks at the frontal problems and challenges through detailed reviews and extended articles covering fundamental properties, characterization and applications of the new high T_c superconductors. The series is aimed at the professional scientist and engineer, as well as at graduate students in physics, chemistry, materials science, solid state electronics and engineering.

In this fourth volume of the "Studies" the articles presented are sufficiently diverse to project many of the unusual facets associated with the broad spectrum of HTSC studies.

Although the basic question of the origin of high T_c has remained disputed, the novel theories and mechanisms of superconductivity form a prominent feature of the present series. Interestingly, the reader will notice two widely differing theoretical approaches appearing in this volume. In Chapter-1, Labbe presents convincing reasons for the unique structural features of HTSC to make the conventional BCS weak-coupling theory a viable choice for the explanation of high Tc and other related properties. In contrast to this is the RVB approach, originated by P.W. Anderson, developed in Chapter-8 by Schmeltzer. A. Nakamura, in Chapter-12, describes superconductivity by what he calls multi-valence resonance condensate.

One of the exciting findings of recent months has been the identification of the electron-type superconductors, different from the so-called hole-type, in the 2-1-4 system. The idea has been well exploited by Rao and his group to form a series of interesting samples of thallium-based cuprates, in Chapter-2, where the sign of the charge carriers can be altered by changing the composition. Bismuth based cuprates are perhaps the most extensively studied compounds today. Their structural chemistry, the effect of lead substitution, thin film and bulk processing, their transport properties etc., are discussed by Takano (Chapter-3), Nakamura and Ogawa (Chapter-5) and Miyazawa et al. (Chapter-9). It is some relief to note (Chapter-5) that, contrary to the general belief, the saturation effect of T_c beyond three Cu-O layers is not so settled yet and further increase in T_c with added layers continues to remain an exciting feasibility for the future. Chapter-14 by Nevriva, is again about the materials processing where the issues discussed are the phase diagrams and crystal growth of YBCO. This cuprate system with rare-earth substitutions has been investigated in depth for critical field and high field magnetization by Kovachev and Vlakhov (Chapter-4) and for thermal conductivity by Jezowski and Klamut (Chapter-11).

The remaining four chapters are essentially technique oriented. While the potential of various high energy spectroscopy techniques for HTSC studies are demonstrated by Balzarotti (Chapter-6) the application of high resolution polarized XPS is discussed by Marcelli (Chapter-13). We should note two other articles in this area in our earlier volumes, namely by Padalia and Mehta on X-ray Photoelectron Spectroscopy in Volume 1 and by Garg on X-ray spectroscopic study of HTSC in Volume 2. In Chapter-7, Boolchand and McDaniel present an extended article on Mössbauer Spectroscopy of HTSC giving particular attention to the question of site-occupancy of Fe in 1-2-3 and 2-1-4 systems. Lastly, Keinonen et al., in Chapter-10, show that the ion beam techniques provide a powerful approach for characterization and analysis of high T_c materials.

To facilitate speedy publication the camera-ready pattern has been followed for the manuscripts. Though the authors may have taken every care to follow a uniform style, some errors including typographical may have inadvertently remained which I have found difficult to rectify in the final stage; it is hoped the reader will overlook them.

I would like to express my thanks to Mr. Frank Columbus, the President, Nova Science Publishers for helpful and efficient cooperation. I am also thankful to my colleagues, Dr. S.K. Agarwal and Dr. C.V. Narasimha Rao for their suggestions and help.

September, 1989 Anant Narlikar

CONTENTS

Contributors to Volume 4 v
Articles Planned for Future Volumes vi
Contents of Volumes 1,2 and 3 xi
Preface xv

CHAPTER-1: WEAK COUPLING THEORY OF HIGH TEMPERATURE SUPERCONDUCTORS 1
J. Labbe, Group de Physique des Solides de l'Ecole Normale Supérieure, Paris Cedex, France

Introduction 1
Structure of the Antibonding d-p Sub-band within a CuO_2 Layer 5
Superconducting Transition Temperature Calculated in the Mean Field Approximation within a CuO_2 Layer 6
Reduction of the Superconducting Transition Temperature by the Critical Fluctuations 9
Small Isotope Effect 10
Discussion of the Stability of the Antiferromagnetic Insulating Phase in Simple Layered Structures 10
Band Structure of Compound with a Multiple Layered Lattice Structure 12
Paramagnetic Susceptibility of the Compounds with Multiple Layered Structures 14
Case of Compounds with Triple Layered Structures 19
References 23

CHAPTER-2: INVESTIGATIONS OF A FEW NOVEL SERIES OF THALLIUM CUPRATE SUPERCONDUCTORS SHOWING POSSIBLE CHANGE IN THE SIGN OF CHARGE CARRIERS WITH COMPOSITION 27
C.N.R. Rao, R. Vijayaraghavan, A.K. Ganguli, V. Manivannan and N.Y. Vasanthacharya, Indian Institute of Science, Bangalore, India

Introduction 27
The $TlCa_{1-x}Ln_xSr_2Cu_2O_{6+\delta}$ Series 28
The $TlSr_{n+1-x}Ln_xCu_nO_{2n+3+\delta}$ Series 32
$Tl_yCa_{1-x}Y_xBa_2Cu_2O_{6+\delta}$ and $Tl_{1-x}Pb_xCaSr_2Cu_2O_{6+\delta}$ 36
Concluding remarks 36
References 38

CHAPTER-3: SOME STRUCTURAL AND CHEMICAL ASPECTS OF THE Bi-Pb-Sr-Ca-Cu-O SYSTEM 39
M. Takano, Kyoto University, Kyoto, Japan

Introduction 39
Phases and their Relations in Bi-Sr-Ca-Cu-O System 40

Phase 2223 49
$ACuO_2$ Crystallizing in Layered Structure 52
References 54

CHAPTER-4: HIGH MAGNETIC FIELD STUDIES OF HIGH TEMPERATURE SUPERCONDUCTORS 59
V.Kovachev and E. Vlakhov, Institute of Solid State Physics, Bulgarian Academy of Sciences, Sofia, Bulgaria

Upper Critical Fields of $RBa_2Cu_3O_{7-x}$ (R=, Nd, Eu) and Ag-doped Ceramics 62
Magnetization of $RBa_2Cu_3O_{7-x}$ (R=Nd, Eu) and Ag-doped $YBa_2Cu_3O_{7-x}$ 75
References 83

CHAPTER-5: FABRICATION OF Bi-Sr-Ca-Cu OXIDE SUPERCONDUCTORS WITH CONTROLLED NUMBERS OF Cu-O LAYERS 87
K. Nakamura and K. Ogawa, National Research Institute for Metals, Tsukuba, Ibaraki, Japan

Introduction 87
Growth of the 2223 Phase in Bi(Pb)-Sr-Ca-Cu Oxide System 88
Growth of the 2223 Phase in Bi(Pb)-Sr-Ca-Cu Oxide Films 91
Synthesis of Multilayered Bi-O/Sr-Ca-Cu Oxide Films 99
Concluding Remarks 109
References 109

CHAPTER-6: CHARACTERIZATION OF HIGH TEMPERATURE SUPERCONDUCTORS BY HIGH EXCITATION SPECTROSCOPIES 113
A. Balzarotti, II Universita di Roma "Tor Vergata", Roma, Italy

Introduction 113
Some Basic Aspects of High Tc Superconductors 114
Extrinsic Surface Properties 117
Intrinsic Bulk Properties 123
Concluding Comments 138
References 139

CHAPTER-7: MÖSSBAUER SPECTROSCOPY OF HIGH TEMPERATURE SUPERCONDUCTORS 143
P. Boolchand and D. McDaniel, University of Cincinnati, Cincinnati, Ohio, U.S.A.

Crystal Structure, Defects and Metal Doping Sites 145

Magnetism and Superconductivity 163
Mössbauer Debye-Waller Factors and Vibrational Density of States 178
Concluding Remarks 189
References 190

CHAPTER-8: RVB AND RELATED THEORIES OF HTSC
D. Schmeltzer, Bar-Ilan University, Israel 199

Introduction 199
The Two Band Model 203
Superconductivity for the Non-Fermi Liquid Case 207
Experimental Implications 213
References 214

CHAPTER-9: FABRICATION OF 107K SUPERCONDUCTING FILMS AND THEIR NORMAL STATE TRANSPORT PROPERTIES 217
S. Miyazawa, M. Mukaida, K. Kuroda and Y. Tazoh,
NTT LSI Laboratories, Kanagawa, Japan

Introduction 217
Preparation of Films 218
Film Thickness Dependence on ρ-T Characteristics 220
Composition Dependence on ρ-T Characteristics 223
X-Ray Diffraction Analysis 227
Magnetic Susceptibility Measurement 229
Hall Measurement 230
Summary 235
References 236

CHAPTER-10: ION BEAM ANALYSIS OF HIGH T_c SUPERCONDUCTORS 239
J. Keinonen, J. Raisanen and E. Rauhala,
University of Helsinki, Helsinki, Finland

Introduction 239
Stopping Power and Ion Beam Induced Damage 241
Rutherford Backscattering Spectrometry 244
Particle Induced X-Ray Emission 251
Nuclear Reaction Analysis 256
Other Low Beam Techniques 258
References 259

CHAPTER-11: THERMAL CONDUCTIVITY OF HIGH TEMPERATURE SUPERCONDUCTORS 263
A. Jezowski and J. Klamut, Polish Academy of Sciences, Wroclaw, Poland

Introduction 263
Thermal Conductivity of High Temperature Superconductors 269
References 305

CHAPTER-12: A NEW THEORETICAL APPROACH TO ELECTRON PAIRING AND SUPERCONDUCTIVITY IN HIGH T_c SUPERCONDUCTORS 311
A. Nakamura, Japan Atomic Energy Research Institute, Naka-gun, Ibaraki, Japan

Introduction 311
Description and Application of the MVRC Model 313
Discussions 331
Conclusion 335
References 336

CHAPTER-13: HIGH RESOLUTION POLARIZED X-RAY ABSORPTION SPECTROSCOPY: SYMMETRY OF THE HOLE STATES IN HIGH-T_c SUPERCONDUCTORS 339
A. Marcelli, INFN-Laboratori Nazionali de Frascati, Frascati, Italy

Introduction 339
Polarized X-Ray Absorption Spectroscopy 342
Determination of the States Induced by Doping at the CuL_3 Edge 352
Correlation between Tc and the Energy Splitting of Polarized Cu2p→3d Transition 359
Summary 366
References 366

CHAPTER-14: PHASE DIAGRAMS IN THE Y-Ba-Cu-O SYSTEM AND CRYSTAL GROWTH OF YBCO SUPERCONDUCTORS 371
Milos Nevriva, Czechoslovak Academy of Sciences, Prague, Czechoslovakia

Introduction 371
Phase Diagram Determination in the Y-Ba-Cu-O System 372
Crystal Growth of $YBa_2Cu_3O_{7-\gamma}$ Superconductors 383
References 392

SUBJECT INDEX 397

WEAK COUPLING THEORY OF HIGH TEMPERATURE SUPERCONDUCTORS

J. Labbe
Group de Physique des Solides de l'Ecole Normale Supérieure,
Paris Cedex, France

1 - Introduction

Many theories of the high T_c superconductors are founded on the hypothesis that the electron-electron correlations are so strong in these materials that, in the absence of dopping or internal charge transfer, they should be Mott insulators. We consider this hypothesis as unlikely for the following reasons.

At first, very strong correlations would arise from a very large repulsive Coulomb energy between electrons within each atom. This would be the case only with very strongly localized atomic orbitals, as for instance the f orbitals in the rare earthes, leading to very narrow energy bands. But in the copper oxides, the d orbitals of copper, or the p orbitals of oxygen, are not so strongly localized, and thus the intra-atomic repulsive Coulomb energy has no reason to be much larger than in the simple transitional metals or their other compounds.

Secondly, it is argued that, in the high T_c copper oxides, the relative strength of the correlations would arise from the smallness of the effective transfer integrals between nearest neighbouring copper ions through the intercalated oxygen ions, as compared to the direct transfer integrals in simple transitional metals. But soft X rays absorption from the 2s level of oxygen (1) shows that, in the high T_c copper oxides, the 2p level of oxygen is not full. This indicates that the electron holes explore as well the 2p states on oxygen as the 3d states on copper (2). Thus, the contribution of the oxygen 2p states to the band structure very close to the Fermi energy is as large as the contribution of the copper 3d states. This gives rise to a partly occupied covalent d-p sub band, with relatively large effective transfer integrals between nearest neighbouring copper ions.

Actually, owing to the larger effective electrical charge of the ion core in copper than in the transitional elements at the begining of the first series, one can expect that the energy of the d level is much lower in copper, in the copper oxides, than for instance in vanadium, in the vanadium oxides, and thus is much closer to the energy of the p level of oxygen. As a result, the effective transfer integral is expected to be much larger, and thus the relative strength of the correlations to be much smaller, in the copper oxides than in the vanadium oxides. Thus, whereas it is reasonable to consider the vanadium oxides as Mott insultators, at least in a large domain of their phase diagram, it seems to be much more unlikely in the case of the high T_c copper oxides.

The observed correlation satellites in Auger (3) and photoemission (4) spectra in the high T_c copper oxides, due to a double ionisation of the copper atoms, make possible a crude estimation of the intra-atomic repulsive energy U between electrons in the d level of copper. One finds that U is roughly equal to 2 eV, and thus is much smaller than the computed band width which is found by many authors (5) as being at least equal to 5 eV. This estimated value of U roughly has the same order of magnitude as the direct transfer integral between copper and nearest neighbouring oxygen sites. As it has been pointed out by Friedel (2), this does not allow the localization of one electron per unit cell, and the more realistic approach is to consider the high T_c copper oxides as Fermi liquids, even if this approximation is more rough than in the simple transitional metals.

Furthermore, a few well established experimental facts, namely the quantization of the flux trapped in a superconducting ring of YBaCuO, measured bv C.E. Gough et al. (6), and the ac-Josephson effect, measured in $La_{2-x}Sr_xCuO_4$ by D. Estève et al. (7), and in YBaCuO by J. Niemeyer et al. (8), provide evidence that, in the superconducting state of these materials, the charge carriers are electron pairs. Moreover, the dc-Josephson effect, observed by J. Niemeyer et al. (8) between YBaCuO and an alloy of Pb with Sn, clearly indicates that these paires are singlet pairs in an s-state.

It is much more difficult to identify the mechanism which gives rise to an attractive interaction between the electrons, leading to their condensation in pairs. Both the physical origin and the strength of this coupling is largely controverted. Very soon after the discovery of the high T_c copper oxides by Bednorz and Müller (9), a widely held opinion was that such large values of the superconducting critical temperature T_c could not be explained by a phonon induced attractive mechanism, at least within the framework of a weak coupling theory of the BCS type (10). This opinion is founded on the well known BCS expression of T_c,

$$k_B T_c = 1.14\, h\omega_D \exp(-1/\lambda) \qquad (1)$$

where ω_D is the Debye pulsation and $\lambda = Vn(E_F)$ is the product of the phonon induced attractive coupling constant between the electrons by the electronic density of states $n(E_F)$ at the Fermi energy E_F. As the BCS theory is a weak coupling theory, which is valid only when the effective coupling constant λ is much smaller than unity, the formula (1) gives a value of T_c which is much smaller than the Debye temperature $T_D = h\omega_D/k_B$, and thus it could hardly explain the large values of T_c which are observed at least in the most recently discovered copper oxides, as for instance $YBa_2Cu_3O_7$, $Bi_2Sr_2CaCu_2O_8$, $Tl_2Ba_2CaCu_2O_8$ or $Tl_2Ba_2Ca_2Cu_3O_{10}$.

But the expression (1) of T_c holds only when the electronic density of states n(E) has a nearly constant value in the whole energy range from $E_F - h\omega_D$ to $E_F + h\omega_D$, as it is actually the case in most of the superconductors with a tridimensional structure, where the variation of n(E) versus E is very smooth in a large neighbourhood of E_F.

On the contrary, in structures with a lower dimensionality, there exist sharp Van Hove singularities in the density of states n(E), and if E_F lies very close to such a singularity, the expression (1) is no longer valid, even in the weak coupling limit. Instead, one finds in that case a new expression of T_c which is much less sensitive to the cut off energy $h\omega_D$ because n(E) is much smaller for E close to $E_F \pm h\omega_D$ than for E close to E_F. And even with a value of ω_D which would be not larger than usually, the calculated value of T_c is largely enhanced by the fact that the number of electronic states which accumulate in the neighbourhood of E_F is much larger than in the tridimensional case.

Also, as the calculated T_c depends on n(E) only through an integral, its value is not very sensitive to the very detailed structure of n(E) within the integration range. Particularly, T_c does not directly depend on $n(E_F)$. This explains why the observed value of T_c remains large in a relatively extended range of the chemical composition of these compounds.

This only small sensitivity of T_c to $n(E_F)$ is in contrast with the sharp dependence of other instabilities on $n(E_F)$, as lattice or magnetic instabilities, which appear as being very sensitive to the chemical composition.

All these peculiarities had been pointed out a long time ago (11) (12) in the case of the A-15 compounds, where the structure of the narrow d band has a quasi-one-dimensional character due to the presence of linear chains of transitional atoms.

The superconducting copper oxides have a layered structure made of weakly coupled CuO_2 planes, or families of CuO_2 planes in the case of the multiple layered structures. And the Fermi energy E_F lies at, or very close to, a logarithmic singularity of n(E). Even in the weak coupling limit, the value of T_c, calculated in the mean field approximation, can be very large because of this singularity. And, as the reduction of T_c by the critical fluctuations is much smaller when the structure is quasi-bidimensional than when it is quasi-one-dimensional, the copper oxides with a layered structure appear as being perhaps the best physical system to get large values of T_c.

When E_F is exactly at the singularity, the bidimensional Fermi line coïncides with a square (13) (14) in the first Brillouin zone, and the dominant instability at low temperature is that one which leads to the onset of an antiferromagnetic insulating phase by the opening of a Slater energy gap all along the Fermi line. But, as the static susceptibility associated with this instability diverges as rapidly as $n^2(E_F)$ when E_F goes into the singularity (15), the antiferromagnetic phase is very rapidly destabilized when E_F is on the contrary just a little shifted from the singularity, by dopping in $La_{2-x}Sr_xCuO_4$, by the internal electronic charge transfer in $YBa_2Cu_3O_{7-x}$ (15) (16), or by the partial removal of the degeneracy of the singularity by the small coupling between nearest neighbouring planes within a family of CuO_2 planes in the multiple layered structures. When the antiferromagnetic phase has vanished, the compound gradually becomes superconducting, as E_F is further shifted, with a value of T_c which can become very large because the shift of E_F, which is needed to destabilize the antiferromagnetic phase, is sufficiently small for keeping E_F close enough to the singularity for the averaged value of n(E), calculated over the whole integration range of the BCS equation for T_c, to be still very large.

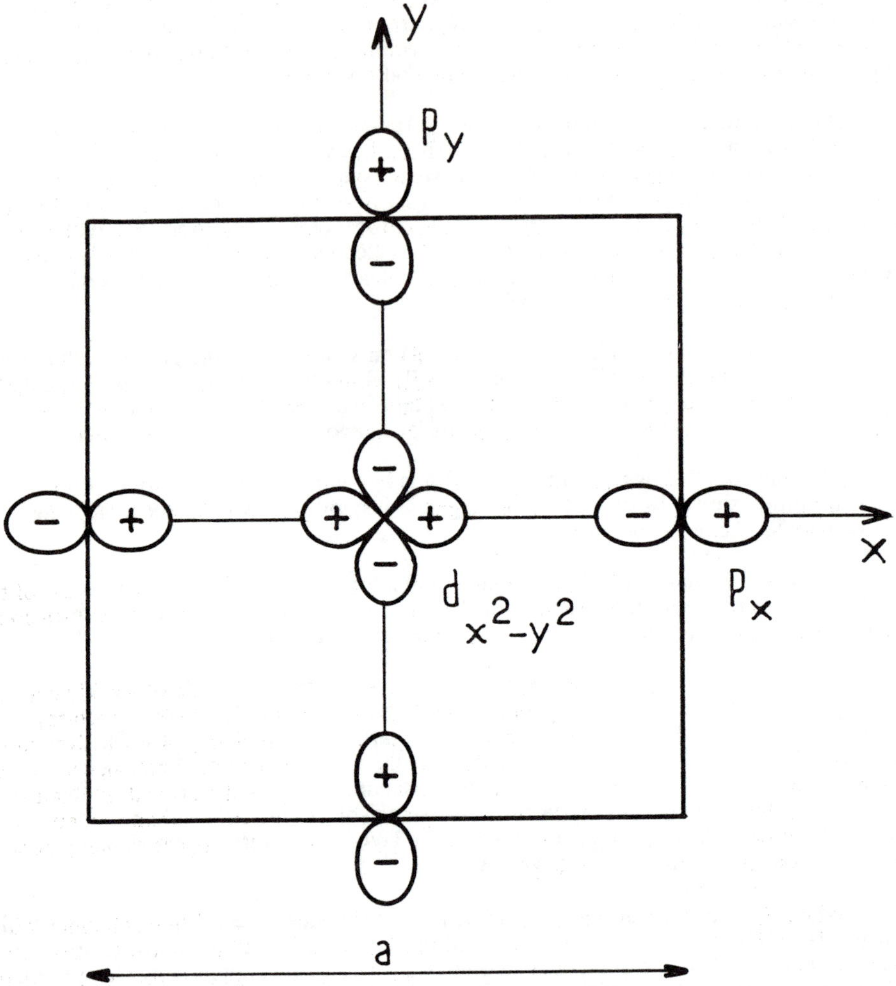

Fig 1

FIG. 1 : Hybridized $d_{x^2-y^2}$ and p_x or p_y orbitals on copper and oxygen sites, respectively, within the square unit cell, with a as the lattice parameter.

The fact that a large value of T_c does not take place immediatly after the vanishing of the antiferromagnetic phase, but needs E_F to be a little more shifted, as it is observed experimentally, provides evidence that the antiferromagnetic fluctuations are not the physical origin of the superconductivity in these materials, but that on the contrary they prevent it. This is one of the reasons why we are much more confident in a phonon induced mechanism of the superconductivity than in a mechanism involving magnetic excitations.

The recent discovery [17] of new high T_c superconducting copper oxides, such as for instance $Nd_{2-x}Th_xCuO_4$ or $Nd_{2-x}Ce_xCuO_4$, where the number of electrons in the partly occupied d-p sub-band increases with x, whereas it decreases with x in $La_{2-x}Sr_xCuO_4$, shows the existence of an electron-hole symmetry in this type of compounds, which is easily understood in an itinerant electron model.

In the following sections we go through detailed calculations. The sections 2 to 6 deal with the case of the high T_c copper oxides having a simple layered structure, and the sections 7 to 9 with the case of those ones having a multiple layered structure.

2 - The structure of the antibonding d-p sub-band within a CuO_2 layer

A simple tight binding analysis of the structure of the d-p sub-band resulting from the strong hybridization of the $d_{x^2-y^2}$ orbitals on copper with the p_x or p_y orbitals on oxygen within a CuO_2 plane, chosen as the xy plane (Fig. 1), leads to the following bidimensional dispersion relation of the partly occupied antibonding d-p sub-band :

$$E_0(\vec{k}) = \frac{1}{2}(E_d + E_p) + \frac{1}{2}\sqrt{(E_d - E_p)^2 + 8\gamma^2 (2 - \cos k_x a - \cos k_y a)} \qquad (2)$$

where γ is the direct transfer integral between the $d_{x^2-y^2}$ and the p_x or p_y orbitals located at nearest neighbouring copper and oxygen sites respectively, E_d and E_p are the intra-atomic energies of these orbitals in the compound, a is the lattice parameter and $\mathbf{k}(k_x,k_y)$ the two dimensional wave vector within a CuO_2 plane.

The bidimensional character is due to the fact that the coupling between two nearest neighbouring CuO_2 planes, or two nearest neighbouring families of CuO_2 planes in the case of the multiple layered compounds, is negligible as compared to the coupling within a plane.

The dispersion relation (2) gives a band structure which could be analysed in terms of a small effective transfer integral $\gamma^* = \gamma^2(E_d - E_p)^{-1}$ between nearest neighbouring copper ions only if the difference $E_d - E_p$ would be much larger than γ. But in that case, the antibonding sub-band would have a dominant d character, with only a very small contribution from the p states on oxygen, and the electron holes in this sub-band would not explore these p states in a sizable way. On the contrary, the strong experimental evidence of the presence of holes as well on the oxygen p states as on the copper d states shows that γ^* is not as small as it is assumed by many authors.

It follows from the dispersion relation (2) that, in the bidimensional reciprocal space, every one of the constant energy lines is characterized by a constant value of the expression $\cos k_x a + \cos k_y a$. To simplify the calculations, it is convenient to replace the expression (2) by its usual simplified version

$$E_0(\vec{k}) \cong -2t\left(\cos k_x a + \cos k_y a\right) \qquad (3)$$

with the origin of the energies chosen at the middle of the sub-band, and where t is a parameter which is not equal to γ^* when $E_d - E_p$ is not much larger than γ, but has a value which must be adjusted so as to give the good size of the band width. Actually, t has the same order of magnitude as the direct transfer integral γ between nearest neigbouring copper and oxygen sites, as far as $E_d - E_p$ is not much larger than γ.

It follows as well from the dispersion relation (2) as from its simplified version (3) that, in the bidimensional reciprocal space, the constant energy line associated with the energy $E = 0$ is a square, the vertices of which coïncide with the middles of the edges of the first Brillouin zone (Fig. 2.a), and the density of states exhibits a logarithmic Van Hove singularity at the energy $E = 0$ (Fig. 2.b).

When E_F is close to the singularity at $E = 0$, one can use for the electronic density of states n(E) per unit cell and per spin the approximate expression [16]

$$n(E) \cong \left(1/2\pi^2 t\right) \mathrm{Ln}\left(16t / |E|\right) \qquad (4)$$

which follows from the dispersion relation (3), and is valid in a large neighbourhood of the singularity.

3 - <u>The superconducting transition temperature calculated in the mean field approximation within a CuO_2 layer</u>

In the BCS theory, T_c is calculated from the well known equation

$$\frac{2}{V} = \int_{E_F - \hbar\omega_D}^{E_F + \hbar\omega_D} \frac{n(E)\, dE}{E - E_F(T_c)} \tanh \frac{E - E_F(T_c)}{2k_B T_c} \qquad (5)$$

where $E_F(T_c)$ is the Fermi energy at the temperature T_c. As E_F lies very close to the singularity at $E = 0$, n(E) can be replaced by its approximate expression (4), which cannot be moved out of the integral in the equation (5) because of its rapid variation with E.

By introducing the approximate expression (4) of n(E) into the equation (5), one gets the following equation for T_c :

$$\frac{4\pi^2 t}{V} = \int_{-u_D}^{u_D} \frac{\mathrm{th}\left(u + \varepsilon\right)}{u + \varepsilon}\left(\mathrm{Ln}\frac{8t}{k_B T_c} - \mathrm{Ln}|u|\right) du \qquad (6)$$

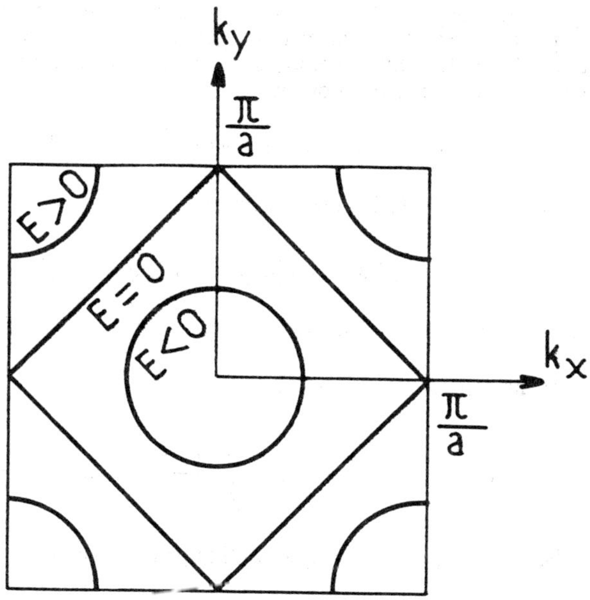

Fig 2 a

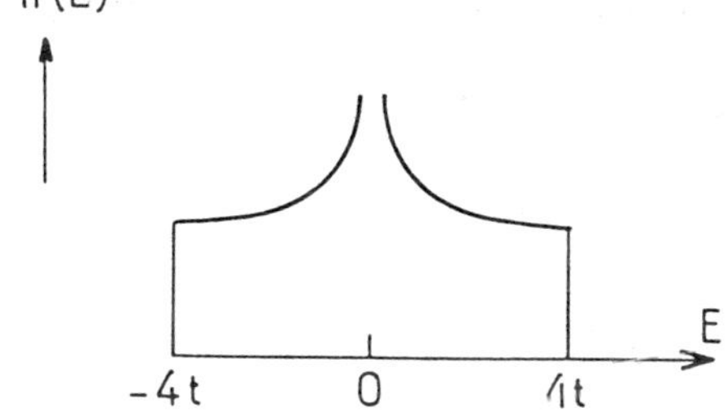

Fig 2 b

Fig. 2 : a) the two dimensional constant energy lines in the first Brillouin zone of the reciprocal lattice of a single CuO_2 layer, where a is the lattice parameter,

b) the corresponding density of states n(E) versus the energy E, where t is the effective transfer integral between nearest neighbouring copper atoms within a layer.

with $u = E/2k_BT_c$, $u_D = h\omega_D/2k_BT_c$, and $\varepsilon = - E_F(T_c)/2k_BT_c$. As long as ε is much smaller than unity, the critical superconducting temperature T_c can be expanded with respect to ε by using the equation (6). In the limit where u_D is much larger than unity, and by retaining only the term of the second order in ε, one obtains :

$$T_c \cong T_{co}\left\{1 - \Gamma\left(E_F(T_{co}) / k_BT_{co}\right)^2\right\} \qquad (7)$$

where the coefficient Γ has the following expression

$$\Gamma = 0.11 \Big/ \left(0.82 + \mathrm{Ln}\left(h\omega_D / 2k_BT_{co}\right)\right) \qquad (8)$$

and where T_{co} is the value of T_c when $\varepsilon = 0$.

When $\varepsilon = 0$, the equation (6) reduces to :

$$\frac{2\pi^2 t}{V} \cong \left(\mathrm{Ln}\frac{8t}{k_BT_{co}}\right)\int_0^{u_D}\frac{\mathrm{th}\,u}{u}\,du - \int_0^{u_D}\frac{\mathrm{th}\,u}{u}\,\mathrm{Ln}\,u\;du \qquad (9)$$

By integrating the two integrals in the equation (9) by parts, with u_D much larger than unity, one gets :

$$\frac{2\pi^2 t}{V} \cong \left(\mathrm{Ln}\frac{8t}{k_BT_{co}}\right)\left(0.82 + \mathrm{Ln}\frac{h\omega_D}{2k_BT_{co}}\right) - \frac{1}{2}\left(\mathrm{Ln}\frac{h\omega_D}{2k_BT_{co}}\right)^2 + 1 \qquad (10)$$

And by retaining only the solution of the equation (10) for which k_BT_{co} is much larger than unity, one finds :

$$k_BT_{co} = 18t\,\exp\left(-\sqrt{C}\right) \quad . \qquad (11)$$

$$C = \lambda^{-1} + \mathrm{Ln}^2\left(h\omega_D/16t\right) \qquad (12)$$

$$\lambda = V/4\pi^2 t \qquad (13)$$

where T_{co} would be the calculated value of T_c in the limit of an exactly half-filled sub-band ($E_F = 0$) if, in that case, the antiferromagnetic insulating phase would not take place instead of the superconducting phase. The expression (13) of the effective coupling constant λ replaces the previous one $\lambda = Vn(E_F)$ of the usual BCS formula (1).

The equations (11) to (13) lead to a value of T_{co} which can easily be very large because of two reasons. The first one is that the prefactor in front of the exponential in the expression (11) is no longer nearly equal to $h\omega_D$, as in the expression (1), but is much larger. The second one is that, in the weak coupling limit where λ is much smaller than unity, it follows from the equations (12), (11) and (7) that C is roughly proportional to $1/\lambda$ and thus T_{co} and T_c to $\exp(-1/\sqrt{\lambda})$, whereas the usual expression (1) would give T_c as being proportional to $\exp(-1/\lambda)$, which is much smaller than $\exp(-1/\sqrt{\lambda})$ when λ is much smaller than unity. This had been pointed out before by Hirsch and Scalapino [19], and a little more recently by Dzaloskinsky [20].

The precise numerical values of the parameters are not well known, and one can only give a rough estimate of the calculated order of magnitude of T_{co} in our model. With for instance $t = 0.5$ eV, $h\omega_D = 0.1$ eV, and $V = 1$ eV, one finds $T_{co} \cong 200$ K.

Furthermore, it follows from the equations (7) and (8) that T_c only slowly decreases when E_F is shifted from the logarithmic singularity at $E = 0$. This arises from the fact that the dominant contribution to the integral in the equation (5) is provided by the whole energy range which roughly extends from $E_F - 2k_BT_c$ to $E_F + 2k_BT_c$. And as long as E_F is smaller than $2k_BT_c$, the singularity of n(E) at $E = 0$ remains within this interval, and T_c keeps a large value. For instance, with the previous numerical values of the parameters, one finds that T_c is reduced by only six per cent when $E_F(T_{co})$ is equal to k_BT_{co}, which corresponds to the introduction of 0.02 holes per unit cell into the sub-band.

4 - The reduction of the superconducting transition temperature by the critical fluctuations

The previous expression (7) of T_c has been derived in the mean field approximation applied to a strictly bidimensional sub-system, consisting of a single CuO_2 plane in the case of a simple layered compound.

If the small coupling between nearest neighbouring bidimensional sub-systems were entirely neglected, the bidimensional critical fluctuations would make vanish the superconducting transition temperature. But Friedel has shown (21) that a very small coupling between nearest neighbouring bidimensional sub-systems is sufficient to strongly reduce the effect of the critical fluctuations. Of course, this coupling suppresses the infinite singularity of n(E), which would exist only in the limit of a perfect bidimensionality, but it is not large enough to appreciably reduce the averaged value of n(E) over the integration range of the equation (5), and thus also the value of T_c calculated in the mean field approximation for a bidimensional sub-system.

Very roughly, the actually tridimensional superconducting transition temperature T_c^{3D} can be calculated from the equation

$$T_c^{3D} \cong T_c / \mathrm{Ln}\, |t/t_\perp| \qquad (14)$$

where t is the previously defined effective transfer integral within a CuO_2 plane, and $t_\perp$ is the coupling parameter between nearest neighbouring bidimensional sub-systems. The ratio $t/t_\perp$ can be estimated from the measured anisotropy of the physical properties of these materials, as the electrical conductivity in the normal state, the plasma frequency, the coherence length, or the penetration depth. Its value is roughly equal to 10. And thus the reduction from T_c to T_c^{3D} by the equation (14) is not larger than by a factor 2 to 3.

Thus, even when taking the critical fluctuations into account, one keeps a large value of the superconducting transition temperature.

5 - The small isotope effect

Even with a phonon induced mechanism of the superconductivity, we expect only a small isotope effect in our model. This is due to the fact that when E_F lies very close to the singularity of n(E) at E = 0, the value of the integral in the equation (5) does not very much depend on ω_D, because the value of n(E) given by the expression (4) very rapidly decreases when E goes from E = 0 to E = $\pm\, h\omega_D$.

The small variation ΔT_c^{3D} of T_c^{3D} resulting from a small variation of $\Delta\omega_D$ of ω_D can be calculated by differentiating the equations (14), (7), (11) and (12). When E_F lies exactly at the singularity ($E_F = 0$), one gets :

$$\Delta T_c^{3D}/T_c^{3D} = \left(\Delta\omega_D/\omega_D\right) / \sqrt{1+\alpha} \qquad (15)$$

with :

$$\alpha = (t/V)\left(2\pi/Ln\left(16t/h\omega_D\right)\right)^2 \qquad (16)$$

It follows from the expression (15) that the isotope effect is expected to be smaller than its usual value in the BCS theory by a factor equal to $\sqrt{1+\alpha}$.

The order of magnitude of the isotope effect observed in $La_{2-x}Sr_xCuO_4$ (22) is explained by our formula (15). In YBaCuO, it seems to be much smaller (23). But it would be very surprising that the mechanism of the superconductivity should be different in these two kinds of compounds. Actually, it is well known that many physical reasons can contribute to reduce the isotope effect in metals or metallic compounds with narrow bands (24). Furthermore, in the high T_c compounds, the relation between the phonon frequencies and the atomic masses is much complicated by the large number of atoms per unit cell. This could lead to a further reduction in the dependence of the superconducting transition temperature on some of the atomic masses.

6. Discussion of the stability of the antiferromagnetic insulating phase in the case of a simple layered structure

The compounds such as for instance $La_{2-x}Sr_xCuO_4$ or $Nd_{2-x}Ce_xCuO_4$ have a simple layered structure, with only one CuO_2 plane per unit cell. When the antibonding dp sub-band is exactly half-filled, the bidimensional Fermi line exactly coïncides with a square in the first Brillouin zone, and its nesting with its translated by one of the four vectors **q** with $q_x = \pm\,\pi/a$ and $q_y = \pm\,\pi/a$ is perfect. As a result, the stable phase in the case of an exactly half-filled sub-band, and at the absolute zero, is the antiferromagnetic (AF) insulating phase, however small the intra-atomic repulsive energy U is. And the existence of this AF phase does not prove that U is large, as many people have seemed to be convinced. On the contrary, the fact that the AF phase is very rapidly destabilized when E_F is a little shifted from the middle of the sub-band seems to prove that U actually is not very large, as we shall see now.

The static bare particle susceptibility, per copper atom and with the wave vector **q**, is given by the well known formula :

$$\chi_0(\mathbf{q}) = \frac{2\mu_B^2}{N} \sum_{\mathbf{k}} \frac{f(\mathbf{k}+\mathbf{q}) - f(\mathbf{k})}{E_0(\mathbf{k}) - E_0(\mathbf{k}+\mathbf{q})} \qquad (17)$$

where μ_B is the Bohr magneton, N the total number of unit cells within each CuO_2 plane, and f(**k**) is the Fermi Dirac occupation factor of the energy level $E_0(\mathbf{k})$.

To discuss the stability of the antiferromagnetic insulating phase, one must consider the case where **q** is one of the four vectors with the components $q_x = \pm \pi/a$ and $q_y = \pm \pi/a$, so that $E_0(\mathbf{k}+\mathbf{q}) = - E_0(\mathbf{k})$. At the absolute zero, and by using for the density of states per unit cell and per spin the approximate expression (16) :

$$n(E) = \frac{1}{2\pi^2 t} \mathrm{Ln} \frac{16t}{|E|} \qquad (18)$$

one obtains from the expression (17) of the bare particule suceptibility :

$$\chi_0(\mathbf{q}) = - \frac{\mu_B^2}{\pi^2 t} \int_{-4t}^{-|E_F|} \frac{dE}{E} \mathrm{Ln} \frac{16t}{|E|} \qquad (19)$$

where E_F is the Fermi energy. By performing the integration, and retaining only the leading term when $|E_F| << 16t$, one gets :

$$\chi_0(\mathbf{q}) = \frac{\mu_B^2}{2\pi^2 t} \left(\mathrm{Ln} \frac{16t}{|E_F|}\right)^2 = 2\mu_B^2 \pi^2 t\, n^2(E_F) \qquad (20)$$

In the case of an half-filled band, one has $E_F = 0$, and $\chi_0(\mathbf{q})$, which here is calculated for $q_x = \pm \pi/a$ and $q_y = \pm \pi/a$, has an infinite value at the absolute zero, with a divergence of the expression (20) which is proportional to $(\mathrm{Ln}|E_F|)^2$, and thus to the square of the density of states $n(E_F)$ at the Fermi energy. Physically, this divergence of the susceptibility when the band is half-filled ($E_F = 0$) arises from the perfect nesting of the squared Fermi line with its translated by the vector **q**.

In the mean field approximation, the static susceptibility of the interacting electron gas is :

$$\chi(\mathbf{q}) = \chi_0(\mathbf{q}) \left\{ 1 - \frac{U}{2\mu_B^2} \chi_0(\mathbf{q}) \right\}^{-1} \tag{21}$$

where U is the intra-atomic Coulomb repulsive energy. Thus, the antiferromagnetic insulating phase is stable as far as the condition

$$U \chi_0(\mathbf{q}) > 2\mu_B^2$$

is satisfied, or equivalently, from the expression (20) of $\chi_0(\mathbf{q})$, as far as the condition

$$|E_F| < 16\, t \exp(-2\pi \sqrt{\frac{t}{U}}) \tag{22}$$

is satified. This explains why the compound $La_{2-x}Sr_xCuO_4$ exhibits an antiferromagnetic ordering within each CuO_2 plane, which excludes the superconductivity in the same plane, as long as the doping rate x is too small [(25)].

7 - The band structure of the compounds with a multiple layered lattice structure

Let us now consider how the previous results are modified in the compounds where the CuO_2 planes are regrouped in families of n planes, the distance between two nearest neighbouring families being much larger than the distance between two nearest neighbouring CuO_2 planes within a family. This is for instance the case of the compounds $Bi_2Sr_2Ca_{n-1}Cu_nO_{4+2n}$, $Tl_2Ba_2Ca_{n-1}Cu_nO_{4+2n}$, and $TlBaCa_{n-1}Cu_nO_{2+2n}$.

The band structure is modified by the small coupling between the nearest neighbouring CuO_2 planes within a family. But as long as we neglect the coupling between distinct famililies, which is much smaller, the wave vector **k** remains parallel to the CuO_2 planes, and the electron gas is still bidimensional. However, we must take into account the fact that each unit cell contains n copper atoms, each one belonging to a distinct plane within a same family, and which are alined along an axis parallel to the z-axis perpendicular to the CuO_2 planes. In the tight binding approximation, a one electron Bloch wave function within each family of CuO_2 planes must be calculated as a linear combination :

$$\psi_{\mathbf{k}}(\mathbf{r}) = \sum_{p=0}^{n-1} A_p\, \phi_{p,\mathbf{k}}(\mathbf{r}) \tag{23}$$

where

$$\phi_{p,\mathbf{k}}(\mathbf{r}) = N^{-1/2} \sum_{n_1 n_2} d(\mathbf{r} - \mathbf{R}_{n_1 n_2} - p\mathbf{b})\, e^{i(k_x n_1 + k_y n_2)a}$$

is the Bloch wave function within the pth CuO_2 plane of a family of n planes, with the wave vector $\mathbf{k}(k_x,k_y)$, $d(\mathbf{r} - \mathbf{R}_{n\,n} - p\mathbf{b})$ is the atomic $d_{x^2-y^2}$ orbital located at the copper atomic site $\mathbf{R}_{n\,n} + p\mathbf{b}$, with $\mathbf{R}_{n\,n} = n_1\mathbf{a}_1 + n_2\mathbf{a}_2$, where n_1 and n_2 are integers, $\mathbf{a}_1$ and $\mathbf{a}_2$ are the vectors of the basis of the squared Bravais lattice within a CuO_2 plane, with $a_1 = a_2 = a$, and the vector **b** is orthogonal to the CuO_2 planes with its length b equal to the distance between two nearest neighbouring CuO_2 planes within a family. The n numerical coefficients A_p in the expression (23) have to be found [26] so that $\psi_{\mathbf{k}}(\mathbf{r})$ is an eigenfunction of the one electron hamiltonian $h = h_0 + h_1$, where h_0 is the tight binding hamiltonian within a CuO_2 plane and h_1 is the coupling between two nearest neighbouring CuO_2 planes within a same family. One has

$$h_0\phi_{p,\mathbf{k}}(\mathbf{r}) = E_0(\mathbf{k})\, \phi_{p,\mathbf{k}}$$

where the expression of $E_0(\mathbf{k})$ is given by the equation (3). Furthermore, in the tight binding approximation, the effective coupling between two copper atoms belonging to two nearest neighbouring CuO_2 in the same family is not negligible only if these two atoms belong to the same unit cell (n_1,n_2), and the matrix elements of h_1 which have to be retained reduce to :

$$< \phi_{p,\mathbf{k}} | h_1 | \phi_{p',\mathbf{k}} > \cong \frac{1}{N} \sum_{n_1 n_2} < d(\mathbf{r} - \mathbf{R}_{n_1 n_2} - p\mathbf{b}) | h_1 | d(\mathbf{r} - \mathbf{R}_{n_1 n_2} - p'\mathbf{b}) > = -\tau$$

with $p' = p \pm 1$ if $0 < p < n-1$, $p' = 1$ if $p = 0$, $p' = n-2$ if $p = n-1$, and where

$$-\tau = < d(\mathbf{r}) | h_1 | d(\mathbf{r} \pm \mathbf{b}) > \qquad (\tau > 0)$$

is the transfer integral between two copper atoms occupying sites related by a translation by the vector **b**. As b is much larger than a, τ is much smaller than t.

The eigenvalues and eigenfunctions of h are easily found for the smallest value of n.

For n = 2 (doublets of CuO_2 planes, as for instance in the compounds $Bi_2Sr_2CaCu_2O_8$ or $Tl_2Ba_2CaCu_2O_8$), one finds the eigenvalues :

$$E^{(\nu)}(\mathbf{k}) = E_0(\mathbf{k}) - (-1)^{\nu}\, \tau \qquad (24)$$

with $\nu = 0$ or 1, and which are associated with the eigenfunctions :

$$\psi_{\mathbf{k}}^{(\nu)}(\mathbf{r}) = 2^{-1/2} \left\{ \psi_{0,\mathbf{k}}(\mathbf{r}) - (-1)^{\nu}\, \psi_{2,\mathbf{k}}(\mathbf{r}) \right\} \qquad (25)$$

The band structure previously found for the uncoupled CuO_2 planes is split by the coupling term h_1 into two sub-bands which have the same structure, but the energies of which are higher (lower) than $E_0(\mathbf{k})$ by an amount τ ($-\tau$) which does not depend on $\mathbf{k}$. Some constant energy lines of these two sub-bands are shown in Fig. 3.a, and the split density of states in Fig. 3.b. The logarithmic singularity, the energy of which is $E = 0$ when the coupling is neglected, is split into new logarithmic singularities the energies of which are $E = \pm \tau$. Thus the coupling strongly modifies the density of states in the neighbourhood of the middle of the band. In particular, for an energy lying in this neighbourhood, the density of states of a doublet of coupled CuO_2 planes is not at all equal to twice the density of states of an isolated plane, in contradiction with one of the hypotheses which were recently introduced to explain the increase of the critical superconducting temperature when the number of CuO_2 planes per unit cell is increases (27).

Obviously, the two split sub-bands do not contain the same number of electrons. When the band of the uncoupled CuO_2 planes is half-filled, the Fermi energy is still $E_F = 0$ when the coupling term h_1 is taken into account, but the sub-band for which $\nu = 1$ is less than half-filled, and the other one, for which $\nu = 0$, is more than half-filled. As a result, the Fermi energy E_F no longer coïncides with the middles of the two new sub-bands where the logarithmic singularities are (Fig. 3.b). Furthermore, the two constant energy lines corresponding to the the energy $E_F = 0$, one in the sub-band with $\nu = 0$, and the other one in the sub-band with $\nu = 1$, no longer coïncide with the square the vertices of which are the middles of the edges of the first Brillouin zone (Fig. 3.a and 3.b). Thus, within each one of the two sub-bands with $\nu = 0$ and $\nu = 1$, there is no longer a perfect nesting of the Fermi line of this sub-band by a vector $\mathbf{q}$ with the components $q_x = \pm \pi/a$ and $q_y = \pm \pi/a$. As a consequence the antiferromagnetic insulating phase is expected to be less stable in the compounds with a double layered structure than in the compounds with a simple layered structure. We shall discuss the stability of the antiferromagnetic phase in the mean field approximation, by calculating the bare particule susceptibility per copper atom of a doublet of coupled CuO_2 planes, and then comparing it to the quantity $2\mu_B{}^2/U$. Of course, this treatment is valid only in the limit where U is not too large.

8 - The paramagnetic susceptibility of the compounds with a multiple layered structure

To calculate this susceptibility, let us come back to the general case of a family of n weakly coupled CuO_2 planes, with n unspecified. By calculating the atomic magnetic moment induced on a copper atom in the pth plane of the family by an external magnetic field

$$B = B_0 \cos(\mathbf{q}.\mathbf{r} + p.\pi)$$

which varies sinusoïdally within each CuO_2 plane, with the amplitude B_0 and the wave vector $\mathbf{q}$ parallel to the plane, and the sign of which changes when going from one plane to a nearest neighbouring one (28), one finds that the bare particule atomic susceptibility of a copper atom in the pth plane of a family of n coupled CuO_2 planes is :

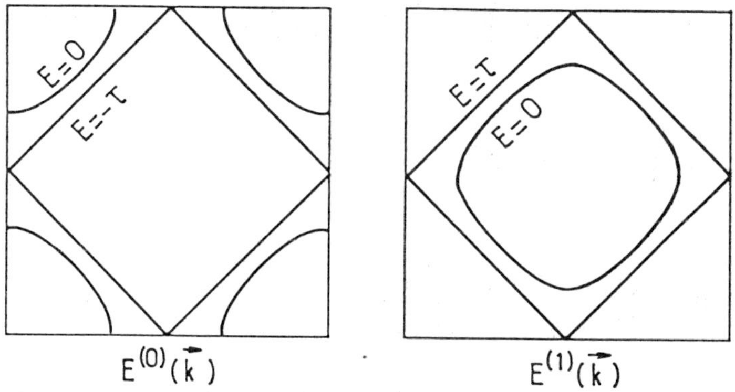

Fig 3 a

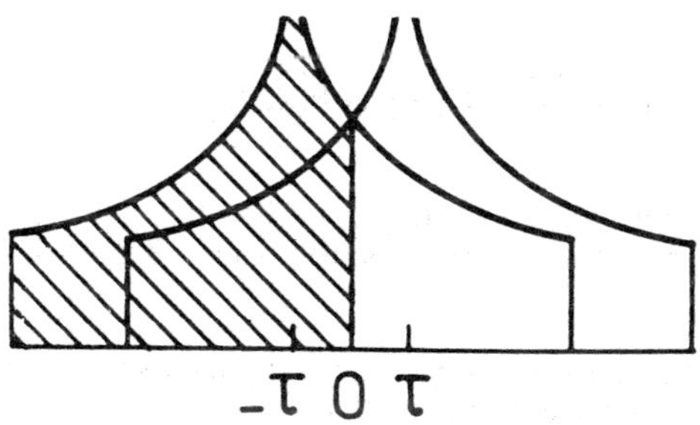

Fig 3 b

Fig. 3 : a) the two dimensional constant energy lines associated with the energies $E^{(\nu)}(\mathbf{k})$ of the two split sub-bands with $\nu = 0$ or 1, in the case of a compound with a double layered structure, and where τ is the small transfer integral between the two CuO_2 layers of a doublet,

b) the corresponding split density of states.

$$\chi^{(n)}_{0,p}(\mathbf{q}) = (-1)^p \frac{2\mu_B^2}{N} \sum_{\nu\nu'} A^{(\nu)}_p A^{(\nu')}_p \alpha_{\nu\nu'} F^{(n)}_{\nu\nu'}(\mathbf{q}) \tag{26}$$

with

$$\alpha_{\nu\nu'} = \sum_{p'=0}^{n-1} (-1)^{p'} A^{(\nu)}_{p'} A^{(\nu')}_{p'} \tag{27}$$

and

$$F^{(n)}_{\nu\nu'}(\mathbf{q}) = \sum_{\mathbf{k}} \frac{f^{(\nu')}(\mathbf{k}+\mathbf{q}) - f^{(\nu)}(\mathbf{k})}{E^{(\nu)}(\mathbf{k}) - E^{(\nu')}(\mathbf{k}+\mathbf{q})} \tag{28}$$

where $f^{(\nu)}(\mathbf{k})$ is the Fermi-Dirac occupation factor of the energy level $E^{(\nu)}(\mathbf{k})$, and $A_p^{(\nu)}$ is the value of the coefficient A_p when the Bloch wave function (23) is an eigenstate associated with the eigenvalue $E^{(\nu)}(\mathbf{k})$ of the hamiltonian h.

Coming back to the case with n = 2, where $A_0^{(0)} = A_1^{(0)} = A_0^{(1)} = -A_1^{(1)} = 2^{-1/2}$, as it can be seen from the expression (25) of the wave function, one gets, from our definition (27) of the coefficients $\alpha_{\nu\nu'}$, $\alpha_{00} = \alpha_{11} = 0$, $\alpha_{01} = \alpha_{10} = 1$, and thus, from the formula (28) :

$$\chi^{(2)}_{0,0}(\mathbf{q}) = \chi^{(2)}_{0,1}(\mathbf{q}) = \frac{\mu_B^2}{N} \left\{ F^{(2)}_{01}(\mathbf{q}) + F^{(2)}_{10}(\mathbf{q}) \right\} \tag{29}$$

Obviously, we find a susceptibility per copper atom which has the same value in the two CuO_2 planes p = 0 and p = 1 of a same doublet, since these two planes are physically equivalent.

When $\mathbf{q}$ is one of the four vectors with the components $q_x = \pm \pi/a$ and $q_y = \pm \pi/a$, the expressions (3) and (24) of the energies show that

$$E^{(1)}(\mathbf{k}+\mathbf{q}) = -E^{(0)}(\mathbf{k}) \tag{30}$$

and the equation (28) leads to :

$$F^{(2)}_{01}(\mathbf{q}) = F^{(2)}_{10}(\mathbf{q}) = -N \int_{-4t-\tau}^{-|E_F|} \frac{n_0(E)\, dE}{2E} - N \int_{-4t+\tau}^{-|E_F|} \frac{n_1(E)\, dE}{2E} \tag{31}$$

where $n_\nu(E) = n(E + (-1)^\nu \tau)$ is the density of states per copper atom and per spin of the νth shifted sub-band, n(E) being the unshifted density of states given by the equation (18). It follows from the equations (29) and (31) that, when $q_x = \pm \pi/a$ and $q_y = \pm \pi/a$, we have :

$$\chi^{(2)}_{0,p}(\mathbf{q}) = -\frac{\mu_B^2}{4\pi^2 t} \sum_{\nu=1,2} \int_{-4t-(-1)^\nu \tau}^{-|E_F|} \frac{dE}{E} \, Ln \frac{16t}{|E+(-1)^\nu \tau|} \qquad (32)$$

which is valid as well in the plane p = 0 than in the plane p = 1.

As previously mentioned, and as far as the material is undoped, the Fermi energy remains equal to zero even when the coupling h_1 is taken into account. Thus, we are interested in the behaviour of the integrals in the expression (32) when E_F goes to zero. Retaining only the leading terms in the limit where $|E_F|$ is much smaller than τ, we find

$$\chi^{(2)}_{0,p}(\mathbf{q}) \cong \frac{\mu_B^2}{2\pi^2 t} \left(Ln \frac{16\,t}{\tau}\right) Ln \frac{16\,t}{|E_F|} \qquad (33)$$

or, from the expression (18) of the unsplit density of states :

$$\chi^{(2)}_{0,p}(\mathbf{q}) \cong \mu_B^2 \left(Ln \frac{16\,t}{\tau}\right) n(E_F) \qquad (34)$$

The comparison of the expressions (20) with the expressions (33) or (34) shows that, when E_F goes to zero, corresponding to half-filling, the susceptibility should diverge much more slowly in the compounds with a double layered structure than in the compounds with a simple layered structure, since in the first ones it diverges only as $Ln|E_F|$, and thus as $n(E_F)$, whereas in the second ones it diverges as $(Ln|E_F|)^2$, and thus as $n^2(E_F)$. This arises from the fact that in the second ones the band is not split into distinct sub-bands, and when it is half-filled, the Fermi line is a square. And the perfect nesting of this square with its translated by a vector **q** with components $q_x = \pm \pi/a$ and $q_y = \pm \pi/a$ includes the singular points at its vertices, which are the middles of the edges of the first Brillouin zone (Fig. 4.a). Thus, the susceptibility diverges as $(Ln|E_F|)^2$ because of the logarithmic Van Hove singularity related to these singular points. On the contrary, in the compounds with a double layered structures, the band is split into two sub-bands, and when it is half-filled, there is a perfect nesting only between the Fermi lines belonging to the two distinct sub-bands, as it can be seen from the relation (30) with $E_F = 0$ (Fig. 4.b), and no longer of each Fermi line with itself [29]. And this perfect nesting excludes the singular points at the middles of the edges of the first Brillouin zone. So that the divergence of the susceptibility is no longer related to the logarithmic Van Hove singularities, but is only a result of the integration of 1/E in the expression (32), and thus it simply behaves as $Ln|E_F|$.

However, if the density of states within an isolated CuO_2 plane actually had an infinite singularity in the middle of the band, we should conclude that the antiferromagnetic phase would exist for exact half-filling ($E_F = 0$) even in the compounds with a double layered structure. But its domain of stability, determined by the new condition

$$|E_F| < 16\,t\,\exp\left(4\pi^2 t \left(U\,Ln \frac{16\,t}{\tau}\right)^{-1}\right) \qquad (35)$$

instead of the condition (22), should be much narrower [30].

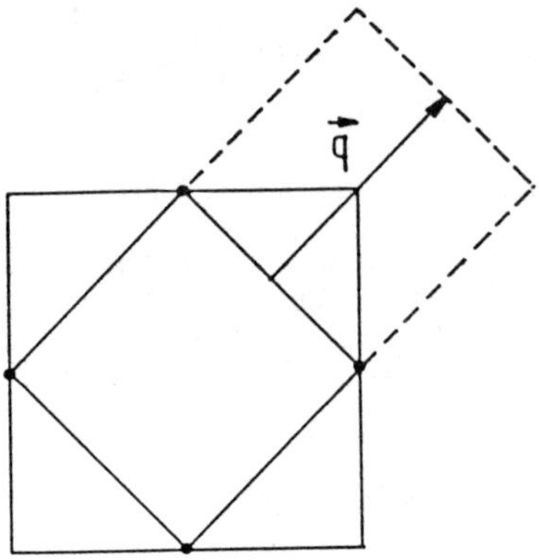

Fig 4 a

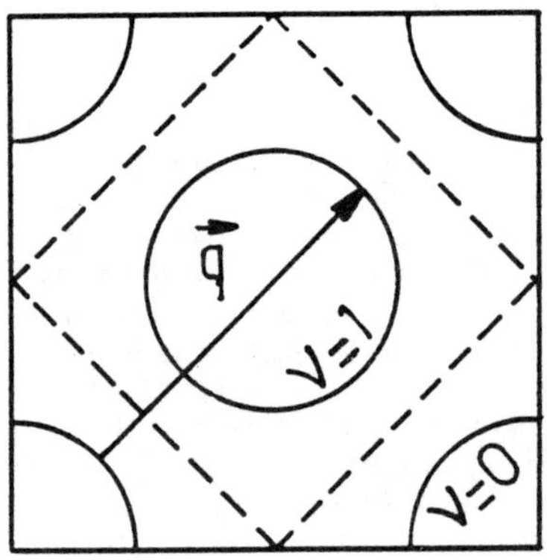

Fig 4 b

Fig. 4 : a) the nesting of the Fermi line by the wave vector **q** with the components $q_x = q_y = \pi/a$, in the case of a compound with a simple layered structure

b) the nesting between the two distinct Fermi lines associated with the two split sub-bands $\nu = 0$ and $\nu = 1$ in the case of a compound with a double layered structure.

But actually, because of the small coupling which also exists between nearest neighbouring doublets of CuO_2, which restores a small but non vanishing three dimensional character of the band structure, but of course is much smaller than the coupling between the two partners of a same doublet of CuO_2 planes, the expression of the density of states is no longer exactly that one given by the expression (18), and $n(E_F)$, calculated before that the splitting is taken into account, is no longer infinite even for exact half-filling ($E_F = 0$), but simply remains large if the three-dimensional character is weak enough. And thus, the condition for the stability of the antiferromagnetic phase could be satisfied when the corresponding susceptibility is proportionnal to $n^2(E_F)$, but no longer satisfied when the same susceptibility is simply proportionnal to $n(E_F)$.

At that point, it is interesting to notice that J. Friedel has shown that the lattice instability with respect to a shear strain which would remove the degeneracy of the logarithmic Van Hove singularity, and which could stem from the proportionality of the corresponding shear modulus to $n(E_F)$, actually does not exist in this model when the weakly three dimensional character of the band structure is taken into account [(15)]. The paramagnetic Pauli susceptibility for $\mathbf{q} = 0$ also is proportionnal to $n(E_F)$, and actually it should not be divergent even for half-filling. Thus, it seems to be clear that, because of the weak three dimensional character of the band structure, only the instabilities related to $n^2(E_F)$ can exist, and thus that the antiferromagnetic phase is non longer stable in the compounds with a double layered structure.

9 - The case of the compounds with a triple layered structure

Let us now consider the case with n = 3 (triplets of CuO_2 planes, as for instance in the compound $Tl_2Ca_2Ba_2Cu_3O_{10}$). The eigenvalues of h are easily found and are

$$E^{(0)}(\mathbf{k}) = E_0(\mathbf{k}) + (\nu - 1)\,\tau\sqrt{2} \tag{36}$$

with $\nu = 0,1$ or 2. They are associated with the eingenfunctions $\psi_{\mathbf{k}}{}^{(\nu)}(\mathbf{r})$ which have the following expressions :

$$\psi_{\mathbf{k}}^{(0)}(\mathbf{r}) = \frac{1}{2}\left\{-\phi_{0,\mathbf{k}}(\mathbf{r}) + \sqrt{2}\,\phi_{1,\mathbf{k}}(\mathbf{r}) - \phi_{2,\mathbf{k}}(\mathbf{r})\right\} \tag{37}$$

$$\psi_{\mathbf{k}}^{(1)}(\mathbf{r}) = \frac{1}{\sqrt{2}}\left\{\phi_{0,\mathbf{k}}(\mathbf{r}) - \phi_{2,\mathbf{k}}(\mathbf{r})\right\} \tag{38}$$

$$\psi_{\mathbf{k}}^{(2)}(\mathbf{r}) - \frac{1}{2}\left\{\phi_{0,\mathbf{k}}(\mathbf{r}) + \sqrt{2}\,\phi_{1,\mathbf{k}}(\mathbf{r}) + \phi_{2,\mathbf{k}}(\mathbf{r})\right\} \tag{39}$$

We see that the band is split into three sub-bands which have the same structure. But whereas the energies of the sub-band with $\nu = 1$ are unshifted, the energies of the sub-bands with $\nu = 0$ or 2 are shifted by an amount which does not depend on **k** and is equal to $-\tau\sqrt{2}$ or $\tau\sqrt{2}$ respectively. Some ones of the constant energy lines are shown in Fig. 5.a, and the split density of states in Fig. 5.b. The logarithmic singularity is split into three new singularities at the energies $E = 0$ and $E = \pm\tau\sqrt{2}$. When the unsplit band is half-filled, the sub-band with $\nu = 1$ remains half-filled, but the sub-band with $\nu = 0$ is more than half-filled, whereas the sub-band with $\nu = 2$ is less than half-filled, the Fermi energy being still equal to $E_F = 0$ (Fig. 5.b).

From the expressions (37), (38) and (39) of the wave functions, one gets

$$A_0^{(2)} = A_2^{(2)} = -A_0^{(0)} = -A_2^{(0)} = \frac{1}{2},\ A_1^{(2)} = A_1^{(0)} = 2^{-1/2},\ A_0^{(1)} = -A_2^{(1)} = 2^{-1/2},\ A_1^{(1)} = 0$$

Then, the definition (27) of the coefficients $\alpha_{\nu\nu'}$, which are symetrical with respect to ν and ν', gives $\alpha_{00} = \alpha_{22} = 0$, $\alpha_{11} = 1$, $\alpha_{01} = \alpha_{12} = 0$, $\alpha_{02} = -1$. And, by using our formula (26), we obtain

$$\chi_{0,0}^{(3)}(\mathbf{q}) = \chi_{0,2}^{(3)}(\mathbf{q}) = \frac{\mu_B^2}{2N}\left\{2F_{11}^{(3)}(\mathbf{q}) + F_{02}^{(3)}(\mathbf{q}) + F_{20}^{(3)}(\mathbf{q})\right\} \tag{40}$$

for the two external CuO_2 planes $p = 0$ and $p = 2$ of a same triplet, and

$$\chi_{0,1}^{(3)}(\mathbf{q}) = \frac{\mu_B^2}{N}\left\{F_{02}^{(3)}(\mathbf{q}) + F_{20}^{(3)}(\mathbf{q})\right\} \tag{41}$$

for the central plane $p = 1$ of the triplet. It is not surprising to find that the susceptibility has not the same value in the central plane than in the two external ones, since physically that plane is not equivalent to the two others.

For $q_x = \pm\pi/a$ and $q_y = \pm\pi/a$, the formula (28) leads to :

$$F_{11}^{(3)}(\mathbf{q}) = -N\int_{-4t}^{-|E_F|}\frac{n_1(E)\,dE}{E} \tag{42}$$

and

$$F_{02}^{(3)}(\mathbf{q}) = F_{20}^{(3)}(\mathbf{q}) = -N\int_{-4t-\tau\sqrt{2}}^{-|E_F|}\frac{n_0(E)\,dE}{2E} - N\int_{-4t+\tau\sqrt{2}}^{-|E_F|}\frac{n_2(E)\,dE}{2E} \tag{43}$$

with $n_\nu(E) = n(E - (-1)^\nu\,\tau\sqrt{2})$. By following the same lines of calculation as in the previous case with $n = 2$, and by retaining the leading terms only, we get :

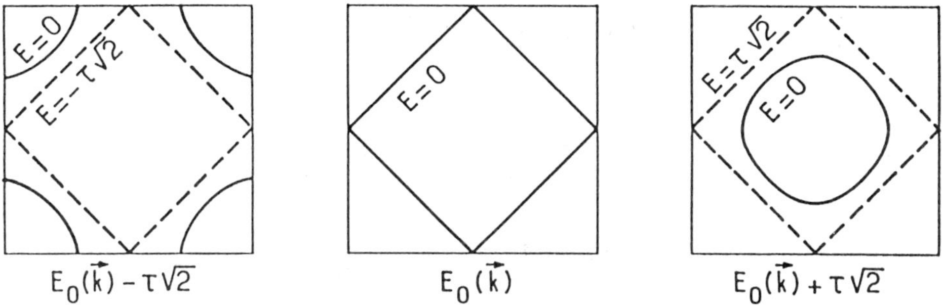

Fig 5 a

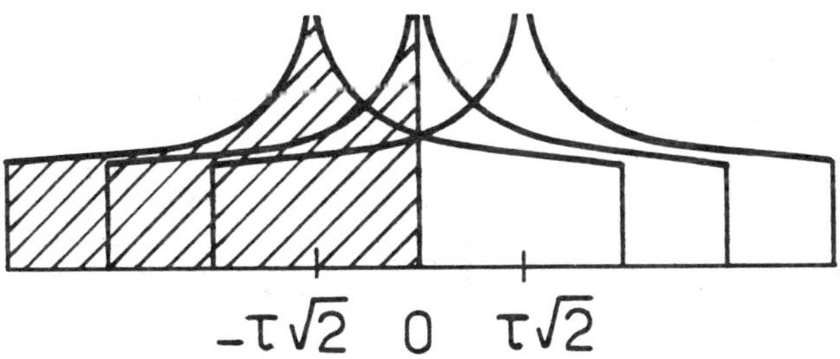

Fig 5 b

Fig. 5 : a) the two dimensional constant energy lines associated with the three distinct split sub-bands in the case of a compound with a triple layered structure, where $E_0(\mathbf{k})$ is the Bloch energy within a single CuO_2 layer in the absence of any coupling between the layers,

b) the corresponding split density of states.

$$F^{(3)}_{11}(\mathbf{q}) \cong \frac{N}{4\pi^2 t}\left(\mathrm{Ln}\frac{16\,t}{|E_F|}\right)^2 = N\pi^2 t\, n^2(E_F) \qquad (44)$$

and

$$F^{(3)}_{02}(\mathbf{q}) = F^{(3)}_{20}(\mathbf{q}) \cong \frac{N}{4\pi^2 t}\left(\mathrm{Ln}\frac{16\,t}{\tau\sqrt{2}}\right)\mathrm{Ln}\frac{16\,t}{|E_F|} = \frac{N}{2}\left(\mathrm{Ln}\frac{16\,t}{\tau\sqrt{2}}\right) n(E_F) \qquad (45)$$

where the last expression is valid in the limit where $|E_F| << \tau$. it follows from the equations (40), (41), (44) and (45) that the expressions of the susceptibility per copper atom are :

$$\chi^{(3)}_{0,0}(\mathbf{q}) = \chi^{(3)}_{0,2}(\mathbf{q}) \cong \mu_B^2 \left\{\pi^2 t\, n^2(E_F) + \frac{1}{2}\left(\mathrm{Ln}\frac{16\,t}{\tau\sqrt{2}}\right) n(E_F)\right\} \qquad (46)$$

in the two external CuO_2 planes p = 0 and p = 2 of a triplet, and :

$$\chi^{(3)}_{0,1}(\mathbf{q}) \cong \mu_B^2 \left(\mathrm{Ln}\frac{16\,t}{\tau\sqrt{2}}\right) n(E_F) \qquad (47)$$

in the central CuO_2 plane p = 1 of the same triplet.

The results (46) and (47) show that the susceptibility is still proportional to $n^2(E_F)$ in the two external planes p = 0 and p = 2 of a triplet of CuO_2 planes, but is proportional to $n(E_F)$ itself in the central plane p = 1 of the same triplet. This is related to the fact that the contribution proportional to $n^2(E_F)$ entirely comes from the unshifted sub-band $\nu = 1$, in which one the Fermi energy $E_F = 0$ still coïncides with the logarithmic Van Hove singularity (Fig. 5.b). And this sub-band is associated with the Bloch wave function given by the equation (38), which is antisymetrical with respect to p - 1, and thus vanishes on the central plane p = 1.

We can conclude from this that the tendency to the antiferromagnetic ordering, which would exclude the superconductivity, is much stronger in the two external planes, with p = 0 or p = 2, than in the central one, with p = 1. And, according to the same kind of arguments as before in the case n = 2, either the domain of stability of the antiferromagnetic phase, with respect to the value of $|E_F|$ in the neighbourhood of $E_F = 0$, is much narrower in the central plane than in the two other ones, and thus it would need a much smaller doping to be destroyed, or it does not exist at all in this plane, because $n(E_F)$, which actually is not infinite, would be not large enough. In the last case, the central plane could be superconducting without any doping, and at the same time the two external planes could exhibit, because $n^2(E_F)$ is much larger than $n(E_F)$, either an antiferromagnetic ordering, or at least large spin fluctuations, which could be detected for instance by nuclear magnetic resonance experiments. It can be noticed that, in the expression of the susceptibility, $n^2(E_F)$ has to be multiplied by a prefactor which is smaller in the expression (46) than in the expression (20) by a factor 2. Thus, the antiferromagnetic phase is expected to be less stable in the two external planes of a triplet of weakly coupled CuO_2 planes in the triple layered structures than in an isolated CuO_2 plane in the simple layered structures, independently of the fact that in the two cases a weak coupling between nearest neighbouring triplets of CuO_2 planes in the first case, or single CuO_2 planes in the second case, is needed to prevent the bidimensional fluctuations to destroye the ordering.

Conclusion

The increase of the value of the superconducting critical temperature T_c, when going from $La_{2-x}Sr_xCuO_4$ to the compounds with a multiple layered structure, could be a consequence of three causes. The first one is the possible absence of any antiferromagnetic ordering in the new compounds, even when they are not doped, or at least the possibly much smaller spin fluctuations than in $La_{2-x}Sr_xCuO_4$, since the domain of stability of the antiferromagnetic phase, if it would exist, should be much narrower. The second one is that the atomic disordering should be smaller in the new compounds, since an external doping seems to be not needed for the onset of the superconductivity in these compounds. The third one is that, as the coupling between the layers is small, and thus also the splitting of the band, the Fermi energy should remain closer to the split Van Hove singularities in the compounds with a multiple layered structure, at least when they are not doped and when the number n of layers in a family is small, than to the single Van Hove singularity in $La_{2-x}Sr_xCuO_4$, in which the doping rate x needed to obtain a high value of T_c is relatively large.

Our theory is a weak coupling theory, since the Coulomb repulsion between the electrons is taken into account only through a self consistent mean field approximation, when calculating the paramagnetic susceptibility. But, within this simple scheme, we have been able to discuss the stability of the antiferromagnetic phase in a very transparent way. And it is not unreasonable to think that the qualitative features of our results could be conserved in a theory which would take the correlations between the electrons into account, since it could be that the Hubbard correlation gap is made less stable, or even unstable, by the splitting of an half-filled band into several sub-bands, each one of them being no longer half-filled. The justification of this statement, which should take into account the mixing of the split sub-bands by the Coulomb repulsion, is an interesting problem.

On the other hand, the author is not convinced that the Coulomb repulsion between the electrons within the same atomic site is as strong as so many authors seem to believe, since the atomic orbitals in these compounds are simply d (or p) orbitals, as in simple transitional metals (or other transitional metal compounds). And one of the distinctive features of the present compounds is that the transfer integrals between the d orbitals of the copper atoms are only effective transfer integrals, through the hybridization of these d orbitals with the p orbitals of the intercalated oxygen atoms, and thus they are smaller than in more usual situations. Also, owing to the layered structure, the number of nearest neighbouring sites of a copper atom is smaller than usually. Thus, it can be thought that, in these compounds, the Coulomb repulsive energy has more or less the same order of magnitude than the one electron terms in the Hubbard hamiltonian, but is probably not much larger. Consequently, it seems to be not more unreasonable to do, as a starting point, the hypothesis of a weak correlation between the electrons than to do the opposite one.

By taking into account the sharpness of the peak which exists in the electronic density of states in the neighbourhood of the Fermi energy, we find that, even in the weak coupling limit, the electron-phonon interaction can give rise to a large value of the superconducting transition temperature. Thus, we see no reason why a phonon induced mechanism would be more unlikely for high T_c superconductivity than the other mechanisms which have been proposed.

Acknowledgements

The author wishes to thank Professor J. Friedel for very stimulating discussions.

REFERENCES

(1) P. Thiry, G. Rossi, Y. Petroff, A. Revcholevski and J. Jegondez, Europhysics Lett. 5, 55 (1988).

(2) J. Friedel, Physica C 153-155, 1610 (1988).

(3) A. Balzarotti, M. De Crescenzi, C. Giovannella, R. Messi, N. Motta, F. Patella and A. Sgarlata, Phys. Rev. B, 36, 8285 (1987).

(4) A. Samsavar, T. Miller, T.C. Chiang, B.G. Pazol, T.A. Friedmann and D.M. Ginzberg, Phys. Rev. B, 37, 5164 (1988).

(5) L.F. Mattheiss, Phys. Rev. Lett., 58, 1028 (1987).

(6) C.E. Gough, M.S. Colclough, E.M. Forgan, R.G. Jordan, M. Keene, C.M. Muirhead, A.I.E. Rae, N. Thomas, J.S. Abell and S. Sutton, Nature 326, 855 (1987).

(7) D. Estève, J.M. Martinis, C. Urbina, M.H. Devoret, G. Collin, P. Monod, M. Ribaud and A. Revcolevschi, Europhys. Lett. 3, 1237 (1987).

(8) J. Niemeyer, M.R. Dietrich and C. Politis, Z. Phys. B 67, 155 (1987).

(9) J.G. Bednorz and K.A. Müller, Z. Phys. B 64, 189 (1986).

(10) J. Bardeen, L.N. Cooper and J.R. Schrieffer, Phys. Rev. 108, 1175 (1957).

(11) J. Labbé, S. Barisic and J. Friedel, Phys. Rev. Lett., 19, 1039 (1967).

(12) J. Labbé, Phys. Rev. 172, 45 (1968).

(13) J.D. Jorgensen, H.B. Schüttler, D.G. Hinks, D.W. Capone, K. Zhang, M.B. Brodsky and D.J. Scalapino, Phys. Rev. Lett. 58, 1024 (1987).

(14) S. Barisic, I. Batistic and J. Friedel, Europhys. Lett. 3, 1231 (1987).

(15) J. Friedel, J. Phys. France 49, 1435 (1988).

(16) R. Combescot and J. Labbé, Phys. Rev. B, 38, 262 (1988).

(17) J.T. Market, E.A. Early, T. Bjornholm, S. Gamaty, B.W. Lee, J.J. Neumeier, R.D. Price, C.L. Seaman, and M.B. Maple, Physica C 158, 178 (1989).

(18) J. Labbé and J. Bok, Europhys. Lett. 3, 1225 (1987).

(19) J.E. Hirsch and D.J. Scalapino, Phys. Rev. Lett. 56, 2732 (1986).

(20) H.E. Dzaloskinsky, Proc. Adriatico Res. Conf. on High Temperature Superconductivity, Trieste (1987).

(21) J. Friedel, J. Phys. France 48, 1787 (1987).

(22) B. Batlogg, G. Kourouklis, W. Weber, R.J. Cava, A. Jayaraman, A.E. White, K.T. Short, L.W. Rupp and E.A. Rietman, Phys. Rev. Lett. 59, 912 (1987).

(23) K.J. Leary, H.C. Loye, S.W. Keller, T.A. Faltens, W.K. Ham, J.N. Michaels and A.M. Stacy, Phys. Rev. Lett. 59, 1236 (1987).

(24) P. Morel and P.W. Anderson, Phys. Rev. 125, 1263 (1962).

(25) W. Kang, G. Collin, M. Ribault, J. Friedel, D. Jérome, J.M. Bassat, J.P. Couture and Ph. Odier, J. Phys. France, 48, 1181 (1987).

(26) In principle, the coefficients A_p should depend on the wave vector **k**. But in our case, they will not, as it will appear later. This arises from the fact that, as the wave vector **k** is parallel to the CuO_2 planes, the phase of the Bloch wave function (8) has the same value on all the copper atoms belonging to a same unit cell, since they are alined along an axis perpendicular to the CuO_2 planes.

(27) J. Bok, Solid State Comm. 67, 251 (1988).

(28) This leads to the susceptibility related to an antiferromagnetic ordering also along the z-axis orthogonal to the planes. Actually, we have also calculated the response to a magnetic field $B = B_0 \cos \mathbf{q}.\mathbf{r}$ inducing a ferromagnetic ordering along the z-axis. But the related susceptibility is found much smaller than the previous one. Thus, such an ordering is less stable, and has not to be considered.

(29) Since in that case $E^{(\nu)}(\mathbf{k}+\mathbf{q}) \neq E^{(\nu)}(\mathbf{k})$ within each sub-band ($\nu = 0$ or 1) and for **k** belonging to the Fermi line of this sub-band.

(30) For instance, with $t = U = 10\,\tau$, one finds that the critical value of $|E_F|$ calculated by the condition (20) is roughly five times smaller than that one calculated by the condition (7), this last one being equal to 0.03 t, corresponding to a number of holes introduced in the half-filled band equal to $x = (0.03/\pi^2 t)\,(1 + \mathrm{Ln}\,(16\,t/0.03))$, which is equal to $x = 0.04$ for $t = 0.5$ eV. From the new condition (20), the corresponding critical value of the number of holes would be much smaller, and equal to $x = 0.01$.

INVESTIGATIONS OF A FEW NOVEL SERIES OF THALLIUM CUPRATE SUPERCONDUCTORS SHOWING POSSIBLE CHANGE IN THE SIGN OF CHARGE CARRIERS WITH COMPOSITION

C.N.R. Rao, R. Vijayaraghavan, A.K. Ganguli, V. Manivannan and N.Y. Vasanthacharya
Indian Institute of Science, Bangalore, India

INTRODUCTION

Cuprates of the general formula $TlCa_{n-1}Ba_nCu_nO_{2n+3}$ are well-characterized high-temperature superconductors [1-4]. It has, however, not been possible to prepare and characterize the corresponding Strontium analogues, $Tl(Ca,Sr)_{n+1}Cu_nO_{2n+3}$. The Sr analogues can be stabilized by partly substituting Tl by Pb as in $Tl_{0.5}Pb_{0.5}Ca_{n-1}Sr_2Cu_nO_{2n+3+\delta}$ [5,6] or Ca by a rare earth as in $TlCa_{0.5}Ln_{0.5}Sr_2Cu_2O_{6+\delta}$ [7]. It has been shown recently that cuprates of the general formula $TlCa_{0.5}Ln_{0.5}Sr_2Cu_2O_{6+\delta}$ exhibit T_cs in the range 60-90K, with the T_c decreasing with the size of the rare earth ion [8]. Varying x in $TlCa_{1-x}Ln_xSr_2Cu_2O_{6+\delta}$ is expected to markedly change the electronic properties. Accordingly, the x=1 members, $TlLnSr_2Cu_2O_{6+\delta}$, are generally insulators. It is possible that a change in the sign of the majority charge carriers can occur in $TlCa_{1-x}Ln_xSr_2Cu_2O_{6+\delta}$ with a change in x. We have investigated several members of the $TlCa_{1-x}Ln_xSr_2Cu_2O_{6+\delta}$ series (T_c, ~ 60-90K) extensively over a range of x as well as with different rare earth substitutions. Besides finding positive thermopower, α, in the $0.25 < x < 0.8$ range and negative thermopower when x = 0.25, suggesting a possible change in the sign of the majority charge carriers, we have noticed a compositionally controlled metal-insulator transition in the normal state.

Most of the Tl cuprates described in the literature contain Ca along with either Ba or Sr. We have explored the possibility of synthesizing cuprate superconductors in Tl-Sr-Cu-O system. We report the successful synthesis and characterization of these cuprates with one or two CuO_2 sheets with T_cs of 40 and 80K respectively. The $TlSr_{2-x}La_xCuO_{5+\delta}$ system shows negative or positive thermopower depending on x. In view of the change in sign of α with composition found in some of these new Tl cuprates, we have examined the thermopower, α, of $Tl_yCa_{1-y}Y_xBa_2Cu_2O_{6+\delta}$ and $Tl_{1-x}Pb_xCaSr_2Cu_2O_{6+\delta}$. The results of these thermopower measurements are indeed interesting and reveal

----

*Contribution No. 647 from the Solid State & Structural Chemistry Unit

**To whom correspondence should be addressed.

certain general features.

THE $TlCa_{1-x}Ln_xSr_2Cu_2O_{6+\delta}$ SERIES

Members of the $TlCa_{1-x}Ln_xSr_2Cu_2O_{6+\delta}$ ($0.25 \leqslant x \leqslant 1.0$) series possess the 1122 type layered structure of $TlCaBa_2Cu_2O_7$ with the P4/mmm space group. In Fig. 1, we show x-ray diffraction patterns of several members

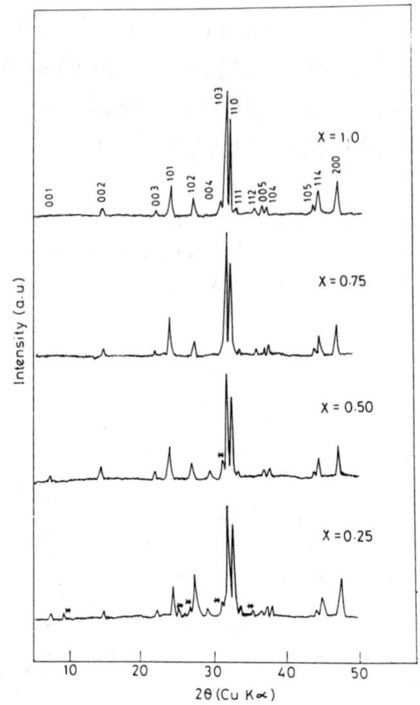

Fig. 1. Powder x-ray diffraction patterns of members of the $TlCa_{1-x}Nd_xSr_2Cu_2O_{6+\delta}$ system. Asterisk denotes impurity phase.

of a series as a function of composition to illustrate that all of them have the same structure. It is generally found that when $x < 0.25$ impurity phases appear, one of them being the 1021 type phase. The $x = 0.5$ composition is always monophasic as shown in Fig. 2. The a parameter increases with x for all the members except Yb; the c - parameter decreases with increase in x. When $x = 0.5$ the a and c parameters as well as the unit cell volume show a minimum around 1.05Å corresponding to Gd^{3+} ion. When Ca is fully substituted by Ln ($x = 1$) all the three parameters increase with the radius of the rare earth ion with a plateau in the middle (around Gd^{3+}). It is possible that there are differences in the oxygen stoichiometry with the change in the Ln ion (for the same x value) causing some changes in the lattice parameters.

In Fig. 3 and Fig. 4, we show the temperature variation of the normalized resistance of $TlCa_{1-x}Eu_xSr_2Cu_2O_{6+\delta}$ and $TlCa_{1-x}Yb_xSr_2Cu_2O_{6+\delta}$ and $TlCa_{1-x}Pr_xSr_2Cu_2O_{6+\delta}$. All these cuprates exhibit high T_c superconductivity in the composition range $0.25 \leqslant x \leqslant 0.75$ and also traverse

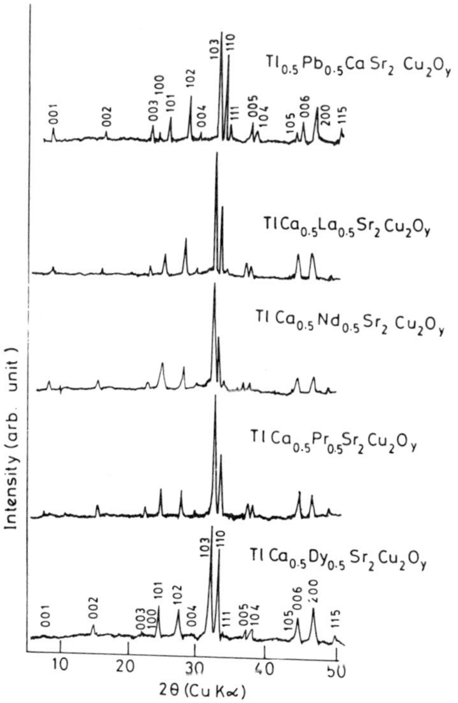

Fig. 2. Powder x-ray diffraction patterns of $TlCa_{0.5}Ln_{0.5}Sr_2Cu_2O_{6+\delta}$ (Ln = La,Nd,Pr and Dy) and $Tl_{0.5}Pb_{0.5}CaSr_2Cu_2O_{6+\delta}$.

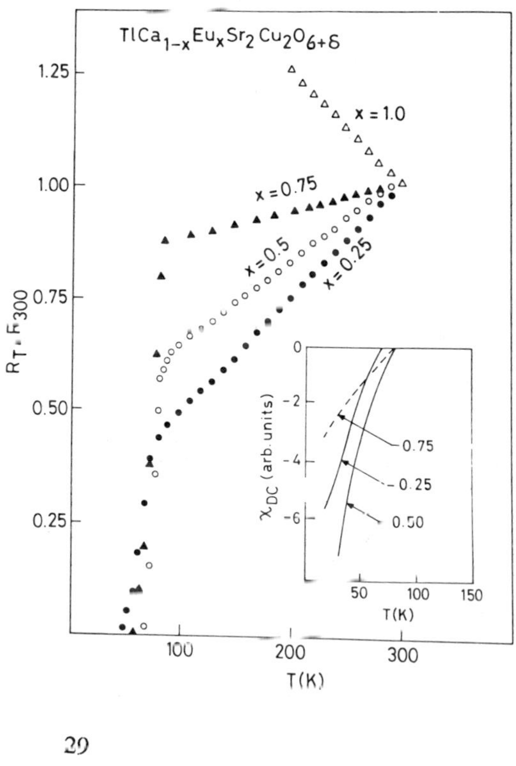

Fig. 3. Plots of the normalized resistivity against temperature for the $TlCa_{1-x}Eu_xSr_2Cu_2O_{6+\delta}$ system. Inset shows the DC magnetic susceptibility data.

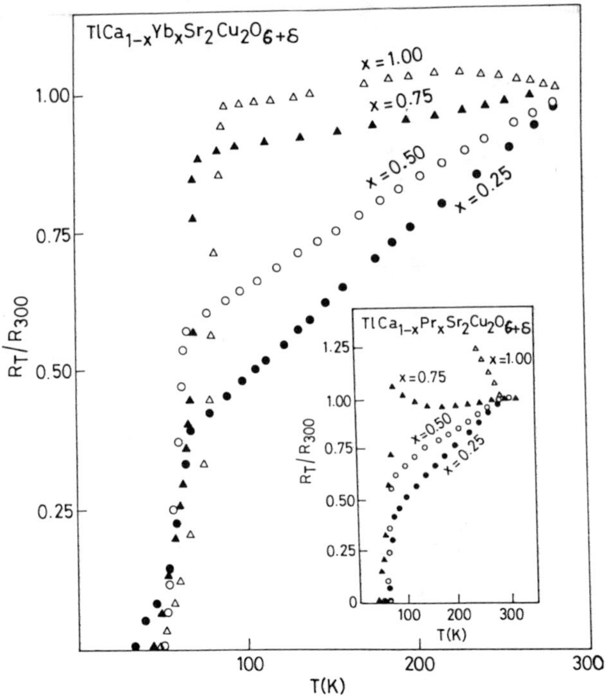

Fig. 4. Plots of the normalized resistivity against temperature for the $TlCa_{1-x}Ln_xSr_2Cu_2O_{6+\delta}$ systems where Ln = Yb or Pr.

a metal-insulator transition in the normal state with change in composition. The metal-insulator transition is possibly associated with the increase in oxygen content accompanying the increase in x. The value of the resistivity at which the change of sign of the temperature-coefficient of resistance occurs is rather low ($\sim 10^{-1}$ ohm. cm). $TlCa_{1-x}Yb_xSr_2Cu_2O_{6+\delta}$ is somewhat unique in that even the x = 1 sample exhibits superconductivity (Fig. 4). The D.C. susceptibility data of the $TlCa_{1-x}Ln_xSr_2Cu_2O_{6+\delta}$ systems confirm the occurrence of superconductivity in these cuprates. We show typical data for the Eu derivatives in the inset of Fig. 3. The superconducting fraction is generally around 20% or less. For a given Ln substitution, the temperature corresponding to the onset of superconductivity is maximum around x = 0.5; the Yb system, however, shows an increase in T_c with increase in x.

In Fig. 5 we show the variation of the T_c of $TlCa_{0.5}Ln_{0.5}Sr_2Cu_2O_{6+\delta}$ as a function of the radius of the ion. We find an increase in T_c with the increase in the radius of Ln^{3+}. This trend possibly arises from the variation in the basicity of the Ln ion or the covalency of the Ln-O bond with the change in the Ln ion. It is also possible that the oxygen stoichiometry (value of δ) changes with the Ln ion to some extent, in turn affecting the T_c value, although the latter explanation would not be consistent with the observed monotonic variation of T_c with Ln.

We have measured the Seebeck Coefficient, α, of several members of these cuprates to determine the sign of the charge carriers.

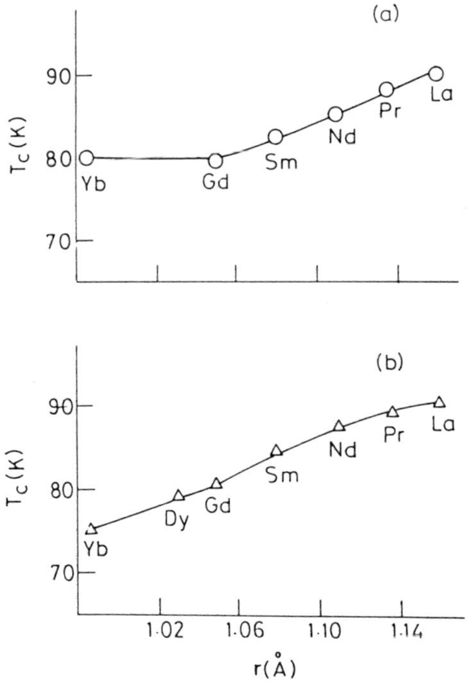

Fig. 5. Variation of T_c with the radius of the Ln ion in $TlCa_{0.5}Ln_{0.5}Sr_2Cu_2O_{6+\delta}$: (a) T_c from susceptibility data; (b) T_c from resistivity data.

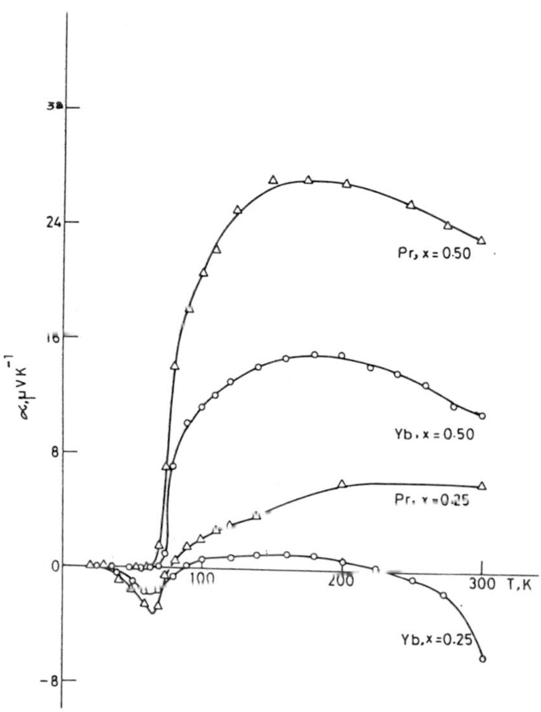

Fig. 6. Temperature-variation of the Seebeck coefficient of $TlCa_{1-x}Ln_xSr_2Cu_2O_{6+\delta}$ (Ln = Yb and Pr).

In general, α is positive at 300K in most of the systems; the magnitude of positive α increases with x, possibly due to the increase in oxygen content with increase in x. What is interesting is that at low temperatures (~ 75K), all the cuprates show a change in sign at x = 0.25. In the Yb system, we find that the x = 0.25 sample shows a negative even at 300K. In Fig. 6, we show the temperature variation of α of a few cuprates. At x = 0.5, all the systems show positive values of α which go to zero at the superconducting transition. When x = 0.25, however, they show negative α at low temperatures and the negative α then goes to zero. These results show that in $TlCa_{1-x}Ln_xSr_2Cu_2O_{7+\delta}$ family of cuprates both holes and electrons are present as charge carriers. It would seem that when x = 0.25, electrons become the majority charge carriers. When x > 0.25, holes seem to take over, in part possibly due to an increase in the oxygen content. The $TlCa_{1-x}Ln_xSr_2Cu_2O_{6+\delta}$ series is, therefore, unique showing negative thermopower at low x which changes over to positive thermopower at higher values of x. What is interesting, however, is that in the 100-250K region the α - T plots always have a positive slope, as if holes are the charge carriers (independent of the sign of α). The change in sign in α around T_c is rather difficult to interpret in a two-band system of this kind.

The Cu states have been investigated by means of Cu K-near edge spectra (XANES) as well as by Cu (2p) core-level spectroscopy. The Cu K-edge spectra clearly show the presence of Cu in the 1+ state in these systems shown by the feature at 8993 eV due to a symmetry related transition. XANES features due to Cu^{2+} are also present in these cuprates. X-ray absorption edge spectra as well as core-level spectra show the rare earth ions to be in the 3+ state.

Cu $2P_{3/2}$ core level spectra show a main peak due to a wellscreened state around 933 eV and a higher bonding energy satellite due to the poorly screened state. It is known that the relative intensity of the satellite with respect to the main peak is a measure of the proportion of the Cu^{2+}-type state present [9,10]. The relative intensity of the satellite in $TlCa_{1-x}Ln_xSr_2Cu_2O_{6+\delta}$ (~ 0.4) is smaller than that in CuO (0.5) suggesting that the proportion of the Cu^{2+} species is smaller in these cuprates. There is some indication that in $TlCa_{1-x}Nd_xSr_2Cu_2O_{6+\delta}$, the proportion of Cu^{2+} is least when x = 0.25. Since the Seebeck Coefficient α becomes negative when x = 0.25, it would appear that the extra electron donated by the rare earth ion is responsible for "electron-superconductivity" as well as for increasing the proportion of the Cu^{1+} state. This probably happens because the increase in oxygen content at low values of x is negligible. At higher x values, however, the increase in oxygen content probably compensates for the electrons donated by Ln.

THE $TlSr_{n+1-x}Ln_xCu_nO_{2n+3+\delta}$ SERIES

We have prepared monophasic compositions of the $TlSr_{2-x}La_xCuO_{5+\delta}$ family ($0.5 \leqslant x \leqslant 1$) possessing the 1021 type of structure by the reaction of the component oxides in a sealed tube at 1170K for 2 hrs. In Fig. 7 we show powder x-ray diffraction patterns of two members of this family. The patterns can be indexed on the basis of a tetragonal cell (space group P4/mmm). The unit cell parameters are: a = 3.762Å, c = 8.923Å for x = 0.5, a = 3.768Å, c = 8.90Å for x = 0.75 and a = 3.78Å, c = 8.893Å for x = 1.00. The c-parameter decreases with the increase in the La content. We have also been able to synthesize the 1021 cuprate with Pr and Nd substituting for Sr. In Fig. 7, we

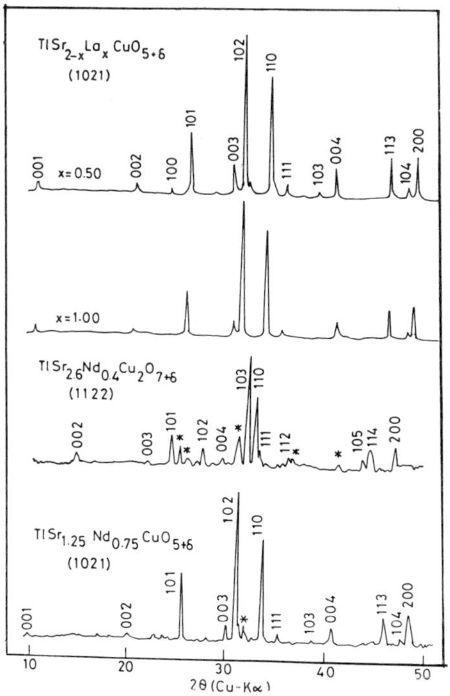

Fig. 7. Powder x-ray diffraction patterns of the $TlSr_{2-x}La_x CuO_{5+\delta}$ (1021) system. Also shown are x-ray patterns of two oxides of the Tl-Sr-Nd-Cu-O system. Asterisk denotes impurity line.

have shown the x-ray diffraction pattern of $TlSr_{1.25}Nd_{0.75}CuO_{5+\delta}$ with a few impurity lines.

Starting from various Tl-Sr-Nd-Cu-O compositions we have been able to prepare the n = 2 member of the new series possessing the 1122 layered structure in 85 ± 5% purity; the 1021 phase [40K T_c] is the impurity phase. In Fig. 7, we have shown an XRD pattern of a 1122 cuprate (~85% pure) obtained from the starting composition $TlSr_{2.6}Nd_{0.4}Cu_2O_y$. This also has a tetragonal structure (space group P4/mmm) with unit cell parameters a = 3.848Å and c = 12.218Å. In both the 1021 and 1122 members of the $TlSr_{n+1-x}Ln_xCu_nO_{2n+3+\delta}$ series, the rare earth ion Ln and Sr occupy the same site. Partial substitution of Tl by Ln is also possible. Based on a study of a variety of compositions in the case of the 1021 derivative, we estimate the replacement of Tl by Ln is likely to be less than 25%.

In Fig. 8, we show the electrical resistivity data of three members of $TlSr_{2-x}La_xCuO_y$ family. The x = 0.5 sample remains metallic down to 20K, but the x = 0.75 and x = 1.0 samples show superconductivity with an onset of temperature around 40K. The x = 1.0 composition shows zero-resistance around 30K. Magnetic susceptibility measurements (inset of Fig. 8) confirm the superconductivity of the 1021 cuprates. The superconducting fraction is around 10% in a favourable composition (such as x = 1.0).

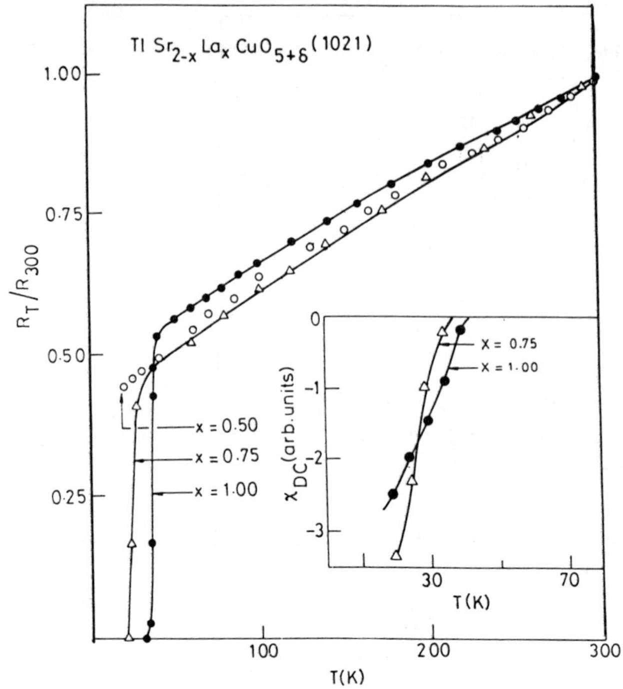

Fig. 8. Temperature-variation of the normalized resistance of $TlSr_{2-x}La_xCuO_{5+\delta}$. Inset shows the temperature variation of the magnetic susceptibility.

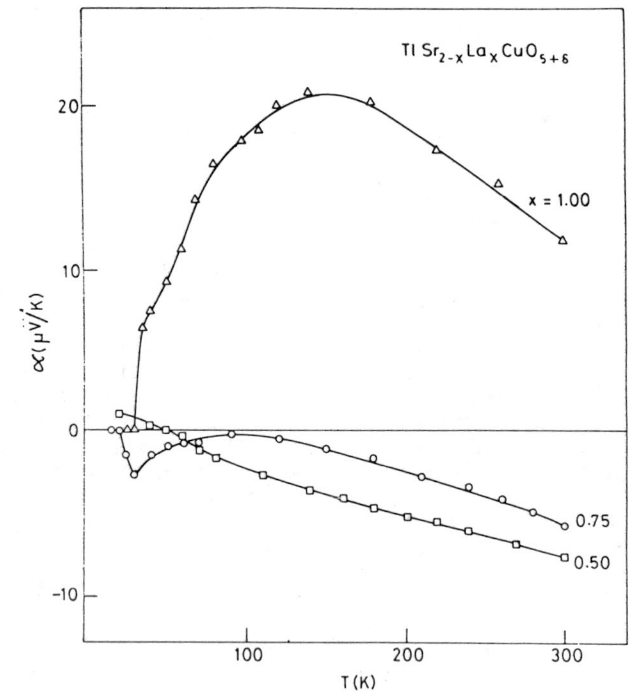

Fig. 9. Temperature-variation of the Seebeck coefficient, α, of $TlSr_{2-x}La_xCuO_{5+\delta}$.

Thermopower measurements on $Tl_1Sr_{2-x}La_xCuO_y$ reveal a fascinating feature of this family. The metallic $x = 0.5$ sample shows a negative Seebeck coefficient, α, from 300K down to 50K and becomes slightly positive at lower temperatures (Fig. 9). In the superconducting $x = 0.75$ sample, α is negative from 300K down to T_c, suggesting that electrons may be the charge carriers. The $x = 1.0$ sample shows positive α from 300K down to T_c. It therefore appears that $TlSr_{2-x}La_xCuO_y$ is another novel series showing a change in the sign of the thermopower with change in composition. Here again the sign of the slope of the α - T plots at relatively high temperatures (100-250K) is negative, suggesting as if electrons are the "real" charge carriers independent of the sign of α.

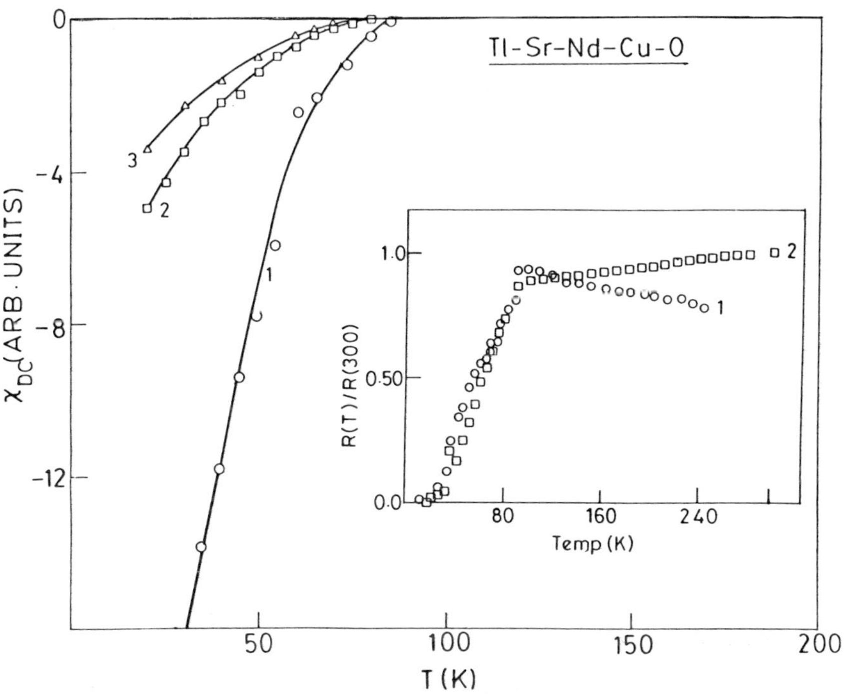

Fig. 10. Magnetic susceptibility data of the 1122 type cuprates obtained from the following starting compositions:
(1) $TlSr_{3.5}Nd_{0.5}Cu_2O_y$; (2) $Tl_{1.2}Sr_{3.5}Nd_{0.5}Cu_2O_y$;
(3) $TlSr_{2.6}Nd_{0.4}Cu_2O_y$. Inset shows the resistivity behaviour of two of the compositions.

The 1021 sample of the Tl-Sr-Nd-Cu-O system also shows superconductivity around 40K. It is really interesting that in the Nd series we could obtain the n = 2 member with the 1122 structure. In Fig. 10 we show the magnetic susceptibility data on the 1122 compounds of Nd obtained from different starting compositions. They indicate onset of superconductivity around 80K. The superconducting fraction in the favourable composition is 10%. Electrical resistivity measurements (Fig. 10) also show a similar onset temperature. These 1122 superconductors show positive thermopower throughout. We have also prepared similar 1122 cuprates of Pr.

$Tl_yCa_{1-x}Y_xBa_2Cu_2O_{6+\delta}$ AND $Tl_{1-x}Pb_xCaSr_2Cu_2O_{6+\delta}$

The results of the different Tl-Ca(Ln)-Sr-Cu-O systems discussed hitherto showing a change in the sign of the thermopower with a change in the composition, prompted us to examine the thermopower of the well-known Tl-Ca-Ba-Cu-O system as well as of the Tl(Pb)-Ca-Sr-Cu-O system. In Fig. 11a we show the thermopower data of $Tl_yCa_{1-x}Y_xBa_2Cu_2O_{6+\delta}$.

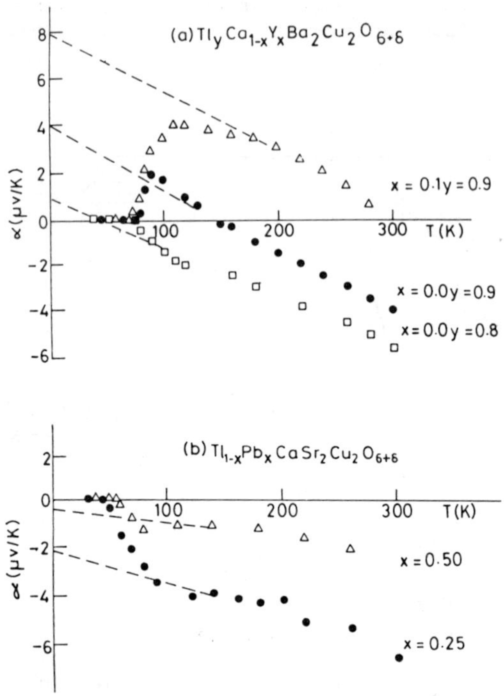

Fig. 11. Thermopower data of (a) $Tl_yCa_{1-x}Y_xBa_2Cu_2O_{6+\delta}$ and (b) $Tl_{1-x}Pb_xCaSr_2Cu_2O_{6+\delta}$.

Here, we see that α is negative from room temperature down to the T_c especially when there is Tl-deficiency and the yttrium content is zero (x = 0.0). Even where α is negative throughout the temperature range, the α - T slope is negative in the high-temperature regime.

Thermopower measurements on $Tl_{1-x}Pb_xCaSr_2Cu_2O_{6+\delta}$ show a similar behaviour (Fig. 11b). All these cuprates show negative α down to the T_c, but the α - T slopes are also negative as if electrons are the charge carriers. It is to be noted however that Pb in this series of cuprates is in the 4+ state as shown by x-ray absorption spectroscopy (Fig. 12). This would suggest as though electrons donated by Pb could also cause negative α.

CONCLUDING REMARKS

The present study has established superconductivity in two new series of superconductors $TlCa_{1-x}Ln_xSr_2Cu_2O_{6+\delta}$ and $TlSr_{n+1-x}Ln_xCuO_{2n+3+\delta}$.

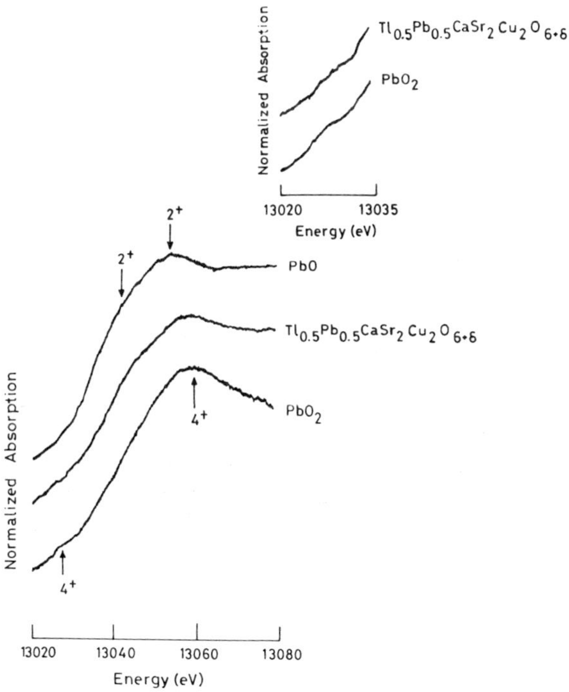

Fig. 12. Pb-L absorption edge of $Tl_{0.5}Pb_{0.5}CaSr_2Cu_2O_{6+\delta}$.

More interestingly, in these two series as well as in $Tl_yCa_{1-x}Y_xBa_2Cu_2O_{6+\delta}$, we see a change in the sign of the thermopower with composition. Such a change in the sign of thermopower would normally be taken as evidence for a change in the sign of the majority charge carriers. Thus, when the sign of the thermopower is negative, especially in the transition region, we would assume electron-superconductivity to be present. At the same time, however, we find that all these cuprates show negative slopes of the α- T plots in the high-temperature regime (100-250K). Such negative α - T slopes are generally considered to be characteristic of hole carriers. This is consistent in the case of $TlCaBa_2Cu_2O_{6+\delta}$ where holes have been generally considered to be responsible for superconductivity. The question that is to be answered is whether in fact, holes are the charge carriers in all these cuprates, independent of the sign of or there is electron-superconductivity at least in the compositions where α is negative throughout as in $Tl_{1-x}Pb_xCaSr_2Cu_2O_{6+\delta}$, where even the extrapolated value of α at zero K is also negative. In the electron superconductor $Nd_{2-x}Ce_xCuO_4$ [9], the thermopower is negative but the α - T slope is negative from 120K to T_c [10]. We recognize that a change in the sign of α in the transition region alone is somewhat difficult to explain in such complex two-band systems where both electrons and holes are present.

ACKNOWLEDGEMENT

The authors thank the Department of Science and Technology, U.S. National Science Foundation and University Grants Commission for support of this research.

REFERENCES

[1] A.K. Ganguli, G.N. Subbanna and C.N.R. Rao, Physica C, 156 (1988) 181.

[2] M. Hervieu, A. Maignan, C. Martin, C. Michel, J. Provost and B. Raveau, J. Solid State Chem. 75 (1988) 212.

[3] S.S.P. Parkin, V.Y. Lee, A.I. Nazzal, R. Savoy, R. Beyers and S.J. La Placa, Phys. Rev. Lett. 61 (1988) 750.

[4] C.N.R. Rao and B. Raveau, Acc. Chem. Res. 22 (1989) 106.

[5] A.K. Ganguli, K.S. Nanjundaswamy and C.N.R. Rao, Physica C 156 (1988) 788.

[6] M.A. Subramanian, C.C. Torardi, J. Gopalakrishnan, P.L. Gai, J.C. Calabrese, T.R. Askew, R.B. Flippen and A.W. Sleight, Science 242 (1988) 249.

[7] C.N.R. Rao, A.K. Ganguli and R. Vijayaraghavan, Phys. Rev. B August issue 1989.

[8] R.Vijayaraghavan, A.K. Ganguli, N.Y. Vasanthacharya, M.K. Rajumon, G.U. Kulkarni, G. Sankar, D.D. Sarma, A.K. Sood, N. Chandrabhas and C.N.R. Rao, Superconductor Science and Technology (in print).

[9] Y. Tokura, H. Takagi and S. Uchida, Nature 337 (1989) 345.

[10] M.E.L. Morales and P.M. Grant, preprint.

SOME STRUCTURAL AND CHEMICAL ASPECTS OF THE Bi-Pb-Sr-Ca-Cu-O SYSTEM

M. Takano
Kyoto University, Kyoto, Japan

INTRODUCTION

The high T_c superconductors in the Bi-Sr-Ca-Cu-O (BSCCO) system [1-3] have "ideal" compositions of $Bi_2Sr_2Ca_{m-1}Cu_mO_z$ (m=1~3), and these crystallize in "ideal" structures in which atomic layers are stacked sequentially as $/BiO/SrO/CuO_2/Ca/---/CuO_2/SrO/BiO/$ [4-8]. More recently, the formation of the m=4 phase was reported [9,10]. T_c varies as 10K, 80K, 110K, and again 110K [10] or 95K [9] as m increases. These superconductors will hereinafter be referred to as 2201, 2212, 2223, and 2234.

Experimental monophasic compositions, however, considerably deviate from these simple cation ratios as typically seen in the 2201 phase to be described below, and their structures are seriously strained as visualized by transmission electron microscopy (TEM) [11]. The strain mode can be expressed in terms of one dimensional (1D), incommensurate modulation. And chemical substitutions like Pb for Bi [12-15] and rare earth elements for Sr [16,17] have been found to influence the modulation considerably. To study the change in T_c in relation to the composition and structure encounters difficulties caused by the complexity in composition and structure and also by the ambiguity of cation valency (Bi^{3+}/Bi^{5+}, Pb^{2+}/Pb^{4+}). Experimentally it is at least evident that oxygen content is important [14,18] as in $La_2CuO_{4+\delta}$ [19] and $YBa_2Cu_3O_{7-\delta}$.

Some other features characterizing the total BSCCO system include the fact that the melting point ranges from 600°C to above 1000°C, strong influences of partial melting on the reaction process, vaporization of metallic elements (especially of Pb) during reaction, and the presence of so many phases surrounding the superconducting phases. These and the complex compositions of the superconductors offer both advantages and disadvantages in the fabrication process.

Intended in this chapter is never a comprehensive collection of a vast number of published reports but is a compact description of some chemical and structural aspects of the BSCCO system which the author and his colleagues have considered important. We have tried to reveal the solid phases and their relations first of all, because we believe that such a study contributes to establishing a sound material scientific base for future developments.

PHASES AND THEIR RELATIONS IN THE Bi-Sr-Ca-Cu-O SYSTEM

Summarized in Fig. 1 are the solid phases formed in the BSCCO system by heating mixtures of Bi_2O_3, $SrCO_3$, $CaCO_3$, and CuO in air [20,21]. The firing temperatures were chosen so that melting did not occur as examined visually. There are so many phases. However, the distribution is not even: In comparison with the Bi-Ca-Cu-O system, there are a variety of ternary oxides in the Bi-Sr-Cu-O (BSCO) system, among which very interesting structural relations have been found.

Bi-Sr-Cu-O (BSCO) System

Four kinds of phases marked A$\sim$D in Fig. 1 are formed in the BSCO system at 840°C in air [21]. The dotted lines connecting CuO, $Bi_{1-x}Sr_xO_z$, and Bi_2CuO_4 indicate the phase relation established at 800°C, but melting occurs at 840°C in this range. And in the area connecting C, D, $SrCuO_2$, and $Sr_3Cu_5O_z$ [22] (or more exactly $Sr_{14}Cu_{24}O_{41}$ as deduced from structural studies [23,24]), two possibilities remain unsettled as shown using a pair of dotted-and-dashed lines. It is possible that there is a further structure in this range at slightly different temperatures. A stands for a solid solution expressed as $Bi_{2+x}Sr_{2-x}Cu_{1+y}O_z$ ($0.1\lesssim x\lesssim 0.6$, $0<y<x/2$), while B does for a compound $Bi_{17}Sr_{16}Cu_7O_y$. At the "ideal" 2201 composition three mixed phases are obtained, A, B, and $Sr_{14}Cu_{24}O_{41}$, while slight deviations toward Bi/Sr>1 and Bi/Cu>2 isolate A and B, respectively. Though these have rather similar compositions, the particles are formed in quite different shapes, plates vs. needles, as compared in Fig. 2.

Solid solution A crystallizes in pseudotetragonal structure with a=b$\sim$5.4Å and c$\sim$24Å (a=b=5.385, c=24.630Å at x=0.125 and y=x/4), in which the BiO, SrO, and CuO_2 layers are developed within the ab planes. The length of the a axis corresponds to (Cu-O-Cu bond length)$\times\sqrt{2}$, and a pair of /BiO/SrO/CuO_2/SrO/BiO/ units are packed along the c axis.

The XRD pattern of compound B was indexed by assuming orthorhombic structure with a=5.425, b=23.254, and c=24.427Å. However, according to TEM studies [25,26] a monoclinic distortion to $b_m\simeq$21Å and $\angle(b_m,c)$=105° is apparent. From the similarity in the lattice dimension and TEM image (Figs. 3(a), (b), and Fig. 5) between A and B, a simplified structural model for $Bi_{17}Sr_{16}Cu_7O_z$ (2z$\simeq$97) has been deduced as shown in Fig. 3(c). The /BiO/perovskite slab /BiO/ unit is sheared into plates as wide as only $\sim$20Å, without being extended two dimensionally as in A. Figure 3(c) illustrates how these plates are packed. There should occur an atomic rearrangement around the shear plane that yields the Bi-rich and

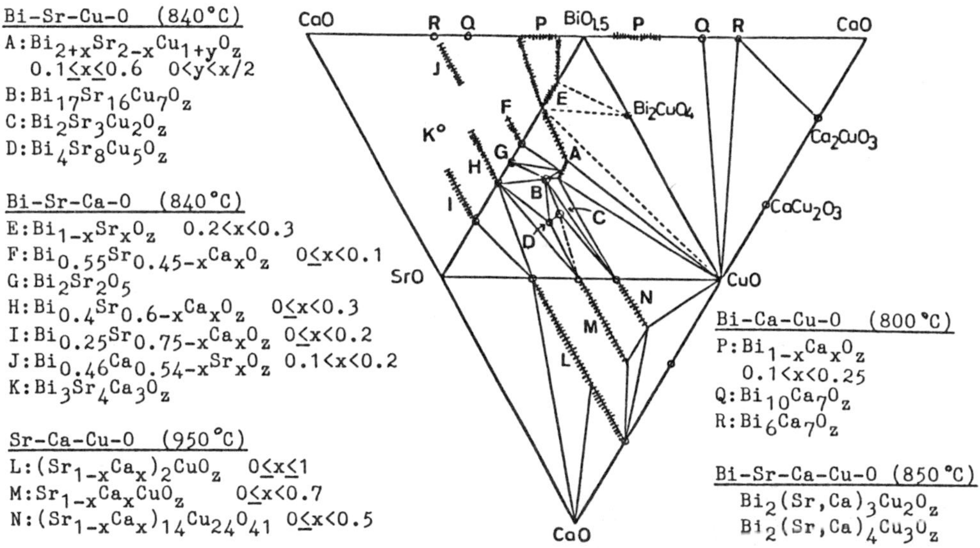

Fig. 1 Phases formed in the Bi-Sr-Ca-Cu-O System [20,21].

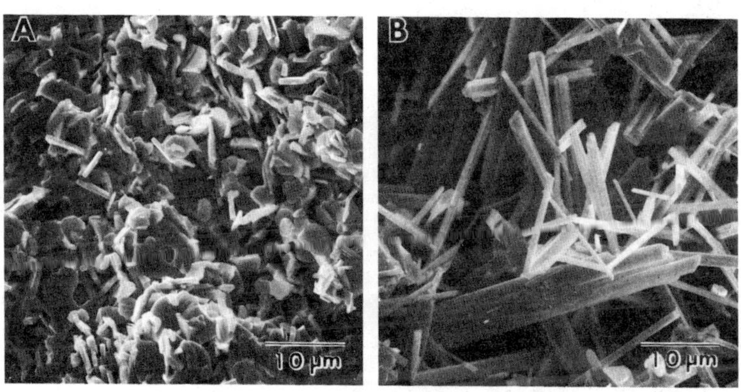

Fig. 2 Particles of $Bi_{2.125}Sr_{1.875}Cu_{1.03}O_z$ (phase A in Fig. 1) and of $Bi_{17}Sr_{16}Cu_7O_z$ (phase B) [21]. The contrast in particle shape should be noticed.

Cu-poor composition. The needle shaped particles are grown along the a axis, in sharp contrast with particles of A flattened along the ab plane.

Another type of interruption of the /BiO/ perovskite slab /BiO/ unit by Sr-Cu-O perovskite layers yields $Bi_4Sr_8Cu_5O_z$ (z=20.5 ideally, Fmmm, z=8, a=5.373, b=33.907, c=23.966Å) as studied by Fuertes et al. [27]. From another viewpoint, this structure may be explained as follows. The Sr-Cu-O perovskite layers cross perpendicular to each other and Bi-O cores fill the empty tubes formed as above. The crossing causes 40% of the Cu atoms to be coordinated in distorted pyramids.

$Bi_2Sr_3Cu_2O_z$ crystallizes in orthorhombic structure with two axial lengths of again 5.396Å and 24.804Å suggesting the presence of the /BiO/ perovskite slab /BiO/ unit. According to TEM studies [25], the b axis is as long as 23Å, along which an incommensurate periodicity is overlapped. Vertical interruption of the above unit by another unit can be seen in the [100] TEM image, but the interruption is done in a more irregular way than in $Bi_4Sr_8Cu_5O_z$. The resulting metallic composition must be complex, not so simple as 2:3:2. It should be remarked that partial substitution of small cations like Ca and Y for Sr induces a transformation into the 2212 type structure [28].

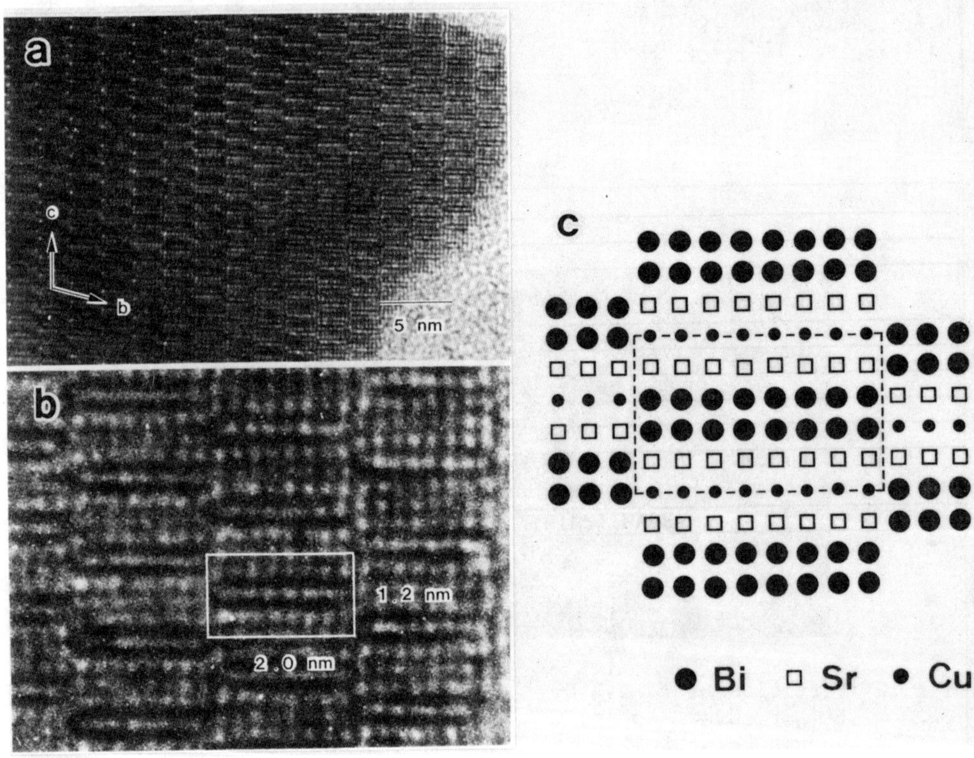

Fig. 3 [100] TEM images (a,b) and a simplified structure model (c) for $Bi_{17}Sr_{16}Cu_7O_z$ [26].

It is now possible to induce the reason why ternary oxides are formed in the BSCO system but not in the BCCO system. The ternary oxides commonly contain the $/BiO/SrO/CuO_2/SrO/BiO/$ unit, though interrupted by other units except for phase A. The SrO and CuO_2 sheets can be ideally matched with each other, if $t=d(Sr-O)/\sqrt{2}d(Cu-O)=1$, where d(Sr-O) and d(Cu-O) stand for the bond lengths. However, t<1 as estimated from the standard ionis sizes. For the combination of Ca-Cu-O, t is further reduced, thus destabilizing the above unit containing Ca^{2+} instead of Sr^{2+}. On the other hand, the fact that $Bi_2Sr_3Cu_2O_z$ does not but $Bi_2Sr_{3-x}M_xCu_2O_z$ (M=Ca or Y) does crystallize in the 2212-type structure indicates that Sr^{2+} is too large to be sandwiched in between a pair of CuO_2 sheets capped by apical oxygen atoms. In support of this, expanded $[CuO_2]_\infty$ sheets having no apical oxygen atoms allow insertion of Sr^{2+} between them as found in the high pressure phase of $SrCuO_2$ (see a later section).

Saggio et al. has studied the subsolidus diagram at 800°C for the BSCO system [29] with and without using lithium carbonate as a mineralizer. They unbelievably did not find phases B, C, and D appearing in Fig. 1 (probably their "2201" phase corresponds to B) but found a pair of compounds with approximate Bi:Sr:Cu ratios of 4:9:1 and 2:7:2. These were formed only when lithium carbonate was used. It would be interesting to study structural relations of these Cu-poor

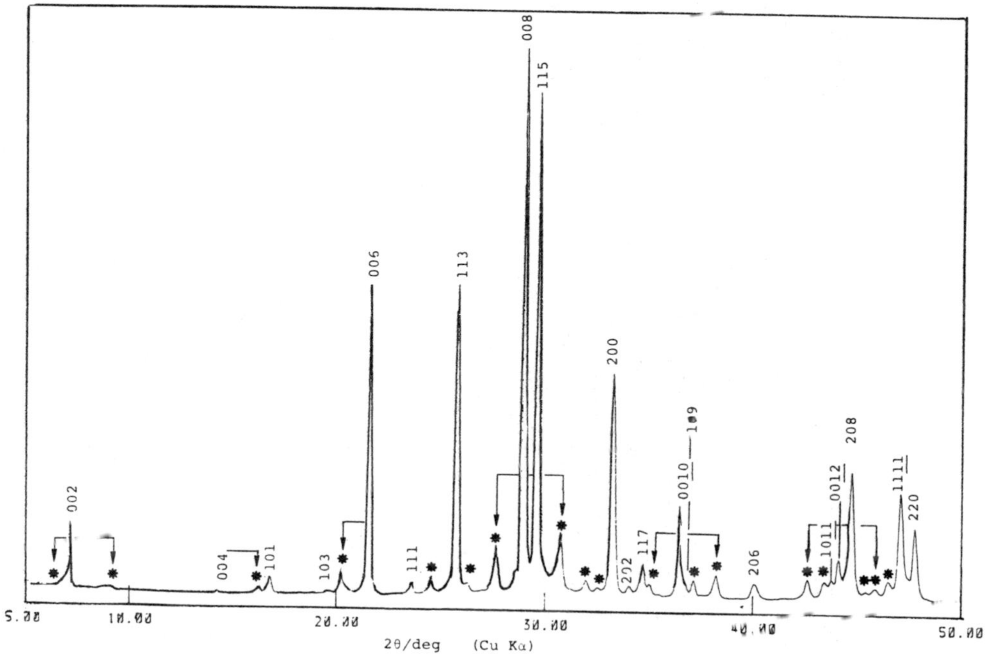

Fig. 4 Powder XRD pattern for $Bi_{2.125}Sr_{1.875}Cu_{1.031}O_z$. The peaks marked with * are satellites due to the one-dimensional modulation. (hkl) indices correspond to the pseudotetragonal cell with a=5.385Å and c=24.630Å.

phases with the others.

Solid solution A, $Bi_{2+x}Sr_{2-x}Cu_{1+y}O_z$, is formed in a considerabl wide range of $0.1 \lesssim x \lesssim 0.6$ and $0 < y < x/2$. If the Cu atoms are assumed t reside in the CuO_2 sheet only, the above composition should be rewritte as $(Bi_{2+x}Sr_{2-x})_{1/(1+y)}\square_{y/(1+y)}CuO_{z/(1+y)}$, where $\square$ stands for a catio vacancy in the Bi and/or Sr sites. Though further studies are needed t reach the final conclusion concerning the ion distribution, an excess o Bi over Sr (plus Ca) has been approved especially for 2212 [6,14,30-32

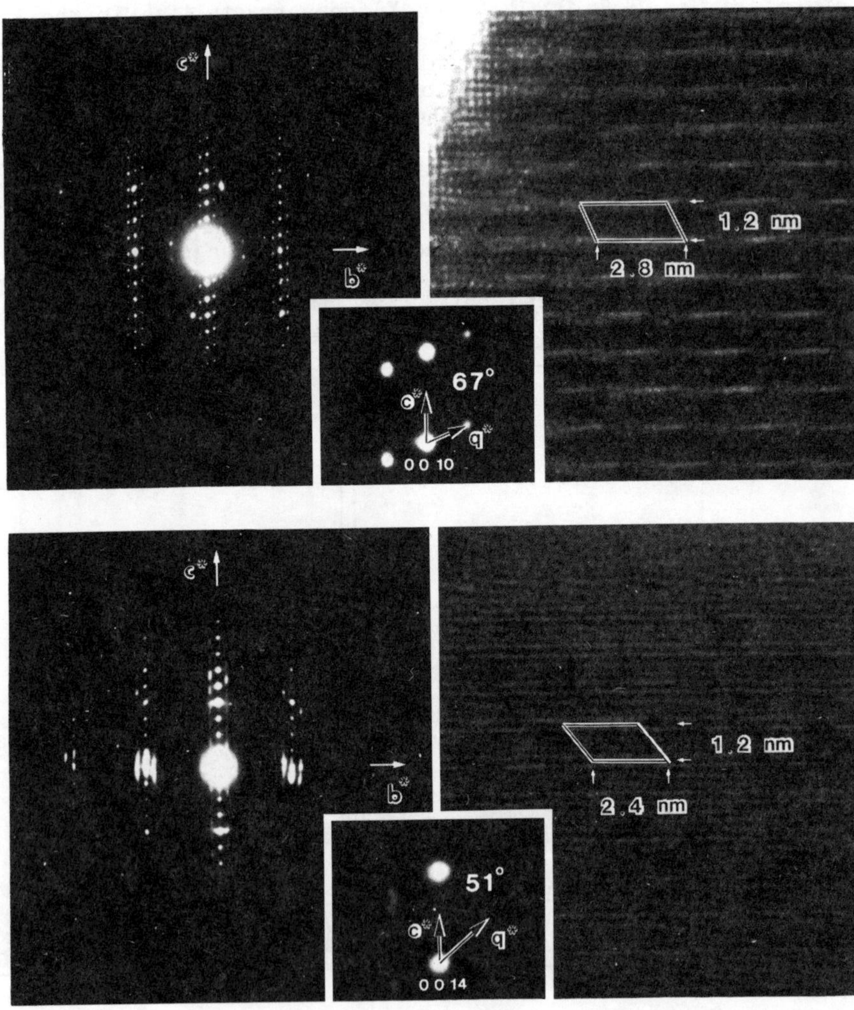

Fig. 5 [100] ED patterns and micrographs showing the compositic dependence of the 1-D modulation. $Bi_{2.1}Sr_{1.9}Cu_{1.03}O_z$ (a) anc $Bi_{2.375}Sr_{1.625}Cu_{1.09}O_z$ (b) [21].

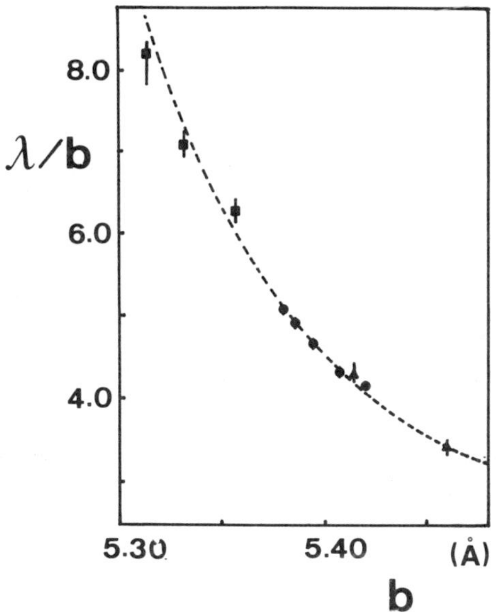

Fig. 6 Hyperbolic relation between λ/b found in the chemically substituted 2201 phase (■:Pb for Bi [26], ●:Bi for Sr [21], and ▲:La for Sr [36,37].

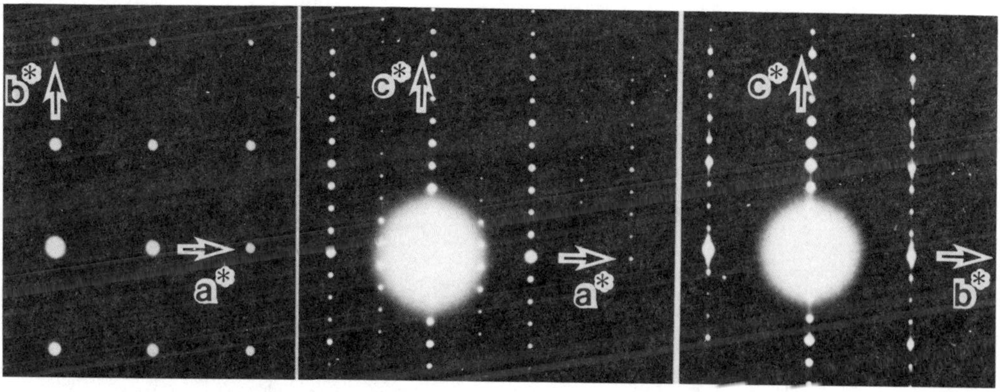

Fig. 7 ED patterns for $Bi_{1.725}Pb_{0.4}Sr_{1.875}Cu_{1.03}O_z$ [26].

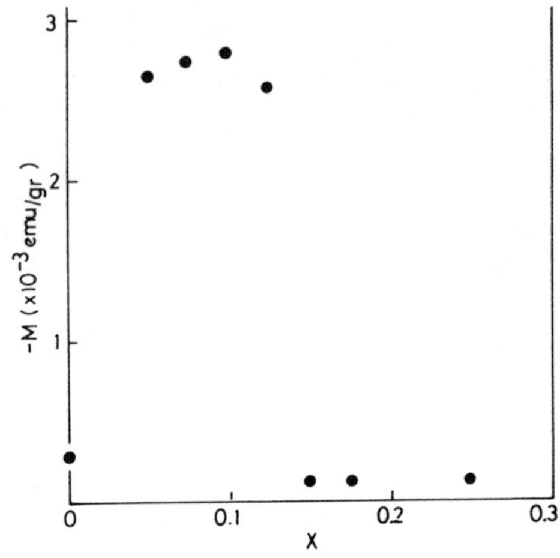

Fig. 8 Composition dependence of the diamagnetic response as a function of $Bi_{2+x}Sr_{2-x}Cu_{1+x/4}O_z$ [21].

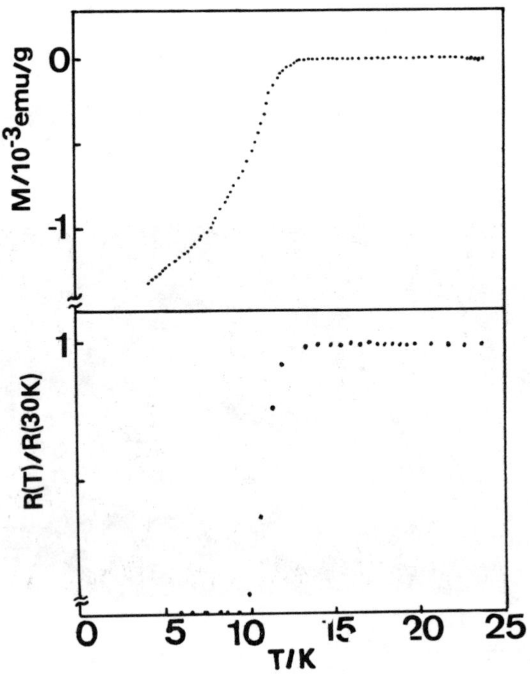

Fig. 9 Superconductivity in the Pb-substituted 2201 phase [26].

and also for Pb-substituted 2223 [13,14]. The Bi:Sr ratio varies from 21:19 to 26:14 in the 2201 phase. As can be seen typically in the powder XRD pattern for x=0.125 and y=0.031 (x/4) (Fig. 4), satellite peaks marked with * due to the 1D incommensurate modulation appear with considerable intensities. The modulation vector changes its magnitude and direction depending upon x (but not remarkably with y). In the [100] ED patterns for x=0.1 and 0.375 in Fig. 5, it is evident that the reciprocal modulation vector q is shorter and, at the same time, makes a larger angle with the c^* axis for x=0.1 than for x=0.375. The corresponding micrographs exhibit the composition dependence in the real space.

The most widely accepted explanation of the 1D modulation is insertion of extra oxygen within each BiO layer, not in-between a pair of BiO layers, the total composition thus being $BiO_{1+\delta}$ (δ>0). The insertion is not random but periodical. The Bi-Bi distance is increased locally as a result, leading to a wavy change in the Bi density. It may be said that a Bi-dilute (oxygen rich) band alternates with a Bi-dense (BiO) band along the b axis in a period of λ [11]. For phase 2201 λ decreases monotonically, as x increases, from ~5b (27Å, x=0.1) to 4.3b (23Å, x=0.5). Phases 2212 and 2223 have similar wave lengths of λ/b ~4.8 [11,15,33]. A very recent Rietveld analysis of powder neutron data [32] characterizes the 2212 phase as NBbmb/1T1, a=5.3957, b=5.3971, c=30.649Å, and q=0.2118b^*. The chemical composition deduced from this analysis is $Bi_{2.3}Sr_{1.4}Ca_{1.3}Cu_2O_9$ (i.e. δ=0.5), and excess Bi atoms substitute at the Sr sites near the oxygen-rich band, probably to compensate the local excess negative charge as suggested in [21] to explain the Bi/Sr>1 composition of the 2201 phase. It should be noticed here that the q vector have different directions in phases 2201 and 2212. It is parallel to the b^* axis in the latter, but not so in the former. The stacking phase angle of the modulated /BiO/ perovskite slab /BiO/ unit is π for the latter, while it is smaller and varies with composition x in the former. The modulation wave pattern shows body-centered orthorhombic symmetry for phases 2212 and 2223, and monoclinic symmetry for the 2201 phase [16,21,33-35].

Another problem in structure analysis is the fact that two different wave lengths seem to be microscopically mixed in a ratio giving an average value of λ/b=4.8 in the 2212 phase [11]. It should now be asked whether the annoying modulation can be removed without destroying superconductivity. In relation to this problem, Hiroi et al. [26] has noticed a specific relation between λ and b. The La-for-Sr substitution [36] and the Pb-for-Bi substitution in the 2201 phase each induces an orthorhombic distortion of the basal plane. Interestingly, the component of the modulation within the basal plane is parallel to the longer axis in the former, while the shorter axis is preferred in the latter. If is platted vs. b for these cases together with the Bi-for-Sr case, in which the basal plane remains essentially tetragonal, a universal hyperbolic relation can be found as seen in Fig. 6. The $\lambda\to\infty$ limit, the vanishment of the modulation, can be realized in $Bi_{2.125-x}Pb_xSr_{1.875}Cu_{1.03}O_z$ at x 0.4. The [100] and [001] ED patterns shown in Fig. 7 exhibit the fundamental spots assigned to the orthorhombic cell with a~5.4Å, b~5.3Å, and c~24Å without accompanying superlattice spots such as seen in Fig. 5. The disappearance of the modulation should make a detailed structure analysis easy, which is in progress.

The substitution of Bi for Sr tends to suppress the superconducting properties as seen in Fig. 8 exhibiting the composition dependence of the diamagnetic response at 4K. In the single phase region, the superconductivity appears only for $0.1 \leq x < 0.15$. Electrical conductivity becomes more and more semiconductive as x increases, though the nominal Cu valency estimated by means of iodometric analysis and a direct oxygen content analysis remains above +2 over a wide range of x. The holes must be trapped in shallow centers. The La substitution shortens the modulation wave length more remarkably than the Bi substitution does [16,37], while T_c is raised up to 25K at $Bi_2Sr_{1.6}La_{0.4}CuO_z$. On the other hand, the Pb substitution can raise T_c by a few degrees as seen in Fig. 9, in which sample the modulation has vanished. Thus, the modulation and the symmetry of the primitive cell, tetragonal or orthorhombic, do not seriously affect the superconductivity.

A very impressive experiment revealing the oxygen's crucial role was reported very recently by Shimakawa et al. [38] for $Tl_2Ba_2CuO_z$ ($z \sim 6$), the 2201 phase in the Tl-Ba-Ca-Cu-O system. Oxygen content can be varied quite reversibly in a range of Δz_0.13, and T_c itself does change reversibly from 0K (normal metal) to 87K, roughly at a rate of $(-\Delta T_c)/\Delta z = -45K/1\%$, as illustrated in Fig. 10. Powder XRD data indicate that the c axis becomes longer almost linearly as T_c increases (z decreases). More microscopically along the c axis, the distance between a pair of TlO layers remains constant, the TlO-BaO layers are separated, while the BaO-CuO_2 distance becomes shorter, as oxygen content decreases. According to Parise et al. [39], excess oxygen atoms in $Tl_{2.0}Ba_{2.0}CuO_{6+\delta}$ are located at interstitial sites between a pair of close packed TlO sheets and the change in δ crucially controls T_c. A similarly strong dependence of T_c upon oxygen content has been reported also for the 2212 phase in the BSCCO system [14,40].

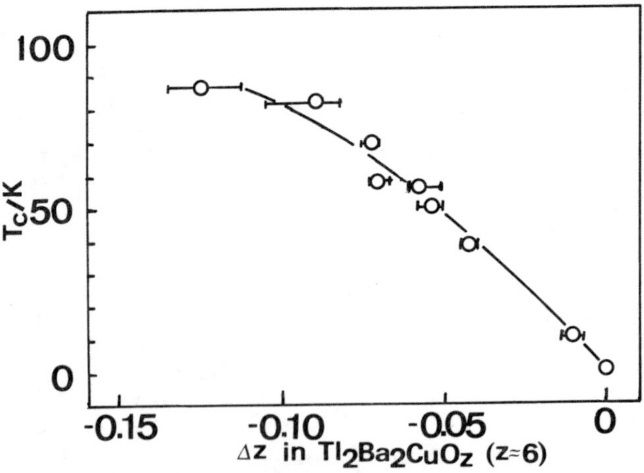

Fig. 10 T_c vs. the change in oxygen content (Δz) in $Tl_2Ba_2CuO_z$ [38].

PHASE 2223

Contrast of the 2223 phase with 2201 is interesting. The modulation wave vector is parallel to the b axis in the 2223 phase as it is in the 2212 phase. On substitution of Pb for Bi, the wave length changes stepwise from $\lambda/b=4.7$ to 8.7 and, at the same time, the modulation wave pattern changes in symmetry from body centered to base centerd orthorhombicity [12]. At a monophasic composition of $Bi_{1.73}Pb_{0.35}Sr_{1.89}Ca_{1.87}Cu_3O_{9.91}$ [13], the latter mode only can be observed, while the two modes are mixed at lower Pb contents [12]. This strongly suggests that the Pb-free and the above mentioned Pb-rich ($Pb/Bi \simeq 1/5$) compositions are different phases and that these can intergrow within a particle as a result of a microscopic fluctuation of the Pb content.

In relation to this, it is interesting to note that annealing at low temperatures, typically 700°C in air, induces a very interesting change in composition and structure. As seen in Fig. 11, a Ca_2PbO_4-like phase containing Bi and Sr also and $(Sr,Ca)_2CuO_3$ precipitate finely on the surface of the original particles [13,18,41]. The XRD and ED patterns of the annealed samples indicate in the sharp peak profiles and the sharp spots that the 2223 phase remains well-crystallized even after the segregation reaction [13,18]. On the other hand, the modulation wave length changes suddenly from the original value of $\lambda/b=8.7$ to 4.7 after the annealing. This change cannot be observed by XRD, because satellite peaks due to the modulation do not appear with detectable intensities.

The stepwise change in λ due to the Pb substitution or the subsequent segregation treatments is qualitatively different from the continuous change found in the pure and substituted 2201 phase. And, though a hyperbolic equation approximates the λ vs. b relation for the 2201 phase, the b parameter for the 2223 phase remains almost constant: $a=b=5.411$, $c=37.12$Å for the original sample with $\lambda/b=8.7$, while $a=b=5.410$, $c=37.14$Å for the annealed sample with $\lambda/b=4.7$ [13,18]. Another difference is that the Pb substitution induces an orthorhombic distortion of the basal plane of the primitive cell for the 2201 phase but not for the 2223 phase. These differences may be considered to result from the effects of the insertion of the Ca and $[CuO_2]_\infty$ layers enhancing the stiffness of the triple CuO_2 layers against distortions from tetragonal symmetry.

The Pb-substituted 2201 phase also shows a similar segregation reaction both on annealing at low temperatures and on electron beam irradiation. The precipitates are always formed on the surface of the original particles, while no orientation relations have been found between the precipitates and the mother phase [26].

These results must be throwing light on the phase relation in this multicomponent system, Bi-Pb-Sr-Ca-Cu-O (BPSCCO), but a systematic understanding would require much more effort.

Partial substitution of Pb for Bi drastically promotes the formation of the 2223 phase as clearly shown for the first time in the XRD and ac susceptibility measurements indicating a volume fraction of ~90% done in this laboratory [42]. This study was done independently of earlier suggestions [6,43] that the Pb substitution helps to attain

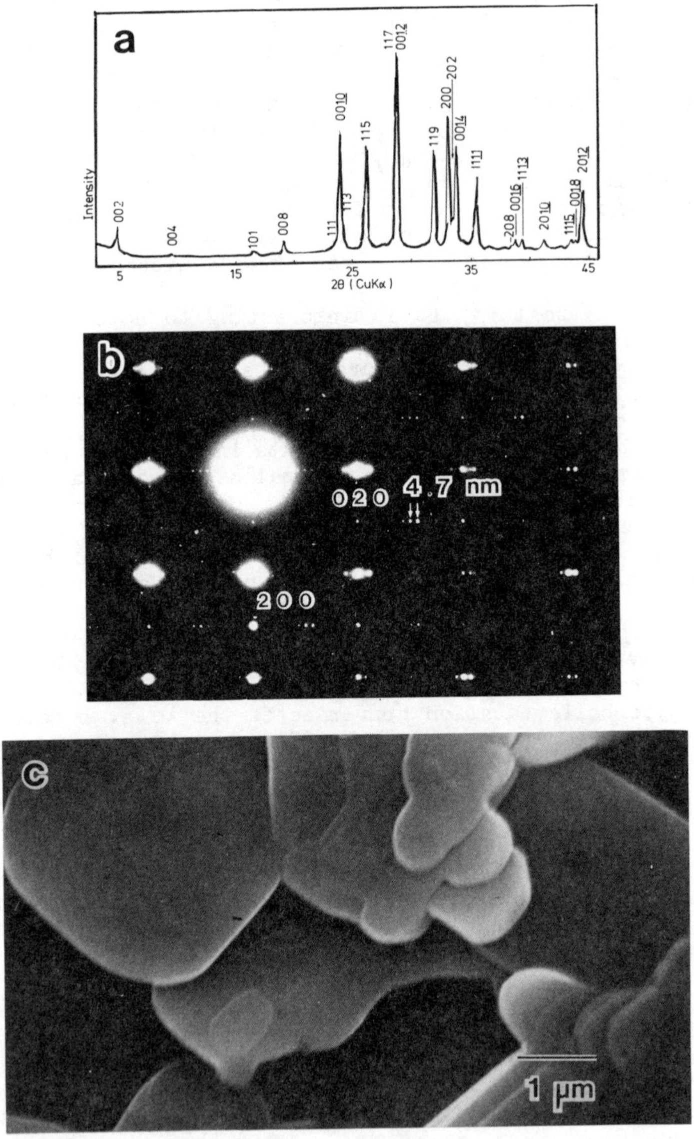

Fig. 11 XRD pattern of $Bi_{1.73}Pb_{0.35}Sr_{1.89}Ca_{1.86}Cu_3O_{9.91}$ (a), the [001] ED pattern (b), and the particle morphology (c) [13,18].

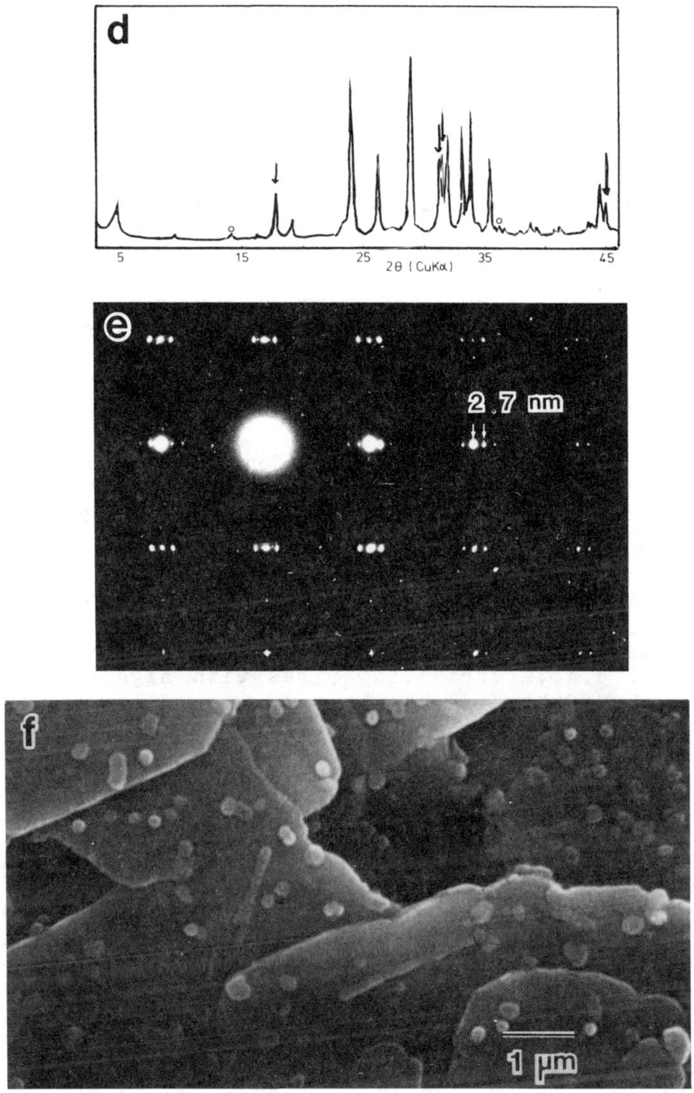

Fig. 11 (continued) Changes after annealing at 700°C in air for the same sample (d~f) [13,18].

the zero resistance, even at a low volume fraction ($\sim$10% [6]). The effects of Pb on the formation process was immediately analyzed by Hatano et al. [44]. They started from powder (a mixture of phase 2212, Ca_2PbO_4, and others) obtained by heating a mixture of appropriate raw materials of oxides and carbonates at 800°C in air and found that the Pb substitution induced at $\sim$840°C a partially molten phase from which the 2223 phase precipitated. The molten phase was found to be rich in Bi and Pb. The temperature range in which the 2223 phase was stabilized was limited to below $\sim$860°C above which the phase tended to be decomposed on melting. The importance of Ca_2PbO_4, formed at much lower temperatures than the 2223 phase formation, was emphasized by Uzumaki et al. [45] also. On the other hand, Endo et al. [46,47] reported that use of a low oxygen partial pressure of 1/13 atm further facilitated the 2223 phase formation.

An interesting and successful way of supplying Pb from the vapor phase was reported very recently by Hayakawa et al. [48]. They prepared an amorphous BSCCO film without Pb and annealed it in a closed capsule together with a pellet of the leaded 2223 phase. Pb was transferred from the pellet to the film and induced the crystallization of the 2223 phase.

Thus, Pb strongly facilitates the formation of the 2223 phase and changes both the wave length and symmetry of the 1D modulation as a result of substitution. The substitution site has been suggested to be the Bi site, at least mainly [49,50].

Finally the author and colleagues would like to stress the following points. There are so many phases in the BPSCCO system, many more than already known, beyond doubt. So, the creation of new reaction processes for the purpose of overcoming weak-links between particles, crucially important to, e.g., fabricating wires with high critical current densities ($\sim$1.7 10^4A/cm^2 at the present stage [51]) and the finding of new phases having higher T_c's are worthy being tried.

$ACuO_2$ CRYSTALLIZING IN LAYERED STRUCTURE

It is interesting to find that $[CuO_2]_\infty$ layers, in which Cu ions are coordinated by oxygen squares, are involved in both the hole-type superconductors with m$\geq$3 in the BSCCO and TBCCO systems and also in "electron"-type superconductors like $Nd_{2-x}Ce_xCuO_4$ [52]. The role of these layers in the latter type of superconductors is unequivocal, because their presence is unique. However, the role is masked in the former by the coexistence of CuO_2 layers capped with apical oxygen ions which can by themselves bear superconductivity as in the 2212 phase. The square planer coordination, in comparison with five- and six-fold coordinations, is expected to allow a considerably different electronic state.

It is interesting to synthesize new phases containing $[CuO_2]_\infty$ layers only and thereby study their chemical and physical properties. It has been found that high pressure treatments induce a transition in $ACuO_2$ (A: alkaline earths) to 2D structure in which layers of A and $[CuO_2]_\infty$ are alternately stacked [53]. As well-known, $BaCuO_2$ and $SrCuO_2$ are stabilized under ambient pressure but do not crystallize in such

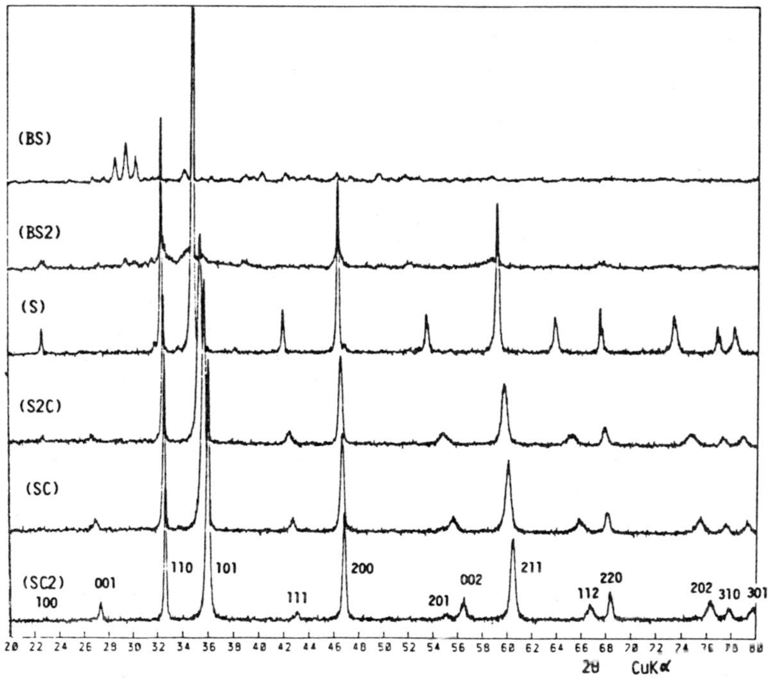

Fig. 12 XRD patterns for $ACuO_2$ treated at 1050°C and 60kb [53].

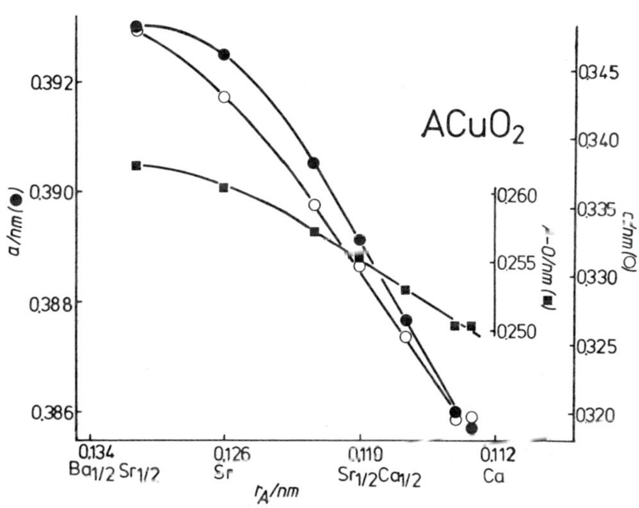

Fig. 13 Composition dependence of the lattice dimensions vs. r_A [53].

layered structure [54,55]. $CaCuO_2$ is not stabilized. This work was initiated from a simple estimation of specific density for $SrCuO_2$: The hypothetical layered structure was suggested to be a high density form. In the course of this study, Siegrist et al. and Yamane et al. reported that even under ambient pressure $Sr_{0.14}Ca_{0.86}CuO_2$ [56] and $Sr_{0.09}Ca_{0.91}CuO_2$ [57] crystallized in the layered structure when these were heated above 1000°C, and space group P4/mmm has been assumed. This tetragonal phase does not appear in Fig. 1, because heating above 1000°C is needed. The composition range can be extended drastically to the proximity of $Ba_{1/3}Sr_{2/3}Cu_2$ under 60kb.

XRD patterns for $Ba_{1/2}Sr_{1/2}CuO_2$ (BS), $Ba_{1/3}Sr_{2/3}CuO_2$ (BS2), $SrCuO_2$ (S), $Sr_{2/3}Ca_{1/3}CuO_2$ (S2C), $Sr_{1/2}Ca_{1/2}CuO_2$ (SC), and $Sr_{1/3}Ca_{2/3}CuO_2$ (SC2) are compared in Fig. 12. Two types of structures are formed, one being the $BaCuO_2$ type for BS and the $[CuO_2]_\infty$-layered type for the other compositions. The (100) and (001) peak intensities strongly depend upon the A composition, reflecting the difference in the atomic scattering factor. The lattice constants and the A-O bond length are plotted vs. the average A radius, r_A, in Fig. 13, where r_{Ca}=0.112, r_{Sr}=0.126 and r_{Ba}=0.142nm are assumed. At low r_A's, both the Cu-O bond length (=a/2) and the A-O bond length are stretched linearly, as r_A increases, at rates of $d(CuO)/dr_A=0.271$ and $d(AO)/dr_A=0.896$. The great difference in the rate manifests the stiffness of the Cu-O bond. The expansion tends, however, to be suppressed for r_A>0.121nm, more strongly for the Cu-O bond. At r_A=0.131nm for sample BS2 the Cu-O bond length seems to be saturated at 0.1965nm. The $BaCuO_2$ type structure sets in for $r_A \geq$ 0.134nm. The serious broadening of the XRD peaks for BS2 seems to indicate poor crystallization of the borderline composition. From $Sr_{0.09}Ca_{0.91}CuO_2$ to this borderline composition the unit cell is expanded by 1.8% along the a axis and by 8.8% along the c axis. The volume contraction of $SrCuO_2$ due to the transition to the layered structure amounts to 9.3%.

The lattice constants extrapolated to $CaCuO_2$ are a=0.3849 and c=0.3160nm. These values are very near the a parameter of and the separation between the $[CuO_2]_\infty$ layers in the TBCCO system [58]. The feasibility of the system to form multi-$[CuO_2]_\infty$ layered structures ($m \leq 6$) may be considered as originating in the good conformity of the a parameter between the $Ca_{m-1}(CuO_2)_{m-2}$ block with the counter building block. Efforts to find new phases based on this study and to dope of $ACuO_2$ with carriers themselves are in progress.

Acknowledgments

The author would like to thank colleagues in Mie Univ., Kyoto Univ., Konan Univ., and Okayama Univ. for allowing him to summarize their results. Discussion with and encouragements by Professor Y. Bando and Emeritus Professor T. Takada are appreciated.

REFERENCES

[1] C. Michel, M. Hervieu, M.M. Borel, A. Grandin, F. Deslandes, J. Provost, and B. Raveau, Z. Phys. B68 (1987) 421.

[2] J. Akimitsu, A. Yamazaki, H. Sawa, and H. Fujiki, Jpn. J. Appl. Phys. 26 (1988) L2080.
[3] H. Maeda, Y. Tanaka, M. Fukutomi, and T. Asano, Jpn. J. Appl. Phys. 27 (1988) L209.
[4] C.C. Torardi, M.A. Subramanian, J.C. Calabrese, J. Gopalakrishnan, E.M. McCarron, K.J. Morrissey, T.R. Askew, R.B. Flippen, U. Chowdhry, and A.W. Sleight, Phys. Rev. B38 (1988) 225.
[5] J.M. Tarascon, Y.Le Page, P. Barboux, B.G. Gagley, L.H. Greene, W.R. Mckinnon, G.W. Hull, M. Giroud, and D.M. Hwang, Phys. Rev. B37 (1988) 9382.
[6] S.A. Sunshine, T. Siegrist, L.F. Schneemeyer, D.W. Murphy, R.J. Cava, B. Batlogg, R.B. Van Dover, R.M. Fleming, S.H. Glarum, S. Nakahara, R. Farrow, J.J. Krajewski, S.M. Zahurak, J.V. Waszczak, J.H. Marshall, P. Marsh, L.W. Rupp, and W.F. Peck, Phys. Rev. B38 (1988) 893.
[7] E. Takayama-Muromachi, Y. Uchida, Y. Matsui, M. Onoda, and K. Kato, Jpn. J. Appl. Phys. 27 (1988) L556.
[8] J.L. Tallon, R.G. Buckley, P.W. Gilberd, M.R. Presland, I.W.M. Brown, M.E. Bowden, L.A. Christian, R. Goguel, Nature 333 (1988) 153.
[9] H. Adachi, S. Kohiki, K. Setsune, T. Mitsuyu, and K. Wasa, Jpn. J. Appl. Phys. 27 (1988) L1883.
[10] C.N.R. Rao, R. Vijayaraghavan, L. Ganapathi, and S.V. Bhat, J. Solid State Chem. 79 (1989) 177.
[11] Y. Matsui, H. Maeda, Y. Tanaka, and S. Horiuchi, Jpn. J. Appl. Phys, 27 (1988) L372.
[12] S. Ikeda, K. Aota, T. Hatano and K. Ogawa, Jpn. J. Appl. Phys. 27 (1988) L2040.
[13] Y. Ikeda, M. Takano, Z. Hiroi, K. Oda, H. Kitaguchi, J. Takada, Y. Miura, Y. Takeda, O. Yamamoto, and H. Mazaki, Jpn. J. Appl. Phys. 27 (1988) L2067.
[14] J.L. Tallon, R.G. Buckley, P.W. Gilberd, and M.R. Presland, Physica C 158 (1989) 247.
[15] R. Retoux, V. Caignaert, J. Provost, C. Michel, M. Hervieu, and B. Vaveau, J. Solid State Chem. 79 (1989) 157.
[16] H.W. Zandbergen, W.A. Groen, F.C. Mijlhoff, G. van Tendeloo, and S. Amelinckx, Physica C 156 (1988) 325.
[17] T. Tamegai, A. Watanabe, K. Koga, I. Oguro, and Y. Tye, Jpn. J. Appl Phys. 27 (1988) L1074.
[18] Y. Takeda, R. Kanno, F. Tanigami, O. Yamamoto, Y. Ikeda, and M. Takano, Physica C, in press.
[19] J.D. Jorgensen, B. Dabrowski, S. Pei, D.G. Hinks, L. Soderholm, B. Morosin, J.E. Schirber, E.L. Venturini, and D.S. Ginley, Phys. Rev. B38 (1988) 11337.
[20] M. Takano, Y. Ikeda, J. Takada, K. Oda, H. Kitaguchi, Y. Miura, Y. Tomii, and H. Mazaki, Proc. MRS Int. Meet. Adv. Mat., Tokyo, May 30-June 3, 1988 (Mat. Res. Soc., Pittsburg, 1989).
[21] Y. Ikeda, H. Ito, S. Shimomura, Y. Oue, K. Inaba, Z. Hiroi, and M. Takano, Physica C 159 (1989) 93.
[22] H. Kitaguchi, M. Ohno, M. Kaichi, J. Takada, A. Osaka, Y. Miura, Y. Ikeda, M. Takano, Y. Bando, Y. Takeda, R. Kanno, and O. Yamamoto, J. Cer. Soc. Jpn. Inter. Ed. 96 (1988) 388.

[23] E.M. McCarron, M.A. Subramanian, J.C. Calabrese, and R.L. Harlow, Mat. Res. Bull. 23 (1988) 1355.
[24] T. Siegrist, L.F. Schneemeyer, S.A. Sunshine, J.W. Waszczak, and R.S. Roth, Mat. Res. Bull. 23 (1988) 1429.
[25] Y. Matsui, S. Takekawa, H. Nazaki, and A. Umezono, Jpn. J. Appl. Phys. 28 (1989) L602.
TEM images have been reported without phase identification. Their phase B probably corresponds to $Bi_{17}Sr_{16}Cu_7O_z$, phase C to $Bi_4Sr_8Cu_5O_z$, phase D to $Bi_2Sr_3Cu_2O_z$ [19].
[26] Z. Hiroi, Y. Ikeda, S. Shimomura, H. Ito, and M. Takano, to be submitted.
[27] A. Fuertes, C. Miravittles, J. Gonzalez-Calbet, M. Vallet-Regi, X. Obradors, and J. Rodriguez-Carvajal, Physica C 157 (1989) 525.
[28] C.C. Torardi, J.B. Parise, M.A. Subramanian, J. Gopalakrishnan,and A.W. Sleight, Physica C 157 (1989) 115.
[29] J.A. Saggio, K. Sujata, J. Hahn, S.J. Hwu, K.R. Poppenlmeier, and T.O. Mason, J. Am. Ceram. Soc. 72 (1989) 849.
[30] A.K. Cheetam, A.M. Chippindale, and S.J. Hibble, Nature 333 (1988) 21.
[31] P.L. Gai and P. Day, Physica C 152 (1988) 335.
[32] A. Yamamoto, M. Onoda, E. Takayama-Muromachi, F. Izumi, T. Isigaki, and H. Asano, preprint.
[33] Y. Matusi, S. Takekawa, H. Nazaki, A. Umezono, E. Takayama-Muromachi, and S. Horiuchi, Jpn. J. Appl. Phys. 27 (1988) L1241.
[34] Y. Matsui and S. Horiuchi, Jpn. J. Appl. Phys. 27 (1988) L2306.
[35] M. Onoda and M. Sato, Solid State Commun. 67 (1988) 799.
[36] J. Darriet, C.J.P. Soethout, B. Chevalier, and J. Etourneau, Solid State Commun. 69 (1989) 1093.
[37] W. Bauhofer, Hj. Mattausch, R.K. Kremer, P. Murugaraj, and A. Simon, Phys. Rev. B39 (1989) 7244.
[38] Y. Shimakawa, Y. Kubo, T. Manako, T. Satoh, S. Iijima, T. Ichihashi, and H. Igarashi, Physica C 157 (1989) 279 and preprint.
[39] J.B. Parise, C.C. Torardi, M.A. Subramanian, J. Gopalakrishnan, A.W. Sleight, and E. Prince, Physica C 159 (1989) 239.
[40] W.A. Groen and D.M. de Leeuw,Physica C 159 (1989) 417.
[41] M. Matsuda, A. Kikuchi, M. Takata, M. Ishii, T. Yamashita, and H. Koinuma, Jpn. J. Appl. Phys. 28 (1989) L940.
[42] M. Takano, J. Takada, K. Oda, H. Kitaguchi, Y. Miura, Y. Ikeda, Y. Tomii, and H. Mazaki, Jpn. J. Appl. Phys. 27 (1988) L1041.
[43] R.J. Cava, B. Batlogg, S.A. Sunshine, T. Siegrist, R.M. Fleming, R. Rabe, L.F. Schneemeyer, D.W. Murphy, R.B. van Dover, P.K. Gallagher, S.H. Glarum, S. Nakahara, R.C. Farrow, J.J. Krajewski, S.M. Zahurak, J.V. Waszcak, J.H. Marshall, P. Marsh, L.W. Rupp Jr, W.F. Peck, and E.A. Rietman, Physica C 153-155 (1988) 560.
[44] T. Hatano, K. Aota, S, Ikeda, K. Nakamura, and K. Ogawa, Jpn. J. Appl. Phys. 27 (1988) L2055.
[45] T. Uzumaki, K. Yamanaka, N. Kamehara, and K. Niwa, Jpn. J. Appl. Phys. 28 (1989) L75.
[46] U. Endo, S. Koyama, and T. Kawai, Jpn, J. Appl. Phys. 27 (1988) L1476.
[47] S. Koyama, U. Endo, and T. Kawai, Jpn. J. Appl. Phys. 27 (1988) L1861.

[48] H. Hayakawa, M. Kaise, K. Nakamura, and K. Ogawa, Jpn. J. Appl. Phys. 28 (1989) L967.
[49] N. Kijima, H. Endo, J. Tsuchiya, A. Sumiyama, M.Mizuno, and Y. Oguri, Jpn. J. Appl. Phys. 28 (1989) L787.
[50] H. Nobumasa, T. Arima, K. Shimizu, Y. Otsuka, Y. Murata, and T. Kawai, Jpn. J. Appl. Phys. 28 (1989) L57.
[51] K. Sato, T. Hikata, H. Mukai, N. Shibuta, and H. Hitotsuyanagi, Proc. MRS Int. Meet. Adv. Mat., San Diego, May 24-29, 1989 (Mat. Res. Soc., Pittsburg, 1989).
[52] Y. Tokura, H. Takagi, and S. Uchida, Nature 337 (1989) 345.
[53] M. Takano, Y. Takeda, H. Okada, M. Miyamoto, and T. Kusaka, Physica C 159 (1989) 375.
[54] R. Kipka and H.K. Müller-Buschbaum, Z. Naturforsch. 32b (1977) 121.
[55] Chr. L. Teske and H.K. Müller-Buschbaum, Z. Anorg. Allg. Chem. 379 (1970) 234.
[56] T. Siegrist, S.M. Zahurak, D.W. Murphy, and R.S. Roth, Nature 334 (1988) 231.
[57] H. Yamane, Y. Miyazaki, and T. Hirai, J. Cer. Soc. Jpn. 97 (1989) 143 (in Japanese).
[58] H. Ihara, Proc. MRS Int. Meet. Adv. Mat., San Diego, May 24-29, 1989 (Mat. Res. Soc., Pittsburg, 1989).

HIGH MAGNETIC FIELD STUDIES OF HIGH TEMPERATURE SUPERCONDUCTORS

V.Kovachev and E. Vlakhov
Institute of Solid State Physics,
Bulgarian Academy of Sciences, Sofia, Bulgaria

The study of high temperature superconductors (HTSCs) in high magnetic fields is of utmost significance for understanding of the macroscopic nature of high temperature superconductivity even the mechanism of its origin is still unclear. Besides their fundamental importance these studies aim a practical task to clarify the potentialities of new superconducting materials and field of their possible application. Here it follows an example. The study of dH_{c2}/dT at $T \to T_c$ and its anisotropy in HTSCs permits to evaluate $H_{c2}(0)$ and the coherent length $\xi(0)$ in directions of the different crystallographic axes. Hereby conclusions could be drawn on tridimentional or quasi-twodimentional nature of superconductivity. At the same time, values obtained for the upper critical field determine the technological potentialities of new materials regarding generation of static magnetic fields.

The investigation of the problem on critical current density in ceramic specimens, single-crystals, epitaxial and oriented films, and other HTSCs in high magnetic fields is also rather important. These studies already revealed the presence of high ($>10^6 A/cm^2$) intragranular currents in the new materials. Simultaneously they confirmed the cause for considerably lower intergranular currents in ceramic HTSCs and their degradation with the field - weak links at the grain boundaries and absence of predominant orientation of crystallites in these materials.

High magnetic field study of HTSCs and the interpretation of obtained data are attended with a number of peculiarities and difficulties evolved from the particular material properties as well as from the experimental technique used. Some of these peculiarities related to HTSCs properties are of essential physical interest, i.e. it is necessary to clarify their nature in order to interpret completely and simply the results of high magnetic field experiment. Typical it is the situation with the superconducting transition broadening in magnetic field and "foot" appearance at the end of the transition, observed by a number of authors during resistivity measurements [1-7].

The transition broadening has been connected with inhomogeneity and polyphase nature of specimens [5], anisotropy of the transport properties and upper critical field [4], fluctuation effects [6] or thermo-activated flux creep [7] while the foot is probably due to the weak links and/or inhomogeneity of specimens [3].

Certain complications appear during high magnetic field study of some HTSCs which contain rare-earth ions with magnetic moment. For instance, it is established that the magnetization curves of $RBa_2Cu_3O_7$ (R=Gd, Dy, Er) deviate from the expected according to Curie-Weiss law [8]. This is probably due to the strong crystal field effects in these materials.

The difficulties evolved from using high impulse magnetic fields should be also mentioned. Eddy currents which are originated in the investigated material influence the magnetization hysteresis particularly in the case of bulk specimens. The latter could evidently affect the magnetic flux and change the hysteresis width, making difficult adequate interpretation of the results.

Objectively existing problems in high magnetic field study of HTSCs some of which were mentioned above, very large difference between specimens prepared by various methods, use of diversity of theoretical models for experimental data interpretation, make the comparison of published in the literature results rather difficult for the present.

Regarding this we considered it necessary to present and comment upon some published data for measured and calculated HTSCs parameters coming from high magnetic field studies (Table 1 - measured values; Table 2 - calculated values).

One can see that data presented in Table 1 were obtained at different definition conditions.Some authors define T_c or $T_c(H)$ as T_c^{onset}, $T_c(0,9Rn)$, $T_c(0,5Rn)$, $T_c(R=0)$ or use variety of extrapolations. Values dH_{c2}/dT differ the same manner depending on the method of T_c determination. Data about the specimens made of the same material vary considerably when obtained at different conditions (see items 9a and 9b, Table 1). In the case of ceramic specimens measurement one did not always account for their different density which leads to an error, for example on specific electrical resistivity determination, etc.

The main problem in experimental data interpretation for upper critical field of HTSCs apparently arises out from the extremely high $H_{c2}(0)$ values in these materials. The extrapolation results for dH_{c2}/dT (at $T \rightarrow T_c$ to $T \rightarrow 0$) highly depend on the adopted theoretical model and corresponding approximations (see Table 2). Higher approximations lead in fact to a little purpose since, for example, spin-orbit scattering parameter in Werthamer, Helfand, Hohenberg theory is introduced as a free parameter and could yield deviations in $H_{c2}(0)$ values more than twice (see items 3 and 4, Table 2). Definitely we always face here the great question if the theoretical models for conventional superconductivity are generally applicable to the new high temperature superconductors.

In the present chapter some experimental results we obtained for the properties of HTSCs in static magnetic fields up to 12-14T and impulse fields up to 40T will be considered. Broadening of the transition in the magnetic field, causes for the transition foot appearance and possibilities for its elimination, H_{c2} values and opportunities for determination of $H_{c2}^{\perp}$ and $H_{c2}^{\parallel}$ out of experimental data for polycrystalline specimens, magnetization curves, the critical current density, and influence of silver additions for diminution of weak links in some rare-earth ceramics were the goal of the present study.

Table 1

No.	Compound	Structure	T_c [K]	$-dH_{c2}/dT$ [T/K]	REF.
1.	$(La_{1-x}Sr_x)_2CuO_4$	Single-crystal	≃30*	≃4 (∥)** ≃0.3(⊥)**	9
2.	$YBa_2Cu_3O_{7-x}$	Single-crystal	88.8+	2.3 (∥)+ 0.46 (⊥)+	10
3.	$La_{1,4}Sr_{0,6}CuO_4$	Polycrystal with ≃ 10% impurity phases	43* 35***	1.48* 0.92***	5
4.	$Y_{0,4}Ba_{0,6}CuO_{3-x}$	Polycrystal with ≃10% impurity phases	93* 89***	1.79***	5
5.	$La_{1,85}Sr_{0,15}CuO_{4-x}$	Polycrystal	≃32***	2.2***	4
6.	$YBa_2Cu_3O_{7-x}$	Polycrystal	≃91***	4.2***	4
7.	$Y_{1,6}Ba_{0,4}CuO_{4-x}$	Multiphase Polycrystal	95.1* 89.5***	5±1* 1.3***	2
8.	$YBa_2Cu_3O_{7-x}$	Single-phase Polycrystal	92.5++	≃3.2++	6
9. a)	$YBa_2Cu_3O_{7-x}$	Polycrystal with <5% impurity phases	92.2***	1.72*** 9.00**	3
9. b)	$YBa_2Cu_3O_{7-x}$	Polycrystal with <5% impurity phases	90.9***	1.14*** 3.9**	3
10.	$YBa_2Cu_3O_{7-x}$	Single-crystal	85-92	4.3 (∥) 0.69 (⊥)	11

* - onset
** - 0.9Rn
*** - 0.5Rn
+ - extrapolated onset
++ - by extrapolation of $(\sigma_{290K}/\Delta\sigma)^2$ and $H_{c2}(T)$
⊥ - normal to the *ab* plain
∥ - parallel to the *ab* plain

Table 2

No.	Compound	$H_{C2}(0)$ [T]	$\xi(0)$ [Å]	Theoretical model	REF
1.	$YBa_2Cu_3O_{7-x}$	140(∥) 29(⊥)	7(∥) 34(⊥)	anisotropic Ginzburg-Landau theory	10
2.	$Y_{1,6}Ba_{0,4}CuO_{4-x}$	320($\lambda_{SO}=\infty$) 130($\lambda_{SO}=0$)	17*** 9***	isotropic Ginzburg-Landau theory	2
3.	$La_{1,85}Sr_{0,15}CuO_4$	32(λ_{SO}= 0) 41(λ_{SO}= 1) 45(λ_{SO}= 2) 48(λ_{SO}= 5) 50(λ_{SO}= ∞)		Werthamer,Helfand, Hohenberg theory	4
4.	$YBa_2Cu_3O_{7-x}$	120(λ_{SO}= 0) 170(λ_{SO}= 1) 195(λ_{SO}= 2) 220(λ_{SO}= 5) 270(λ_{SO}= ∞)			
5.	$La_{1,4}Sr_{0,6}CuO_4$	45* (H_P=79) 23***(H_P=64)	23* 32***	Ginsburg-Landau theory	5
6.	$Y_{0,4}Ba_{0,6}CuO_{3-x}$	110***(H_P=164)	14***		
7.	$YBa_2Cu_3O_7$	400(∥)	3.6(∥) 22(⊥)	anisotropic Ginzburg-Landau	11
		64(⊥)	2(∥) 22(⊥)	thermodynamic fluctuation effects	

* - onset data
*** - midpoint data
λ_{SO} - spin-orbit scattering parameter; H_P - Pauli limit

I.UPPER CRITICAL FIELDS OF $RBa_2Cu_3O_{7-x}$ (R=Y, Nd, Eu) AND Ag-DOPED CERAMICS

I.1.Introduction

A lot of investigations on the HTSCs critical magnetic fields of single-crystals [9,11,12,13], polycrystals [2,14,15,16], and orientated thin films [17,18] are known. These studies show the strong anisotropy of the critical fields of metaloxide compounds. The investigations of (Y_{123}, La_{214}, Eu_{123}, Gd_{123}) and other single-crystals give $-dH_{C2}^{\parallel}/dT$ (R=0.5Rn) values in the range of 1.95-3.3T/K and $-dH_c^{\perp}/dT$ (R=0) in the range of 0.23-0.9T/K. The anisotropy parameter determined according to the angle dependence of H_{c2} for Y_{123} by Worthington et al.[11] is approximately 10, and according to [12] for the same material this value is approximately 5. The corresponding $-dH_{c2}/dT$ (R=0.5Rn) for polycrystalline ceramic spesimens Y_{123} varies from 0.6 T/K [14] to 5.8T/K [15], and for $-dH_{c2}/dT$ (R=0) from 0.5T/K [19] to 4T/K [20].

The resistivity transitions R(T) in magnetic field broaden considerably for polycrystalline 123 specimens, even more for Bi-ceramics [21], and as well as for single-crystal ones [11,12,13]. Magnetic field application do not influence in practice only the transition width of $(La_{1-x}Sr_x)_2CuO_4$ single-crystal [14]. Study of the factors influencing the transition broadening and determination of methods to control it is of great interest.

The HTSCs bulk specimens prepared by ceramic technology represent polycrystals with random oriented anisotropy axes of the separate crystallites. For the crystallites oriented with the *ab* plain parallel to H the superconducting transition occurs at higher temperature compared to that for the crystallites of another orientation. At further decrease of specimen temperature more and more crystallites turn superconducting and at $H<H_{c2}^{\perp}$ all crystallites could go to superconducting state.Percolation theory, assuming existance of metallic contacts between separate crystallites, permits to predict the shape of resistivity transition R(T) for polycrystalline specimens [1,22]. Glazman et al.[22] suggest a procedure for determination of $dH_{c2}^{\parallel}/dT$ and $dH_2^{\perp}/dT$ for single-crystal blocks (crystallites) composing the polycrystalline material using experimental data for $R(T)_H$ and $H_{c2}(T)$ for this material. Dmitrenko et al. [23] have experimentally investigated polycrystalline specimens obtained of a melt (with direct metallic contacts between crystallites) and ascertained good agreement with the theory [1,22]. The same authors [23] have proposed and successfully approbated a method for experimental data interpretation of polycrystalline specimens containing weak links.This circumstance affords an opportunity to acquire information for the critical magnetic fields of single-crystal blocks (crystallites) investigating $R(T)_H$ of polycrystalline specimens prepared by various methods and having different microstructure.

The aim of this part of the work is to study the critical magnetic fields in $RBa_2Cu_3O_{7-x}$ (R = Y, Nd, Eu), as well as $RBa_2Cu_3O_{7-x}$(R = Y,Eu) with addition of Ag prepared by ceramic technology.

I.2.Experimental

. The Y_{123}, Nd_{123}, and Eu_{123} specimens are prepared of high purity rare-earth oxides R_2O_3,$BaCO_3$, and CuO. The calcination was carried out in the temperature range of 850-900°C for 6-12 h in air. The sintering was accomplished in oxygen atmosphere at temperature of 880-980°C for 6-20 h with intermediate grinding and pressing in some of the cases. The specimens were tablets with 10-27 mm in diameter. The pressure to make into tablets varied in the of range 100-500 MPa. The silver addition was introduced as Ag_2O to the initial composition (precursor). In some cases Ag_2O was added to the stoichiometric 123 amount of the other components, in the other as a substituting part of initial raw materials.

The regime of preparation for Ag-doped samples did not differ a lot of that for undoped material. A special feature was that the sintering has been carried out at temperatures not exceeding 950°C to prevent uncontrollable flow of elementary silver out. Obtained specimens were characterized using X-ray diffraction analysis, scanning electron microscopy and X-ray microprobe analysis [25]. Resistivity measurements were carried out on samples with typical dimensions 10x2x2 mm, cut out of the tablets, using standard four-probe technique. The electrical contacts were made by means of: 1)silver paste at room temperature, 2)thermally processed silver coating with $R_{\square}=1\div3.10^{-4}$ ohm.cm, and 3)eutectic In-Ga for critical current measurements.

The constant magnetic field was produced up to 12T by Bitter type magnet. An impulse type solenoid was used for conducting the investigations in the fields up to 40T for a period of 10 ms in liquid helium or liquid nitrogen. For all experiments the direction of the applied magnetic field H was normal to the measuring current I ($H\perp I$).

The temperature measurements of the resistance in the range of 4.2-300K were carried out in a helium vapors cryostat. The temperature was monitored using calibrated thermocouples Cu+Fe/Cu and Cu+Fe/NiCr with accuracy of 0.02K.

The R(T) dependence was taken at constant magnetic field varying the temperature. Cooling down and warming up rates of the container with the investigated specimen were established in such a manner that R(T) were identical in both directions within accuracy of less than 0.1K.

Specimens $YBa_2Cu_3O_{7-x}$ doped with silver are polyphase [24,25]. They contain 123 phase mainly. Silver, in the form of metallic ingredients, is randomly distributed between crystallites of 123 phase and 211 phase. Silver presence was occasionally detected by X-ray microprobe analysis in the element composition of 123 crystallites, which are usually with an elongated form, and sometimes like sticks or pins. In 211 phase however no traces of Ag were detected.

I.3.Results

R(T) dependences of $YBa_2Cu_3O_{7-x}$, $EuBa_2Cu_3O_{7-x}$, and $NdBa_2Cu_3O_{7-x}$ for different magnetic fields H are similar. R(T) for Y_{123} are shown in Fig.1. The transition width monotonously enlarges with the field increase. At the beginning of the transition (T_c^{onset}) the shift is negligible compared to this in the end (T_c^{end}). The existence of a foot in the function $R(T)_{H=const}$ could be seen. Further temperature reduction leads to linear section and a "tail" of R(T) approximating slowly R=0.

Fig.2 shows the dependences $R(T)_{H=const}$ for specimen doped with Ag, (D62.2) obtained from Y_{123} precursor with 3 mol% Ag_2O substitution instead of $BaCO_3$. An essential peculiarity of the silver doped specimens is the absence of the foot for D62.2, or strong decrease in the foot size for specimen Eu20 prepared from Eu_{123} precursor with addition of 20 mol% Ag_2O.

It is known [16,23] that the foot on R(T) dependence in magnetic field is probably due to the presence of weak links between separate crystallites. The critical magnetic field and critical current of the weak links are considerably lower compared to those for the crystallite volume. The latter leads to foot appearance when magnetic field is

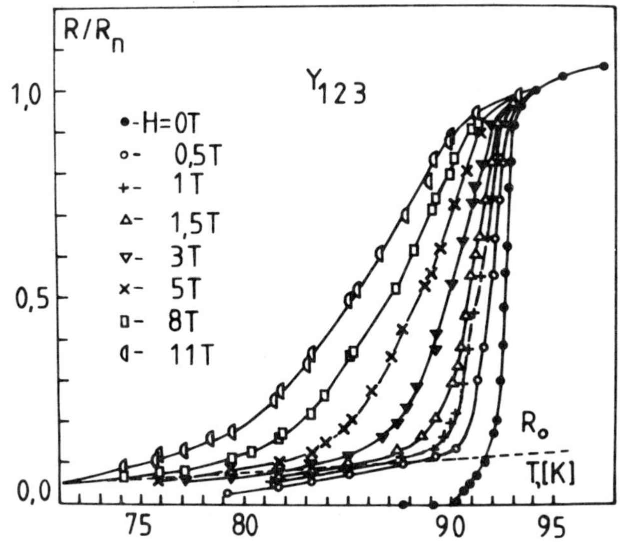

Fig.1 Resistance ratio R/Rn vs. temperature T for $YBa_2Cu_3O_{7-x}$ (Y_{123}) at various magnetic fields as indicated on the figure. The "foot" value Ro is shown by the deshed linc.

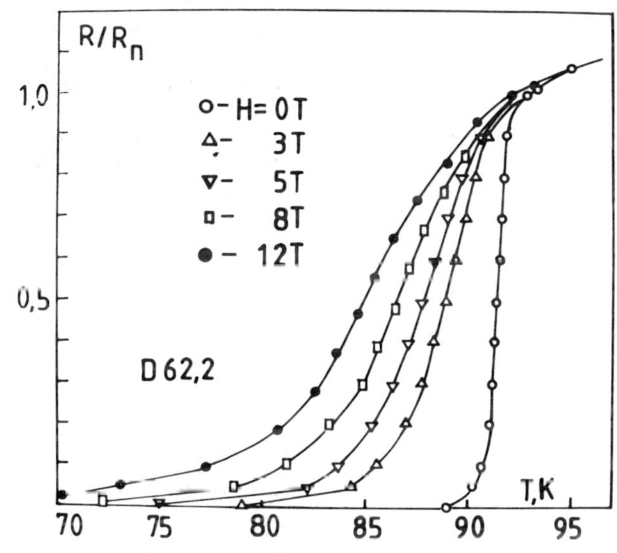

Fig.2 Resistance ratio R/Rn vs. temperature T for Ag-doped $YBa_2Cu_3O_{7-x}$ (D62.2) at various magnetic fields as indicated on the figure.

applied or higher density of the measuring current is used.

Fig.3 shows the dependence R(H) at T=77.35K obtained for Eu20 in the impulse magnetic field. This material has the lowest T_c and the broadest transition at H=0 (see Table I.1) among the other materials investigated. Nevertheless, its resistivity at H=40T is lower than Rn i.e., the specimen still contains a part in superconducting state.

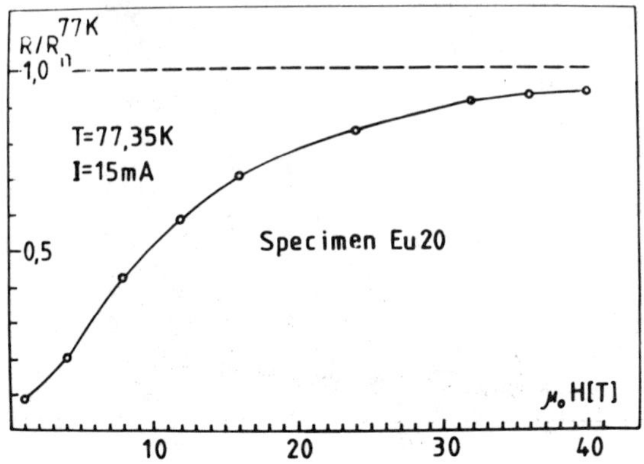

Fig.3 Resistance ratio R/Rn vs. magnetic field $\mu_0 H$ at 77.35 K for Ag-doped $EuBa_2Cu_3O_{7-x}$ (Eu20).

Fig.4 illustrates the $H_{c2}(T)$ dependence obtained at different parameter [R(T)-Ro]/[Rn-Ro], where Ro is the foot value for Y_{123} specimen. The use of parameter [R(T)-Ro]/[Rn-Ro] instead of R/Rn provides a possibility, according to [23], to eliminate the weak links influence and to avoid the ambiguity, due to the polycrystalline specimen microstructure. The dependence $H_{c2}(T)$ at [R(T)-Ro]/[Rn-Ro] could be considered as consisting of two linear sections, namely at H<2T and H>3T, which are smoothly connected. The similar relations for other investigated specimens are not shown in the chapter graphically to shorten the presentation, but some more important parameters for the further discussion are given in Table I.1.

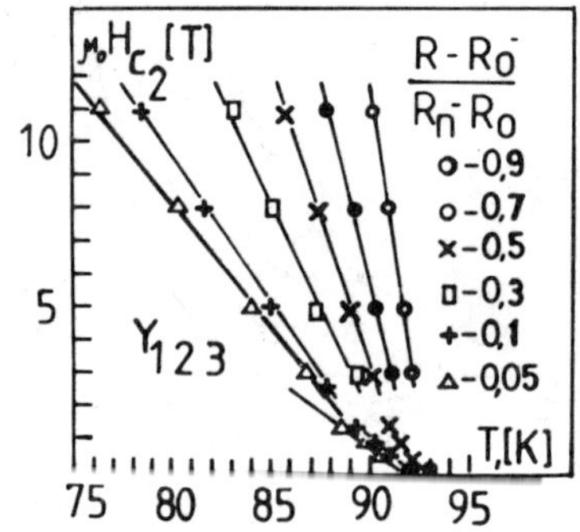

Fig.4 Upper critical field H_{c2} vs. temperature T of Y_{123} for different [R(T) - Ro]/[Rn - Ro] ratio as indicated on the figure.

Table I.1

Specimen	T_c [K]	ΔT_c [K]	R(T)-Ro/Rn-Ro	$-dH_{c2}/dT$ (H>3T) [T/K]	$-dH_{c2}/dT$ (H<2T) [T/K]
Eu_{123}	93.8	1.25	0.05 0.10 0.50	0.910 - 1.960	0.500 0.647 -
Nd_{123}	93.3	2.85	0.05 0.10 0.50	0.687 - 1.810	0.550 0.620 -
D62.2	91.4	1.10	0.05 0.10 0.50	0.850 - 2.160	0.477 0.550 -
Eu20	87.1	4.90	0.05 0.10 0.50	0.680 - 1.167	0.289 0.354 -

I.4.Discussion

From here presented R/Rn=f(T) relations in magnetic fields up to 12T for $RBa_2Cu_3O_{7-x}$ we shall pay more detailed attention to these of $YBa_2Cu_3O_{7-x}$.The compound $YBa_2Cu_3O_{7-x}$ is a subject of predominant numbers of investigations of HTSCs and the most comprehensive information on its superconducting properties, normal state parameters, and structure is now available in the literature.

As it has already been mentioned the transition foot is an evidence for a weak links presence due to Josephson-like behavior of interfaces between the crystallites. Ceramic technology for specimen preparation provides relatively low density and random orientation of crystallites that leads to very low current density compared to that of single-crystals [26,27].

The beginning of the foot at different magnetic fields determine the level of Ro. At this resistance value the percolation limit of the links with metallic behavior between the crystallites is already overpassed and the resistivity is defined by the weak links number, their critical superconducting characteristics and the flux creep effects [28,29,34] at certain H, J and T. The Ag addition to $RBa_2Cu_3O_{7-x}$,however, leads to a complete disappearance of the foot for D62.2 or to its considerable decrease for Eu20. Indeed on one hand the silver addition is situated mainly as separate phase between the grains providing for metallic character of the links [25], and on the other [24] leads to a decrease of the transition width ΔT even below 1K in zero magnetic field.

The foot disappearance in the temperature dependence of resistivity in magnetic field should be interpreted simply as an elimination of the weak links as a result of Ag addition. The disappearance of the weak links in [23] is gained by using melt preparation technology which increases the density of the material. The investigation of critical current density $J_c(H)$ of the specimens doped with Ag showed considerable increase of J_c compared to this of pure compound $YBa_2Cu_3O_{7-x}$ made by ceramic technology [24,30, see also the second part of this chapter]. Values for J_c of Ag doped specimens [24,30], however, are far lower than this for textured bulk samples $\simeq 10^4$ A/cm^2 [31] probably due to the random orientation of crystallites in Ag-doped material.

Two independent methods [23] for determination of the derivative $dH_{c2}^{\perp}/dT$ characterizing the critical field of single-crystal blocks consisting the polycrystalline specimen at H normal to the *ab* plain are known. According to Glazman et al.[22] in case of a strong anisotropy for homogeneous polycrystalline specimens (i.e. without local fluctuations of T_c in the specimen) in the range of linear part of R(T) dependence the following equation is valid:

$$1 - \frac{R}{Rn} = 3\,\frac{T_c(0) - T}{H} \cdot \frac{dH_{c2}^{\perp}}{dT} \tag{1}$$

Here the correlation between resistivity (R) and the relative part of the volume in superconducting state (c) was used. This equation is valid at small values of c:

$$1 - R/Rn \simeq 3c \tag{2}$$

When the specimen is inhomogeneous the equation (1) is valid only in high magnetic fields H which meet the requirements:

$$H \gg \Delta T\,\frac{dH_{c2}^{\perp}}{dT} \tag{3}$$

where ΔT is the maximum deviation of the transition temperature in different parts of the specimen.

The equation (1) is modified and approbated as well for the case of polycrystalline specimen having weak links [23] as follows:

$$1 - \frac{R(T) - Ro}{Rn(T) - Ro} = 3 \cdot \frac{T_c(0) - T}{H} \cdot \frac{dH_{c2}^{\perp}}{dT} \tag{4}$$

where Ro is the foot in the R(T) dependence,

Rn is determined by extrapolation of the linear part of the resistivity dependence,

$T_c(0)$ is midpoint of the transition to superconducting state at H=0.

Figs 5-9 show the experimental results for specimens Y_{123}, Nd_{123}, Eu_{123}, D62.2, and Eu20 at H > 3T introduced in the form:

$$1 - \frac{R(T) - Ro(T)}{Rn(T) - Ro(T)} = f\left(\frac{T_c(0) - T}{H}\right) \tag{5}$$

A satisfactory fit of the linear part of the universal equation (4) in the range $0.2 < \{1 - [R(T) - Ro(T)]/[Rn(T) - Ro(T)]\} < 0.8$ to the

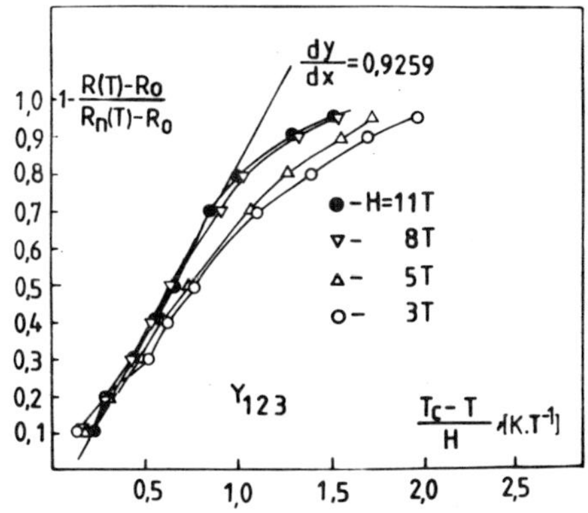

Fig.5 Dependence $1-[R(T)-Ro]/[Rn(T)-Ro]$ vs. $[T_c-T]/H$ of Y_{123} at various magnetic fields. The derivative dy/dx of the linear part of the experimental curves for H = 8T and H = 11T is shown.

experimental data for $RBa_2Cu_3O_{7-x}$ (R = Y,Nd,Eu) in magnetic fields H = 8T and 12T is observed. At lower fields the deviation of the universal equation is considerable and using equation (3) the maximum difference in the local transition temperatures ΔT could be evaluated. Thus in principle the degree of inhomogeneity of the specimen could be estimated. When H in (3) is introduced for 7-8T we obtain ΔT = 22.7, 28.8 and 24 K for Y_{123}, Nd_{123} and Eu_{123} respectively. These values are considerably higher than the experimentally determined transition width in zero field ΔT (1.40,2.85, and 1.25K respectively). The value of ΔT at H=0 do not reflects the real scattering of the local critical temperature of the

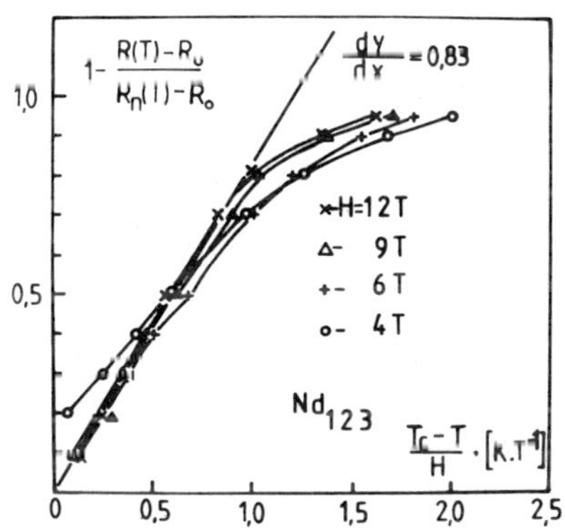

Fig.6 Dependence $1-[R(T)-Ro]/[Rn(T)-Ro]$ vs. $[T_c-T]/H$ of $NdBa_2Cu_3O_{7-x}$ (Nd_{123}). Derivative dy/dx for H=6-12 T is shown.

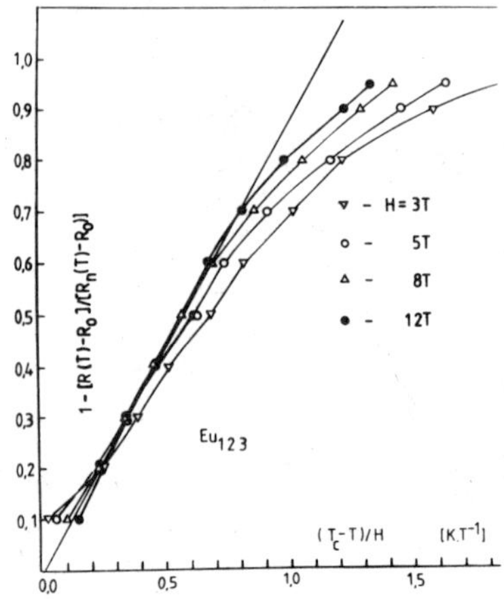

Fig.7 Dependence 1-[R(T)-Ro]/[Rn(T)-Ro] vs.[T_c-T]/H of $EuBa_2Cu_3O_{7-x}$ (Eu_{123}).

material, since the resistivity becomes zero when percolation value C_P=0.17 for the continuum is reached [32], but not at the complete disappearance of resistivity in all parts of the specimen.

The values for $dH_{c2}^{\perp}/dT$ determined according to equation (4) are given in Table I.2. For the specimen D62.2 even in fields of 8-12T there is not satisfactory fit to the universal equation, but the linear sections for 8T and 12T are almost parallel (dy/dx ≃ 1.086), which corresponds to $dH_{c2}^{\perp}/dT$ = 0.362 T/K. For Eu20 the linear parts fit in more narrow interval 0.1 < {1 - [R(T) - Ro(T)]/[Rn(T) - Ro(T)]} < 0.4 compared to this for undoped 123 material.

The second independent method for $dH_{c2}^{\perp}/dT$ determination uses the dH_{c2}/dT experimental data for polycrystalline specimen. If the specimen

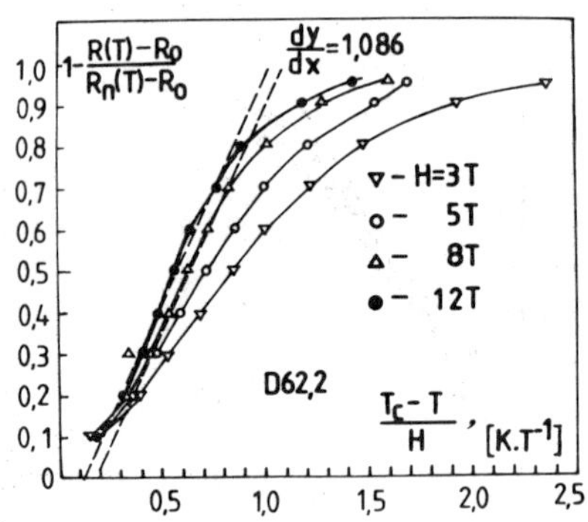

Fig.8 Dependence 1-[R(T)-Ro]/[Rn(T)-Ro] vs.[T_c-T]/H of D62.2. Derivetives dy/dx for H=8T and H=12T are shown.

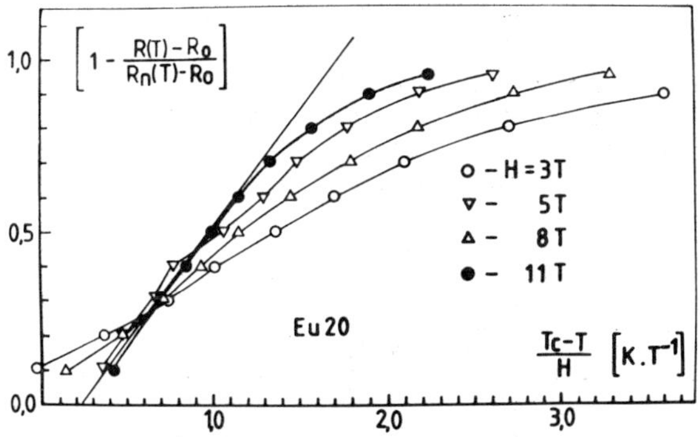

Fig.9 Dependence 1-[R(T)-Ro]/[Rn(T)-Ro] vs.[T_c-T]/H of Eu20.

is homogeneous one and does not have non-superconducting impurities in case of strong anisotropy ($dH^{\parallel}_{c2}/dT \gg dH^{\perp}_{c2}/dT$) the next equation is valid:

$$\left.\frac{dH_{c2}}{dT}\right|_{R=0} = \frac{1}{C_P} \cdot \frac{dH^{\perp}_{c2}}{dT} \simeq 6 \frac{dH^{\perp}_{c2}}{dT} \qquad (6)$$

For inhomogeneous specimens the percolation value is different. According to [15] in the case of strong anisotropy this value is determined by the averaged along all directions $dH_{c2}(\theta)/dT$ derivative and is equal to $2dH^{\perp}_{c2}/dT$. In low magnetic fields ($H\to0$) it could be written:

$$\left.\frac{dH_{c2}}{dH}\right|_{H\to0} = 2 \frac{dH^{\perp}_{c2}}{dT} \qquad (7)$$

As it was mentioned above two linear sections of $H_{c2}(T)$ at H<2T and H>3T smoothly connected could be seen for Y_{123} in Fig.4. The values for $(dH_{c2}/dT)_{H<2T}$ and $(dH_{c2}/dT)_{H>3T}$ for other investigated specimens are presented in Table I.1. These results are in qualitative agreement with the conclusions for inhomogeneity of the specimens. The ratios of $(H_{c2}/dT)_{H>3T}$ to $(dH_{c2}/dT)_{H<2T}$ are presented in Table I.2. Obtained values are in the range of 1.25 - 2.36. According to (6) and (7) the ratio of $H_{c2}(T)$ slopes should be equal to 3. Thus considerably smaller values come from the experimental data regarding the theoretical prediction for $(dH_{c2}/dT)_{H>3T}/(dH_{c2}/dT)_{H<2T}$.

The values for $dH^{\perp}_{c2}/dT$ obtained by these two independent methods are given in Table I.2.

Data analysis in Table I.2 shows satisfactory agreement for the results of $dH^{\perp}_{c2}/dT$ for $RBa_2Cu_3O_{7-x}$ (R = Y, Eu, Nd) obtained from the two independent methods. Standing, on this we consider that the theoretical method suggested by Glazman [22] adequately reflects the H_{c2} behavior of single-phase specimens $RBa_2Cu_3O_{7-x}$.

However, values of $dH^{\perp}_{c2}/dT$ calculated by these two methods for specimens doped with Ag differ substantially. It could be assumed that the evaluation according to equation (4) is more inaccurate and overvalued. The universal relation for Eu20 is valid in more narrow range of 1-[R(T)-Ro]/[Rn(T)-Ro] (Fig.9), and for D62.2 even at 8T and 12T there is not fit to the linear sections (see Fig.8). On the other hand, the addition of silver, which leads to multiphase material and presence of metallic silver in the pores between granules of 123 and 211 phases, could influence the percolation ratio to differ from that accepted in [32].

Table I.2

Specimen	$\left.\dfrac{(dH_{c2}/dT)_{H>3T}}{(dH_{c2}/dT)_{H<2T}}\right\vert_{R\to 0}$	$-dH^{\perp}_{c2}/dT$*	$-dH^{\perp}_{c2}/dT$**
Y_{123}	1.58	0.308	0.240-0.320
Eu_{123}	1.83	0.292	0.250-0.323
Nd_{123}	1.25	0.277	0.275-0.310
D62.2	1.78	0.362	0.238-0.275
Eu20	2.36	0.220	0.145-0.177

* - calculated according equation (4).
** - calculated according equation (7) for H<2T and [R(T)-Ro]/[Rn(T)-Ro], equal to 0.05 and 0.1 respectively.

According to [22] the derivative $dH^{\parallel}_{c2}/dT$ ($H\parallel ab$) for polycrystalline material could be obtained by the following equation:

$$T_c^{ons}(0) - T_c^{ons}(H) = H \cdot \left(\frac{dH^{\parallel}_{c2}}{dT} \right)^{-1} \tag{8}$$

Figs 10 and 11 show the dependence $T_c^{ons}(0) - T_c^{ons}(H) = f(H)$ for specimens Y_{123}, D62.2, and Eu20. The onset transition temperature $T_c^{ons}(H)$ was determined equal to the temperature of the cross point of the extrapolated normal resistivity and the tangent to the $R(T)_{H=const}$ dependence in its section [R(T)-Ro]/[Rn-Ro]>0.5.For the specimens Y_{123} and D62.2 this dependence is linear and $dH^{\parallel}_{c2}/dT$ is estimated to be equal to -6.54 and -5.71 T/K respectively. For the specimen Eu20 (Fig.11) two sections could be distinguished. The first one (0<H<5T) with linear dependence and $dH^{\parallel}/dT$ = -1.58 T/K, and the second one (H>5T) with $dH^{\parallel}_{c2}/dT \simeq$ -2.7 T/K. The relevant evaluation of $dH^{\parallel}_{c2}/dT$ for Nd_{123} and Eu_{123} is difficult to be done for intermediate values of the magnetic field due to the insufficient accuracy in $T_c^{onset}(H)$ determination. The assumption that the error in slope determination at H=0 and H=12T is minimum yields $dH^{\parallel}_{c2}/dT$= -7 T/K and -15 T/K for Nd_{123} and Eu_{123} respectively.

Obtained according to (8) values for $dH^{\parallel}_{c2}/dT$ are overvalued compared to those measured on Y_{123} single-crystal (criterion 0.5Rn) -3.3 T/K [12] and on single-crystal Eu_{123}[13]~-5T/K. Comparable, however, are the results for $H^{\parallel}_{c2}/dT$ measured by Sun et al. [7] on $ErBa_2Cu_3O_{7-x}$ highly orientated thin film.

The established differences between theoretically predicted [22] ratio of the linear sections of $H_{c2}(T)$ at R=0 or R=Ro for specimens with weak links and experimentally obtained (see Table I.2) could be

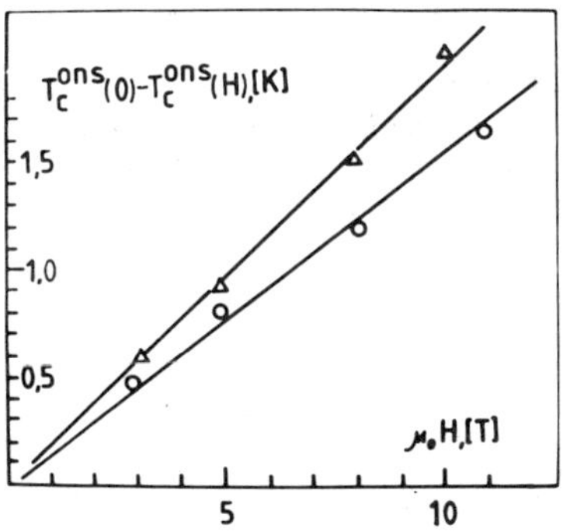

Fig.10 $T_c^{onset}(0) - T_c^{onset}(H)$ vs. magnetic field $\mu_0 H$ for D62.2(△) and Y_{123}(O).

conditioned by the influence of a number of factors, which are not taken into account in [22]. The latter are thermodynamic fluctuations influence close to T_c, as well as the flux creep effects.

Iwata et al.[34] registered one of the most narrow superconducting transitions of HTSCs in single-crystal Eu_{123}, namely ΔT_c = 0.25 K at $T_c(0.5R_n)$ = 94.8 K. However, the applied magnetic field H=5T in this case also leads to considerable broadening of the transition ($\Delta T \simeq 4$ K). The broadening of the transition in a single-crystal specimen could not be determined by the percolation effects since there are not differently oriented crystallites as in the case of polycrystalline specimens.

There is actually a case of practically unchanged transition width for zero field and H=8T, described by Hidaka et al.[9] for single-crystal $(La_{1-x}Sr_x)_2CuO_4$. Transition broadening for HTSCs single-crystals in magnetic field probably is determined by the thermal activated flux creep [28,29,34]. For HTSCs with $T_c \simeq 90$ K the flux creep is considerable, but at $T_c \simeq 30$ K it is highly suppressed, which probably explains such a behavior of $(La_{1-x}Sr_x)_2CuO_4$ in magnetic field [9].

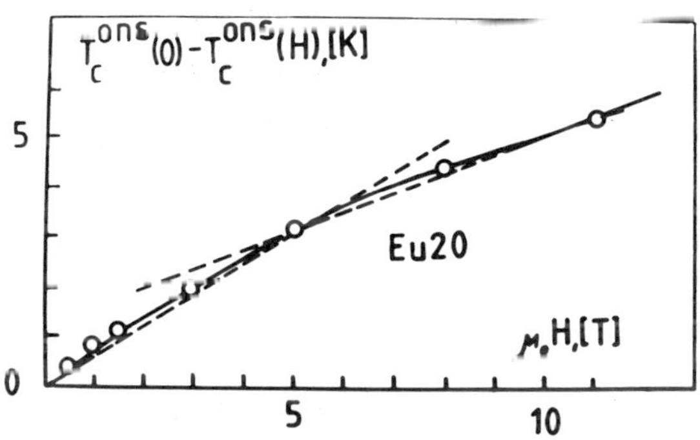

Fig.11 $T_c^{onset}(0) - T_c^{onset}(H)$ vs. magnetic field $\mu_0 H$ for Eu20).

The model for flux creep [28] predicted $H_{C2} \simeq (T_C-T)^{1,5}$, which drove us to use logarithmic co-ordinates for the $H_{C2}=f(T_C-T)$ dependence. The results for specimens Y_{123} and D62.2 (Fig.12) and Eu_{123} (Fig.13) correspond to R=Rn and R=0 (for D62.2) or R=Ro. Closest to 1.5 are the exponents for Y_{123}, 1.59 and 1.545 respectively, while for the Ag-doped D62.2 they are higher (1.815 and 1.68). For Eu_{123} exponents lower than 1.5 are observed, 1.43 and 1.347 respectively. The analysis of these results requires more information than available for the investigated materials.

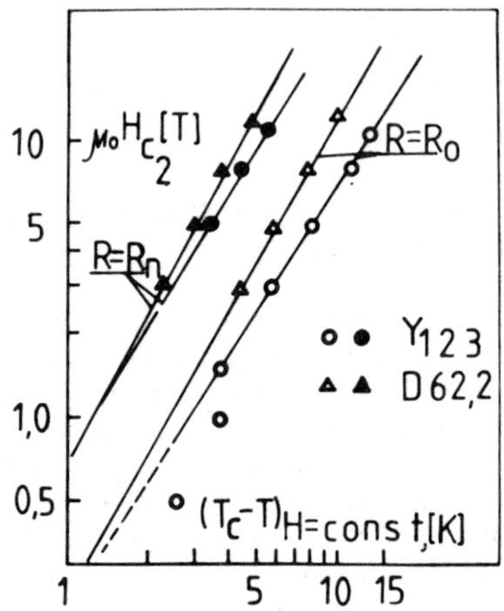

Fig.12 Upper critical field H_{C2} vs. $[T_C-T]_{H=const}$ at R=Rn(▲,●) and R=Ro (△,o) in logarithmic scale for Y_{123} (o,●) and for D62.2 (△,▲).

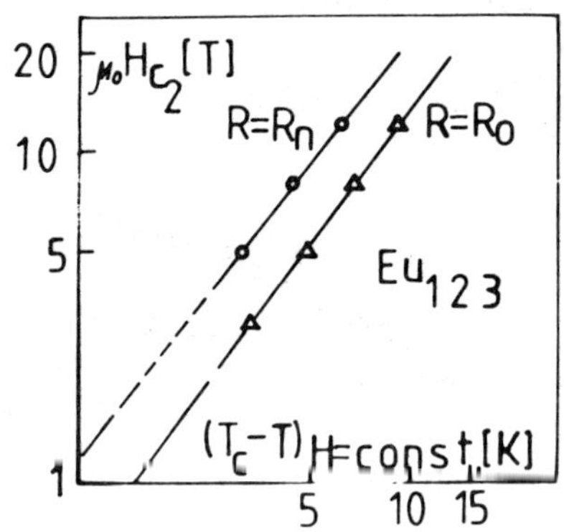

Fig.13 Upper critical field H_{C2} vs. $[T_C-T]_{H=const}$ at R=Rn (o) and R=Ro (△) in logarithmic scale for Eu_{123}.

I.5.Conclusions

The results presented and discussed here lead to two base conclusions:

1.The transition broadening in magnetic field for $RBa_2Cu_3O_{7-x}$ (R = Y, Eu, Nd) ceramics could be considerably reduced by Ag doping. It could be expected for other HTSCs ceramics.

2.Using the experimental data $R(T)_{H=const}$ obtained for polycrystalline HTSCs it is possible to estimate the critical field derivatives $dH^{\perp}_{C2}/dT$ and $dH^{\parallel}_{C2}/dT$ for single-crystal blocks constituting the ceramic material $RBa_2Cu_3O_{7-x}$ (R = Y, Eu, Nd).

II.MAGNETIZATION OF $RBa_2Cu_3O_{7-x}$ (R=Nd,Eu) AND Ag-DOPED $YBa_2Cu_3O_{7-x}$

II.1.Introduction

The magnetization of $RBa_2Cu_3O_{7-x}$ (R is rare-earth) in high magnetic fields has been studied mainly regarding two aspects:

1)Investigation of the antiferromagnetism at ≤1K in the compounds containing rare-earth ion with magnetic moment. Studies of the origin of antiferromagnetic ordering phenomenon - crystal field effects; weak superexchange interaction via intervening oxygen atoms surrounding the magnetic ion; magnetic dipolar interaction. Investigation of the rare-earth ion ground states in HTSCs. These studies are carried out in possibly highest magnetic fields and low enough temperatures [35,36.37,38].

2)Investigation of the magnetic flux penetration and distribution and the behaviour of macro- (intergranular) and microscopic (intragranular) shielding currents. Investigation of the magnetization hysteresis, flux creep, pinning and critical currents in HTSCs. These studies are carried out in the entire temperature range below 77K and high enough fields [39-42].

The results presented in this part of the work concern mainly the second group of problems [43].

The magnetic moment investigations of bulk and powdered ceramic specimens show practically identical results. This is an opportunity for some authors to represent bulk specimens as a complex of quasi-insulated grains. The magnetization of these grains is mainly determined by the intragranular critical current density J_{CG}, their dimensions, shape and structure, as well as their superconducting parameters T_C, H_{C1}, H_{C2} and anisotropy. When the crystallite anisotropy axes are randomly oriented they should be considered as exhibiting different J_{CG}, H_{C1} and H_{C2} at certain orientation of the magnetic field H.

According to Boan critical state model [44] the magnetic moment hysteresis is caused by the induced superconducting current. The critical current of a cylindrical specimen is determine by the following dependence:

$$M_{IR} = \frac{\mu_0 R J_C}{3} \qquad (9)$$

where M_{IR} is the irreversible part of M,
R is the cylinder radius.

After the estimation from equation (9) in [36] for $YBa_2Cu_3O_{7-x}$ the intragranular current density is $J_{CG} \simeq 1.5 \div 3 \cdot 10^{10}$ A/m^2 (4.2K) in the fields up to 29T. However, the intragranular critical current density for the same material given in [40] is $J_{CG} = 8.9 \cdot 10^{10}$ A/m^2 at H=0.3T and $J_{CG} = 5 \cdot 10^{10}$ A/m^2 at H=1T.

The aim of this part of the work is to investigate the magnetization of bulk polycrystalline specimen $RBa_2Cu_3O_{7-x}$ with magnetic rear-earth ion (Nd) or non-magnetic one (Y,Eu), as well as the magnetization of Ag-doped $YBa_2Cu_3O_{7-x}$ in magnetic field up to 14T, and to compare the calculated values for J_{CG} with experimental results of J_C, directly obtained by resistivity measurements in high magnetic field.

II.2.Experimental and results

The studied specimens are prepared by ceramic technology as reported in [24] and Ag-doped according to the method described above. The magnetization was measured using a Faraday type magnetometer with sensitivity of 10^{-3} emu/g. Specimens are cooled down to the appropriate temperature in zero magnetic field. The increase of the field to its maximum value (14T) has been carried out in successive steps of 0.2-1T. From 14T the field was reduced down to 13.5T and after that in successive steps of 1-0.2T down to zero.

Figure 14 shows the dependence M(H) for $EuBa_2Cu_3O_{7-x}$ and $NdBa_2Cu_3O_{7-x}$ at increasing ($H^{\uparrow}$) and decreasing ($H_{\downarrow}$) fields and temperature T=4.21K. Fig. 2 shows the same dependence at temperature T=77.35K.

Figs.16 and 17 illustrate the results for M(H) investigation of Ag-doped $YBa_2Cu_3O_{7-x}$ at temperatures of 4.21K and 77.35K respectively. Specimens D20.2 (0.2 mol% Ag_2O), D6.3 (1 mol% Ag_2O) and D21.2 (5 mol% Ag_2O) were prepared substituting the appropriate amount Y_2O_3 with Ag_2O in the initial composition of the 123 precursor. Specimen D41.1 was prepared by substituting of corresponding amount of $BaCO_3$ by 1 mol% Ag_2O.

The dependencies M(H) for specimens D63.2, D62.2, Y20 at T = 4.21K and T=77.35K are presented in figs. 18 and 19 respectively. Specimens D63.2 and D62.2 were obtained by substitution of 3 mol% Y_2O_3 and 3 mol% $BaCO_3$ by Ag_2O respectively. Specimen Y20 is prepared by adding of 20 mol% Ag_2O to the stoichiometric precursor Y_{123}.

The critical current density for D6.3 and D41.1 specimens measured by resistivity method using four-probe technique at T=4.2K and 66K is shown in Fig. 20.

II.3.Discussion

As known, at the initial increase of the field from zero the diamagnetic signal is linearly proportional to the field due to complete Meissner effect. Further ($H \geq H_{C1}$), the magnetic flux penetrates in to the specimen and the M(H) dependence deviates from the linear one.

Using our experimental data it is difficult to determine the

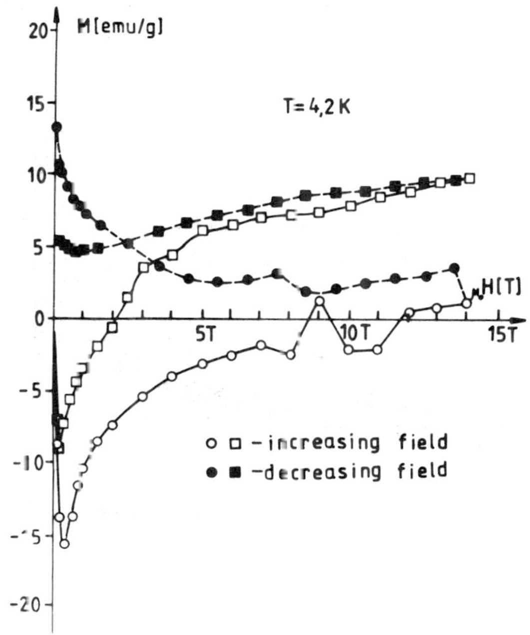

Fig.14 Magnetization M vs. magnetic field μ_0H for Nd_{123} (□,■) and Eu_{123} (○,●) at temperature T=4.21K.

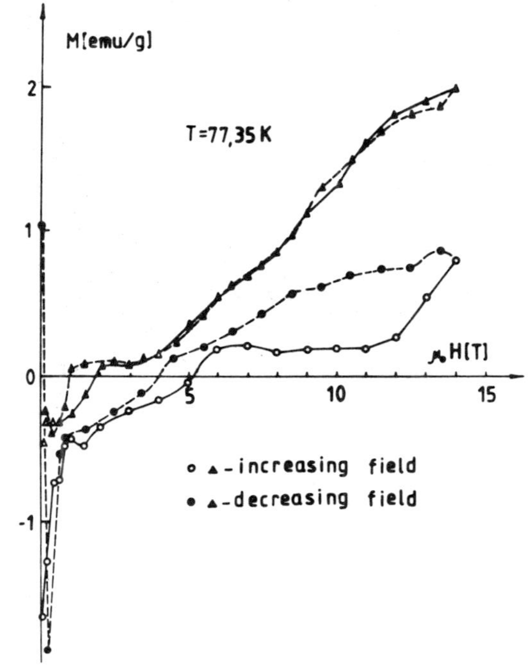

Fig.15 Magnetization M vs. magnetic field μ_0H for Nd_{123} (△,▲) and Eu_{123} (○,●) at temperature T=77.35K.

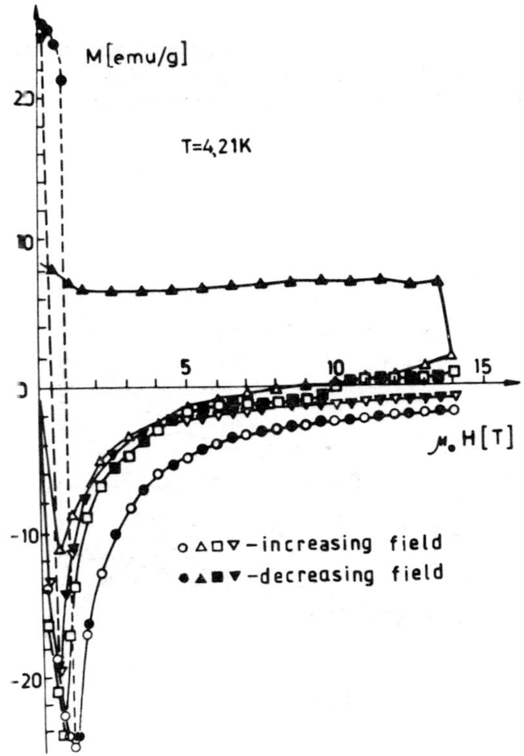

Fig.16 Magnetization M vs. magnetic field $\mu_o H$ for D6.3 (○,●); D20.2 (▽,▼); D21.2(□,■); and D41.1 (△,▲) at temperature T=4.21K.

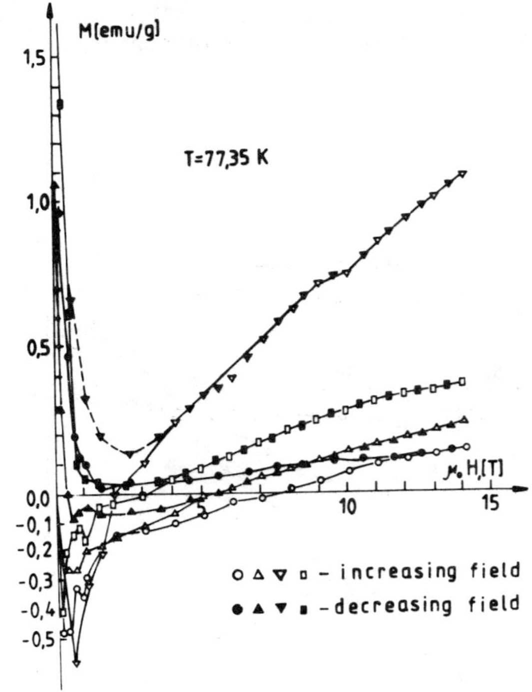

Fig.17 Magnetization M vs. magnetic field $\mu_o H$ for D6.3(○,●); D20.2(□,■); D21.2(△,▲); and D41.1 (▽,▼) at temperature T=77.35K.

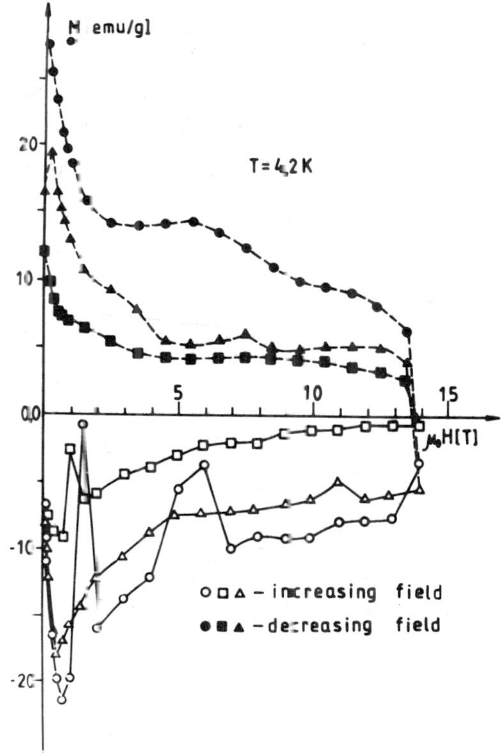

Fig.18 Magnetization M vs. magnetic field μ_0H for D62.2(△,▲);D63.2(□,■); and Y20(o,●) at temperature T=4.21K.

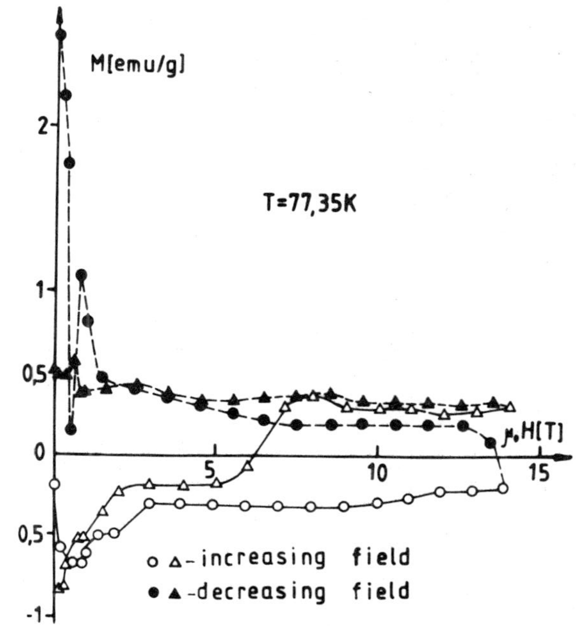

Fig.19 Magnetization M vs. magnetic field μ_0H for D63.2 (o,●) and Y20 (△,▲) at temperature T=77.35K.

field of first flux penetration for the materials studied. Even the M(H) dependences up to H=0.4T seem rather close to the linear one (see Fig.14 for instance) but the scale is very small to be positive about.

The detail study of HTSCs ceramics magnetization and the induction distribution in the material bulk at low fields and T=4.2K [39] shows that macroscopic demagnetization current density is of the order of 10^3 A/cm^2 up to $H \simeq 5 \cdot 10^{-3}$ T. In the fields $5 \cdot 10^{-3}T < H < 1.3 \cdot 10^{-2}T$ it is $\leq 10^2$ A/cm^2, while at $H > 1.3 \cdot 10^{-2}T$ the weak links between crystallites are destroyed and only intragranular demagnetization currents remain to exist. For bulk polycrystalline specimens Y_{123} and Gd_{123} Kupfer et al. [42] have established that the intergranular current $J_{CJ}(H=0) \simeq 10^2 - 10^3$ A/cm^2 has Josephson-like behaviour (junction current via weak links) only in very low fields. In a field higher than 0.3T the intergranular current in these specimens is almost independent of H, and its value becomes negligible $\simeq$ 10 A/cm^2. The intragranular current J_{CG}, however, has very high value of $10^6 - 10^7$ A/cm^2 and depends slightly on the magnetic field up to 12T.

In our case the situation must be similar. At the field of the largest diamagnetic signal there is some penetration of magnetic flux in to the specimen bulk (a deviation of M(H) from linearity is observed). Thus, the diamagnetism is mainly due to intragranular shielding currents and possibly due to local intergranular currents via superconducting grain boundaries [42] at the given field and temperature.

The behaviour of M(H) dependences on further increase of the field is related in principle to the magnetic flux penetration in to the separate crystallites and decrease in the intragranular diamagnetism.

In the case of HTSCs compounds containing rare-earth ions (in our case this is $NdBa_2Cu_3O_{7-x}$, see Figs. 14 and 15) a transition from diamagnetic to paramagnetic state is observed at certain fields and temperatures, in spite of some superconducting phase remain existing. It should be noted that similar transition to paramagnetic state could be observed also in compounds with rear-earth with no magnetic moment and this is due to the impurity magnetic phases of another origin - spin magnetism of Cu^{++} ions. For example, such a behaviour we have observed in polyphase Y-Ba-Cu-Pt-O due to the presence of Y_2BaCuO_x and $YBa_2Cu_3O_{6.3}$ paramagnetic phases. These data were obtained by both the magnetic and heat capacity measurements [45,46].

Let us consider some of magnetization peculiarities of the investigated specimens at liquid helium temperature. For the specimens prepared by substitution of Y_2O_3 by Ag_2O in the nominal composition of the precursor the M(H) dependence is almost completely reversible in increasing and decreasing field H>0.4-1T. At lower fields the magnetic moment hysteresis increases sharply up to a value of $\Delta M \simeq 25$ emu/g (H=0) for specimens D20.2 (0.2 mol% Ag) and D6.3 (1 mol% Ag). Such a behaviour could be possibly interpreted as determined by the lack of effective pinning centers inside the grains and at the same time the existence of considerably high pinning at the grain boundaries of Ag-doped material. Another feature observed is that the diamagnetic signal maximum occurs at higher magnetic fields compared to those for undoped specimens. The localization of diamagnetic signal maximum for D20.2, D21.2 and D6.3 specimens corresponds to a field H = 0.4, 0.6 and 1.0T respectively, while for undoped specimens Y_{123} [40] it is $\simeq$0.3T. The displacement of diamagnetic signal maximum to higher fields shows most likely that intergranular shielding currents between some separate grains or groups of grains still exist in the fields approximately 1T.

The diamagnetic signal for D20.2 and D6.3 specimens is also observed at maximum H values, which comes to show that the upper critical field dH_{c2} of the grains exceeds 14T at T=4.2K.

The M(H) dependence for Eu_{123} and Nd_{123} specimens shown in Fig.14 reveal the existance of hysteresis ($\Delta M \neq 0$) in the whole range studied. Its value for Eu_{123} is maximum (ΔM=24 emu/g) at 0.4T. The maximum value for Nd_{123} is two times lower, ($\Delta M \simeq 13$ emu/g), and at H>3T this hysteresis is $\Delta M \leq 2$ emu/g. As mentioned above the paramagnetic contribution of Nd added to the diamagnetic signal leads to a sign change of M at considerably lower values of the field $H \simeq 2T$.

Fig. 18 shows that some of Ag-doped specimens have considerable hysteresis ΔM in the whole range up to 14T. One can see at H>1T some rather strange peculiarities of M(H) for D63.2 and Y20, when the diamagnetic signal sharply deceases (almost down to zero for Y20) and on further increase of the field its value is almost recovered. Similar behaviour Y20 has also at H=4÷6T. These features (assuming they do not arise out of some unknown for us experimental error) could be hardly interpreted without additional information about the behaviour and the interplay of intergranular and intragranular shielding currents in the specimens. Information for intragranular currents, as known could be derived via investigation of M(H) dependence of powdered samples with no contact between separate grains. Another opportunity to make difference between inter- and intragranular currents could come from the investigation of specimens with tubular configuration [47].

M(H) for D62.2, shown on the same figure , does not manifest similar peculiarities and has considerable hysteresis up to H=14T, which at H>5T is almost independent on the field.

D41.1 specimen, which is prepared in the same way as D62.2, namely by substitution of $BaCO_3$ with Ag_2O in the nominal precursor composition, exhibits less but still noticeable hysteresis (ΔM=5÷16 emu/g).

Direct measurement of the transport superconducting current by resistivity four-probe technique could help for the interpretation of ΔM(H) dependence. Such measurements we performed [24] with specimens D41.1 and D6.3 at temperatures of 4.2K and 66K in liquid helium and pumped out liquid nitrogen. The results are shown in Fig.20. In the course of these experiments "a training effect" was observed, the origin of which is not clear for the present. The critical current for specimen D6.3 went up after the maximum field of H=14T was initially applied. The critical current increase was considerable ≃35% (at T=4.2K) and≃20%(atT=66K).

The dependencies J_c(H) for both specimens have different behaviour. For D6.3 a considerable decrease of J_c(≃4 times) can be seen on the field increase up to 1.5T, but relatively slight dependence exists in higher fields. This behaviour correlates to the large ΔM observed at H<1T in Fig.16.

A slight, almost linear dependence of J_c(H) is observed for specimen D41.1, and its critical current at H>0.5T exceeds that for the specimen D6.3. This material shows improved pinning at 4.2K and 66K compared to D6.3 as shown in Fig.20 and the insert.

Using the critical state model we have estimated the characteristic size of the shielding current circuits for D41.1 material at H=5T and

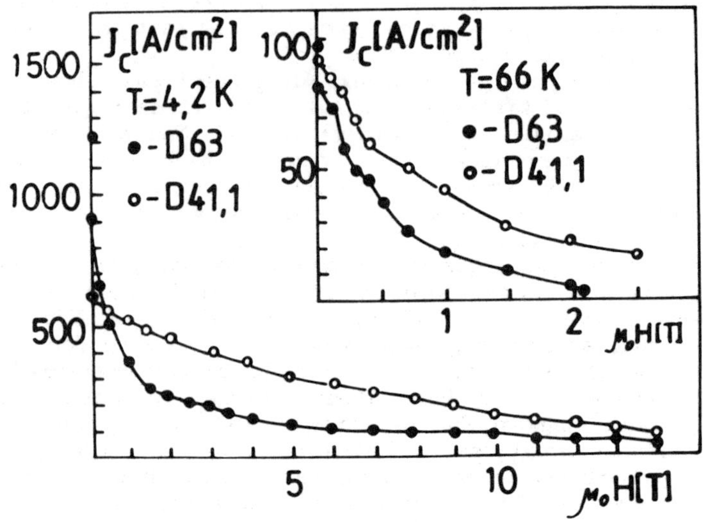

Fig.20 Transport critical current density J_c vs. magnetic field $\mu_0 H$ for D6.3 (•) and D41.1 (o) at temperatures T=4.2K and T=66K (insert).

T=4.2K. From J_c(H) data of this material at (Fig.20) we have taken J_c=310 A/cm^2 and from its M(H) dependence (Fig.16) the value of $\Delta M(5T) \simeq 7$ emu/g. The equation (9) gives d=3 mm which is of the order of the specimen size instead of the order of the grain size. This result is a little surprising but on the other hand it is in agreement with Kupfer's data [42] for the existance of almost field independent intergranular currents in HTSCs ceramics.

According to the results obtained there is a reason to consider that certain substitution of Ag_2O for $BaCO_3$ could improve the intergranular pinning of $YBa_2Cu_3O_{7-x}$. We must note, however, that different way Ag-doped specimens exhibit rather different behaviour.

It is interesting to note that the magnetization changes its sign for sample D41.1. This effect is probably due to the existing of paramagnetic phase Y_2BaCuO_5 (211) in this sample. The presence of 211 phase has been detected by X-Ray analysis and is also supported by the magnetic susceptibility measurements.

Let us follow shortly the M(H) dependence at liquid nitrogen temperatures. One can note right away the reduction more than one order of magnitude for M and hysteresis ΔM. This decrease leads to a larger slope of M(H), which for specimen with no magnetic rear-earth ion is determined by the existance of 211 phase. It is worth to note that peculiarities similar to those of specimens D63.2 and Y20 (see Fig.18) are observed at 77K too for D20.2 and D21.2 (see Fig.17). Surprising is the Eu123 behavior, which shows increase of the hysteresis ΔM at H>6T. Radically different is the behavior of the specimen Y20, which has $\Delta M \leq 1$ emu/g up to H=6T, and at H>6T the hysteresis disappears completely.

II.4.Conclusion

We showed that a radical difference exists for the dependences M(B) and J_c(H) for specimens prepared in different way of Ag doping. Silver most probably influenced the kinetics of the phase formation in the compound. This in turn reflects on the quantity as well as on the characteristics of the separate phases and via these on the physical properties of the specimens. There is a reason to consider that in some cases substitution or addition of Ag_2O could improve the pinning of $RBa_2Cu_3O_{7-x}$ (R is rare-earth).

It will be of great interest to distinguish inter- and intragranular current contribution in M(H) dependencies of Ag-doped HTSCs, to establish the causes for the observed pinning improvement and to find a method to control it.

Acknowledgements

The authors would like to thank the colleagues with International Laboratory for High Magnetic Fields and Low Temperatures in Wroclaw, Poland for their collaboration in magnetic field measurements.

This work has been completed with the financial support of the Bulgarian Committee for Science under contract No.706.

REFERENCES

[1] D.O.Welch, M.Suenaga and T.Asako, Phys.Rev. B36, (1987) 2390.

[2] T.P.Orlando, K.A.Delin, S.Foner, E.J.McNiff Jr., J.M.Tarascon, L.H.Green, W.R.McKinnon, and G.W.Hool, Phys.Rev B36, (1987) 2394.

[3] J.J.Neumeier, Y.Dalichaouch, R.R.Hake, B.W.Lee, M.B.Maple, M.S.Tarikachvili, and K.N. Y ang, R.P.G u ertin, and M.V.Kuric, Physica C. 152, (1988) 293.

[4] P.L.M.Van Bentum, H.Van Kempen, L.E.C.Van de Leemput, J.A.A.J. Pernboom, L.W.M.Schreurs, and P.A.A.Teunissen, Phys.Rev.B36, (1987) 5279.

[5] K.Murata, H.Ihara, M.Tokumoto, M.Hirabayashi, N.Terada, K.Senzaki, and Y.Kimura, Japan.J.Appl.Phys.26, (1987) L473.

[6] J.S.Moodera, P.M.Tedrow, and J.E.Tkaczyk, Phys.Rev.B36, (1987) 8329.

[7] J.Z.Sun, K.Char, M.R.Hahn, T.H.Geballe, N.A.Kapitulnik, Appl.Phys. Lett. 54, (1989) 663.

[8] K.Nakao, K.Tatsuhara, N.Miura, S.Uchida, H.Takagi, T.Wada, S.Tanaka, J.Phys.Soc.Japan 57, (1988) 2476.

[9] Y.Hidaka, Y.Enomoto, M.Suzuki, M.Oda, and T.Murakani, Japan.J.Appl. Phys. 26, (1987) L377.

[10] T.K.Worthington, W.J.Gallagher, and T.R.Dinger, Phys.Rev.Lett. 59, (1987) 1160.

[11] T,K,Worthington, W.J.Gallagher, D.L.Kaiser, F.H.Hodzberg, and T.R. Dinger, Physica C153-155, (1988) 32.

[12] Y.Iye, T.Tamegai, T.Sakakibara, T.Goto, N.Miura, H.Takeya, and H.Takei, Physica C152-155, (1988) 27.

[13] Y.Tajima, M.Hikita, A.Katsui, Y.Hidaka, T.Iwata, and S.Tsurumi, J.Crystal Growth 85, (1987), 665.

[14] F.G.Aliev, N.B.Brandt, V.V.Moshchalkov, and G.A.Chesnokov, Problemi visokotemperaturnoi sverhprovodimosti, Sverdlovsk, Part I, (1987) 50.
[15] K.Rogacki, C.Shklovski, A.Y.Zaleski, et al., Preprint Polish Acad. Sci., Inst.Low Temp.and Struct.Res., Wroclaw, (1987) 1337.
[16] V.G.Barjakhtar, V.M.Pan, A.G.Popov et al., Phisica niskih temperatur 13, (1987) 823.
[17] S.V.Gaponov, N.V.Il'in,M.A.Kal'jagin,E.B.Kluenkov, M.D.Strikovskii, and L.M.Fisher, Pisma v ZETPh 48, (1988) 155.
[18] P.England, T.Venkatesan, X.D.Wu, A.Inam, M.S.Hedge, L.T.Greeks and H.G.Craighead, Appl.Phys.Lett. 53, (1988) 2336.
[19] M.Nakao, M.Nemoto, R.Yanasa et al., Jap.J.Appl.Phys. 26, (1987) 794.
[20] K.Noto, H.Murita, K.Watonabe et al., Jap.J.Appl.Phys. 26, (1987)802.
[21] H.Kumakura, H.Shimizu, K.Takahashi, K.Togano, and H.Maeda, Jap.J. Appl.Phys. 27, (1988) L668.
[22] L.I.Glazman, A.E.Koshelev, A.G.Lebed, Pisma v ZETPh Prilovenie 46, (1987) 148.
[23] I.M.Dmitrenko, A.M.Glukhov, A.S.Zaika, A.S.Pokhila, et al., Phisica niskih temperatur 14, (1988) 1045.
[24] V.T.Kovachev, E.S.Vlakhov, K.A.Nenkov, V.A.Lovchinov, D.P.Lepkova, and S.Djambazov, Proc.ICEC 12, Southampton, UK (1988) 1026.
[25] Y.Dimitriev, B.Samuneva, Y.Pirov, E.Gattef, Y.Ivanova, V.Dimitrov, E.Kashchieva, V.Kovachev, and E.Vlakhov, Poc.ICEC 12, Southapmton, UK (1988) 1022.
[26] T.R.Dinger, T.K.Worthington, W.J.Gallagher, and R.L.Sandstrom, Phys.Rev.Lett. 58, (1987) 2687.
[27] Y.Enomoto, T.Muracami, T.Suzuki, and K.Moriwaki, Jap.J.Appl.Phys. 26, (1987) 1248.
[28] Y.Yeshurun, and A.P.Malozemoff, Phys.Rev.Lett. (in press).
[29] Y.B.Kim, C.F.Hempstead, and A.R.Struad, Phys.Rev.Lett. 9, (1962) 306.
[30] P.N.Peters, R.C.Sisk, E.W.Urban, C.Y.Huang, and M.K.Wu, Appl.Phys. Lett 52, (1988) 2066.
[31] M.M.Waldrop, Science 238, (1987) 1656.
[32] B.I.Shklovskii, and A.L.Efros, Elektronnie svoistva legirovannikh poluprovodnikov, Nauka, Moskva, (1979) 416.
[33] T.Iwata, M.Hikita, Y.Tajima, and S.Tsurumi, J.Crystal Growth, (1987) 661.
[34] D.Dew-Hughes, Cryogenics 28, (1988) 674.
[35] Z.Fisk, J.D.Thompson, S.W.Cheong, R.M.Aikin, J.L.Smith, and E.Zirngiebl, J.Magnetism an Magnetic Materials 67, (1987) L139.
[36] L.W.Roeland, F.R.De Boer, Y.K.Huang, A.A.Menovsky, and K.Kadowaki, Physica C 152, (1988) 72.
[37] K.Kadowaki, Y.K.Huang, M. van Sprang, H.P. van der Meulen, J.Q.A. Koster, et al., Int.J.Mod.Phys. 1, (1987) 525.
[38] K.Kadowaki, H.P. van der Meulen, J.C.P.Klaase, M.van Sprang, J.Q.A. Koster, et al., Physica B 145, (1987) 260.
[39] V.M.Dzugutov, and L.M.Fisher, Phisica twerdogo telo 30, (1988) 2148.
[40] M.Polak, F.Hanic, I.Hlasnik, M.Majoros, F.Chovanec, I.Horvath, L.Krempasky, et al., Physica C 156, (1988) 79.
[41] S.K.Malik, C.V.Tomy, Ram Prasad, M.C.Soni, Ashok Moham, and S.K. Gupta, Solid State Comm. 65, (1988) 581.
[42] H.Kupfer, I.Apfelstedt, R.Flukiger, C.Keller, R.Meier-Hirmer, B.

Runtsch, A.Turowski, U.Wiech, and T.Wolf, Cryogenics 28,(1988) 650.
[43] E.S.Vlakhov, V.T.Kovachev, K.A.Nenkov, T.Mydlarz, Int.J.Mod.Phys. (to be published).
[44] C.P.Bean, Phys.Rev.Lett 8, (1962) 250.
[45] V.T.Kovachev, E.S.Vlakhov, K.A.Nenkov, V.A.Lovchinov,M.M.Gospodinov, A.K.Stojanova, D.A.Dimitrov, M.Czyczek, and T.Mydlarz, Int.J.Mod. Phys. B 1, (1987) 223; ibid 479.
[46] V.T.Kovachev, B.M.Terzjiska, H.Madge, D.A.Dimitrov, and E.S.Vlakhov, Proc.ISCMP-5, Varna, (1988) (World Sci.Pabl.Co.,Singapore,in press).
[47] F.J.Eberhardt, A.D.Hibbs, and A.M.Campbell, Cryogenics 28, (1988) 681.

FABRICATION OF Bi-Sr-Ca-Cu OXIDE SUPERCONDUCTORS WITH CONTROLLED NUMBERS OF Cu-O LAYERS

K. Nakamura and K. Ogawa
National Research Institute for Metals, Tsukuba, Ibaraki, Japan

I. INTRODUCTION

Since the discovery of multiphase Bi-Sr-Ca-Cu-O superconductors[1], it becomes soon clear that these phases can be distinguished by the number of CuO layers[2-4], n, sandwiched by two BiO double layers. Thus the chemical compositions of these compounds are

$$Bi_2Sr_2Ca_{n-1}Cu_nO_{2n+4} \quad (n=1-3) \qquad (1)$$

Hereafter these phases are abbreviated as the 22(n-1)n phase. The correlation between the superconducting transition temperature, Tc, and the number of CuO layers, n, is also established. i.e., Tc increases with increasing n from 1 to 3 as 8, 80 and 107 K, respectively[1,5]. Although Tc varies significantly with the small variation of each component within the solid-solubility range[6], much effort has been extended to increase the volume fraction of the n=3 phase, i.e., the 2223 phase[7-10]. The difficulty to produce the monophasic 2223 compound lies in the fact that this phase grows in a very narrow temperature range around 875 3 when annealed in air(=20% O_2)[11]. Thus the amount of the 2223 phase produced in early studies was less than 20 % even if a correct cationic ratio 2223 was chosen for the starting oxide mixture[12]. Recently several processes have been proposed for the efficient growth of the 2223 phase, i.e., long term annealing[8], Pb addition[9], and annealing under low oxygen partial pressure[10]. Of these most efficient process is the Pb addition[9]. The first part of this chapter (section II) describes the growth of the leaded 2223 phase. Emphasis is put on the sintering temperatures and atmosphere, which have closely related to the exothermic peaks observed in the DTA analysis[11,13].

When the Pb addition method is applied to enhance the fraction of the 2223 phase in thin films, various problems related to the ease of Pb evaporation[14], the requirement of a long-term annealing[15] and the formation of impurity phases such as Ca_2PbO_4[8] are found. The ease of Pb evaporation from the leaded film suggests the possibility of a reversible reaction concerning the doping of Pb into the film from the vapor phase. The second part of this chapter (Section III) shows the formation of the 2223 phase under a controlled Pb potential[16].

The possibility of introducing more than three CuO layers between the two BiO double layers are the motive force for the third part of this report. When a conventional single target sputtering or a co-evaporation from multi-evaporation sources is applied, the control of the number of CuO layers is extremely difficult or almost impossible even if we choose an appropriate target composition to yield a correct cationic ratio in the deposited film[17]. However, it becomes recently clear that the multi-layer deposition technique is the most promising method to produce a single phase with a desired number of CuO layer, i.e., desired c dimension[18-21]. The third part of this chapter (Section IV) will describe the control of c dimension by means of multilayer deposition technique[19]. The superconducting properties and the thermal stability of these films will be also described.

II. Growth of the 2223 Phase in Bi(Pb)-Sr-Ca-Cu Oxide System. -Effects of Oxygen Partial Pressure and Annealing Temperature.-

Takano et al[9] have reported an enhanced effect of Pb addition on the formation of the 2223 phase in the Bi-Sr-Ca-Cu-O system. The optimum initial compositions they reported were 0.7(0.3):1:1:1.8 and 0.96(0.24):1:1:1.6 (hereafter we denote the order for Pb containing case as Bi(Pb):Sr:Ca:Cu). The Bi to Pb ratios 0.7:0.3 and 0.96:0.24 mean that Bi is partially substituted for Pb in this system. The sintering was made at 1118 K for 244 h. The sintering temperature 845 C is by about 30 C lower than that of the Pb free Bi-Sr-Ca-Cu-O, i.e., 875 C. Later, Endo et al[10] reported the effect of oxygen partial pressure and pointed out that the leaded 2223 phase grew most efficiently under 1/13 atm oxygen partial pressure.

The sample preparation is as follows[13]. The mixed oxide powder having nominal composition 0.96:0.24:1:1:1.6 was calcined twice at 800 C for 12 h in air and then pressed into a pellet of 11 mm in diameter. The sintering was carried out at temperatures between 823 to 875 C for 60 h and cooled to room temperature in the furnace. The flowing oxygen partial pressure was varied from 1 to 0.007 atmosphere.

The differential thermal analysis(DTA) was performed in a $Ar-O_2$ mixed gas flow with various oxygen partial pressure. The powdered sample of 600 mg in weight in an aluminum cell attached on a Pt-PtRh thermocouple was heated up to 800 C and held for 2 h, and the temperature was decreased down to 750 C. After this pre-treatment, DTA curve was measured by increasing temperature from 750 to 900 C with a heating rate of 37 deg/h. The pre-treatment mentioned above and the deliberately slow heating rate are the most important experimental conditions to obtain reproducible and reliable data.

Figure 2.1 shows the annealing temperature dependence of the XRD intensities of the (115) peak in the 2212 and 2223 phases. The

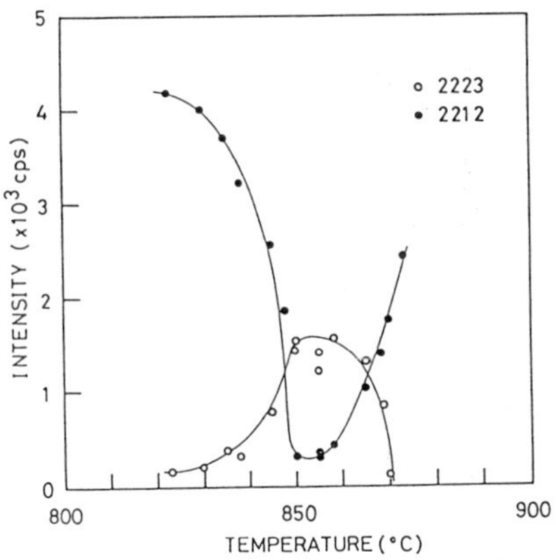

Fig. 2.1. Annealing temperature dependence of XRD intensities of the (115) line of the 2212 and 2223 phases. The pellets were sintered for 60 h [11].

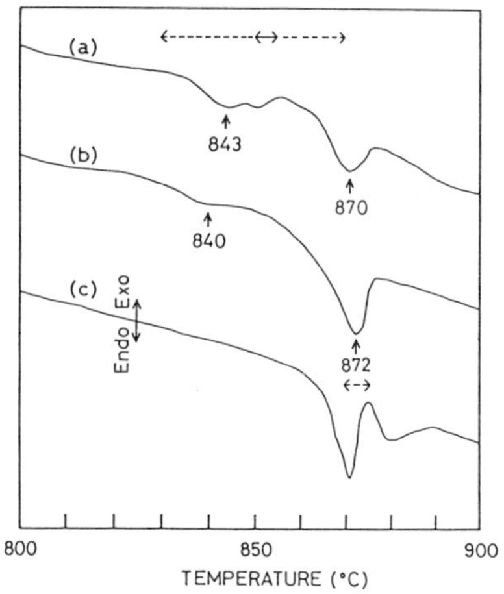

Fig. 2.2. DTA curves of (a) calcined powder (mainly composed of the 2212 phase), (b) annealed for 850 C for 60 h (mainly the 2223 phase) and (c) the Pb-free calcined powders. The heating rate is 0.6 C/min. The temperature ranges for the 2223 phase formation are shown with arrows [11].

temperature range where the 2223 phase is formed lies between 835 to 869 C. The optimum sintering temperature is seen to lie between 850 to 855 C, where the volume fraction of the 2223 phase exceeds 90 %. The increase in the fraction of the 2223 phase shows good correlation with the decrease in the fraction of the 2212 phase, indicating that the 2223 phase grows from the 2212 phase. The temperature range where the 2223 phase grows, i.e., 835 to 869 C. is lower and wider than that of the Pb free Bi-Sr-Ca-Cu-O system, i.e., 872 to 877 C. Figure 2.2 shows the DTA curves of the calcined powder (before sintering) mainly comprising the 2212 phase (curve a) and of the pellet which was sintered at 850 C for 60 h and mainly comprising the 2223 phase (curve b). Both curves have two endothermic peaks between which the 2223 phase grows most efficiently. The higher endothermic peak can be assigned to the decomposition reaction of the 2212 and 2223 phases into the high melting point compounds and the low melting point liquid phase.

Curve (c) in Figure 2.2 shows the DTA curve of the Pb free compound. two endothermic peaks between which the 2223 phase in the Pb free system grows most efficiently are observed around 875 C. The narrow temperature range, 10 C, may be the main cause of the difficulties found in the formation of the 2223 phase from the Pb free oxide powder.

Figure 2.3 shows the DTA curve of the calcined sample ((a) in Fig. 2.2) measured in a flowing $Ar-O_2$ mixture of various oxygen partial

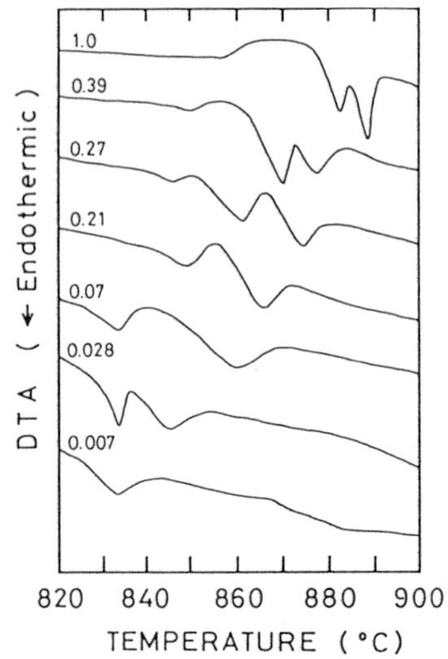

Fig. 2.3. DTA curves of the calcined powder examined in the various oxygen partial pressure from 1 to 0.007 atm. The heating rate is 0.6 deg/min [13].

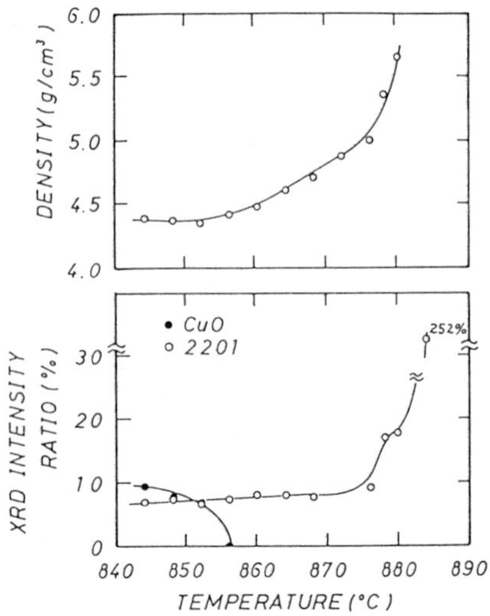

Fig. 2.4. Annealing temperature dependence of the XRD line intensities of (111) of CuO and (105) of the 2201 phase relative to that of the (105) of the 2212 phase. The sample were quenched into a liquid nitrogen bath [13].

pressure. From this figure one sees that at higher oxygen partial pressure, three endothermic peaks exist between 820 to 900 C. The three endothermic peaks are denoted as Peak I, Peak II and Peak III in the order of increasing temperature. At lower O_2 pressure Peak I disappears. These endothermic peak temperatures decrease monotonously with decreasing O_2 partial pressure. The chemical species existing at each endothermic peak were determined by means of XRD for the samples annealed at various temperatures between 840 to 885 C for 1 h in 1 atm O_2 and then quenched to liquid nitrogen. Various chemical species including the 2201, 2212, CuO and Ca_2PbO_4 are found to exist in the quenched samples. The fraction of these phases decreases or increase with varying the quenching temperature but the main phases are the 2212 and 2201. In Figure 2.4 the dependence of XRD intensities of the (111) line of CuO and the (105) line of the 2201 phase relative to the (105) of the 2212 phase on the quenching temperature are shown. The peak position of each endothermic peak is shown by arrow. It is shown that the CuO lines disappear above Peak I and the (105) line of the 2201 phase shows an abrupt increase above Peak II. The density of the sample increased from 4.8 to 5.8 with increasing temperature from 870 to 880 C indicating the existence of liquid phase above this temperature range (Fig. 2.4-a).

On the basis of this relationship and the density and appearance of the quenched samples, Peak I, II and III can be assigned to the

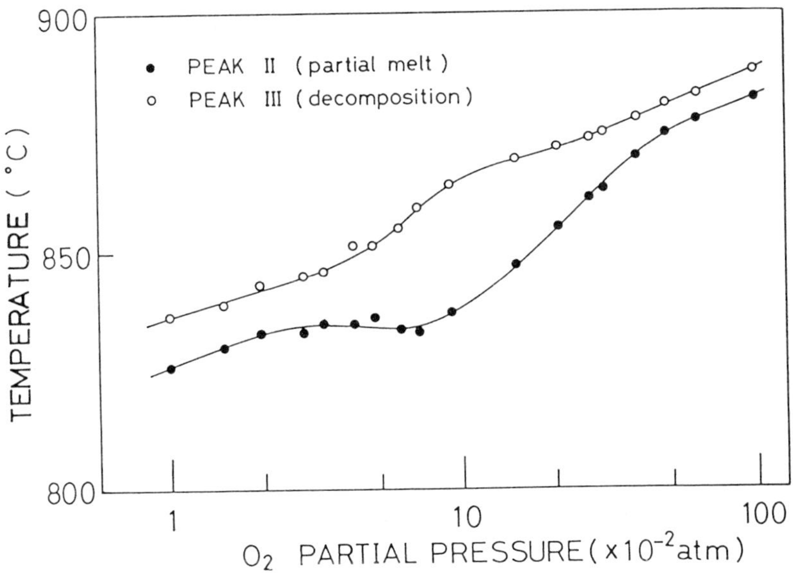

Fig. 2.5. Oxygen partial pressure dependence of the peak temperature for endothermic Peak II and Peak III, corresponding to the partial melting and decomposition of the calcined powder, respectively [13].

endothermic peak arising from the melting of CuO, partial melting of the calcined compounds, and its decomposition, respectively. The 2223 phase grows most efficiently around Peak II or in the temperature range sandwiched by Peak I and II, where the partial melting state exists (see Fig. 2.1). The 2223 phase grows from the 2212 phase by taking up the necessary elements such as Cu and Ca, from the partially melted liquid phase,which offers as the fast diffusion path.

In Figure 2.5, the temperature range where the endothermic peaks , Peak I and II, appear in the DTA curves are plotted against O_2 partial pressure. Both peak temperatures decrease with decreasing O_2 partial pressure. But the gradients are not the same. Consequently, the difference between the two peaks, where the 2223 phase grows most efficiently, becomes maximum at about 0.07 atm, showing that the optimum partial pressure lies around 0.07. In summary, the results of DTA and XRD measurement on the Bi(Pb)-Sr-Ca-Cu-O system confirms the temperature range in which the partial melting occurs, and in which the 223 phase grows most efficiently. The broadening of the temperature range with decreasing the oxygen partial pressure has a maximum around 0.07 atm oxygen partial pressure.

III. Growth of the 2223 Phase in Bi(Pb)-Sr-Ca-Cu Oxide Films under a Controlled Pb Potential.

For bulk leaded pellet, the loss of Pb from the pellet by annealing in air is not so decisive as to affect the formation of the 2223 phase. For example, for the oxide mixture having the cationic ratio 0.95(0.25):1:1:1.6 , the loss of Pb during the calcination and sintering at 850 C for 230h amounts to about 33 %[11]. However, in the case of thin oxide film, the larger surface to volume ratio results in the complete evaporation of Pb from the film and consequently, suppresses the formation of the 2223 phase. Figure 3.1 shows the change in Pb content for the Pb containing sputter-deposited film (16(3.7):20:21.5:39 in the cationic ratio) when annealed at 850 C in air. As can be seen from the figure, the Pb content decreases very rapidly with annealing time and becomes about 50 % of its initial value within 10 min and almost undetectable after annealing for 30 min. The XRD of the film thus annealed consists of almost the 2201 phase with a trace amount (ca. 1/10) of the 2223 phase. Taking into account of the Pb evaporation, Hakuraku et al[14] have prepared amorphous films from a target with a large Pb content, i.e., 2.7:2:2:2:4 in the cationic ratio, and have obtained films with a large volume fraction of the 2223 phase by annealing the amorphous film in air at 850 C for 15 h. However, no detailed information on the amount of Pb in the sputtered film and the post annealed one has been reported.

The fact that Pb is easily evaporated from the film strongly suggests that when Pb free Bi-Sr-Ca-Cu oxide film is annealed under a constant Pb potential, the concentration of Pb must increase so as to balance the Pb potential between the vapor and solid phases. The most convenient method to control the Pb chemical potential in the solid phase is to adjust Pb concentration in a single or multi-component metallic oxide having a large solid solubility for PbO. The chemical potential in the vapor phase can be simply controlled by equilibrating the oxide in a closed gold capsule. Among many possible combination, Bi(Pb)-Sr-Ca-Cu-O system is assumed to be most favorable since the solid solubility of Pb in this system is large and the Pb chemical potential can be easily

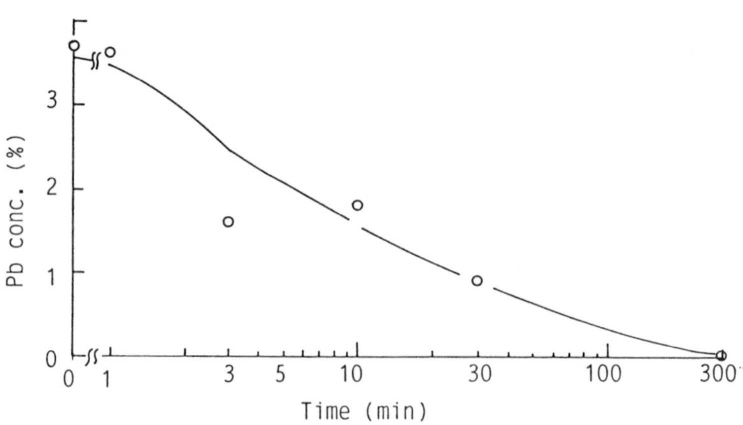

Fig. 3.1. Variation of Pb content vs time of annealing at 855 C in air.

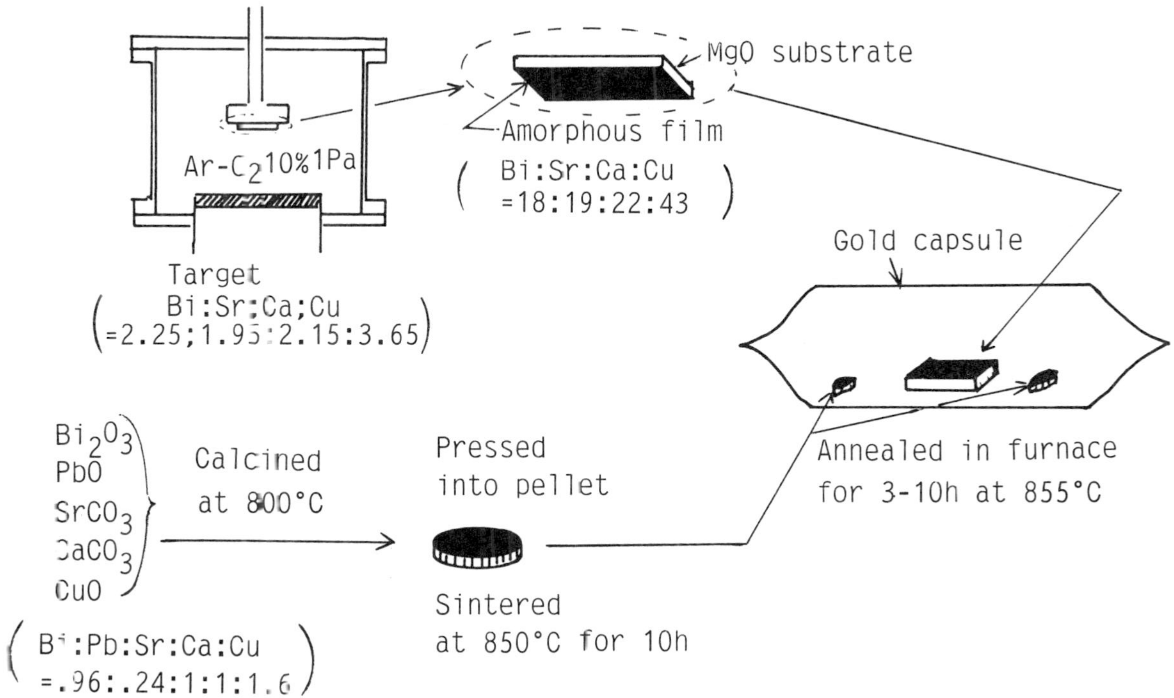

Fig. 3.2 A flow chart for Pb doping processs.

controlled by substituting Pb for Bi. When the volume of Pb source oxide is larger than the Pb free film and when the two solid phases are confined in a closed gold capsule having smaller volume (smaller volume in this case means that the amount of Pb in the vapor phase must be smaller than that in the solid phase), the Pb potential in the Pb source phase can be approximated as constant.

Figure 3.2 shows a schematic drawing of our simple Pb doping process. Firstly the Pb free amorphous film is sputter-deposited onto MgO substrate. The composition of the Pb free sputtered film is 16:19:22:43. As the Pb source phase, we have chosen sintered Bi-Pb-Sr-Ca-Cu oxide having a cationic proportion of 0.96(0.24):1:1:1.6 following the composition by Takano et al[9]. This sintered pellet is crushed into pieces of almost 0.1 g in weight. A piece of the crushed Pb source and the Pb free film (5x5mm) having 200 nm to 1 um thick(thus the net amount of oxide film is about 0.3 to 0.1 mg, which is less than 0.3 % of the source pellet) are confined in a closed gold capsule (Φ20×50mmℓ). This capsule is then inserted into a box furnace. The annealing temperatures are 820 to 870 C for 30 to 40 h.

Figure 3.3 shows how the Pb concentration in the two solid phases (Pb source and film) becomes closer with time after annealing at 855 C. With increasing annealing time the Pb concentration in the film increased and approached to that in the Pb source pellet. It should be noted that the Pb concentration of the annealed film is independent of the initial crystalline state of the film, i.e., amorphous or crystalline 2212 phase. As will be shown later, however, the initial crystalline state of the film has a decisive effect on the growth of the 2223 phase in the film.

It should be emphasized that the growth of the 2223 phase depends not only on the Pb content, annealing temperature and annealing time but also on the manner of heat treatment, i.e., heating and cooling rate, and on the thickness of the films. Therefore in the present case the following three runs of heat treatment are applied as shown schematically in Figure 3.4. In Run A the as-sputtered amorphous film is initially annealed at 850-870 C for 1 h and furnace cooled to room temperature.

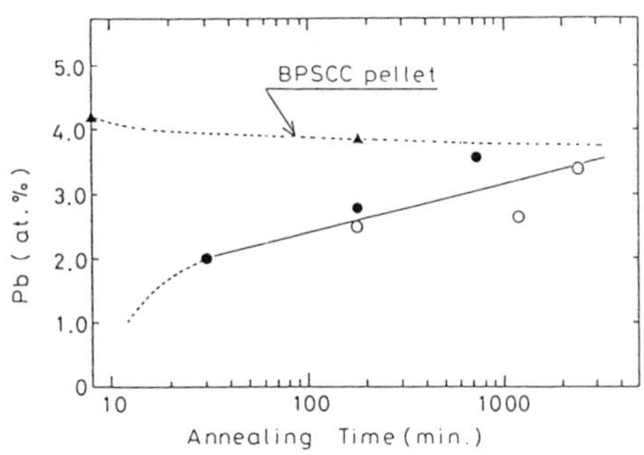

Fig. 3.3. The annealing time dependence of Pb concentration for the film (1.8 um thick) kept at 850 C. ○:slow heating, ●:rapid heating [16].

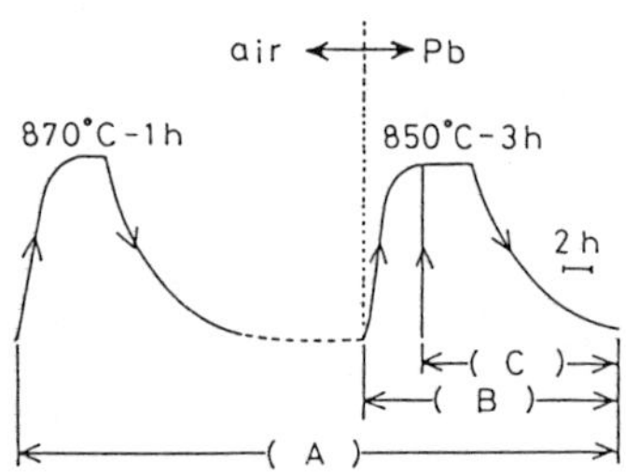

Fig. 3.4. A flow chart showing typical three runs of heat treatment.

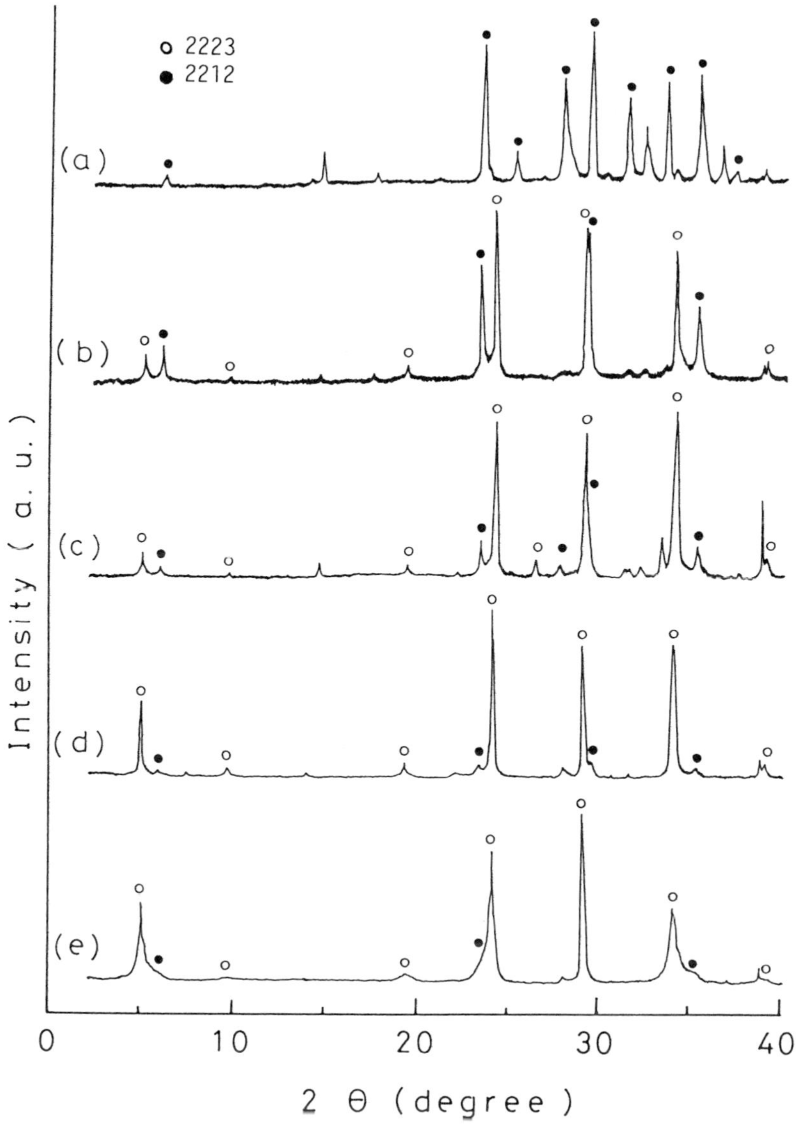

Fig. 3.5. XRD patterns of Bi(Pb)-Sr-Ca-Cu-O films after annealing in P atmosphere at 850 C for 3 h. o:the 2223 phase. ●:the 2212 phase (a):Run A (1.8μm thick), (b):Run B (1.8μm thick), (c):Run C (1.8μ thick), (d)Run C (0.23μm thick) and (e):Run B (0.23μm thick) [16].

After this pre-treatment, the film is then confined in a gold capsule together with a piece of Pb source pellet and furnace heated in the same manner as Run B. The difference in Run B and C is that in Run C the sample contained in the gold capsule is rapidly inserted into the furnace kept at a desired annealing temperature.

Figure 3.5 shows the XRD patterns of the films obtained by the heat treatment denoted by Run A-C. Although the Pb concentration in the film is almost identical, the diffraction patterns differ markedly depending on the initial crystalline state, In Run A, the 2212 phase is initially formed by the pre-heat treatment in air. In this case the 2223 phase does not grow even after the heat treatment in Pb atmosphere. Contrary to this, when the initial crystalline state is amorphous, the 2223 phase grows efficiently but the volume fraction of the 2223 phase depends on the history of heat-treatment and the thickness of the film as shown in Fig. 3.5 (b)-(e). For the thicker film (1.8 um thick), the growth of the 2223 phase depends strongly on the heating rate. This is closely related to the intermediately formed crystalline state, crystalline grain size and orientation. The largest volume fraction of the 2223 phase is found in the thinner film (0.23 um thick) as shown in Figures 3.5 (d) and (e). In this case, the heating rate affects the width of diffraction lines.

Figure 3.6 shows the XRD patterns of the films (1.8 um thick)

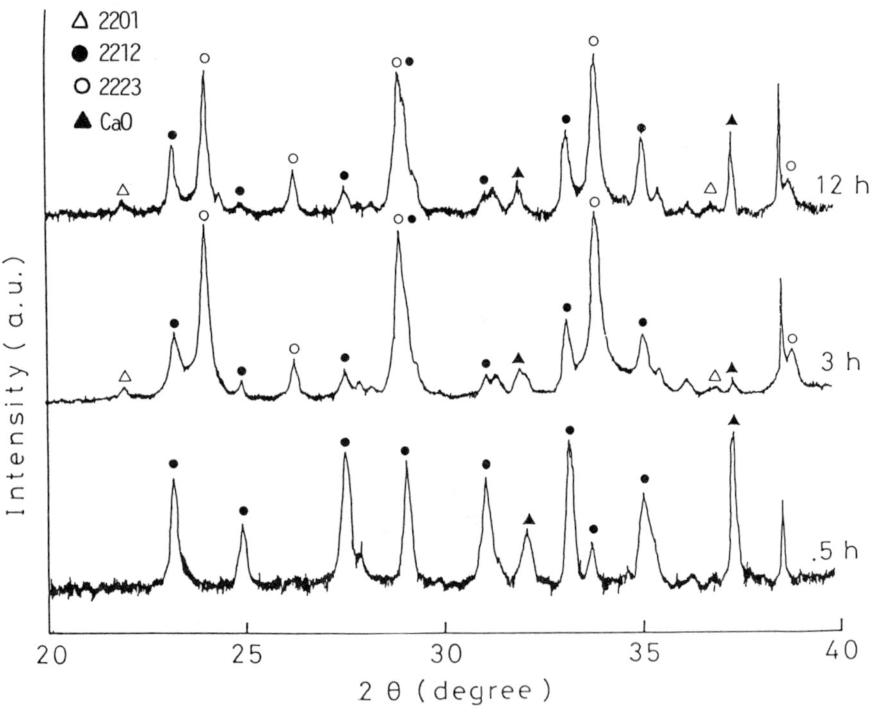

Fig. 3.6. XRD patterns of the films rapidly heated (Run C) and kept for 0.5,3 and 12 h, respectively [16].

rapidly heated and kept at 850 C for 0.5, 3 and 12 h. The most significant difference between the rapidly heated (Run C) and slowly heated (Run B) case lies in the grain orientation of the the 2212 phase which is initially formed under Pb atmosphere. The XRD pattern for the film rapidly heated and annealed at 850 C for 0.5 h shows the less oriented 2212 because relatively strong (117) and (115) lines can be observed other than (001) lines. The intensity ratio L(115) to L(0012), where L(hkl) denotes the hkl intensity of the 2212 phase, is a measure of the grain orientation and is 1.5 which is about one half of that for the randomly oriented powdered sample but is larger than that for the highly oriented one. In addition to the less-oriented 2212 phase, a fair amount of impurity phases such as Ca_2CuO_3 and CaO can be observed. These impurity phases, with the aid of Pb from the vapor phase, must be supplied to the 2212 phase to form the 2223 phase. The XRD patterns, 850x3h and 850x12h, show the growth of the 2223 phase during the annealing, but at the same time we can observe that the relative intensity of the impurity phases decreased appreciably, indicating the existence of the following reaction,

$$2212 + (0021,\ 0010,\ 0001) \longrightarrow 2223 \qquad (2)$$

At the same time, the ratio L(115)/L(0012) decreases to 0.3, indicating that the randomly oriented 2212 grains disappeared. These results show that the 2223 phase grows most efficiently from the initially formed randomly oriented 2212 phase.

Figure 3.7 summarized the above mentioned XRD observation. The fraction of the high Tc phase is represented by the intensity ratio of the (0014) line, i.e., H(0014) to H(0014)+L(0012), where H denotes the 2223 phase. Contrary to the rapidly heated one, the intermediately formed 2212 phase is sufficiently oriented, and the growth of the 2223

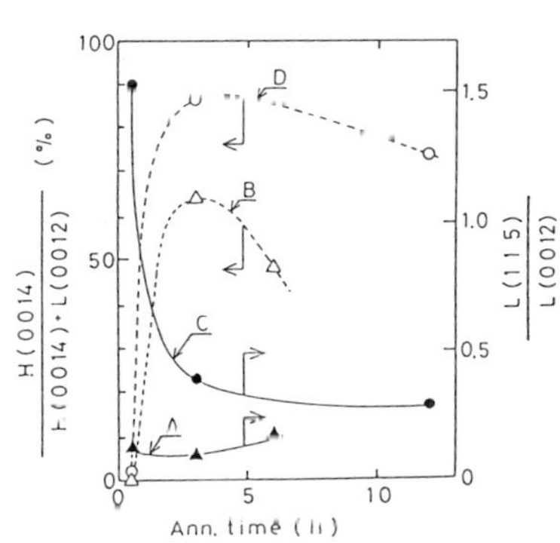

Fig. 3.7. Annealing time dependence of H(0014)/ H(0014)+L(0012) and L(115)/L(0012) when annealed in Pb atmosphere at 850 C. A and B: slow heating. C and D: rapid heating [16].

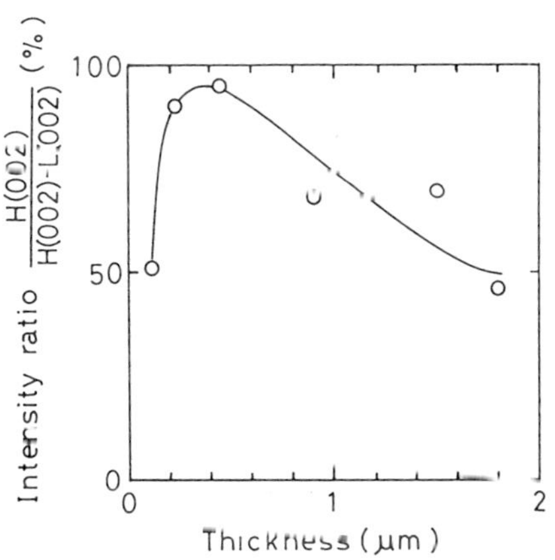

Fig. 3.8. Film thickness dependence of the fraction of the 2223 phase after annealing at 850 C for 3 h (Run B).

phase is not so remarkable as that for the rapidly heated case. This observation is consistent with the fact that the 2223 phase shows minimal growth from the well crystallized 2212 phase which is formed before the Pb treatment as shown in Run A case.

The dependence of the 2223 phase growth on the initial thickness of the amorphous film is shown in Figure 3.8 for the slow heating case. The fraction of the 2223 phase shows maximum around 4500 A thick. This indicates that the diffusion of Pb into the film is not the growth rate controlling step in this system, but the shape factor such as grain size and orientation of the 2212 phase and the distribution of impurity phases as CaO and Ca_2CuO_3 plays an important role.

The growth of the 2223 phase vs. annealing temperature is shown in Figure 3.9. Similar to the Pb doped material, the temperature range in which the 2223 phase grows becomes relatively broad owing to the addition of Pb from the vapor phase, Above 870 C, no 2223 phase can be detected even after a long-term annealing and the 2201 phase grows significantly indicating that the complete decomposition of the 2223 and 2212 phases occurred above this temperature. The optimum annealing temperature deduced from this figure is 855 to 850 C, where 60 to 95 % of the 2223 phase grows depending on the heating rate and thickness of the film. These observation suggest a similar growth mechanism operating in the films as well.

Figure 3.10 shows resistivity-temperature curves for the same films shown in Fig. 3.5(a) to (e). Because film (a) consists only of the 2212 phase, the resistivity curves shows a low Tc transition. Although the volume fraction of the 2223 phase for (b) to (e) is almost the same, Tc end of the thick film rapidly heated (Run C) is lower by about 35 K than that of films (b), (d) and (e). This difference can be accounted for by the difference in the distribution of the 2212 phase in the crystal

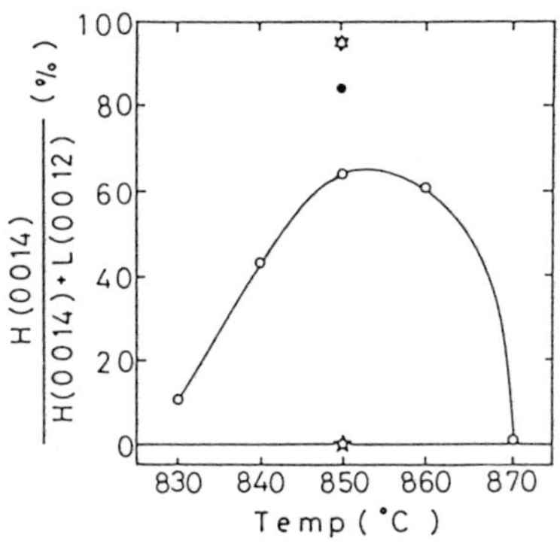

Fig. 3.9 The anneling temperture dependence of H(0014) /[H(0014)+L(0012)] for the films(1.8um thick) annealed at 830 to 870 C(—o—). The symbols ☆, ●, △ and ▽ correspond to (a), (c) (d) and (e) in Fig.3.5 , respectively.

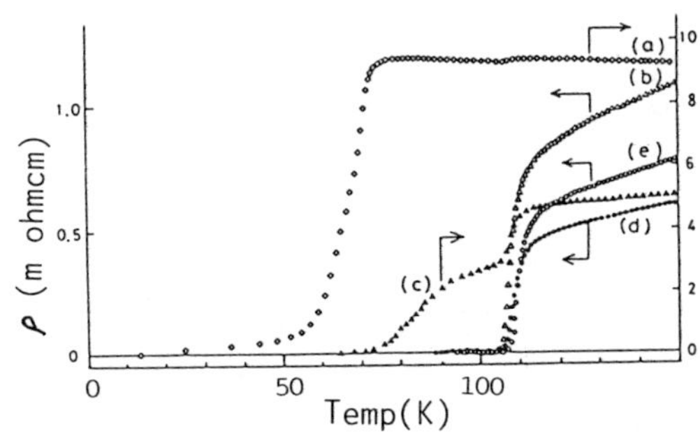

Fig. 3.10. The temperature dependence of resistivity curves for films (a)-(e) shown in Fig. 3.5.

grains of the dominant 2223 phase. In summary, a considerable amount of the 2223 phase can be obtained by annealing the Pb free Bi-Sr-Ca-Cu oxide films in an appropriate Pb potential. The initial crystalline state of the films, the heating rate and the thickness of the films affect the fraction of the 2223 phase in the annealed films.

IV. Synthesis of Multilayered Bi-O/Sr-Ca-Cu Oxide Films Having Desired Modulation Wave Length along the Growth Direction.

So far we are mainly concerned with the increase in the volume fraction of the high Tc phase, i.e., 2223 vs 2212 phases. These phases can be obtained as a thermally stable phase at a given temperature range. The oxygen partial pressure and Pb chemical potential also affect the thermal stability of these phases. In many cases, however, the intergrowth of these phases are often observed by the HRTEM observation [22-25]. The intergrowth of many phases with various different $CuCaO_2$ layers suggests that the free energy against the number of $CuCaO_2$ layers have no sharp minimum. These considerations lead to a possibility of synthesizing thermally metastable phase with n>3. In this chapter we show our recent investigation on this field.

Figure 4.1 shows a schematic diagram of the bitarget sputtering apparatus equipped with two magnetron cathodes. The targets are 75 mm in diameter metallic Bi and sintered Sr-Ca-Cu-O with a cationic ratio 2:2:2-3. To enhance the oxidation of the film during the sequential deposition, Ar-10%O_2 gas is blown directly to the substrate through a circular shaped oxygen nozzle. The total pressure during the sputtering is 1Pa but the oxygen partial pressure at the substrate should be larger than the averaged O_2 pressure. The modulation wave length Λ, defined as the thickness of the repeated unit sandwiched with two Bi double layers, was adjusted by varying the deposition time for the Sr-Ca-Cu-O layer, usually 16 to 36 s, while keeping that for the Bi-O double layer constant as shown schematically in Figure 4.2. The total number of repeated layers was 50 to 100 in the present case. When the substrate

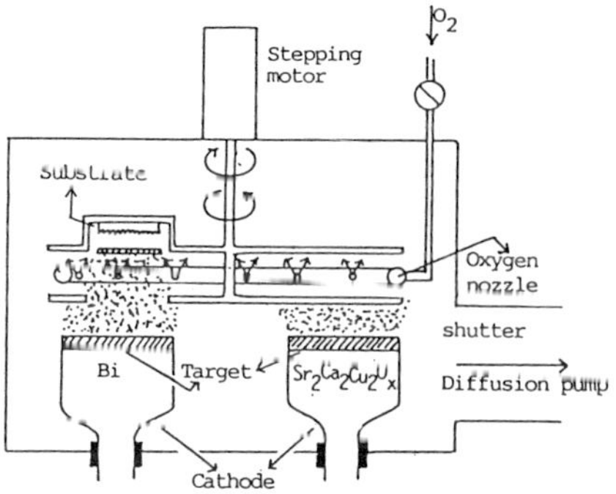

Fig.4.1. Sputtering system geometry for the synthesis of the Bi-Sr-Ca-Cu-O/B-O/····· multi-layered film.

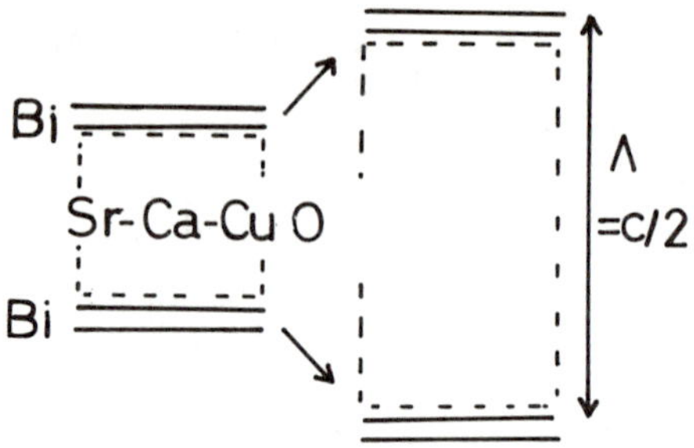

Fig. 4.2. A schematic diagram showing how to control the modulation wavelength along the growth direction. At lower Ts the films become amorphous. At Ts higher than 600 C the arrangement of the Sr-Ca-Cu-O layer becomes ordered to construct a thermodynamically stable perovskite structure.

temperature, Ts, was lower than 400 C, the as-deposited films showed diffraction pattern characteristic of alternate amorphous layers, i.e., consisting of a sharp superstructure diffraction lines located at lower angles and a halo at higher angles, i.e., 3-7 and 20-40 (degree in 2θ) in CuKα radiation. Above Ts>600 C, all diffraction lines can be identified as those from the layered perovskite phases.

Figure 4.3 shows the XRD patterns of the multilayered films deposited at Ts=650 C. The number 22(n-1)n (n=1-5) in the figure shows that the film came close to the simulating 22(n-1)n structure. All diffraction patterns can be assigned to (00ℓ). In synthesizing these multilayered films, for example, the deposition time for one Bi double layer was kept at a constant value of 9 s, while fixed the deposition time for the Sr-Ca-Cu-O layer varied as 15, 20, and 30 s for n=3, 4 and 5, respectively. When the deposition times for the Sr-Ca-Cu-O is fixed at one of these discrete values, the observed diffraction pattern of the resultant films can be assigned to one of the layered perovskites as shown in the figure. However, when the deposition time for the Sr-Ca-Cu-O layer is intermediate between these discrete values, the diffraction linewidth becomes broader around the sharp lines.

The modulation wave length and the composition of the resultant layered films are listed in Table 1. The ideal compositions estimated from the 22(n-1)n structure are also shown for comparison. It is emphasized that the composition of the thin film (1200-2500 A thick) were determined by means of EDX analysis and consequently several % errors must exist. Even though taking into account of uncertainty in the concentration, the deviation from the ideal composition is significant. For the 2201 phase, 20 % Ca substitute for Sr indicating that the solubility of these system is extremely large, i.e., chemical disorder concerning the arrangement of Sr and Ca must exist in the present film. The existence of such a disorder in the chemical composition is a characteristic of this film and is closely related to the superconducting properties of the layered films as will be shown later.

In considering the thermal stability of the multilayered film, it is important to consider how the crystallization of the film during the deposition depends on the substrate temperature Ts. At Ts lower than 400 C, the as-deposited film is amorphous and converts into the 2212 phase

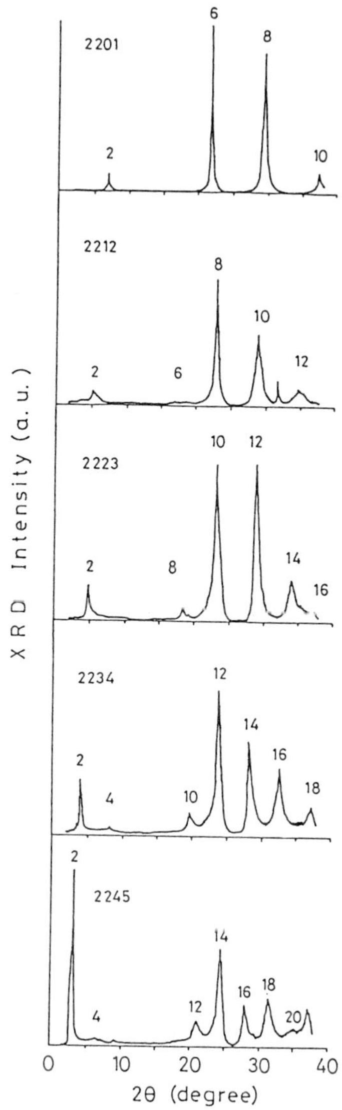

Fig. 4.3. XRD patterns of the as-deposited films (Ts=650 C) simulating the 2201, 2212, 2223, 2234 and 2245 phases (from up to down), respectively. Numbers attached to the diffraction lines represent ℓ of (00ℓ) [19].

Table 1. Modulation wavelength, Λ, and composition of the as-deposited films [19].

Structure	Λ(Å)	Composition (%)			
		Bi	Sr	Ca	Cu
2201	12.0	36.5	15.0	20.7	27.8
2212	15.6	23.6	18.9	22.9	34.6
2223	18.8	26.7	18.9	24.5	29.9
2234	21.6	18.7	20.6	25.8	35.0
2245	25.8	13.8	23.3	29.1	33.8

when annealed above 800 C. Figure 4.4 shows the XRD patterns of the film having designed modulation period of 18 A corresponding to the 2223 structure. From this figure, one sees that the XRD linewidth becomes narrower with increasing the substrate temperature. At Ts=665 C, the linewidth of the as-deposited film is very narrow which is almost equal to that of the Pb doped film having Tc (zero) = 106 K. Although the XRD line width of the as-deposited film (Ts=665 C) is very narrow, Tc end, in many cases is lower than 30 K, and sometimes shows semiconductor-like behavior.

When these as-deposited films were annealed at 760 C for 1 h in air, all the films deposited at different Ts exhibit superconductivity. For the film deposited at Ts=600 C, the somewhat broad XRD of the 2223 phase completely converts into sharp XRD lines of the 2212 phase, indicating that the as-deposited 2223 phase was re-crystallized completely to give the 2212 phase. Contrary to this, the somewhat broader XRD lines of the 2223 phase deposited at Ts=645 does not change appreciably by the post-annealing. A detailed examination of Fig. 4.4 shows that the linewidth becomes rather broader by annealing, indicating that a clear intergrowth structure developed in the annealed film. The most significant difference between the two is that the annealed films exhibit Tc zero at ca. 50 K, whereas the as-deposited films show semiconductor-like behavior

As shown in Fig. 4.4., the linewidth, diffraction angle and relative line intensity of the (00ℓ) lines do not change after annealing at 760 C. This indicates that the as-deposited film is structurally stable when deposited at Ts=665 C. If some kind of rearrangement of cations, such as ordering of Ca and Sr atoms or substitution of Sr for Bi occurred during annealing, i.e, the relative intensity of (00ℓ) line should change appreciably. The invariability of the relative diffraction line intensity strongly suggests that the interlayer diffusion did not occur during the annealing process since the interlayer atomic rearrangement changes the XRD structure factor of each layer. In spite of the structural stability of as-deposited films, the superconducting properties changed appreciably after the post-annealing; Tc increased from 0- 20 K to 66 K (760x1 h) and 82 K(810x1 h). One of the possible

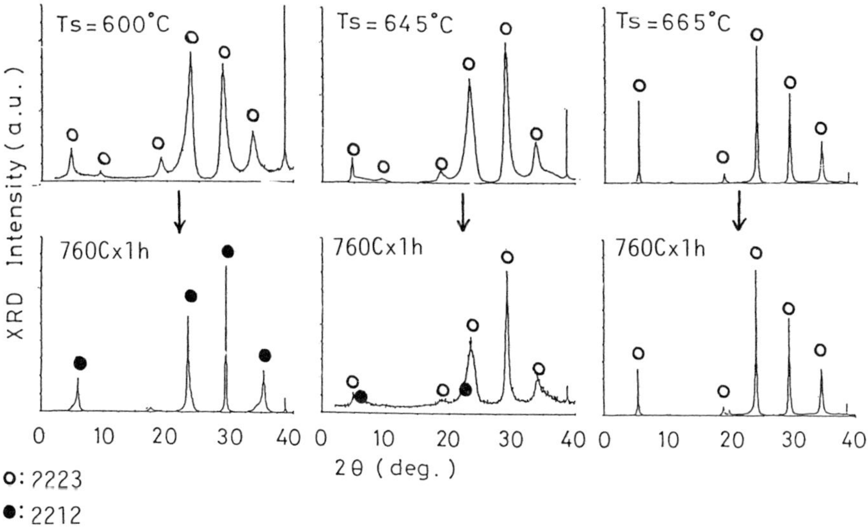

Fig. 4.4. Effects of substrate temperature, Ts, on the crystallinity and thermal stability of the as deposited films. The change in the XRD patterns after annealing at 760 C for 1 h in air are also shown in the lower part of the figure. When Ts=600 C, the 2223 phase completely decomposed into the 2212 phase with annealing. ●: Λ=15.6 A o: Λ=18.8 A

explanation of this increase in Tc is that some ordering occurred within each layer plane (00ℓ) since XRD intensity is not affected by the atomic rearrangement within the XY plane.

When the artificially layered phase has the 2234 structure (Λ =21.6 A), this phase is not stable as compared with the phases with n<=3. Consequently, precipitation of phases with n<=3 is expected to occur when annealed. The XRD of the 2234 film annealed at 760x3h, 810x1h and 850x1h, respectively, are shown in Figure 4.5. The XRD pattern of the film annealed at 760 is almost equal to that of the as-deposited film, but the RT resistivity dropped by one order of magnitude and exhibits superconducting transition around 4 K. However when the film was annealed at 810 to 850 C in air, the 2212 and 2201 phases are observed to precipitate from the layered 2234 phase as shown in Fig. 4.5 (b) and (c). No additional phases arising from the impurity phases, however, have been observed here. In spite of the precipitation of phases with n=2 and 1, the XRD angle and relative intensity of the layered 2234 phase keep their original values suggesting that no structural change occurred in the 2234 phase during annealing.

The high resolution SEM images of the as-deposited and the annealed film are shown in Figure 4.6. The surface of the as- deposited film is completely smooth, which is in contrast with the rough surface of the sputter-deposited film from a single target source or the film annealed

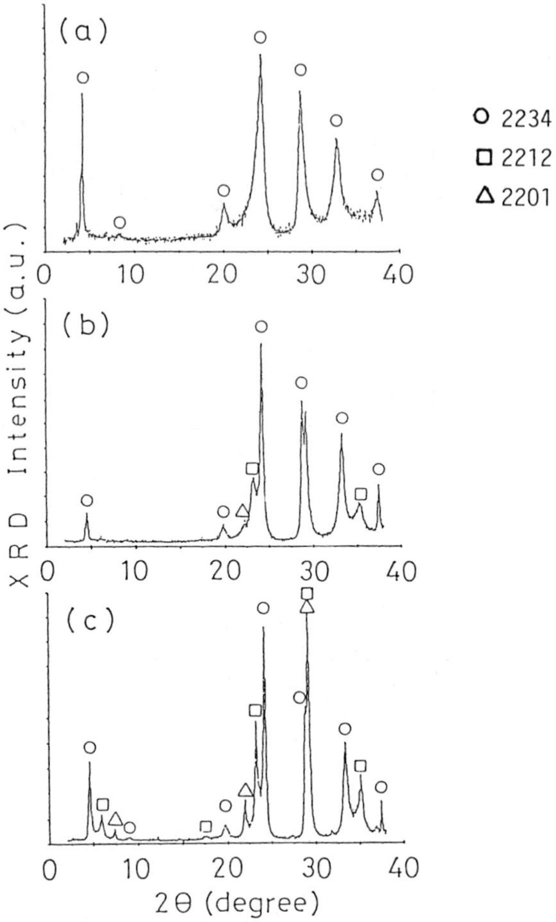

Fig. 4.5. XRD patterns of the multilayered films with Λ =21.6 A after annealing at (a)730 Cx3h, (b)810x1h and (c)850x1h. ○:c=43A, □:c=31A,△ :c=24A phase respectively [19].

from the amorphous state. From the SEM images of the annealed sample, we see at least two phases precipitate from the layered 2234 phase. From the XRD patterns in Fig. 4.6, the precipitated phases can be identified as the 2212 and 2201 phases, respectively. The scanning XPS observation also indicates that Ca rich phase also precipitates on the surface of the layered 2234 phase. These precipitated phases have a plate like shape and seem to grow epitaxially with their a-b plane parallel to the layered film. These plate like crystals do not contact each other as schematically shown in Figure 4.7. From these observations it seems likely that only the underlayered 2234 phase is electrically continuous through out the film and consequently the phase responsible for the

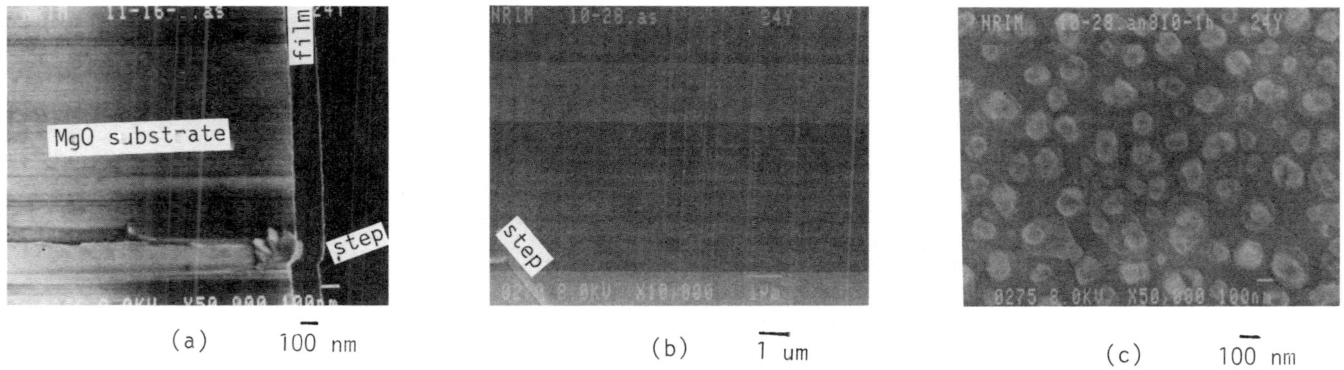

Fig. 4.6. HRSEM images of the as-deposited (a,b) and annealed(c) films with designed period of Λ =21.6 A. (a) cross-sectional view. (b) surface view. and (c) surface view (760x1h) [19].

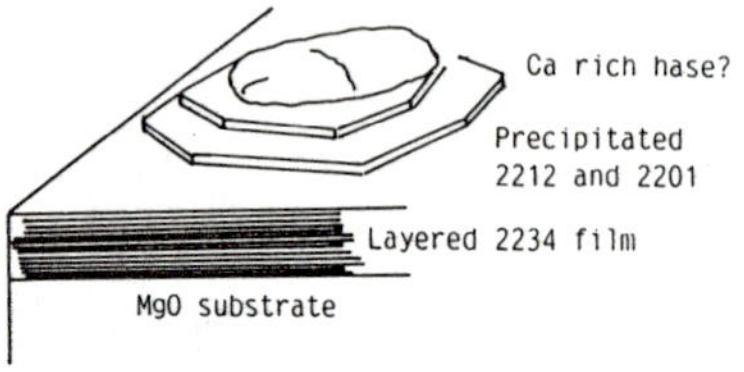

Fig. 4.7. A schematic diagram showing the precipitation of the plate-like 2201 crystals on the surface of the layered 2234 phase . Ca rich phase also precipitates on the surface of the 2201 and 2212 phases.

superconductivity of the annealed film must still be the layered 2234 phase.

The resistivity vs temperature curves of the 2234 phase of the films annealed at various temperatures are shown in Figure 4.8. The decrease in the resistivity after annealing at 730 C shows that some kind of structural ordering occurred during the annealing process. The increase in Tc for the films annealed above 810 C could be accounted by the increase in chemical ordering by supplying excess chemical components to the growing 2212, 2201 and Ca rich phases but no essential change in the structure of the 2234 phase occurred as described above. Tc onset of the annealed films lies around 115 K and Tc end lies around 65 K. Tc end of 65 K is higher than of the layered 2223 film deposited at Ts=645 C and annealed at 760 C (Tc end= 50 K) but lower than that deposited at Ts=665 C and annealed at 760-810 C for 1h (66-82 K).

It is interesting to discuss on the correlation between the number of CuO layers, n, and Tc. For the $Tl_2Ba_2Ca_{n-1}Cu_nO_{2n+4}$ (=22(n-2)n) and $TlBa_2Ca_{n-1}Cu_nO_{2n+4}$ (=12(n-1)n) systems, Ihara has derived empirical equations between the number of CuO layers,n, and Tc as [26],

$$Tc= 83n\exp(-n/4) \quad \text{for } 12(n-1)n \qquad (3)$$

$$Tc=115n\exp(-n/3) \quad \text{for } 22(n-1)n \qquad (4)$$

For the Bi-Sr-Ca-Cu-O system the number of CuO layers obtained by the thermal equilibrium method was limited to 3. Adachi et al have synthesized the layered 2234 phase by applying the sequential deposition method and have obtained superconductivity by annealing the as-deposited film at 850 C for 5h. The Tc onset of the film was 90 K and end around 75 K. They concluded that Tc zero of the 2234 phase should be 90 K, by about 16-8 K lower than that of the 2223 phase. However, the annealing temperature was so high and the annealing time was so long as to prevent the interdiffusion between the film and substrate and the recrystallization of the film. From the fact that Tc of the 2223 phase from the multilayer deposition technique is by about 30 to 60 K lower than that obtained from the thermal equilibrium method, it is reasonable to consider that Tc of the layered 2234 phase should be more than 30 to 60 K larger than the observed value, 50-75 K. Therefore we consider that Tc vs number of Cu layer problem still remains unsolved.

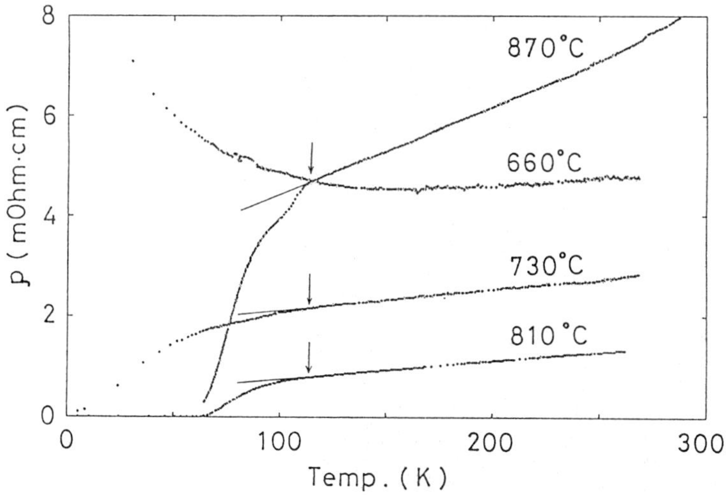

Fig. 4.8 Resistivity vs temperature curves for the films of Λ=21.6A after annealing at 660 Cx3h, 730x3h, 810x1h and 850x1h, respectively [19].

So far we have shown that the XRD pattern of the as deposited and annealed films show no essential change except for that the linewidth becomes narrower in the annealed film. In order to make clear what kind of ordering occurred during the annealing process in more detail, we have examined high resolution (HR) TEM observation for the 2234 films.

The HRTEM images of the as-deposited and annealed films having the 2234 structure (Λ=21.6 A) are shown in Figure 4.9 [27]. Similar to the bulk BSCCO, double dark dotted lines corresponding to the double Bi-O layers which are generally parallel to the MgO substrate are observed for the both films. In the case of as-deposited film, however, the lines are frequently bent, tilted, disconnected and terminated. The thickness of Bi-O double layers/perovskite sandwich can be assigned to 21.6 A, in agreement with the XRD. However, other layer thicknesses 12, 15, 18 and 25 A can be frequently observed in the as-deposited film whereas less frequently observed in the annealed one. The most significant difference between the two is that the modulated structure with wavelength of 25.8 A along the b axis can be clearly seen in the annealed film and not in the as-deposited one. In the annealed film, portions where modulation can be and can not be seen in the figure correspond to the a- and b- plane of the crystal. Thus the film is composed of mosaic crystals with a grain size of a few nm.

Comparison of the cross-sectional TEM images of the as-deposited and the annealed films shows that the formation of a clear modulation structure along the b axis seems to be the most probable cause for the development of superconductivity of the annealed multilayered film. The modulation structure is considered to arise either from the substitution

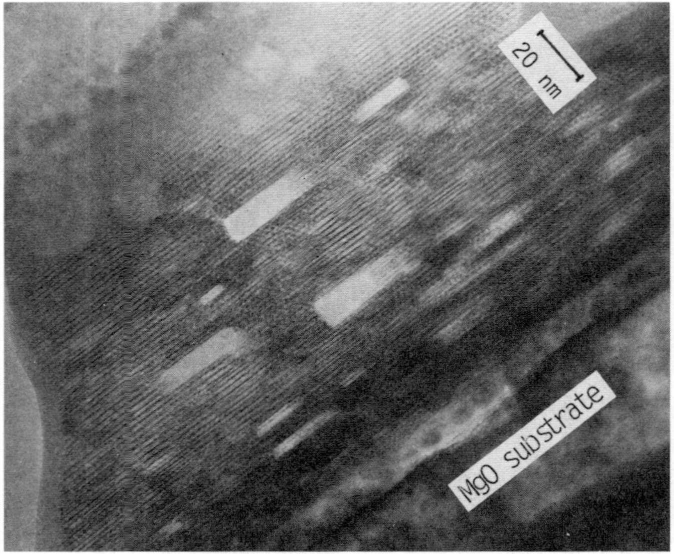

(a) annealed at 810°Cx1h

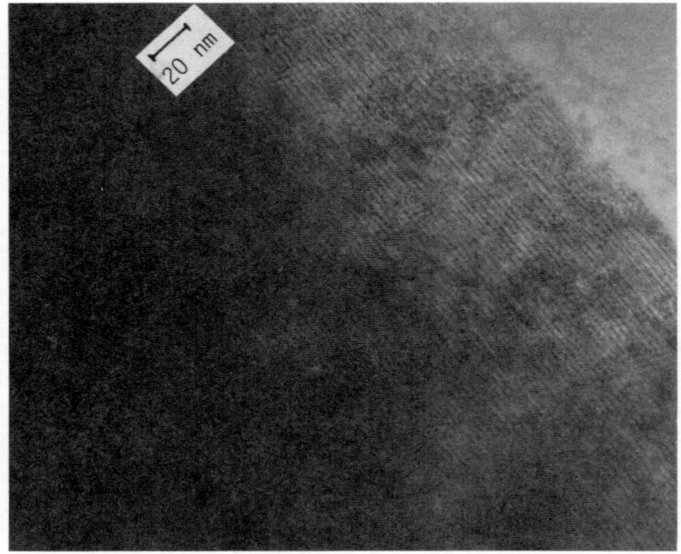

(b) as-deposited (Ts=650°C)

Fig. 4.9. Cross-sectional images of the HRTEM for tha annealed (a) and as-deposited films[27].

of Sr for Bi [28,29] or from the ordering of excess or deficient oxygen atoms in the Bi-O double layers [30,31]. If the interlayer atomic rearrangements occurred during the post annealing, the relative intensity of the XRD (00ℓ) lines must vary since the structure factor $F_{00} = f_n exp(2 i z_n)$ must change, where f_n and z_n are the atomic scattering factor and z coordinate of the nth atom in the layer, respectively. The XRD patterns before and after the post annealing, however, shows no sign of interlayer atomic rearrangement as shown in Figs. 4.4 and 4.5. Several probable explanations for this phenomenon are possible;(1) Oxygen rearrangement including oxygen ordering in the bi-O double layers (2) Absorption or desorption of oxygen by which most favorable hole concentration is realized and (3) Rearrangement of cations within each layers (z coordinate does not change by such a rearrangement). Our recent experimental results that a in situ post annealing in a low oxygen pressure (several torr) and low temperatures (450 C for 30 min) improve the superconducting properties of the as-deposited films strongly suggest the importance of oxygen atoms for the development of superconductivity in the post annealed films.

V. Concluding Remarks

Through the investigation of the way how to control the structure of BSCCO, the multilayer deposition technique is convinced to be the most promising method to obtain a desired number of CuO layers between the two BiO double layers. The superconducting properties of the as-deposited films are not satisfactory at present. However, the large solid solubility in the multilayered films shows the possibility to synthesize thermodynamically non-equilibrium phases in which the carrier concentration and interatomic distances can be controlled to obtain better superconducting properties.

The authors would like to acknowledge members of our group, Drs. S. Ikeda and T. Hatano of NRIM, Mr. J. Sato (on leave from Hitachi Cable, Co. Ltd.), Dr. K. Aota (on leave from Citizen Watch, Co. Ltd.), Mr. H. Hattori (on leave from Sumitomo Electric Industries, Co. Ltd.) and Mr. H. Hayakawa (on leave from Yaskawa Electric MFG., Co. Ltd.) for their contribution to our recent investigations.

References

[1] H.Maeda, T.Tanaka and M.Fukutomi and T.Asano, Jpn. J. Appl. Phys. 27 (1988) L209.
[2] R.M.Hazen, L.W.Finger, R.J. Angel, C.T.Prewitt, N.L. Ross. C.C. Hadidiacos, P.J. Heaney, D.R. Veblen, Z.Z.Sheng, A.Elali, A.M.Herman, Phys. Rev. Lett. 60 (1988) 1657.
[3] J.M. Tarascon, Y.Le. Page, F. Barboux, B.B. Bagley, L.H. Greene, W.R. Mckinnon, G.W. Hull, M. Girond and D.M. Hwang, Phys. Rev. B37 (1988) 9832
[4] E.T-Muromachi, Y. Uchida, Y. Matsui, M. Onoda and K. Kato, Jpn. J. Appl. Phys. 27 (1988) L556.

[5] J. Akimitsu, A. Yamazaki, H. Sawa and H. Fujiki, Jpn. J. Appl. Phys., 27 (1988) L209.
[6] T. Sato, T. Yoshitake, Y. Kubo and H. Igarashi, Appl. Phys. Lett. 53 (1988) 1213.
[7] T. Kajitani, M. Hirabayashi, M. Kikuchi, K. Kusaba, Y. Shono, N. Kobayashi, H. Iwasaki and Y. Muto, Jpn. J. Appl. Phys. 27 (1988) L1453.
[8] H. Nobumasa, K. Shimizu, Y. Kitano and T. Kawai, Jpn. J. Appl. Phys. 27 (1988) L846.
[9] M. Takano, J. Takada, K. Oda, H. Kitaguchi, Y. Miura, Y. Ikeda, Y. Tomii and H. Mazaki, Jpn. J. Appl. Phys. 27(1988) L1041.
[10] U. Endo, S. Koyama and T. Kawai, Jpn. J. Appl. Phys. 27 (1988) L1476.
[11] T. Hatano, K.Aota, S.Ikeda, K.Nakamura and K. Ogawa, Jpn. J. Appl. Phys. 27 (1988) L2055.
[12] M.Uehara, Y.Asada, H.Maeda and K.Ogawa, Jpn. J. Appl. Phys. 27 (1988) L665
[13] K. Aota, H. Hattori, T. Hatano, K. Nakamura and K. Ogawa, submitted to Jpn. J. Appl. Phys.
[14] Y.Hakuraku, Y. Aridomi and T.Ogushi, Jpn. J. Appl. Phys. 27 (1988) L2091.
[15] H. Nasu, T. Kato, S.Makida,T.Imura and Y.Osaka, Jpn. J. Appl. Phys. 27 (1988) L1669.
[16] H.Hayakawa, M.Kaise, K.Nakamura and K.Ogawa, Jpn. J. Appl. Phys. 28 (1989) L967
[17] Several methods to obtain films with a large fraction of the 2223 phase using single target evaporation were published including, H. Itozaki,K.Higaki, K.Harada, S.Tanaka, N. Fujimori and S. Yazu, Proc. !st. Inter. Symposium on Superconductivity,(Nagoya, Aug. (1988)).
[18] H.Adachi, S.Kohiki, K.Setsune, T.Mitsuyu and K.Wasa, Jpn. J. Appl. Phys. 27 (1988) L1883.
[19] K.Nakamura, J.Sato, M.Kaise and K.Ogawa, Jpn. J. Appl. Phys. 28 (1989) L437.
[20] T. Kawai, M.Kanai and S.Kawai, Abstracts of ISTEC Workshop on Superconductivity (Oiso, Feb. (1989)).
[21] J. Fujita, T.Tatsumi, T.Yoshitake and H.Igarashi, Appl. Phys. Lett. 54 (1989)234.
[22] S.Ikeda, H.Ichinose, T.Kimura, T.Matsumoto, H.Maeda, Y.Ishida and K.Ogawa, Jpn. J. Appl. Phys. 27 (1988) L999.
[23] Y.Matsui, S.Takekawa, H.Nozaki, A. Umezono, E.T-Muromachi, S.Horiuchi, Jpn. J. Appl. Phys. 27 (1988) L1241.
[24] S.Iijima, T.Ichinose, Y.Shimkawa. T.Manako and Y.Kubo, Japn. J. Appl. Phys. 27 (1988) L1054.
[25]S.Adachi, H.Hirano, Y.Takahashi, O.Inoue and S.Kawashima, Jpn. J. Appl. Phys. 28 (1989) L209.
[26] H.Ihara, R.Sugise, T.Shimomura, M.hirabayashi, N.Terada, M.Jo, K. Hayashi and M. Tokumoto, Extend Abstracts of Japan-China Oxide High Tc Superconductors Symposium P88 (April 9-12,(1989) Tsukuba).
[27] S. Ikeda, J. Sato and K. Nakamura, Jpn. j. Appl. Phys. 28 (1989) in press.
[28] S.Horiuchi, H.Maeda, Y.Tanaka and Y.Matsui, Jpn. J. Appl. Phys. 27 (1988) L1172.
[29] H.Nobumasa, T.Arima, K.Shimizu, Y.Otsuka, Y.Murata and T.Kawai, Jpn. J. Appl. Phys. 28 (1989) L187.

[30] H.W.Zandbergen, W.A.Groen, F.C.Mijlhoff, G.Van Tedeloo and S.Amelinckx, Physica C 156 (1988) 325.
[31] C.C.Torardi, J.B.Parise, M.A.Subramanian, J.Gopalakrishnann and A.W.Sleight, Physica C 157 (1989) 115.

CHARACTERIZATION OF HIGH TEMPERATURE SUPERCONDUCTORS BY HIGH EXCITATION SPECTROSCOPIES

A. Balzarotti
II Universita di Roma "Tor Vergata", Roma, Italy

1.INTRODUCTION

In the course of the unprecedented research effort which followed the discovery of high-temperature superconductivity by Bednorz and Müller [1], virtually all the available experimental spectroscopic methods and techniques have been used to elucidate the electronic properties of these materials. Albeit much has been learned during three years of intense activity, we have to admit that the present understanding of some non trivial issues is still unsatisfactory. As an example, it is curious that, while a widespread consensus exists on the pairing of oxygen holes as responsible for the superconducting state, the core-level O1s photoemission spectrum is still among those less firmly established. One reason for these difficulties arises from material problems due to the complexity of the phase diagrams of these multinary superconducting compounds with coexisting phase regions [2] and from the process of annealing with scarce equilibration of the copper-containing phases with the oxygen partial pressure, which results in oxygen non-stoichiometry [3]. These problems became gradually less important after the isolation of the superconducting phases and the progress in crystal growth and characterization. Single crystals of high crystalline perfection and textured epitaxial films deposited on oriented substrates can nowadays be prepared which have allowed a detailed monitoring of the electronic states at the Fermi level (E_F) and their temperature dependence in $Bi_2Sr_2CaCu_2O_{8+\delta}$ [4] by angle-resolved ultraviolet photoemission spectroscopy. These measurements indicate a superconducting gap $\Delta(0)$ of 25-30 meV, i.e., $2\Delta(0)/kTc \approx 8$ for T_c=90 K, twice as large as the BCS value 3.52. This value classifies the high temperature superconductors (HTSC) as strong-coupling superconductors.

Another source of experimental uncertainty is the imperfect characterization of the real surfaces analyzed by the high-energy spectroscopies. By real surfaces - in contrast to ideal - we mean surfaces prepared *ex situ* and then inserted in the ultrahigh-vacuum (UHV) chamber after an indefinite exposure to the ambient. This unavoidably results in the formation of a layer whose contamination depends on several factors such as surface reactivity and morphology, time of exposure etc. which are difficult to quantify. This means that the surface composition is different from the bulk composition and since most high-excitation spectroscopies like x-ray photoemission (XPS), soft-x-ray absorption (XAS), electron energy loss (EELS), Auger (AES) and Bremsstrahlung Isochromat (BIS) are surface sensitive, many of these techniques will give results representative of the surface more than of the bulk. This is particularly noticeable for ceramic superconductors containing atomic species like Cu whose formal valence can be simply modified by changing the oxygen content. If one considers the fact that several details of the oxygen core-level spectra of copper and nickel oxides have yet to be understood [5], one realizes the importance of characterizing a "clean" surface to record the genuine electronic response of these superconductors.

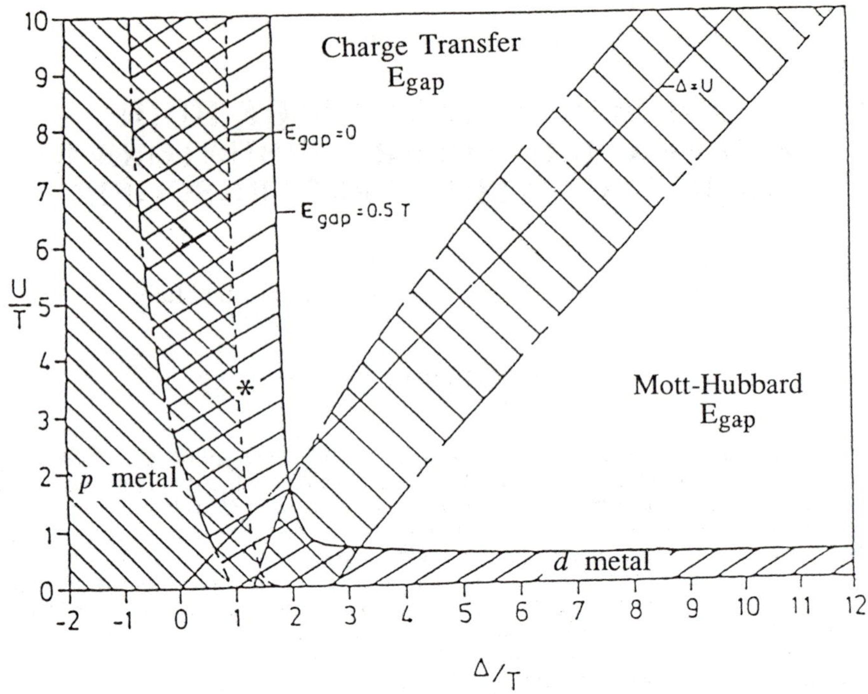

Fig.1

ZSA phase diagram for transition-metal compounds. The asterisk denotes CuO, according to the data of Ref.[42]. Adapted from Zaanen et al. [7].

The electronic properties of the new HTSC have been reviewed recently [6]. Here we emphasize those aspects which are surface related and attempt to rationalize, when possible, the results reported so far by the principal core-level spectroscopies by considering the spectral modifications of polycrystalline samples, films and single phase crystals of $La_{2-x}Sr_xCuO_4$ (hereafter referred to as La-Sr-Cu-O or 2-1-4), $YBa_2Cu_3O_{7-\delta}$ (Y-Ba-Cu-O or 1-2-3) and $Bi_2Sr_2CaCu_2O_{8+\delta}$ (Bi-Sr-Ca-Cu-O or 2-2-1-2) due to surface decomposition, impurity phases, excess oxygen and cationic nonstoichiometry. We try to separate the intrinsic properties from the effects of contamination, a comparison being made with single crystals peeled (for Bi-Sr-Ca-Cu-O) or fractured in high vacuum (for Y-Ba-Cu-O). We review here XPS, Auger, BIS and EELS measurements done in our laboratory and include some unpublished material.

2. SOME BASIC ASPECTS OF HIGH-T_c SUPERCONDUCTORS

One of the earliest and most accepted evidences obtained from photoemission and other high-energy spectroscopies applied to HTSC is that the correlation effects, i.e., local electron-electron interactions determine to a large extent the electronic structure of these materials. Valence-band and core-edge satellites, multiplet structure and band shifts are all signatures of a localized behaviour.

The original proposal by Zaanen et *al.*. [7] that the metallic character of various transition-metal compounds can be explained by considering, besides the Mott-Hubbard charge fluctuations $d_i^n d_j^n \leftrightarrow d_i^{n-1} d_j^{n+1}$, another charge fluctuation $d_i^n L^0 \leftrightarrow d_i^{n+1} L^{-1}$, where L^{-n} (or $\underline{L}^n$) denotes n ligand holes, may be applied to understand the core and valence level spectra of the Cu-based superconductors. In these materials the energy Δ necessary to transfer electrons from Cu to O is much less than the Coulomb repulsion U_{dd} between two *d* holes, thus stabilizing a divalent Cu^{2+} ($3d^9$) configuration and a filled O $2p^6$ band. A charge transfer gap is then formed between the O *2p* band and the Hubbard Cu *3d* bands. This picture is appropriate to charge-transfer undoped semiconductors rather than to Mott-Hubbard insulators as illustrated by the well-known Zaanen-Sawatzky-Allen (ZSA) phase diagram shown in Fig.1.

Doping the material with oxygen holes has at least two important consequences. First, hole states of dominant O 2p character are formed in the gap thus giving metallic character to the compound. Second, the valence band is partially unfilled which makes the local screening of a second hole created by the photoemission process an effective mechanism which yields resonances and shake up satellites. Many features observed in the valence band photoelectron spectra are similar for different superconducting oxides and they can be resonantly enhanced when the photon energy is tuned at the onset of the appropriate atomic core level excitation. The enhancement arises from the coupling between the direct transition and the decay channels opened by the excitation of the core electron.

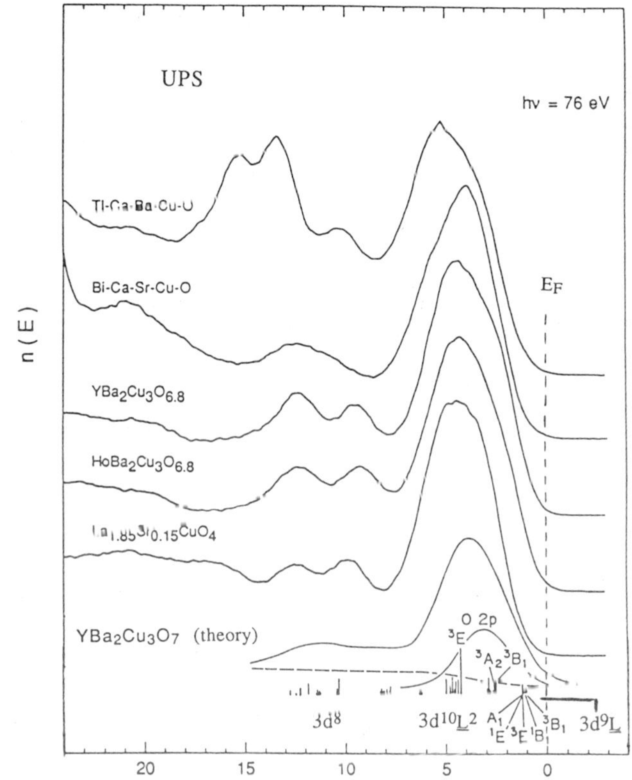

Fig.2

Valence-band photoemission spectra of five high-temperature superconductors. The theoretical curve is the cluster calculation of Fujimori [9]. Adapted from Kurtz [8].

In Fig.2 the energy distribution curves measured on five different superconductors by Kurtz et *al.* [8] at photon energies hν=76 eV using synchrotron radiation are shown. By tuning the photon at 74 eV, the energy needed to excite the Cu 3*p* core levels, the two features at 12.5 eV and 9.5 eV are enhanced which assigns them to multiplet-split Cu satellites.

If we regard the Cu 3*d* ground state as an admixture of $|3d^9\rangle$ and $|3d^{10}\underline{L}\rangle$ states, the local configurations of a small cluster that result from the photoemission process are $3d^8$, $3d^9\underline{L}$ and $3d^{10}\underline{L}^2$. They are shown schematically in Fig.3, in which we distinguish between ligand holes on the same oxygen atom $3d^{10}\underline{L}^2$ or the nearby atoms $3d^{10}\underline{L}\underline{L}$. The $3d^9\underline{L}$ state corresponds to the valence band emission. The $3d^8$ (Cu^{3+}) state interferes with the same state produced by the super Coster-Kronig Auger decay of the intermediate Cu $\underline{3p}3d^{10}$ state. From the energy difference between the satellites and the valence band states, as provided by the result of a configuration-interaction (CI) calculation [9], included for comparison in Fig.2, one estimates a correlation energy, U_{dd}, for the Cu *d* holes of about 6 eV. The 9.5 eV satellite resonates with the excitation of the O 2*s* level as well. This resonant excitation is explained as an interference between the direct excitation and the resonant decay of the excited oxygen core level $\underline{2s}\,3d^{10} \rightarrow 3d^{10}\underline{L}\underline{L}$, where $\underline{2s}$ refers to the O 2*s* hole. Thus the 9.5 eV satellite is associated with two O 2*p* holes on different, neighboring ligands. The correlation energy for these interacting holes, U_{pp}, is approximately 4.5 eV. In Sect. 4.1 we discuss other shake up features associated with core level excitations.

Hole doping in HTSC occurs in the $Cu\text{-}O_2$ planes through different charge compensation mechanisms which depend on the different crystal structure and on the presence of fluctuating valence elements. Charge fluctuations are not restricted to one sublattice but they occur concomitantly on the Cu and O lattices in 1-2-3 and 2-1-4 superconductors and on Cu and Bi sublattices in the 2-2-1-2 superconductor. In the case of the antiferromagnetic insulator La_2CuO_4, chemical substitution of trivalent La for divalent Sr in the LaO layers gives $La_{2-x}Sr_xCuO_4$ which is superconducting for x>0.04 with a maximum T_c near 40 K for $x \approx 0.15$. $YBa_2Cu_3O_{7-\delta}$ is doped by forming oxygen vacancies in the $Cu\text{-}O_3$ chain (i.e., with a decrease in δ) and below $\delta \approx 0.7$ the system transforms from an antiferromagnetic insulator to a metal which is superconducting for temperatures lower than 93 K. A charge transfer gap is then formed between the O 2*p* band and the upper Hubbard Cu 3*d* band. For Bi-Sr-Ca-Cu-O and the newest $Tl_2Ba_2CaCu_2O_8$ (2-2-1-2) system with critical temperatures in excess of 100 K, the BiO and TlO layers play the same role as the Cu-O chains in the 90 K material. However, these systems are multiphased and layered and their stacking sequence is not always maintained giving rise to stacking faults [10] and incommensurate modulation [11]. These structural imperfections allow retention of excess oxygen - presumably in the BiO layers - and thus it is difficult to determine to what extent the loss of the excess oxygen or the charge fluctuations at the Cu site or the crystalline defects act to modify T_c.

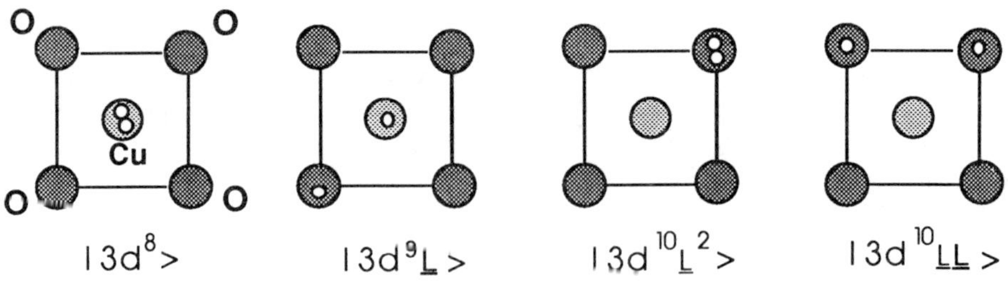

Fig.3

Simplified illustration of the localized excitations of the $Cu\text{-}O_2$ cluster.

3. EXTRINSIC SURFACE PROPERTIES

The surface sensitivity of high-energy photoelectron spectroscopies which probe the electronic properties within the escape depth of photoelectrons (typically 5 to 20 Å), needs surface cleanliness. However, in the case of the oxide superconductors, which are usually prepared *ex situ* and then analyzed in UHV environment, surface degradation occurs because of the low chemical stability of the surface exposed to air for an undetermined length of time. There have been several reports on the surface reactivity of HTSC [12-15]. Particularly studied is Y-Ba-Cu-O that is known to be extremely sensitive to moisture. The $YBa_2Cu_3O_7$ superconducting phase reacts with the atmospheric water according to the reaction

$$2\ YBa_2Cu_3O_7 + 3\ H_2O \leftrightarrow 3\ Ba(OH)_2 + 5\ CuO + Y_2BaCuO_5 + 1/2\ O_2, \qquad (1)$$

resulting in the formation of a surface containing the "green" insulating phase Y_2BaCuO_5 besides Ba hydroxide and Cu oxide. For polycrystalline pellets, where grains and intergrain regions are present, particularly in low-density samples ($\leq 80\%$ of the theoretical density), the surface composition can differ considerably from that of the bulk. Scraping or filing the surface in UHV reduces the intergrain regions and fractures each crystalline grain into multiple grains [16], thus restoring fresh surfaces of bulk stoichiometry. Nevertheless, significant parts of the scraped surfaces contain grain boundaries which can be contaminated with impurity phases like barium carbonate and barium hydroxide, unreacted species or compounds formed during the sintering process. For instance, if the temperature is well above 921 °C in the liquid sintering process of Y-Ba-Cu-O, the grains are covered by a $BaCuO_{2+x}$ phase approximately 20 Å thick [16]. Superconducting films, deposited by sputtering or laser ablation techniques on oriented substrates, as yttrium-stabilized zirconia, MgO, and $SrTiO_3$, are usually cleaned by Ar^+ sputtering, followed by annealing in O_2 at 400-600 °C [17] or heated in vacuum at moderate temperatures [18]. All these treatments are, however, rather risky due to the reduction of copper from Cu^{2+} to Cu^{1+} occurring under ionic bombardment [19], amorphisation of the bombarded area, as revealed by microprobe analysis [20], and possible oxygen losses [18] at low pressures.

Another source of complication could arise from the inhomogeneity of the samples resulting from the standard method of oxidation. If, for instance, a sintered sample is cooled down from elevated temperatures at a constant partial pressure of oxygen, the sample surface and its grain boundaries become more oxidized with respect to the bulk of the grains. In order to maintain a constant stoichiometry, the samples must be equilibrated with their surrounding atmosphere by decreasing the equilibrium oxygen partial pressure with decreasing temperature. A careful study conducted on $YBa_2Cu_3O_x$ [3] has demonstrated that the constant stoichiometry cooling process produces homogeneous samples of well defined x and good superconducting properties.

From the above considerations it appears that HTSC single crystals are the most suited for surface analysis because their crystalline perfection is greater and their stoichiometry closer to the ideal one. However, apart from problems related to structural perfection (twinning, stacking, vacancies, superstructures) [21-22], obtaining a reproducibile (i.e., independent of the cleaning procedure) clean surface of known chemical composition is not trivial even for single crystals. Crystallographically ordered clean surfaces of La-Sr-Cu-O single crystals have been recently obtained by annealing in oxygen or air and by ion bombardment and annealing in UHV[23]. This is made possible by the stability of the $La_{1-x}Sr_xCuO_4$ surfaces against oxygen losses and their scarce reactivity with residual gases. The same is not true for $YBa_2Cu_3O_7$ crystals for which the surface composition is strongly dependent upon the cleaning procedure. Y-Ba-Cu-O single crystals are usually of small dimensions with respect to those of La-Sr-Cu-O and this does not facilitate their cleaning by mechanical methods in vacuum. Moreover, even for *in situ* cleaved high-quality single crystals [24] the surface (oxygen) stoichiometry is not always guaranteed [25]. From the point of view of chemical surface stability Bi-Sr-Ca-Cu-O and Tl-Ba-Ca-Cu-O single crystals are certainly better, due to their stability in the presence of water [14,26] and other reagents [15]. In addition, Bi-Sr-Ca-Cu-O single crystals are layered and thus easily cleaned in UHV by simple stripping thin layers with adhesive tape. Occasionally the cleave occurs along planes containing intercalated impurities [27].

Because oxygen is the main constituent present in the surface layer, it is not surprising that little agreement exists on what the intrinsic O 1*s* core-level photoemission spectrum should be. As remarked in Ref.[28], it is almost certain that most published spectra are not characteristic of the clean material but reflect an unknown condition of the surface. The majority of authors find two peaks at binding energies of about 528.5±0.3 eV and 531.0±0.3 eV whose relative intensity depends strongly on cleaning treatments, environmental conditions and quality of the samples. For the more chemically stable $La_{1-x}Sr_xCuO_4$ and $Bi_2Sr_2CaCuO_{8+x}$ compounds, there is new accumulating evidence that the intrinsic O 1*s* spectrum consists of a single peak at 528.5 eV while the 531 eV peak is due to extrinsic surface effects [29-34]. The 531/528.5 ratio in $Bi_2Sr_2CaCuO_{8+\delta}$ tends to decrease as the surface cleans up. This is evident if one follows the evolution of the O 1*s* line of the $Bi_2Sr_2CaCuO_{8+x}$ polycrystalline material during the cleaning of the surface by thermal heating shown in Fig.4 (left). It is clear that a mild heating up to 145°C gets rid of more than half of the high-binding energy component and that no further changes take place between 145°C and 290 °C. The spectral weight of the 531 eV component is about 20% of the 528 eV peak and could result from nonequivalent (structurally or chemically) oxygens [30] or from residual contamination. It should be noted that the O 1*s* spectrum of CuO pellets exhibits a very small shoulder around 531 eV [35] which is attributed to absorbed water or oxygen molecules [36].

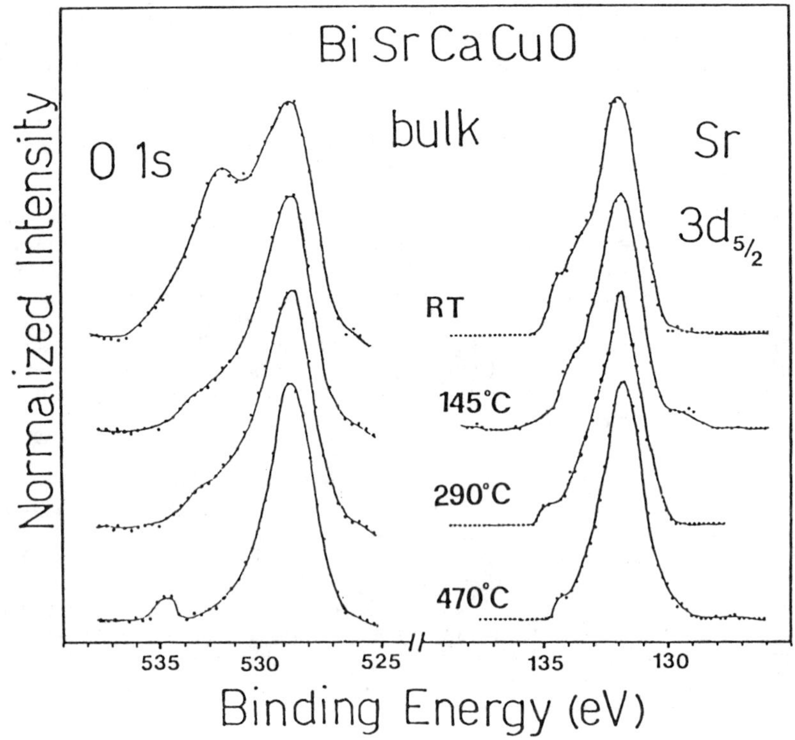

Fig.4

The effect of heating in vacuum on the $Bi_2Sr_2CaCu_2O_{8+\delta}$ ceramic at the O 1s and Sr $3d_{5/2}$ core levels. The $3d_{5/2}$ component has been isolated assuming a spin-orbit splitting of 1.7 eV and a statistical ratio $3d_{5/2}/3d_{3/2}$ =1.5. Notice the replica in the O 1s spectrum at 470 ℃.

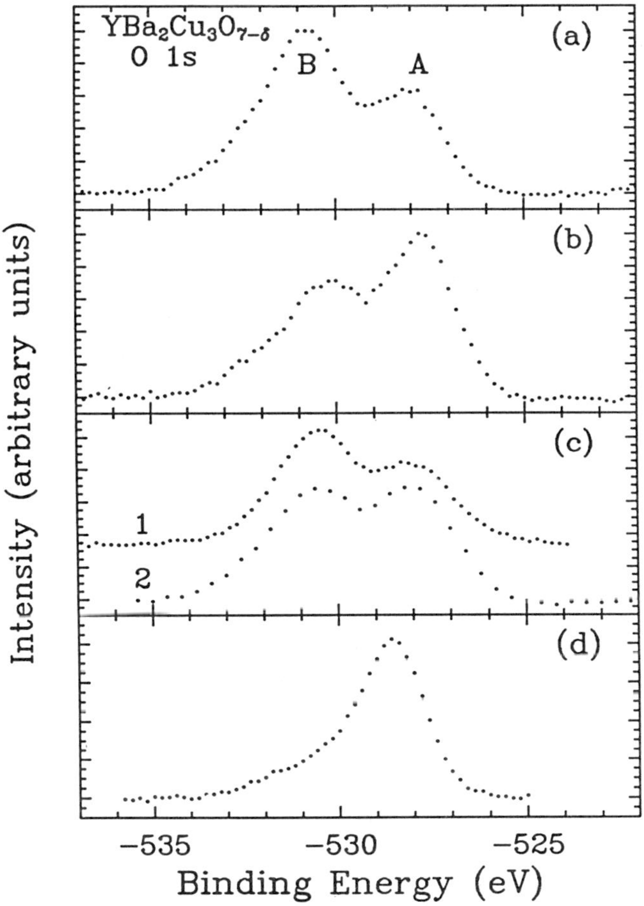

Fig.5

O 1s XPS spectrum of $YBa_2Cu_3O_{7-\delta}$.
(a) bulk sample, after exposure to air
(b) - bulk sample, freshly scraped in UHV
(c) - same as (b) but for increasing grazing angles (1→2)
(d) - film, cleaned by electron bombardment.

Cleaning the surface by electron bombardment reduces the 531 eV peak but the Cu 2*p* spectrum is modified as well. In particular, the satellite-to-main peak ratio (see below) decreases by about 30% and the main peak becomes narrower. These changes are indicative of copper reduction due, probably, to the evolution of the off-stoichiometric oxygen in the terminal BiO plane [21]. More complex is the problem of surface cleaning encountered with $YBa_2Cu_3O_7$. Nominally cleaned bulk samples scraped or fractured in vacuum exhibit a double peaked O 1*s* structure with a substantial component at higher binding energies. A third, weak structure is sometimes reported around 535 eV, but it is generally not found in the best surfaces. The relative intensity of the two components depends strongly on surface conditions [35]. In Fig. 5 we show the O 1*s* spectrum of a sintered sample exposed to oxygen before (panel a) and after (panel b) surface scraping in vacuum (~$6x10^{-11}$ Torr).

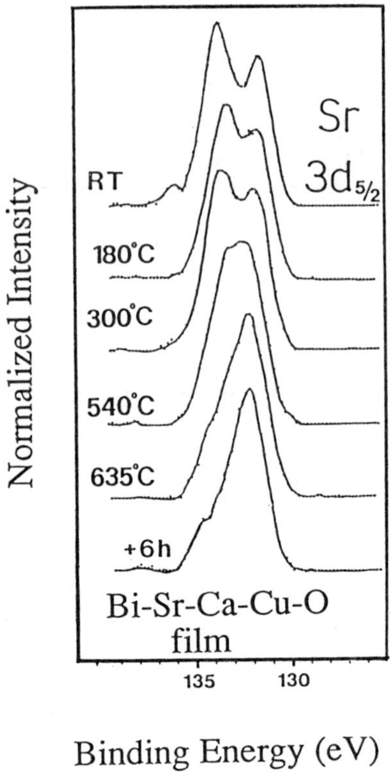

Fig.6

The effect of heating in vacuum on the Sr $3d_{5/2}$ component of Bi-Sr-Ca-Cu-O films deposited by laser ablation.

The ratio of the peak intensity is inverted indicating that a strong modification of the surface conditions has occurred. Further scraping does not modify the spectrum appreciably. It can be easily verified by decreasing the offtake angle ($\Theta < 30°$ from the surface) that peak B increases with respect to A - see panel (c), curve labeled (1) - suggesting that B originates mainly from the uppermost layers (5-10 Å from the surface). We conclude that in polycrystalline $YBa_2Cu_3O_7$ samples the average stoichiometry of the clean surface is represented by a double peak structure as depicted in panel (b). This, however, does not represent the cleanest surface that one can prepare. As a matter of fact, we have measured [37] the O 1*s* spectrum on well-characterized superconducting films (~2000 Å thick) grown by laser ablation on yttrium-stabilized zirconia after cleaning the surface by a mild electron bombardment. The result shown in panel (d), indicates clearly that peak B is strongly reduced and the spectrum closely resembles those of the other HTSC. In contrast to Bi-Sr-Ca-Cu-O, the Cu 2*p* and the Ba 4*d* spectra are not appreciably modified, indicating that stoichiometric oxygen does not evolve under electronic bombardment. It can therefore be concluded that Y-BaCu-O films are more stable than Bi-Sr-Ca-Cu-O films under electronic bombardment. Since on surfaces of high-density (95%) ceramics fractured in UHV the O 1*s* spectrum is single [38], the comparison of (b) with (d) suggests that peak B arises probably from residual contamination present in the intergrain regions of the kind discussed above. Removal of the high-binding energy component after electron-gun operation has been also reported by van der Marel *et al.* [39].

The Sr and Ba *3d* spectra for $Bi_2Sr_2CaCu_2O_{8+x}$ and $YBa_2Cu_3O_7$, respectively, are other sensitive probes of the surface degradation since they are strictly correlated with the O 1*s* spectrum.

In bismuth cuprates crystals the Sr 3*d* spectrum consists of a single spin-orbit doublet [31,32] with peaks at 132.2 and 133.9 eV and intensity ratio lower than the statistical weight (1.17 against 1.5). In thin films exposed to atmosphere each spin-orbit line doubles and the entire spectrum becomes complex. A given component of the spin-orbit doublet can be isolated by numerical stripping, as done in Fig. 6 for the $3d_{5/2}$ component, and its evolution followed during surface cleaning by heating the film in vacuum [34]. It is seen that the subpeak disappears at temperatures close to 540°C, concomitant with the decrease of its oxygen partner, and the doublet finally merges into a single peak. It is interesting to note the difference between adsorbed oxygen, which is eliminated by moderate heating without affecting the Sr spectra (bulk case in Fig.4), and the bound oxygen which is removed at high temperature in the films. In both cases the clean surface displays the characteristic Sr 3*d* spectrum of the single crystal.

In $YBa_2Cu_3O_7$, the intrinsic Ba photoemission spectra are more difficult to establish. The correlation between O 1*s* and Ba 3*d* core-level spectra has been studied by Flavell and Egdell on ceramic surfaces [40] by collecting electrons at different angles from normal to grazing incidence. They found that the contaminant component of the Ba 3*d* spectrum is confined to a thin layer on top of the intrinsic material and disappears after *in situ* oxygen annealing. In such conditions the O 1*s* spectrum has a spectral profile similar to that of the case (b) in Fig.5 which they consider to be intrinsic to the 1:2:3 phase. A single Ba $3d_{5/2}$ line is measured on high-density sintered powders [38].

The Ba 4*d* photoemission measured on clean thin film surfaces [37] and shown in Fig.7 (c), consists of two spin-orbit lines $4d_{5/2}$ at 87.5 eV and $4d_{3/2}$ at 90.0 eV.By stripping the $4d_{5/2}$ component (right panel of Fig.7), one sees that it is a single line with a small residual high-binding energy tail (less than 10% of the 87.5 eV intensity) overlapping the $4d_{3/2}$ component.

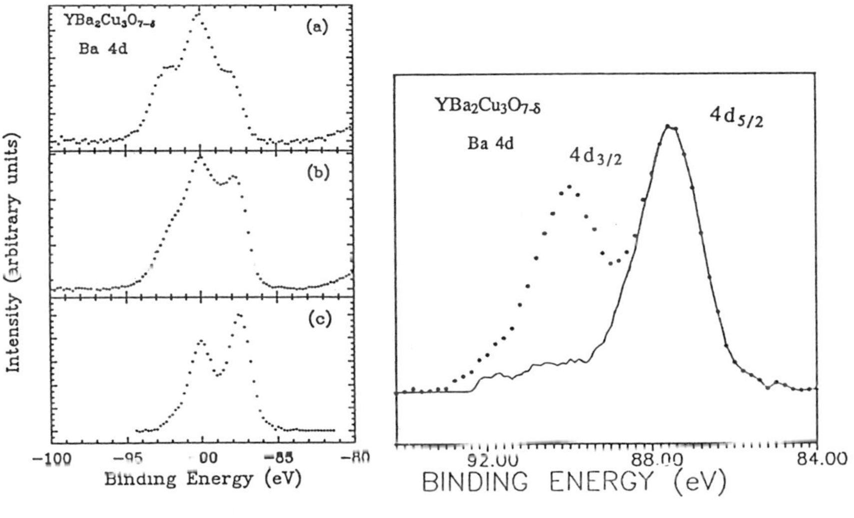

Fig.7

Left: Ba 4d XPS spectra of Y-Ba-Cu-O.
(a) - bulk sample, after exposure to air
(b) - bulk sample, freshly scraped in UHV
(c) - film cleaned by electron bombardment
Right: stripping of the Ba $4d_{5/2}$ *component in the film showing a single line with a residual contamination overlapping the* $4d_{3/2}$ *line.*

Such a tail is substantial in sintered samples exposed to air (a) and reduces in intensity after scraping in UHV (panel b) but does not disappear. It is responsible for the three-component spectrum shown in (a) and (b). As for O 1*s*, the extra tail feature increases going to grazing angles [37], indicating its predominant surface origin. Among the remaining core states, the Cu 2*p* emission around 933 eV is the most studied. Since Cu states strongly hybridize with oxygen states, changes of the oxygen stoichiometry, formal valence and bonding configuration are expected to modify Cu emission. For the Cu $2p_{3/2,5/2}$ core-level excitation the local final-state screening of the 2*p* hole at the Cu site is modified by the surrounding charge density due to oxygens. Extrinsic oxygen-rich phases due to surface degradation tend to broaden the well-screened $2p_{3/2}$ component at 932.8 eV ($\underline{2p}3d^{10}$ final state) and to reduce the strength of the shake-up satellite ($\underline{2p}3d^{9}$ final state) centered around 942 eV. In terms of Cu formal valence the decrease of the satellite intensity is simply described as a reduction of copper from Cu^{2+} to Cu^{1+}. Since the ionic configuration of monovalent Cu is $3d^{10}$, the satellite should be absent for complete reduction of copper. The linewidth of the 932.8 eV component is affected as well in both Bi-Sr-Ca-Cu-O [34] and Y-Ba-Cu-O [35] by extrinsic reasons. Taking for reference the satellite-to-main peak intensity ratio and linewidth of CuO which are 0.53-0.55 [41,42] and 2.9 eV [30], respectively, one finds 0.48 for Bi-Sr-Ca-Cu-O [34], and 0.35 for Y-Ba-Cu-O [35] and La-Sr-Cu-O [43]. The linewidth is always larger (≈4 eV) in superconductors than in cupric oxide. It should be noted that the asymmetric broadening of the main line is not all due to extrinsic features. Some authors [35, 44,45] have reported a shoulder on the high-binding energy side of the Cu $2p_{3/2}$ line of Y-Ba-Cu-O which is attributed to the $3d^9\underline{L}$ ground state of a formal Cu^{3+} in the Cu-O chains, as discussed in Sect.4.1.

A strong reduction of copper to the nominal Cu^{1+} configuration is induced by exposing the surface of UHV scraped Y-Ba-Cu-O pellets to hydrogen at about 10^{-4} Torr for a few minutes [37], as shown in Fig.8. Besides the strong decrease of the satellite, we observe an overall shift (≈2 eV) of the spectrum towards lower binding energies and a sharpening of the $2p_{3/2}$ line. For short exposures, Cu reduction is limited to the surface volume, since further scraping reverts the spectrum back to the original condition.

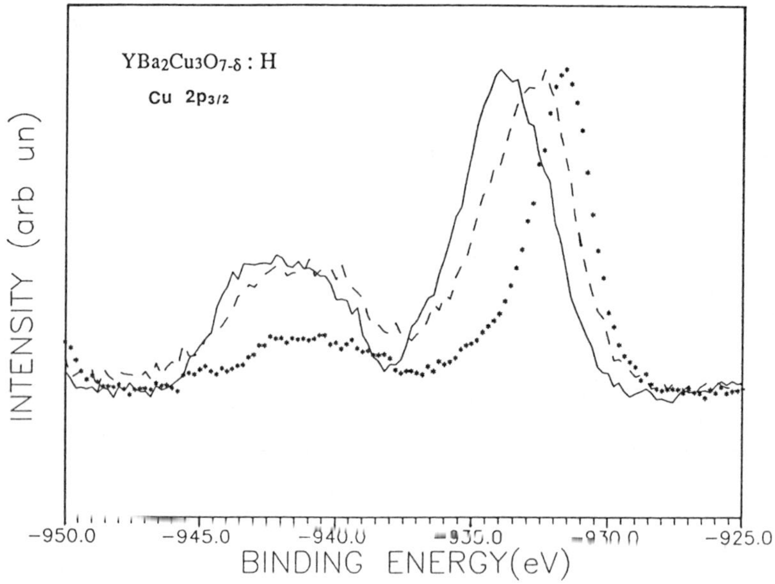

Fig.8

Reduction of Cu^{2+} to Cu^{1+} at the Cu $2p_{3/2}$ spectrum of Y-Ba-Cu-O after exposure to hydrogen at 10^{-4} Torr for 5 min. Full curve: before hydrogenation; dotted curve: after hydrogenation; dashed curve : scraped after hydrogenation.

4. INTRINSIC BULK PROPERTIES

4.1 Photoemission Spectra

From the analysis made in the previous section, it is clear that core-level photoemission spectra representing the bulk properties of the superconductor are quite difficult to establish. As a guideline one could say that single crystals and textured thin films well oxidized and unexposed to the atmosphere display the simplest lineshape for a given edge. Single-phase sintered materials have in general a more complex behavior approaching that of single crystals for dense samples.This points to the importance of controlling the amount of grain-boundary and integranular phases, their distribution and purity to achieve high-critical currents in polycrystalline samples. It should be said that the quality of HTSC single crystals is still unsatisfactory and therefore it is difficult at present to estimate the degree of spectral distortion due to surface and volume contamination. Since the electronic effects responsible for superconductivity are tiny, it might not be wise to rule out any indication of such effects before a clear demonstration of their extrinsic nature is achieved. We will discuss below the photoemission spectra of HTSC which are considered to be representative of the superconducting bulk phase.

1) Cu 2p edge- There is little doubt left that the Cu $2p_{1/2,3/2}$ XPS spectrum is indicative of a dominant divalent state for copper. The spectra of all superconductors $La_{1-x}Sr_xCuO_4$, $YBa_2Cu_3O_{7-\delta}$ and $Bi_2Sr_{2-x}Ca_{1-x}Cu_2O_{8+\delta}$ look very similar to each other and to that of CuO [34,46] and display clear signs of strong electronic correlations. As a prototype clean surface, we show in Fig. 9 the Cu $2p_{1/2,3/2}$ energy distribution curve measured on a superconducting film of $YBa_2Cu_3O_{7-\delta}$ ($\delta \approx 0.1$) at normal (panel a) and grazing incidence (b). The lineshape is almost insensitive to the offtake angle and this is taken as a proof of a good surface.

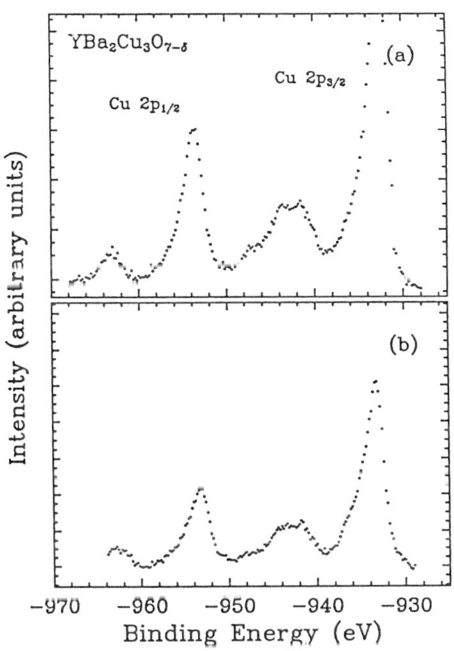

Fig.9

Cu 2p x-ray photoemission of a clean film of Y-Ba-Cu-O at (a) normal and (b) grazing incidence.

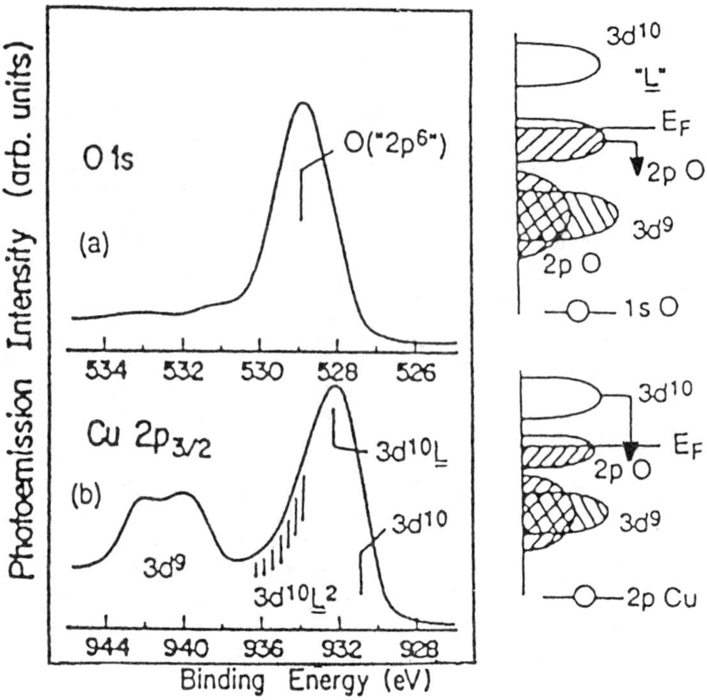

Fig.10

Core level O 1s and Cu $2p_{3/2}$ XPS spectra of Bi-Sr-Ca-Cu-O single crystals and possible level diagrams in presence of the core hole. Adapted from Wendin [6].

The Cu $2p$ main line is clearly asymmetric and the shake up satellite is intense and structured.In general a strong satellite is observed when the core hole lifetime is long, i.e., the screening charge takes a relatively long time to be transferred from oxygens to copper. This is indicative of a bonding charge distribution displaced towards oxygen in the ground state (ionic bonding) and of modest hybridization for the Cu-O orbitals [6]. In the sudden approximation [47] the position of the $3d$ final state is at the frozen (unrelaxed) energy, but its actual position depends on the relaxation mechanism which involves a number of unknown (number and symmetry of the initial and final states, local geometry, hybridization etc.).

A simplified description of this complex situation has been widely employed to analyze the core-level spectra of copper-based HTSC [35,43, 48,49]. It starts from a single isolated small cluster Cu-O_n (n=4) to represent the central Cu atom surrounded by four oxygen ligands, and local correlations are treated within an impurity Anderson model, much on the lines adopted to describe copper dihalides [50]. In the cluster model the O $2p$ band is treated as a set of molecular orbitals derived from the oxygen ligand orbitals and the intraatomic interactions are neglected. For the undoped material, the ground state of the neutral cluster embedded in an effective medium in an extended CI scheme is a superposition of $|3d^9\rangle$ and $|3d^{10}\underline{L}\rangle$ states. The interaction between these two configurations is described by the hybridization integral $T=\langle d^9 | H | d^{10}\underline{L}\rangle$ and the Cu $3d \rightarrow$ O $2p$ charge tranfer energy $\Delta = \langle d^{10}|H|d^{10}\rangle - \langle d^9|H|d^9\rangle$ is a measure of the bonding-antibonding separation in the ground state. The hole doped superconductor contains the additional configurations $3d^8$, $3d^9\underline{L}$ and $3d^{10}\underline{L}^2$ with a variable number of O 2p holes. The $3d^8$ configuration can be excluded from the basis set since the high-correlation energy of the Cu d-electrons moves this state to about 10 eV excitation energy relative to the Fermi level [51]. The divalent-like charge transfer state $|3d^9>$ gives the major contribution because of the substantial copper-oxygen hybridization. The occupation of the $|3d^{10}\underline{L}^2>$ ground state will always be small because of the double excitation on the ligand. The resulting ground state of the cluster is approximated by $\Psi g=\alpha|3d^9>+\beta|3d^{10}\underline{L}>+\gamma|3d^9\underline{L}>$. Charge neutrality requires that electrons are exchanged between the cluster and the surrounding medium ; a delocalized **k** electron could be added to the $|3d^9\underline{L}>$ to preserve the number of holes in the cluster.

Removal of a core ($\underline{c}$=$\underline{2p}$) electron will suddenly switch on a 2p core hole and will project the system into a two-hole final state configuration $\Psi e=\alpha'|\underline{c}3d^9>+\beta'|\underline{c}3d^{10}\underline{L}>+\gamma|\underline{c}3d^9\underline{L}>$. The empty $|3d^{10}\underline{L}>$ initial state is pulled down locally with respect to the $|3d^9>$ state by the Coulomb interaction U_{cd} between the core-hole and the 3d electron and may be filled by charge transfer from the O 2p band. The resulting, locally screened state, has the configuration $\underline{2p}\,3d^{10}\underline{L}$ and constitutes the main peak while the locally unscreened $|\underline{2p}\,3d^9>$ state is the shake up satellite seen in Fig.9. In the metallic state the ligand hole must be screened. For integer occupation of the ligand orbitals $|d^{10}\underline{L}>$, one has the three possibilities $\underline{L}^n$, n=0,1,2. The skake down $|\underline{2p}\,3d^{10}>$ and the shake up $|\underline{2p}3d^{10}\underline{L}^2>$ states contribute to the asymmetry of the main peak. The unscreened state $|\underline{2p}3d^9\underline{L}>$, which should be located close to the satellite, decays into the screened $|\underline{2p}3d^{10}\underline{L}^2>$ state and it is not observed. Its energy coincides also with that of the $2p_{3/2}$ component in $NaCuO_2$ photoemission spectrum [45]. A schematic energy level diagram illustrating the above situation is presented in Fig.10 (b).

The final states eigenvalues of the excited cluster can be simply calculated by diagonalizing the Hamiltonian 3x3 matrix:

$$\begin{vmatrix} -E & T & 0 \\ T & -E-U_{cd}+\Delta & T' \\ 0 & T' & -E-U_{cd}+U_{pp} \end{vmatrix} = 0, \qquad (2)$$

where $T'=<3d^{10}\underline{L}|H|3d^{10}\underline{L}^2>$ and U_{pp} is the Coulomb (screened) repulsion between the O 2p holes. If the hole are delocalized in a continuum of states (the Cu-O planes are metallic), U_{pp} will be small and the $|\underline{2p}\,3d^{10}\,\underline{L}>$ and $|\underline{2p}\,3d^{10}\underline{L}^2>$ nearly degenerate [52]. In this case the two-hole multiplet will broaden in a Doniach-Sunjic tail [6,53] with an asymmetry index α. Since the oxygen partial density of states at the Fermi level is found to be small in these superconductors, the line asymmetry is not negligible [54]. When non superconducting or insulating phases are mixed to the superconducting one, the electronic screening is reduced and the weight of the $|\underline{2p}\,3d^{10}\underline{L}^2>$ increases. A distinct peak or a large shoulder will appear on the high energy side of the Cu $2p_{3/2}$ peak [35,44,45,55] as illustrated in Fig.11 for the case of $YBa_2Cu_3O_7$ [35]. Thus the appearance of a sharp structure on the high energy side of the 933 eV peak and of its spin orbit partner might indicate a partially conducting or insulating phase, such as the intergranular phase in sintered samples or a phase(s) originating from surface contamination or off-stoichiometry in the films. The intensity of this structure could also be a function of doping through the concentration dependence of vacancies.

In Fig.12 we plot the calculated stoichiometric ratio $[Cu^{3+}]/[Cu^{2+}]=(1-2\delta)/(2+2\delta)$ as a function of the oxygen deficiency δ (broken curve) for $YBa_2Cu_3O_{7-\delta}$ pellets and compare it with the experimental area ratio obtained from the analysis of the Cu 2p x-ray photoemission spectrum [35].

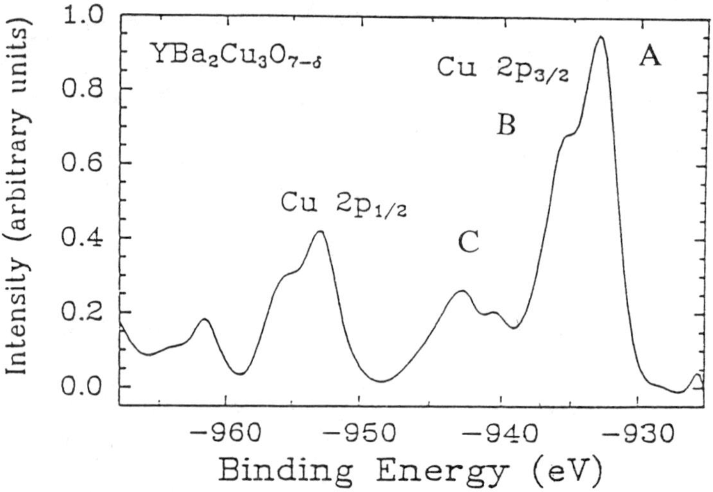

Fig.11

Cu 2p XPS spectrum of Y-Ba-Cu-O pellets. Notice the peak on the high-binding energy side of the spin-orbit doublet.

It is apparent that an average over-concentration of formal Cu^{3+} of 30% is needed to bring the experiment and theory in agreement suggesting that an excess oxygen-rich phase could be present on the scraped surface exposed to the x-rays. Changes of the satellite-main peak ratio as a function of temperature has been reported and correlated with T_c [44,45]. We believe that temperature effects must be be taken with caution, particularly when the 531 eV peak in the O 1*s* spectrum is clearly observed. We remark that a decrease of the copper-oxygen hybridization could also be observed from the valence band spectra but this information is not always available.

The correlated parameters T,T', Δ, U_{pp} , U_{cd} as well as the number of electrons in the occupied ground state $n_d = 9 \cos^2\theta + 10 \sin^2\theta$, where $\tan(2\theta)= 2T/\Delta$ [50] can be calculated by fitting the experimental satellite-main line intensity ratio and energy separation. In Table 1 a typical set of these parameters extracted from the Cu 2*p* spectra of Y-Ba-Cu-O and Bi-Sr-Ca-Cu-O is shown. It is interesting to note the lack of hybridization (T' = 0), in which case the Cu^{2+} and Cu^{3+} formal states are decoupled, and the small U_{pp}=1.5-2.0 eV Hubbard repulsion suggesting a screened interaction between the O 2*p* holes. The hybridization integral T, 2.5 eV, is appreciable in all cases, in agreement with the large covalence of the Cu-O bond estimated from valence band photoemission spectra.

2) O 1*s* edge- As stated above for Cu 2*p*, the O 1*s* spectrum of cleaned superconducting films and single crystals, Fig. 5 (d), consists of a single peak A and it is remarkably similar for all superconductors. Peak B is found consistently in the pellets. If we regard the peak B as completely spurious, we must exclude large correlation effects in the final state at the oxygen site.

Table I

SAMPLE	ΔE_{AB}(eV)	ΔE_{AC}(eV)	I_C/I_A	T	T'	Δ	U_{pp}	U_{pd}	n_d
$YBa_2Cu_3O_{6.85}$	2.8±0.6	10.7±1.0	0.35±0.2	2.8	0	0.0	2	8.5	9.5
	(2.83)	(10.2)	(0.29)						
$YBa_2Cu_3O_{6.80}$	2.4±0.5	9.7±0.6	0.32±0.2	2.7	0	0.0	1.5	8.1	9.5
	(2.32)	(9.70)	(0.28)						
$Bi_2Sr_2CaCu_2O_{8+\delta}$ film	0	8.7±0.3	0.33±0.2	2.1	0	0.0	0	7.5	9.5
		(8.60)	(0.34)						
$Bi_2Sr_2CaCu_2O_{8+\delta}$ bulk	0	8.7±0.3	0.48±0.2	1.5	0	0.0	0	8.2	9.5
		(8.73)	(0.49)						

Typical electronic structure parameters (in eV) for Y-Ba-Cu-O and Bi-Sr-Ca-Cu-O as derived from the analysis of the Cu 2p XPS spectra. ΔE are the binding energy positions and I_C/I_A the satellite-to main peak intensity ratios according to Fig.11.

As noted by Wendin [6], the electronic screening of the 1*s* orbitals is already complete in the ground state, and therefore the $\underline{1s}2p^63d^9(O^{2-})$ final-state configuration is the most probable, while the shake up configurations $\underline{1s}2p^63d^{10}\underline{L}$ $(O^-)$and $\underline{1s}2p^63d^{10}\underline{L}^2(O)$ occur with much more low probability. To the latter states, however, could (but less convincingly) be associated the residual intensity around 531 eV. In this case the assignment is controversial being associated with different types of planes [31,56], singly negative ions [35,44], dimers [57,58] and peroxitons [59]. A possible level diagram leading to the observation of a single 528.5 eV structure is shown in Fig. 10(a), where the valence O 2*p* states are locally pulled down by the oxygen core hole.

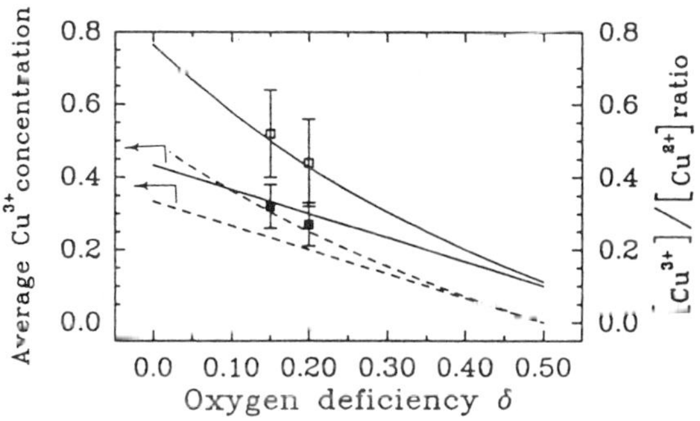

Fig.12

Average [Cu^{3+}]/[Cu^{2}] ratio (right-hand scale) and average concentration of Cu^{3+} ions (left-hand scale) estimated from the Cu 2p spectra as a function of the oxygen vacancy concentration δ (squares). Dashed lines: calculated neglecting oxygen charge fluctuation (d=0); solid lines: calculated including oxygen charge fluctuation (d=0.33).

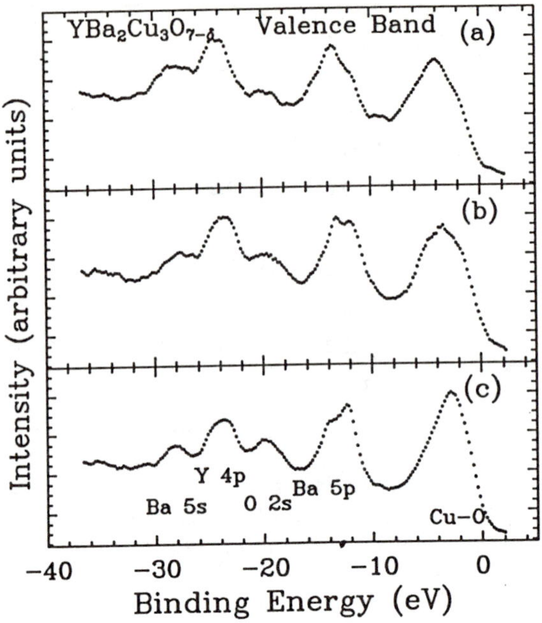

Fig.13

Valence band XPS of Y-Ba-Cu-O
(a) - Bulk sample exposed to air
(b) - bulk sample freshly scraped in UHV
(c) - thin film cleaned by eletron bombardment.

3) Valence band - In Fig.13 we present the valence-band x-ray photoemission spectra for $YBa_2Cu_3O_{7-\delta}$ measured, using the Al $K_{\alpha 1,2}$ x-ray line as excitation source (1486.6 eV), on (a) pellets exposed to the atmosphere, (b) pellets cleaned by scraping in UHV, (c) thin films cleaned by electron bombardment.We notice a narrowing of the Cu-O hybrid states and of the Ba 5p emission which is better resolved with respect to (a).The largest spectral distortion in dirty samples is observed for the spin-orbit split Ba-5p structure centered around 13 eV which is barely resolved in (a) and in previously published data [39]. Moreover, the valence band centroid in the case (c) is shifted slightly towards E_F relative to the cases (a) and (b). The density of states at E_F is low, as reported by many authors, and this is believed to be due to the strong electron-electron correlation. Similar results have been found recently on pellets [38]. The small peak at 9-10 eV, which corresponds to an analogous structure observed by ultraviolet photoemission and attributed to a two-hole O $2p^4$ resonance, is strongly reduced in cleaned samples, which raises the question of its intrinsic character [25]. The XPS spectrum is dominated by the Cu 3*d* emission, whereas at low photon energies the O 2*p* cross section is enhanced relative to that of Cu. The main valence-band features of $YBa_2Cu_3O_7$ are rather well accounted for by a CI calculation applied to a $(CuO_5)^{8-}$ pyramidal cluster which includes the apex oxygen on the Ba plane [9]. The theoretical analysis suggests that doping creates holes on the BaO *p*-like orbitals of symmetry σ_z and on the *p* $\sigma_{x,y}$ orbitals in the CuO_2 plane. When the energy difference $\Delta\varepsilon_p$ between the O in-plane and off-plane is larger than 1.7 eV, the local triplet state $^3B_1$ has the lowest excited state energy for the doped system. The near-degeneracy with the singlet $^1B_1$ state reduces the ferromagnetic coupling to values small enough to not disrupt Cooper pairing of holes in the strong-coupling

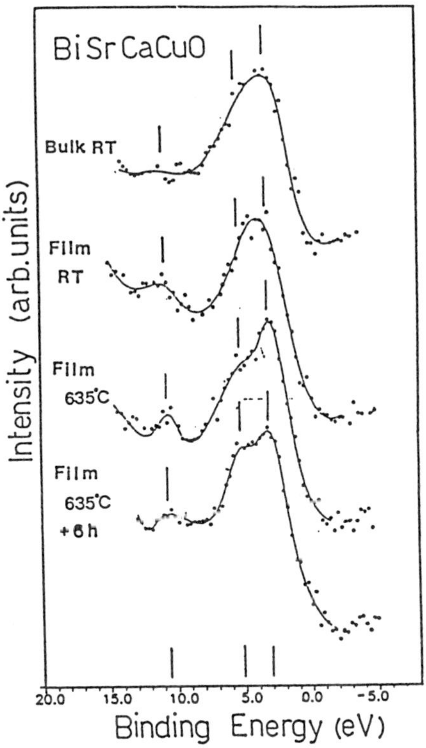

Fig.14

Valence-band XPS of Bi-Sr-Ca-Cu-O films deposited by laser ablation after heating in vacuum at different temperatures. The top spectrum refers to bulk samples. Background not removed.

regime. A calculation confined to the Cu-O_2 plane oxygens [60] finds the singlet state $^1A_1$ (of symmetry x^2-y^2) closest to E_F for U>Δ and the triplet state $^3B_1$ (of symmetry $3z^2$-r^2) at higher binding energy with a large oxygen hole-Cu $3d^9$ antiferromagnetic exchange coupling. Experiments give different results which support either p_z [61] and p_z and $p_{x,y}$ [62] character. The BaO-CuO_2 plane separation involving states hybridization through the intervening apical oxygen is important in view of the strong dependence of T_c upon the Cu(1)-O(4) distance.

Fig.14 shows the valence band emission for bulk and films of Bi-Sr-Ca-Cu-O grown by laser ablation [34]. Two distinct features of the Cu $3d$ -O $2p$ band clearly show up in the film heated in vacuum at 635 °C. The peak at 3 eV from E_F is undoubtedly associated with O since it decreases after prolonged heating suggesting that the copper-oxygen hybridization is diminished by possible formation of oxygen vacancies The peak at 5 eV has a preferential contribution from copper orbitals, as shown by Takahashi et *al.* [4]. The emission at 10.5 eV could be due to Bi 6s emission [31,63] superimposed on the Cu d^8 final-state emission.

At the same emission we find a contribution from a replica of the O 2*s* peak at 20 eV due to the Al $K_{\alpha 3,4}$ source line. Cluster calculations [31] describe quite well the valence spectrum with a set of parameters U, Δ, and T similar to those used for $YBa_2Cu_3O_7$. According to core-level electron-energy-loss [64] and polarized x-ray photoabsorption [62] measurements, the holes have $p_{x,y}$ symmetry, i.e. they lie in the plane of the CuO_2 squares.

4.2 On-site Correlations and X-Ray Auger Spectra

The Auger decay of excited core electrons leaving two-holes in the valence band is a direct probe of the effective Coulomb hole repulsion U_j^{eff} on the same atomic site j. U_j^{eff} describes an effective interaction between holes which involves both intraatomic (readjustment of the atomic charge) and extraatomic relaxation (screening) and tends to the Hubbard on-site U_j value for large U_j. The kinetic energy E_{kin} of an Auger transition $C_\alpha VV$ involving two electrons of the same valence band V is a function of U_j^{eff}

$$E_{kin}(C_\alpha VV)= E_b(C_\alpha) - 2E\ (V) - U_j^{eff}, \qquad (3)$$

where $E_b(C_\alpha)$ and E(V) are the binding energies of the core level C_α and of the centroid of the valence band, respectively.

A convenient way of distinguishing superconducting from nonsuperconducting phases is that of using the modified Auger parameter β defined as

$$\beta= E_{kin}(C_\alpha VV)+ E_b(C_\alpha), \qquad (4)$$

which is independent of charging effects and on the choice of the zero energy scale. Using the Wagner representation [65] and the Cu L_3VV Auger parameter of a few binary copper compounds [66], it is possible to see that β for the superconducting phase lies in between that of CuO and Cu_2O, while for non- (or partially) superconducting phases β is well below that of Cu_2O. In Fig. 15 open and full symbols refer to measurements done on a few bulk superconducting ceramics and films of YBa-Cu-O and Bi-Sr-Ca-Cu-O with cleaned or "dirty" surfaces, respectively.

If the electron correlation is strong, i.e. U_j^{eff} is larger than the valence bandwidth W, the Auger line shape deviates from the self-convolution of one-electron valence-band density of states [67]. This is the case for the Cu 2*p* level of HTSC.[73]. For the O 2*p* holes which are delocalized in the valence band, a realistic value for $U_{p,p}$ would be 4-5 eV to be compared with an oxygen bandwith of about 2 eV. It should be noted that U_j^{eff} is not necessarily the same as that entering Eq.(2) since in the latter case the holes could not sit on the same site and then their screening should be different.

A simple analytical expression for the two-hole Auger spectral function valid for filled nondegenerate bands has been derived [67]:

$$D(E)=\frac{D_o(E)}{(1- U_j^{eff} F(E))^2+(\pi\ U_j^{eff} D_o(E))^2}, \qquad (5)$$

where $D_o(E)$ is the self-convolution of the one-particle valence band density of states and F(E) is the Hilbert transform of $D_o(E)$. U_j^{eff} can be adjusted to fit the experimental position of the Auger spectrum. For unfilled bands, Eq.(5) can still be used [68], provided the number of holes per valence state is much smaller than 1 and U_j^{eff} is not too large. For most HTSC the number of Cu 3d holes is 0.4-0.5 and U_j^{eff}=6-8 eV, so this condition is approximately satisfied.

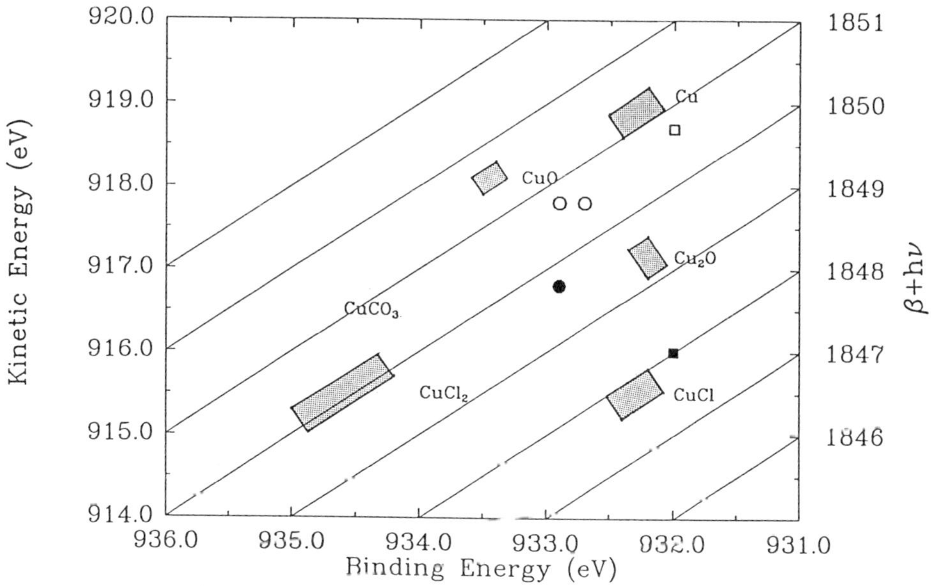

Fig.15

Two-dimensional Wagner plot showing the Auger parameter β for some copper compounds (rectangles) and superconductors (symbols).

1) Cu $L_{2,3}VV$- The Cu $L_{2,3}VV$ spectrum measured on a clean superconducting film of $YBa_2Cu_3O_{7-\delta}$ is shown in Fig.16 using the Al 1486.6 eV line. The main structure peaks at 918 eV and the entire spectrum resembles that of CuO [42,49] but it is more broadened towards low kinetic energies. This peak is associated with the final state configuration $3d^8\underline{L}$ originating from the $\underline{2p}\ 3d^{10}\underline{L}$ initial state of Cu^{2+}[69]. In the atomic limit the $3d^8\underline{L}$ multiplet contains the states ^1S, ^1D, ^1G,^3P, ^3F, of which the ^1G is the most intense state [70]. The enhanced Auger cross section between 925-930 eV (not shown in Fig.16) occurs in the region of the $3d^9\underline{L}^2$ final-state configuration coming from the $\underline{2p}\ 3d^{10}\underline{L}$ initial state which is separated by about U_{dd}-Δ from the $3d^8\underline{L}$ state.The $\underline{2p}\ 3d^9$ initial state ends up as the $3d^7$ final state which is expected to be found on the low kinetic energy side of the main peak, at an energy distance approximately $2U_{dd}$-$U_{p,d}$.

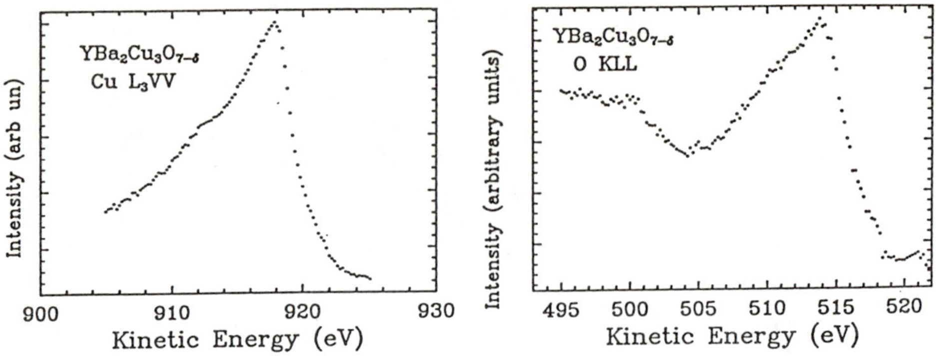

Fig.16

Cu $L_{2,3}VV$ and O 1s KLL Auger spectra of clean superconducting films of Y-Ba-Cu-O. Background not removed.

The problem of evaluating U_{dd} is not trivial, since a source of uncertainty in Eq.(3) is the centroid of the valence band. To make a quantitative comparison, we have used in Eq.(5) two different projected density-of-states calculations to fit the Auger lineshape. The calculated U_{dd} values are ≈ 2 eV and $\approx$ 7 eV from the data of Ref. 71 and 72, respectively. The large discrepancy is attributed to the different position of the centroid of the valence band with respect to E_F provided by these calculations. The largest correlation energy is more reliable because it is closer to that of other estimates [51] and to the values measured for CuO [42,73]. Using U_{dd}=7 eV and $U_{p,d} \approx$ 8 eV, derived from the analysis of the Cu $2p$ photoemission, we find the $3d^7$ transition around 912 eV kinetic energy in the region of the shoulder in Fig.16.

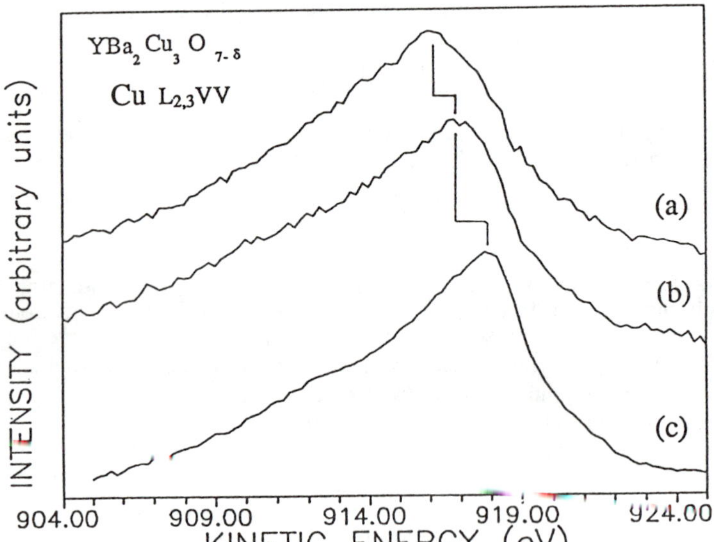

Fig.17

Spectral modification of the Auger $L_{2,3}VV$ spectrum under electron bombardment for Y-Ba-Cu-O thin films of increasing surface cleaniliness from (a) to (c).

However, the $L_2L_3M_{4,5}$ Coster-Kronig-preceded Auger transitions ending up as $3d^7$ states should nearly coincide with the $3d^7$ final state [39]. The modification of the Auger lineshape as a function of surface cleaning for a superconducting $YBa_2Cu_3O_{7-\delta}$ film can be followed in Fig.17. Curve(a) represents the Auger $L_{2,3}VV$ lineshape before cleaning.The main peak is shifted to about 916 eV kinetic energy and a shoulder is clearly visible around 918 eV. Curve (b) refers to a partially clean surface while in curve (c) the cleaning process is complete. We notice that the 916 eV peak is gradually removed while the 918 eV peak and the $3d^7$ peaks increase. Thus the 916 eV feature may be associated to an oxygen-rich phase or to surface degradation. By comparison with the Auger CuO spectrum [42], which shows the main peak at 916 eV, we speculate that this extra phase is nonsuperconducting. Evidence of the 916 eV peak has been reported in ceramics pellets scraped in vacuum [57,59,73] and its correlation with the 531 eV O $1s$ structure was stressed [35].In presence of excess O in the surface layer or in the intergrain regions, the 916 eV structure could be due to the $\underline{2p}3d^9\underline{L} \rightarrow \underline{2p}3d^9\underline{L}^2 \rightarrow 3d^8\underline{L}^2$ transition. Another assignment of this structure to a Cu^{1+} species has been repeatedly reported and its temperature (reversible) dependence was dicussed in terms of dimerized holes [57,59,74]. In view of the above results and of the observation that the Cu $2p$ satellite-to-main peak ratio decreases slightly - from 0.25 to 0.21- after cleaning, such an assignment should be considered with care.The Cu $L_{2,3}VV$ spectrum of $Bi_2Sr_2CaCu_2O_{8+\delta}$ films and polycrystalline crystals [32,34] is similar to that of $YBa_2Cu_3O_{7-\delta}$, as shown in Fig.18, and consists of the $3d^8\underline{L}$ transition at 918.7 eV in bulk samples and 918.5 in the film. The Auger spectrum of this material shows less signs of surface degradation than $YBa_2Cu_3O_{7-\delta}$. A incipient shoulder at lower kinetic energy (2.3 eV from the main peak) is only seen in the film and it is quickly washed out by heating up the film to 540 °C.

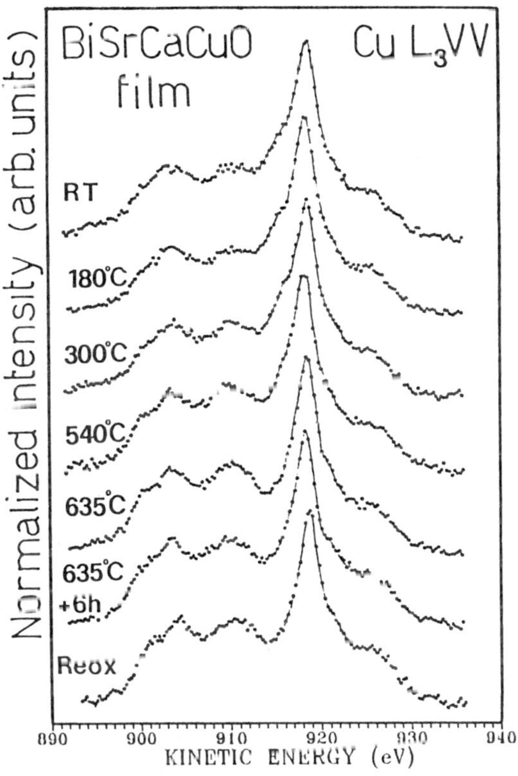

Fig.18

The effect of heating on the Cu L_3VV Auger spectrum of Bi-Sr-Ca-Cu-O films.

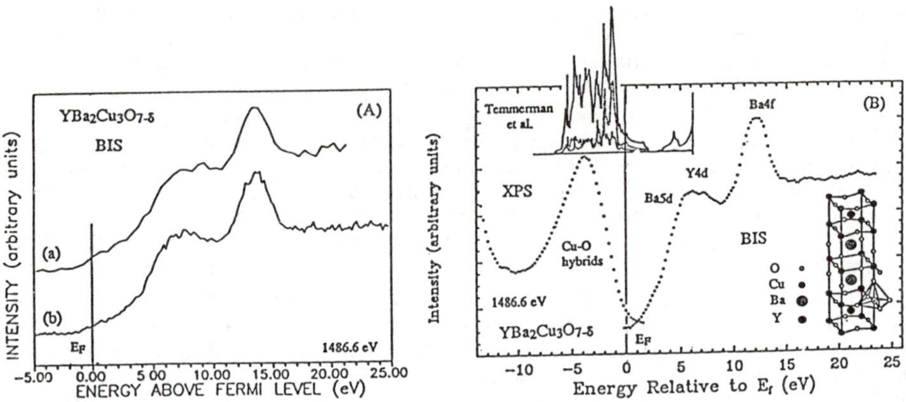

Fig.19

(A) - Effect of different exposures to the electron beam on the BIS spectra of Y-Ba-Cu-O. (a) first scan; (b) 10 h of exposure (sample current ≈300 μA).
(B) - XPS and BIS spectra showing the valence and conduction band regions, respectively. The calculated one-electron density of states is taken from Temmerman et al. [72].

As for $YBa_2Cu_3O_{7-\delta}$, the $3d^7$ transition occurs at 6-8 eV lower kinetic energies. Some stuctures peaked at 900.6 and 903.6 eV, seen also in $Pb_2Sr_2Ca_{1-x}Lu_xCu_3O_{8+\delta}$ [75], are observed but their origin is uncertain. A possible assignment [34] is to L_3M-$M_{4,5}M_{4,5}M$ three-hole final states transitions ranging between 16.5 eV in atomic copper [70] and 5 eV in metallic copper [76] from the main peak.

2) O KVV- In the broad O $KL_{2,3}L_{2,3}$ Auger spectrum of Y-Ba-Cu-O, shown in Fig.16, we can distinguish two components at 514.0 eV and around 508-510 eV kinetic energy, while a single structure at 512±0.5 eV is measured in CuO [42,69]. The most intense component corresponds to the $^1D_2$ state of the configuration $2s^22p^4$ of the O^{2-} ion and the other multiplets $KL_1L_1(^1S)$, $KL_1L_{2,3}$ ($^1P_1$, ^3P) and $KL_{2,3}L_{2,3}$ ($^1S_0$, ^3P) are not resolved. Using Eq.(3), and taking ≈4 eV for the oxygen valence band centroid and 528 eV for the O 1*s* binding energy, we estimate U_{pp}≈5 eV and U_{pp}≈14 eV for the 514.0 eV and 508-510 eV Auger structures, respectively. Since U_{pp}=5 eV is about as large as the valence bandwidth, i.e. $U_{pp}/W \approx 1$, the oxygen holes can be considered as strongly correlated particles. U_{pp}≈14 eV is large but typical of non-transition metal oxides [77] and indicates intratomic screening in the O 2p band due to the large polarizability of the O^{2-} ion. The intensity of the 508-510 eV component is correlated to that of peak B of Fig.5 and increases on going from normal to grazing incidence emission [37], suggesting an origin in part from surface oxides and/or excess or chemisorbed oxygen.

4.3 Bremsstrahlung Isochromat Spectroscopy (BIS)

BIS spectra measured on a $YBa_2Cu_3O_7$ thin film with incident electrons of 1500 eV and with the monochromator set for detection at 1486.6 eV are displayed in Fig.19 (A) before (a) and after (b) 10 h of exposure to the electron beam (≈300 μA current). Dramatic changes of the relative intensities of the structures are not observed indicating that the surface conditions do not change appreciably. Some loss of intensity takes place in the region of the Ba 5*d* transitions around 9 eV which is correlated to changes in the oxygen concentration on the Ba plane after prolonged exposure to the beam. In Fig.19 (B) the BIS spectrum is juxtaposed

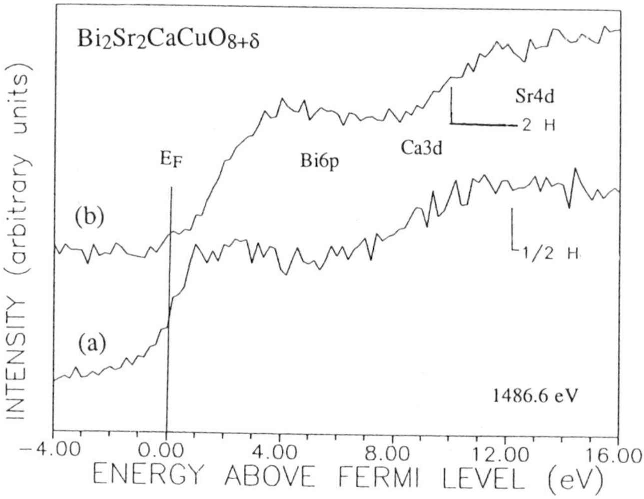

Fig.20

Effect of different exposures to the electron beam on the BIS spectra of Bi-Sr-Ca-Cu-O films.

with the valence band x-ray photoemission spectrum to emphasize the weak emission at E_F. By comparison with the results of van der Marel et *al.*[39], the agreement is good except for a less intense peak at ≈2.5 eV in our case. The spectrum looks very similar to the inverse photoemission spectrum measured by Meyer et *al.* [38] for incident electron energies larger than 36 eV above E_F. The sensitivity of the states near E_F to oxygen indicates some O 2*p* character for them. The largest emission is due to the Ba 4f states around 13-14 eV, while the features between 6 and 9 eV above E_F reflect emission mainly from Y 4*d* and Ba 5*d* levels [38,78,79].

BIS spectra measured under the same experimental conditions on a film of $Bi_2Sr_2CaCu_2O_{8+\delta}$ for two different times of exposure to the electronic beam are shown in Fig.20. In comparison to $YBa_2Cu_3O_{7-\delta}$, the spectra are strongly modified indicating instability of the sample surface under the beam. Considerable spectral changes are reported in inverse photoemission spectra taken on single crystals depending upon the preparation conditions [27].

4.4 Electron Energy Loss Spectra

The electron energy loss spectroscopy in transmission mode is not surface sensitive and therefore the problems of preparing surfaces representing bulk properties are largely avoided. If the incident high-energy (typically higher than 100 keV) electron beam propagates along the **c** axis of an uniaxial crystal, the momentum transfer **q** is predominantly perpendicular to the motion of the electrons and the loss spectrum gives information on the component of the dielectric complex tensor $\epsilon_\perp(\mathbf{q},\omega)$ parallel to the **a**, **b** plane [80].The non-relativistic energy loss rate W(**q**,ω) for the momentum **q** and the loss energy hω is given approximately by

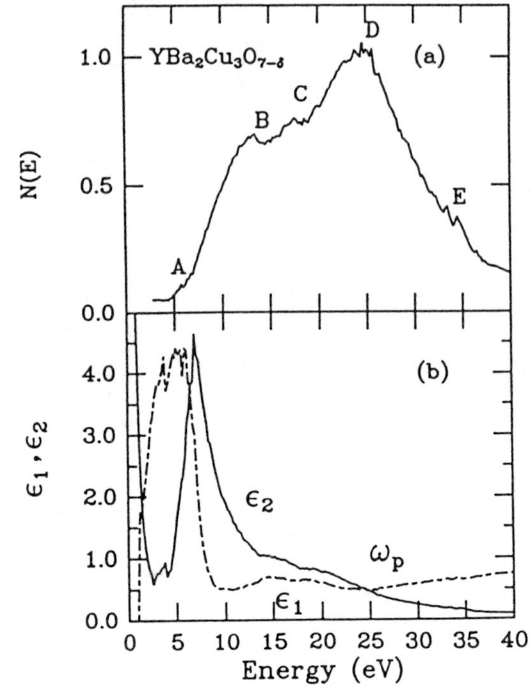

Fig.21

(a) - Experimental loss function measured in reflection mode on Y-Ba-Cu-O pellets after background subtraction; (b) - Components of the dielectric tensor derived from the Kramers-Kronig analysis.

$$W(\mathbf{q},\omega) \approx -\frac{8\pi e^2}{\hbar q_\perp^2} \, \mathrm{Im} \frac{1}{\varepsilon_\perp} . \qquad (6)$$

With the Kramers-Kronig relations the real part of $1/\varepsilon_\perp$ and then $\varepsilon_\perp(\mathbf{q},\omega)$ are calculated. They can be directly compared with those obtained from optical measurements. The beam must cross the samples and therefore thin films 500-1000 Å thick are needed. The loss function of $Bi_2Sr_2CaCu_2O_8$ films peeled off from large single crystals was measured by Nucker et *al*. [81] in the energy range 0.2 to 48 eV. They found a plasmon at 0.7 eV which disperses quadratically with q and it is caused by the collective motion in the O 2*p* band of the nearly free ($m^*/m \approx 1$) self-doping holes present in the Bi-O planes. A series of interband transitions to empty conduction states - starting from 1.2 eV upwards and involving Sr 4*d*, O 2*s* , Ca 3*p* and Bi 5*d* core levels - are also measured. No temperature dependence for either interband transitions or plasmon losses have been measured down to 30 K, indicating that changes of the electronic structure as a function of temperature are within the experimental resolution of 0.15 eV.

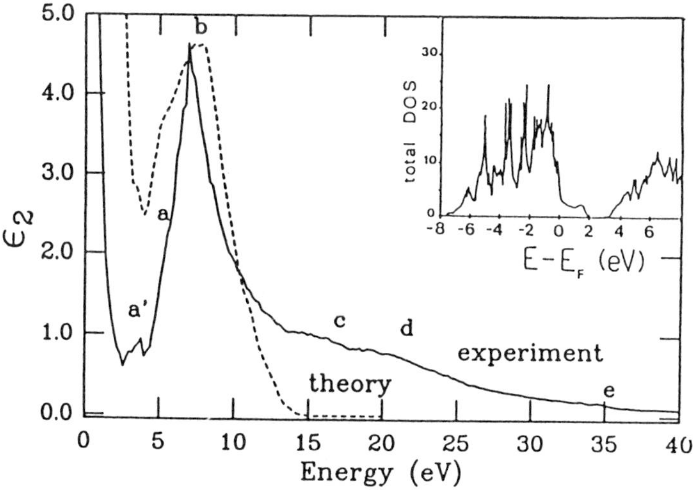

Fig.22

Experimental (full line) and theoretical (dotted line) ε_2 function calculated from the non-direct transition model using the density of states from Ref.[72].

In reflection mode thin samples are no longer required but the surface conditions are very stringent, as for the optical reflectivity, and much care must be exercized to avoid extrinsic features being measured. This is particularly true in the case of sintered Y-Ba-Cu-O samples. In Fig.21 (a) we show a typical loss function $-\mathrm{Im}(1/\varepsilon_{\perp})$ measured between 5 and 40 eV on pellets of $YBa_2Cu_3O_{7-\delta}$ after removal of zero loss and multiple losses contributions [82]. The components ε_1 and ε_2 of the dielectric tensor are represented in Fig.21 (b). A bulk plasmon is measured around 25 eV where $\varepsilon_1 \approx \varepsilon_2$ are small. Interband transitions marked A to E, are observed and are assigned with reference to the x-ray photoemission and BIS data of Fig.11(c) and Fig.19 (B), respectively. To make the assignment quantitative, one can use published valence and conduction band partial densities of states, ρ_v and ρ_c, and adopt the simplified non-direct transition model [83] to calculate ε_2, i.e.:

$$\varepsilon_2(\omega) = (M_{cv}{}^2/\omega^2) \int_0^{\infty} d\omega' \, \rho_v(\omega) \, \rho_c(\omega-\omega'), \qquad (7)$$

where $M_{cv}{}^2$ is the (constant) matrix element. The partial densities of states which contribute to the total density are Cu 3*d* and O 2*p* for the valence band and O 3*s* , Cu 4*s* , Y 4*d*, and Ba 5*d* for the conduction band. The convolution of these features up to 10 eV show features which agree with the experimental peaks a', a, and b, as shown in Fig.22. The measured transitions and their attribution are summarized in Table 2. In contrast to Bi-Sr-Ca-Cu-O, we have observed [82] small temperature effects between 80K < T < 300K as a shift of the main ε_2 peak toward lower energies at low temperatures and we have speculated an increased concentration of holes in the O 2*p* band. However, in view of possible contamination of the sample surface during cooling in UHV, independent confirmations on good single crystals are desirable before a serious interpretation of this feature is attempted.

Table II

N(E)	ε_2	assignment
-	(a') 3.5	Cu 3d-O 2p → E_F
-	(a) 5.5	Cu 3d-O 2p → s,d (4.0)
(A) 6.0	-	-
-	(b) 8.0	Cu 3d-O 2p → s,d (6.5)
(B) 13.5	-	-
-	(c) 16.0	Cu 3d-O 2p → Ba 4f Ba 5p → E_F
(C) 17.5	-	-
-	(d) 20.5	O 2s → E_F Ba 5p → Ba 5d
(D) 25.0	-	valence electron plasmon
(E) 34.0	(e) 34.0	Y 4p → Y 4d

Energy position, referred to the elastic peak, of the loss structures as labeled in Fig. 21, and of ε_2 as labeled in Fig.22.

At oblique incidence the parallel component of the momentum transfer $q = \omega/v = k_0E/2E_0$, where v, k_0 and E_0 are the velocity, momentum and energy of the incoming electrons is non vanishing. At the primary beam energy E_0= 170 keV, the momentum transfer parallel to the **c** axis is 0.41 and 0.72 $Å^{-1}$ at the O1*s* and Cu 2*p* edges, respectively [64]. In these conditions the loss along the **c** axis is measured and important information on the symmetry of the unoccupied oxygen ligand hole states has been obtained in 1-2-3 and 2-2-1-2 superconductors [64]. We do not discuss these results here since they have been illustrated in detail elsewhere [6]. We only stress that these measurements, as well as polarized light x-ray absorption studies [61] on single crystals, may provide a high-quality characterization of HTSC, provided material problems are properly overcome.

5. CONCLUDING COMMENTS

We have shown, with regard to the high-excitation spectroscopies, that the new high-temperature superconductors present severe effects of degradation both at the surface and for a substantial distance inside the material, which alter their electronic properties within the sampling depth of a few tens of Angstroms typical of the high-energy techniques. These effects depend strongly on the stability of the compounds, cleaning procedures and on the measuring probes and are of different nature and intensity in sintered pellets with respect to films and single crystals but they can rarely be ignored. The quality of the experimental data is steadily improving as the methods of preparation and characterization of the samples become better. At present, the best spectroscopic results have been obtained on highly-textured epitaxial thin films or good single crystals measured in ultrahigh vacuum conditions without exposure to the ambient. It is hoped that better electronic characterization of this interesting class of compounds will help to select the correct theoretical model(s) among the several proposed and will guide in a short time to identify the key features responsible for the high-temperature superconductivity.

Acknowledgements - I am strongly indebted to all my collaborators M.De Crescenzi, N.Motta, F.Patella, F.Rochet, M.Sastry and A.Sgarlata for their valuable help and the clarifying discussions during joint work on the subject. This work has in part been supported by the CISM-GNSM.

REFERENCES

[1] J.G.Bednorz and K.A.Müller, Z.Phys. B **64**, 189 (1986).
[2] R.S.Roth, J.R.Dennis, and K.L.Davis, Adv. Ceram. Mat. 2, 303 (1987).
[3] H. Verweij and W.H.M.Bruggink, J.Phys.Chem. Solids **50**, 75 (1989);
[4] T.Takahashi, H.Matsuyama, H.Katayama-Yoshida, Y.Okabe, S.Hosoya, K.Seki, H.Fujimoto, M.Sato, and H.Inokuchi, Nature (London), **334**, 691 (1988); Phys. Rev.B **39**, 6636 (1989); J.M. Imer, F.Patthey, B.Dardel, W.D.Schneider, Y.Baer, Y.Petroff, and A.Zettl, Phys.Rev.Letters **62**, 336 (1989); R.Manzke, T.Buslaps, R.Claessen, and J.Fink, Europhys.Letters (to be published); C.G.Olson, R.Liu, A.-B.Yang, D.W.Lynch, A.J.Arko, R.S.List, B.W.Veal, Y.C.Chang, P.Z.Jiang, and A.P.Paulikas, Science (to appear).
[5] K.Wandelt, Surf.Sci. Rep. **2**, 1 (1982), pag.37.
[6] G.Wendin, Physica Scripta T **27**, 31 (1989); K.C.Hass, In *Solid State Physics* , ed. by H.Ehrenreich and D.Turnbull (Academic, New York) **42**, 213 (1989); W.E.Pickett, Rev. Mod. Phys. **61**, 433 (1989).
[7] J.Zaanen, G.A.Sawatzky and J.W.Allen , Phys.Rev.Letters **55**, 418 (1985).
[8] R.L.Kurtz, S.W.Robey, R.L.Stockbauer, D.Mueller, A.Shih, L.Toth, A.K.Singh,and M.Osofsky, Proc. 1st Iberian Vacuum Meeting, Vacuum (to be published).
[9] A.Fujimori, Phys.Rev.B**39**, 793 (1989).
[10] J.M.Tarascon, W.R.McKinnon, P.Barboux, D.M.Hwang, B.G.Bagley, L.H.Greene, G.W.Hull, Y.Le Page, N.Stoffel, and M.Giroud, Phys. Rev. B, in press.
[11] M.A.Subramanian, C.C.Torardi, J.C.Calabrese, J.Gopalakrishnan, K.J.Morrisey, T.R.Askew, R.B.Flippen, U.Chowdhry, and A.W.Sleight, Science **239**, 1015 (1988); P.Bordet, J.J.Capponi, C.Chaillout, J.Chenavas, A.W.Hewat, E.A.Hewat, J.L.Hodeau, M.Marezio, J.L.Tholence, and D.Tranqui, Physica C **156**, 189 (1988); Y.Le Page, W.R.McKinnon, J.M.Tarascon, and P.Barboux, Phys.Rev.B (to be published).
[12] R.L.Barns and R.A.Laudise, Appl. Phys.Lett. **51**, 1373 (1987).
[13] K.Frase, E.G.Liniger, and D.R.Clarke, Adv.Ceram.Mater. **2**, 698 (1987).
[14] K. Yoshikawa, M.Yoshida, and M. Nakano, Jpn.J.Appl.Phys. **27**, L2324 (1988).
[15] H.K.Liu, S.X.Dou, A.J.Bourdillon, and C.C. Sorrell, Supercond. Sci. Technol. **1**, 194 (1988).
[16] F.Stucki, P.Bruesch, and Th. Baumann, Physica C **156**,461 (1988).
[17] J.E.Blendell, C.K.Chiang, D.C.Cranmer, S.W. Freiman, E.R.Fuller Jr., E. Drescher-Kraisicka, W.L.Johnson, H.M.Ledbetter, L.H.Bennet, L.J.Swatrzendruber, R.B.Marinenko, R.L.Myklebust, D.S.Bright, and D.E.Newbury, *Chemistry of High Temperature Superconductors,* D.L.Nelson, M.Stanley Whittingham, and T.F.George, eds. Am. Chem.Soc., Washington, D.C., (1987).
[18] W.K.Ford, C.T.Chen, J.Andersen, J.Kwo, S.H.Liou, M.Hong, G.V.Rubernacker, and J.E.Drumheller, Phys.Rev.B**37**, 7924 (1988).
[19] T.H.Fleish and G.I.Mains, Appl.Surf.Sci. **10**, 51 (1982).
[20] C.H.Chen, L.F.Schneemeyer, S.H.Liou, M.Hong, J.Kwo, H.S.Chen, and J.V.Waszczak, Phys.Rev. B **37**, 9780 (1988).
[21] H.W.Zanderbergen, W.A.Groen, F.C.Mijlhoff, G.van Tendeloo, and S.Amelinckx,Physica C **156**, 325 (1988).
[22] B.Raveau, C.Michel, M.Hervieu, and J.Provost, Rev. Solid State Science, **2**, 115 (1988).
[23] V.A.Grazhulis, G.A.Emel'chenko, and A.M.Ionov, Physica C **159**, 404 (1989).
[24] M.Ting, N.G.Stoffel, Q.B.Chen, D.LaGraffe, P.A.Morris, W.A.Bonner, G.Margaritondo, and M.Onellion , Phys.Rev. **38**, 897 (1988).
[25] See,e.g., G.Wendin in Ref.6 for a discussion of the 9.5 eV valence band in Y-Ba-Cu-O.

[26] R.L.Stockbauer, S.W.Robey, R.L.Kurtz, D.Mueller, A.Shih, A.K.Singh, L.Toth, and M.Osofsky, Proc. AIP Conf. on thin Film Processing and Characterization of HTSC, Atlanta, Georgia ,October 1988, to appear.
[27] W.Drube, F.J.Himpsel, G.V.Chandrashekhar, and M.W.Shafer, Phys.Rev.B (submitted1989).
[28] For a partial list of references on the O 1s photoemission spectra, see J.C.Fuggle, J.Fink and N.Nücker, Proc. Adriatico Res.Conf.,Trieste, Wordl Sci. Publ.,Singapore, (1988), Ed. S.Lundqvist, E.Tosatti, M.P.Tosi, and Yu Lu, pag. 649.
[29] T.Takahashi, F.Maeda, H.Katayama-Yoshida, Y.Okabe, T.Suzuki, A.Fujimori, S.Hosoya, S.Shamoto, and M.Sato, Phys. Rev.B**37**, 9788 (1988).
[30] P.Steiner, S.Hüfner, A.Jungmann, S.Junk, V.Kissinger, I.Sander, W.R.Thiele, N.Backes, and C.Politis, Physica C **156**, 213 (1988).
[31] A.Fujimori, S.Takekawa, E.Takayama-Muromachi, Y.Uchida, A.Ono, T.Takahashi, Y.Okabe,and H.Katayama-Yoshida, Phys.Rev.B **39**, 2255 (1989).
[32] F.U.Hillebrecht, J.Fraxedas, L.Ley, H.J.Trodahl, J.Zaanen, W.Braun, M.Mast, H.Petersen, M.Schaible, L.C.Bourne, P.Pinsukanjana, and A.Zettl, Phys.Rev.B **39**, 236 (1989).
[33] H.M.Meyer III, D.M.Hill, J.H.Weaver, D.L.Nelson, and C. T.F.Gallo, Phys.Rev.B **38**, 7144 (1988).
[34] F.Rochet, A.Balzarotti, M.De Crescenzi, N.Motta, M.Sastry, F.Patella, J.Perrière, F.Kerhervé, G.Hauchecorne, M.Froment, Physica C **159**, 447 (1989).
[35] A.Balzarotti, M.De Crescenzi, N.Motta, F.Patella, and A.Sgarlata, Phys. Rev. **38**, 6461 (1988).
[36] P.Steiner, S.Hüfner, A.Jungmann, V.Kinsinger, and I.Sander, Z.Phys.-Cond.Matter B **74**, 173 (1989).
[37] A.Balzarotti *et al.* (to be published).
[38] H.M.Meyer III, D.M.Hill, T.J.Wagener, Y.Gao, J.H.Weaver, D.W.Capone II, and K.C. Goretta, Phys. Rev. B **38**, 6500 (1988).
[39] D. van der Marel, J.van Elp, G.A.Sawatzky, and D.Heitmann, Phys.Rev.B **37**, 5136 (1988).
[40] W.R.Flavell and R.Egdell, Supercond.Sci. Technol. **1**, 118 (1988).
[41] F.Parmigiani and G.Samoggia, Europhys.Lett. **7**, 543 (1988).
[42] J.Ghijsen, L.H.Tjeng, J.van Elp, H.Eskes, J.Westerink, G.A.Sawatzky, and M.T.Czyżyk , Phys.Rev.B **38**, 11322 (1988).
[43] Z.X.Shen, J.W.Allen, J.J.Yeh, J.S.Kang, W.Ellis, W.Spicer, I.Lindau, M.B.Maple, Y.D.Dalichaouch, M.S.Torikachvili, J.Z.Sun, and T.H.Geballe, Phys.Rev. B **36**, 8414 (1988).
[44] T.Gourieux, G.Krill, M.Maurer, M.F.Ravet, A.Menny, H.Tolentino, and A.Fontaine, Phys.Rev.B**37**, 7516 (1988).
[45] P.Steiner, S.Hüfner, V.Kinsinger, I.Sander, B. Siegwart, H.Schmitt, R.Schulz, S.Junk, G.Schwitzgebel, A.Gold, C.Politis, H.P.Müller, R.Hoppe, S.Kemmler-Sack, and C.Kunz, Z.Physik-Cond.Matter B**69**, 449 (1988).
[46] N.Nücker, J.Fink, B.Renker, D.Ewert, C.Politis, P.J.W.Weijs, and J.C.Fuggle, Z.Phys.-Cond.Matter B**67**, 9 (1987).
[47] More details on the relaxation mechanisms in solids can be found, e.g., in D.C.Langreth, Phys.Rev. B **1**, 471 (1970) and references therein.
[48] A.Fujimori, E.Takayama-Muromachi, Y.Uchida, and B.Okai, Phys.Rev. B**35**, 8814 (1987).
[49] J.C.Fuggle, J.P.W.Weijs, R.Schoorl, G.A.Sawatzky, J.Fink, N.Nücker,, J.P.Durham, and W.M.Temmermann, Phys.Rev. B**37**, 123 (1988).
[50] G.van der Laan, C.Westra, C.Haas, and G.A.Sawatzky, Phys. Rev.B **23**, 4369 (1981), J.Zaanen, C.Westra, and G.A.Sawatzky, Phys.Rev.B **33**, 8060 (1986).
[51] M.R.Thuler, R.L.Benbow, and Z.Hurich, Phys. Rev. B**26**, 669 (1982).
[52] With these notations the $|2\underline{p}3d^{10}\underline{L}^2>$ should be more accurately written $|2\underline{p}3d^{10}\underline{LL}>$ to indicate that doping holes might not reside on the same oxygen atom.
[53] S.Doniach and M.Šunjić, J.Phys.C **3**, 285 (1970).
[54] A.Balzarotti, M.De Crescenzi and L.Incoccia, Phys.Rev.B **25**, 6349 (1982).
[55] W.R.Flevell and R.G.Egdell, Phys.Rev.B **39**, 231 (1989).
[56] J.H.Weaver, H.M.Meyer III, T.J.Wagener, D.M.Hill, Y.Gao, D.Peterson, Z.Fish, and A.Arko, Phys.Rev.B**38**, 4668 (1988).

[57] D.D.Sarma,K.Sreedhar, P.Ganguly, and C.N.R. Rao, Phys. Rev.B **36**, 2371 (1987).
[58] B.Dauth, T.Kachel, P.Sen, K.Fischer, and M.Campagna, Z.Phys.-Cond.Matter B **68**, 407 (1987).
[59] B.K.Chakraverty, D.D.Sarma, and C.N.Rao, Physica C **156**, 413 (1988).
[60] H.Eskes and G.A.Sawatzky, Phys.Rev.Lett. **61**, 1415 (1988).
[61] A.Bianconi, M.De Santis, A.Di Cicco, A.M.Flank, A.Fontaine, P.Lagarde, H.Katayama-Yoshida, A.Kotani, and A.Marcelli, Phys.Rev. B**38**, 7196 (1988).
[62] P.Kuiper, M.Grioni, G.A.Sawatzki, D.B.Mitzi, A.Kapitulnik, A.Santaniello, P.De Padova, and P.Thiry, Physica C **157**, 260 (1989).
[63] H.M.Meyer III, D.M.Hill, J.H.Weaver, D.L.Nelson, and C.F.Gallo, Phys.Rev.B **38**, 7144 (1988).
[64] N.Nücker, H.Romberg, X.X.Xi, J.Fink, B.Gegenheimer, and Z.X.Zhao, Phys.Rev. B **39**, 6619 (1989).
[65] C.D.Wagner, L.H.Gale, and R.H.Raymond, Anal. Chem.**51**, 466 (1979).
[66] *Handbook of X-ray Photoelectron Spectroscopy*, Ed.by G.E.Mullenberg, Perkin-Elmer Corp. pag.171.
[67] M.Cini, Solid State Commun. **24**, 681 (1977); G.A.Sawatzky, Phys.Rev.Lett.**39**,504 (1977).
[68] M.Cini and C.Verdozzi, Nuovo Cimento **9**,1 (1987).
[69] A.Balzarotti, M.De Crescenzi, N.Motta, F.Patella, and A.Sgarlata, in "*Auger Spectroscopy and Electronic Structure* ", Springer Series in Surface Science, Ed.by K.Wandelt, G.Mondio, and G.Cubiotti, (Springer Verlag, Heidelberg, 1989), in press.
[70] S.Aksela and J.Sivonen, Phys.Rev.A**25**, 1243 (1982).
[71] L.F.Mattheiss and D.R.Hamann, Solid State Comm. **63**, 395 (1987).
[72] W.Temmerman, Z.Szotek, P.J.Durham, J.M.Stocks, P.A.Sterne, J.Phys.F: Metal Phys. **17**, L319 (1987).
[73] A.Balzarotti, M.De Crescenzi, C.Giovanella, R.Messi, N.Motta, F.Patella, and A.Sgarlata, Phys.Rev.B **36**, 8285 (1987).
[74] D.D.Sarma and C.N.Rao Solid State Commun. **65**, 47 (1988).
[75] C.N.Rao, V.Bhat, R.Nagarajan, G.Ranga Rao, and G.Sankar, Phys. Rev.B **39**, 9621 (1989).
[76] E.Antonides, E.C.Janse, and G.A.Sawatzky, Phys.Rev.B**15**, 4596 (1977).
[77] P.Légaré, G.Maire, B.Carriere, and J.P.Deville, Surf. Sci. **68**, 348 (1977). See also the compilation in Ref.20.
[78] T.J.Wagener, Y.Gao, J.H.Weaver, A.J.Arko, B. Flandermeyer, and D.W.Capone II, Phys.Rev. B **36**, 3899 (1987).
[79] J.A.Yarmoff, D.R.Clarke, W.Drube, U.O.Karlsson, A.Taleb-Ibrahimi, and F.J.Himpsel, Phys.Rev.B **36**, 3967 (1987)
[80] A.Balzarotti and M.Piacentini, *Electronic Structure and Electronic Transitions in Layered Materials*, Ed.by V.Grasso, D.Reidel Publ.Co., pag.289 (1986).
[81] N.Nücker, H.Romberg, S.Nakai, B.Scheerer, J.Fink, Y.F.Yan, and Z.X.Zhao, Phys.Rev.,(submitted 1989).
[82] A.Balzarotti, M.De Crescenzi, N.Motta, F.Patella, and A.Sgarlata, Solid State Commun. **68**, 381 (1988).
[83] F.Wooten, *Optical Properties of solids*, Academic Press, New York, (1972).

MÖSSBAUER SPECTROSCOPY OF HIGH TEMPERATURE SUPERCONDUCTORS

P. Boolchand and D. McDaniel
University of Cincinnati, Cincinnati, Ohio, U.S.A.

The discovery [1-4] of high-temperature superconductors (HTSC) has generated a sense of excitement and challenge in the solid state sciences. It has touched inter-related disciplines such as physics, chemistry, geology, materials science, and electrical engineering. The high-transition temperatures, coupled with novel processing methods being used to prepare these materials in the form of wires and thin-films offers the promise of applications both in microelectronics and magnet technology. However, for this promise to become a reality, a better understanding of these materials will have to emerge. Even though nearly two years have passed since the announcement of the 90K superconductor, the basic mechanism of the phenomenon continues to present the major challenge in this field [5-6].

Crystal structures of HTSC [7-10] involve large unit cells in which point defects (oxygen vacancies, impurity atoms), line defects (twin-planes) and inter-granular impurity phases are pervasive. Soon after isolation of (Y or RE)$_1$$Ba_2Cu_3O_{7-\delta}$, (abbreviated as 1-2-3) as the 90K superconducting phase, the sharp oxygen stoichiometry dependence (δ) of the superconductive transition temperatures (T_c) was noted [12] (Fig. 1). Indeed the connection between structural features of these oxides and their physical properties such as magnetism, specific heat, elasticity, superconductivity, electronic and vibrational density of states represents an area of intense current activity [13] largely because it holds the key to understanding the materials. It is likely that as better quality materials emerge from new processing methods, advances in both applications and the basic science of these materials will emerge and set the stage for isolating the origin of superconductivity.

Local probes of solids such as EXAFS, NMR, NQR, μSR, Mössbauer and Raman scattering are all likely to play a significant role in understanding these materials. There are several reasons. In these large unit cell crystals, superconductivity appears to be localized onto specific structural elements - CuO_2 planes. Furthermore, impurity-atom alloying is known to drastically alter superconducting properties either because these act as electrical dopants (both donors [14] and acceptors [15-17] are known) or because these stabilize a given crystalline phase. Narrow windows of atomic radii differences and electronegativity differences between cations appear to be critical to realization of high-

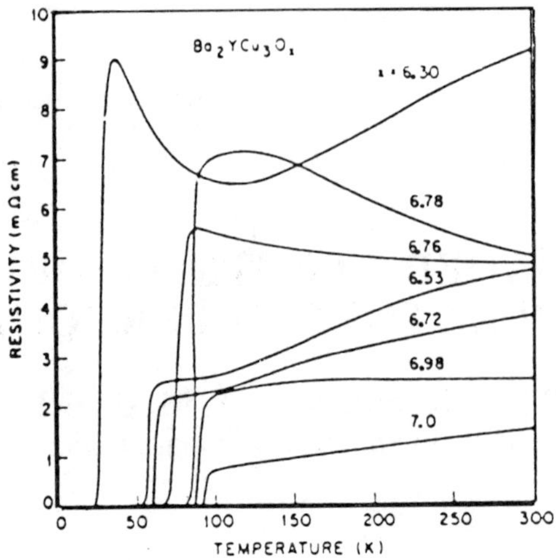

Fig. 1a T-dependence of Resistivity for $Ba_2YCu_3O_x$ samples at indicated oxygen stoichiometry x. Figure taken from ref. 12.

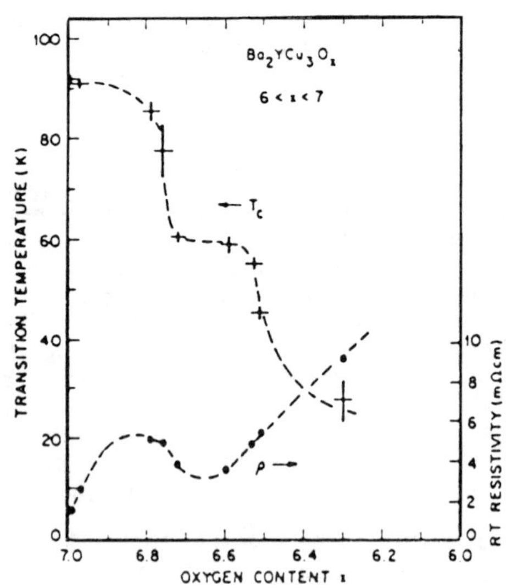

Fig. 1b Superconductive Transition temperature T_c and Room Temperature resistivities at a function of oxygen content x for $Ba_2YCu_3O_x$ samples from Fig. 1a. Figure taken from ref. 12.

T_c [18]. In either case basic issues concerning impurity site location, charge state, spin state and local density of vibrational states come up. It is in this context Mössbauer spectroscopy can serve as a valuable microscopic probe of these oxides.

In this chapter we illustrate these areas with specific examples, some taken from published literature and others from our own work. This chapter is structured as follows. In Section I we discuss the correlation between microstructure, defects and metallic doping sites; Section II focuses on magnetism and superconductivity and in the concluding section III we describe experiments on Mössbauer Debye-Waller factors that have shed light on phonon-instabilities observed in the present oxides.

I. CRYSTAL STRUCTURES, DEFECTS, AND METAL DOPING SITES

After a brief description of the crystal structures of HTSCs, and defects native to these structures, we proceed to discuss chemical doping effects. Our thrust is to review different types of experiments that illuminate the issue of Fe-doping sites in $YBa_2Cu_3O_{7-\delta}$, a subject of some current interest. We will conclude this section with very specific site assignments that can form the basis of discussions for magnetic effects in section II.

Crystal Structures

Crystal structures of the two families of cuprate superconductors based on La_2CuO_4 (Fig. 2) and $YBa_2Cu_3O_{7-\delta}$ (Fig. 3) have been established by x-ray [11,19,20] and neutron Bragg diffraction [9,10,21]

La_2CuO_4 crystallizes in an orthorhombicly distorted version [9] of the K_2NiF_4 structure [22] which may be visualized as made up of CuO_2 planes with strong intraplanar but weaker interplanar interactions. Cu^{2+} cations possess a coordination number (CN) of 6. The Cu coordination is not cubic, however, because the Cu-O bond lengths in the CuO_2 planes are distinctly shorter than Cu-O bond lengths perpendicular to the planes. In La_2CuO_4, an orthorhombic distortion results in the symmetry axis of the CuO_6 polyhedra being canted with respect to the c-axis. La replacement by a few atomic percent of Sr or Ba leads to stabilization of the tetragonal K_2NiF_4 structure in which the symmetry axis of the CuO_6 polyhedra becomes co-linear with the c-axis.

$La_2CuO_{4-\xi}$ prepared by reacting La_2O_3 and CuO at 1100°C in air invariably forms as an oxygen deficient ($\xi \neq 0$) material which is found to be a semiconducting antiferromagnet [23,24] with a Neel temperature T_N = 235K. Tetragonal La_{2-2x} (Sr or $Ba)_{2x}CuO_{4-\xi}$ samples on the other hand display an onset of superconductivity at 2x=0.07 for Sr and at 2x=0.05 for Ba. Pristine $La_2CuO_{4-\xi}$ samples display filamentary superconductivity [25] upon oxygen loading (ξ<0) or fluorination as reported in some recent publications [26].

$YBa_2Cu_3O_{7-\delta}$ samples at an oxygen content of ~7 (i.e. δ~0) crystallize in an orthorhombic structure (a=3.8158(1)Å, b=3.8822(1)Å, c=11.6737(3)Å) [27] which may be described as a triple cell perovskite with oxygen vacancies. Moving along the c-axis one encounters a sequence of planes: Cu(1)-O, Ba-O, Cu(2)-O, Y, Cu(2)-O, Ba-O, in the unit cell. Neutron scattering results [28] show that oxygen vacancies are largely ordered: in the Cu(1)-O plane the O(5) oxygen sites (along the a-axis) are vacant while the O(4) oxygen sites (along b-axis) are occupied leading to Cu(1)-O(4)-Cu(1)-chains running parallel to the b-axis. In the Y-plane, the oxygen sites are vacant leading to Cu(2)-

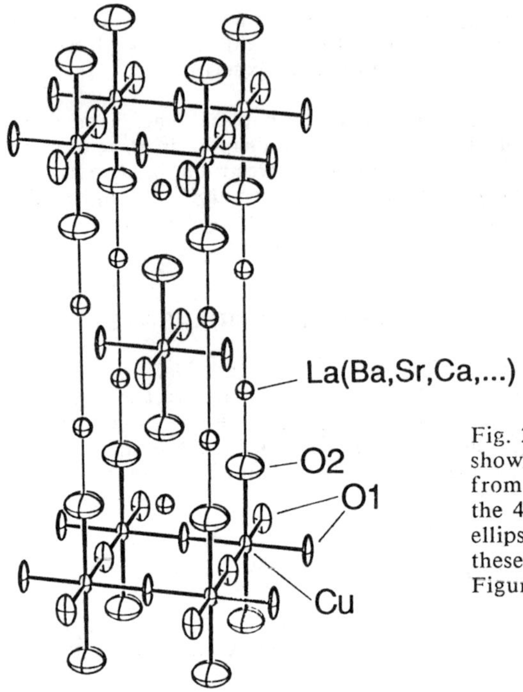

Fig. 2 Crystal structure of $La_{2-x}Ba_xCuO_4$ showing the CuO_2 planes which are displaced from each other so that the O atoms along the 4-fold axis do not join Cu atoms. The ellipsoid of vibrational amplitude for these O atoms is appreciably larger. Figure is taken from ref. 9.

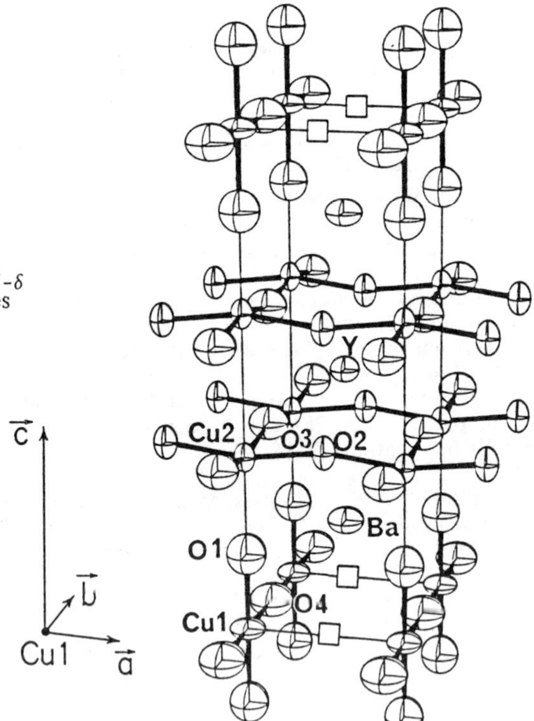

Fig. 3 Crystal structure of $YBa_2Cu_3O_{7-\delta}$ showing the pair of buckled CuO_2 planes for every CuO_3 chain in the unit cell. Note that the O(4) -Cu(1) -O(4) linear chains run along the b-axis. Figure adapted from ref. 10. O(5) vacancies indicated by squares.

Cu(2) contacts. These oxygen vacancies lead to a stoichiometry of 7 in the orthorhombic cell. The unit cell consists of two CuO_2 planes for every CuO_3 chain. Note that the Cu coordination in chains is nearly square-planar (CN=4; two O(4) sites and two O(1) sites as nearest neighbors (nn) while that in planes is square pyramidal (CN=5; two O(2) sites, two O(3) sites and an O(1) apex site as nn).

Defects

Oxygen vacancies are native defects in the present perovskites. Solid state reaction of finely powdered Y_2O_3, $BaCO_3$, CuO in air at 950°C leads to the formation of tetragonal $YBa_2Cu_3O_{7-\delta}$, which can exist over a broad range of oxygen stoichiometries $0.5<\delta<1.0$. At $\delta=1$, tetragonal $YBa_2Cu_3O_6$ consist of Cu(1) sites that are 2-fold coordinated to O(1) nn along the c-axis with both the O(4) and O(5) sites vacant [29]. At higher oxygen stoichiometries $1.0>\delta>0.5$, the tetragonal phase is achieved by equal but random O(4) and O(5) site occupancies, as revealed by neutron scattering.

The 90K superconducting orthorhombic phase is thought to be formed upon oxygen soaking of the tetragonal $YBa_2Cu_3O_6$ phase at modest temperatures (400°C) in three distinct steps [30]. The first step consists of an oxygen diffusion process which leads to an increase in the oxygen content of the tetragonal phase to a threshold value of about 6.5. The transformation of tetragonal $YBa_2Cu_3O_{6.5}$ to orthorhombic $YBa_2Cu_3O_{6.5}$ is thought to be martensitic in character involving a shear motion of a large number of atoms quite spontaneously and leads to coherent (110) twin planes in the daughter phase. Further diffusion of oxygen in the orthorhombic phase increases both the oxygen content and the transition temperatures (T_c) as noted earlier (Fig. 1).

The plateau in T_c at 60K over an oxygen content in the range $6.50<7-\delta<6.70$ (see Fig. 1) is ascribed to a new orthorhombic phase [11]. The structure of this phase has been recently isolated [31] by High Resolution Electron Microscopy (HREM) and indicates a modulation of the lattice structure of the 90K phase: specifically in the a-direction every other O(4) oxygen site is vacant. This leads to a doubling of the interchain separation and the "a" cell length parameter. In principle such a structure should possess a stoichiometry of $YBa_2Cu_3O_{6.5}$. Partial occupancy of the vacant O(5) sites is thought to be responsible for the persistence of the 60K superconducting phase to higher oxygen contents (up to 6.70). Because of the high mobility of oxygen in the 1-2-3 phase, oxygen vacancy ordering is sensitive to the thermal treatment of samples. In particular the tetragonal to orthorhombic transition can be shifted up in oxygen stoichiometry by use of a higher annealing temperature. This means that mere specification of oxygen content is not sufficient to completely specify the material since the type of oxygen vacancy ordering prevailing does influence the physical properties of the material.

$BaCO_3$, $BaCuO_2$ and Y_2BaCuO_5 (green phase) are the non-superconducting impurity phases [32,33,34] that can appear at grain boundaries of sintered 1-2-3 samples prepared the usual way using oxides and $BaCO_3$ as the starting materials. Departures from the 1-2-3 stoichiometry and overprocessing will invariably precipitate these impurity phases that can be detected by Raman Scattering, Infrared Transmission and Auger or x-ray photoelectron spectroscopy. The presence of these impurity phases at grain boundaries is thought to act as weak links (Josephson junctions) in polycrystalline materials drastically lowering transport critical currents in spite of the fact that intragrain critical currents (probed in magnetization experiments) are several orders of magnitude higher. Samples of superconducting 1-2-3 phase made using carbon free precursors (such as nitrates) in novel processes involving solution routes such as aerosols [35,36] and sprays in rather controlled conditions will undoubtedly assist in clarifying the underlying materials issues. Furthermore, the recent fabrication of Ag: 1-2-3 composites [37,38] may provide not only a way to replace intergrain weak-links by a good metal such as Ag, but also to pin fluxoids in these type II superconductors and minimize the detrimental dissipation effects associated with melting of the flux lattice [39,40].

An intrinsic structural feature of the orthorhombic phase is the existence of twin-planes. Transmission or scanning electron-micrographs of this phase display rather striking parallel streaks in a grain and these are manifestations of (110) twin-planes [41]. Such planes parallel to the c-axis are thought to provide a means to release strain energy by a static deformation in the growth of the larger sized orthorhombic ($a \neq b$) cell from the tetragonal one. Twinning as a structural defect in the superconducting phase is more than of passing interest. There are extensive theoretical discussions in the literature [42] to suggest that twinning may induce T_c enhancement. Perhaps some of the T_c enhancement in the newly discovered oxides may be twinning related [43,44], a speculation for which decisive experimental evidence has yet to emerge.

Metal Doping Effects

Metal dopants (Fe, Co, Ni, Zn, Ga, Al) directed at Cu sites in the 1-2-3 phase, systematically alter the superconducting properties and crystalline structure [45,46]. The role of such dopants has been at the focus of much basic research in these materials largely because such studies provide a controlled way to alter the structural properties of these oxides. This is essential to understand the nature of superconductivity in these perovskites with large unit cells. Such studies were the first to demonstrate [47] that it is the CuO_2 planes and not the CuO_3 chains in the orthorhombic phase, that are the significant structural elements that support superconductivity. Divalent Ni and Zn dopants apparently replace Cu^{2+} in the CuO_2 planes and therefore are far more effective in lowering T_c than the trivalent Fe, Co, Al dopants that largely substitute in chains. Furthermore, Fe and Co doping in the 1-2-3 phase has attracted much interest because these transition metal dopants, do not lower T_cs as drastically as previously noted in the low-temperature superconductors. Samples of $YBa_2(Cu_{1-x}Fe_x)_3O_{7-\delta}$ display superconductivity over a rather large alloying concentration range $0<x<0.15$ (Fig. 5) with a crystallographic phase change from orthorhombic ($x=0$) to tetragonal ($x \gtrsim 0.025$) taking place as a function of alloying concentration x [46,48]. Remarkably, the transition temperatures T_c remain continuous across this structural transition. The oxygen content ($7-\delta$) of the Fe-doped samples increases [45] as a function of x, indicating that the charge state of this metal dopant is certainly higher than the charge state of the host Cu cations.

Mössbauer spectroscopy. In principle, this method provides a powerful microscopic probe [49] of Fe dopant sites in $YBa_2(Cu_{1-x}Fe_x)_3O_{7-\delta}$ samples. In practice the lineshapes have revealed an unusually rich hyperfine structure [50-82] indicating that multiple sites are involved. Disentangling the possibilities has proved to be difficult because the spectra change drastically both as a function of Fe concentration x as well as oxygen content $7-\delta$. Blue and co-workers examined the spectral lineshapes as a function of oxygen content at a fixed Fe-doping level. To change the oxygen stoichiometry of the samples, Blue and co-workers vacuum annealed their virgin sample ($\delta \simeq 0$) at a pre-determined annealing temperature T_a for 3 hours and established the oxygen content by a thermogravimetric analysis The spectra (Fig. 4a) display qualitative changes in lineshapes as a function of oxygen content and, furthermore, these

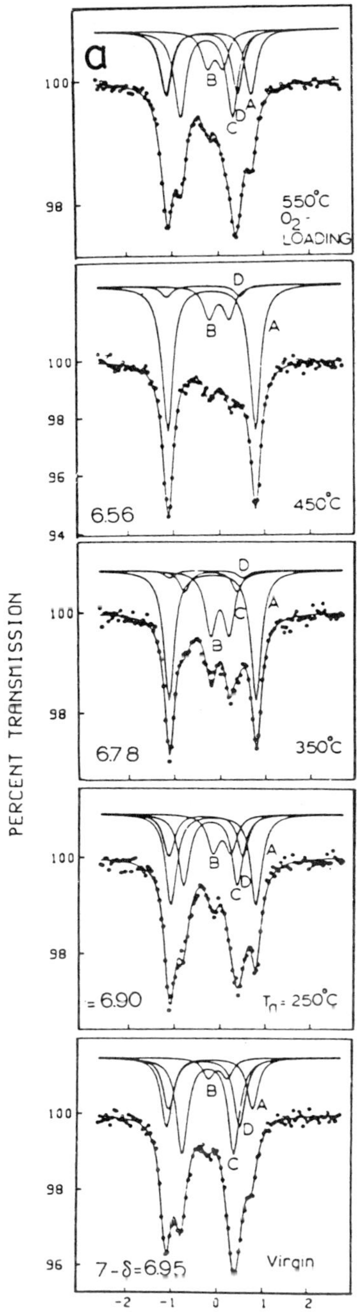

Fig. 4a (left) shows spectra of virgin and thermally annealed virgin $YBa_2(Cu_{0.985}Fe_{0.015})_3O_{7-\delta}$ samples. Thermal annealing was carried out in vacuo at indicated temperatures for 3 hours. Figure taken from ref. 50.

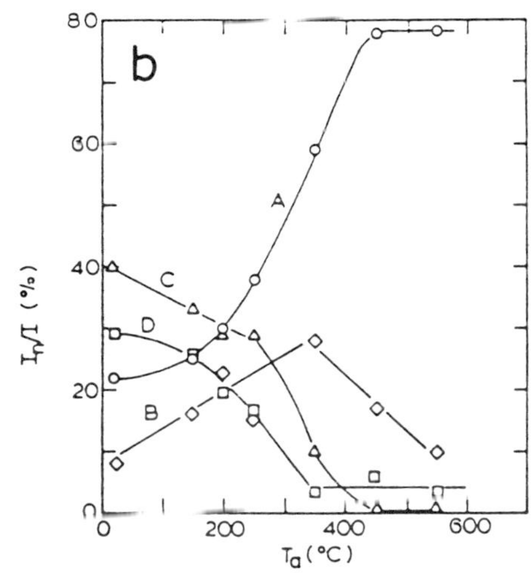

Fig. 4b shows the Mössbauer site intensity ratio I_n/I deconvoluted from the spectra of Fig. 4a plotted as a function of thermal annealing temperature T_a. Figure taken from ref. 50.

changes can be reversed upon oxygen absorption. To analyze the lineshapes, Blue and co-workers found it to be sufficient to include 4 sites (A, B, C, D), each a doublet, displaying no measurable intensity asymmetry (I_+/I_-) and having the same linewidth (Γ_{obs}=0.28 mm/s). Table I summarizes the isomer-shift and quadrupole splitting parameters deduced from a deconvolution of the virgin sample ($\delta \simeq 0$) spectrum, along with isomer shifts and quadrupole splittings of Fe^{4+} charge state reported in several prototypical oxides. The table also includes saturation (T=0K) magnetic fields for the various sites taken from results of other workers.

Table 1

Host	Isomer Shift* (mm/s)	Quadrupole Splitting (mm/s)	Saturation Magnetic Field (Tesla)	Ref.
$BaFeSi_4O_{10}$	+ 0.75	0.51	-	a
$BaFeO_3$	+ 0.03	0.0	24.0	b
$SrFeO_3$	+ 0.0	0.0	33.0	c
$Sr_{1/2}R_{3/2}Li_{1/2}Fe_{1/2}O_4$	-0.2 ~ -0.1	1.0~1.3	13~22	d
$YBa_2(Cu_{0.985}Fe_{0.015})_3O_{7-\delta}$:				e
Site A	+ 0.05(2)	1.90(5)	21.5[f]	
Site B	+ 0.20(2)	0.41(1)	50(2)[f]	
Site C	- 0.01(1)	1.14	24.3[f]	
Site D	- 0.11(2)	1.60	?	

*relative to Fe metal at 300K

a. M.G. Clark, M. Bancroft, A.J. Stone, J. Chem. Phys. 47 (1967) 4250. This mineral, Gillespite, contains high-spin Fe^{2+} in a square planar environment. The shift of Fe^{2+} increases ongoing to a tetrahedral environment and with an increase in CN. Although Fe^{2+} appears to be ruled out for the Fe sites in the Y-based 1-2-3, the assignment of an oxidation state from the isomer shift alone is not unambiguous.

b. T. Ichida, J. Sol. St. Chem. 7(1973) 308; Y. Takeda, M. Shimada, F. Kanamaru, M. Koizuuni, J. Sol. St. Chem. 7(1973) 229.

c. P.K. Gallagher, J.B. MacChesney, D.N.E. Buchanan, J. Chem. Phys. 41 (1964) 2429.

d. M.F. Thomas, G. Demazeau, M. Pouchard, P. Hagenmuller, Sol. State Commun. 39 (1981) 751; J.A. Birch and M.F. Thomas, J. Phys. C 19 (1986) 7071.

e. C. Blue et al. ref. 50

f. Q.A. Pankhurst et al. ref. 51.

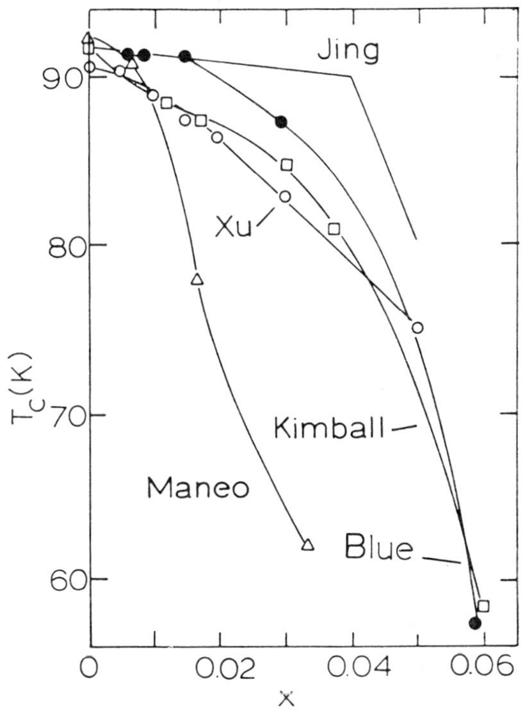

Fig. 5 T_c as a function of Fe concentration x in $YBa_2(Cu_{1-x}Fe_x)_3O_{7-\delta}$ samples taken from the work of several groups. Figure is taken from ref. 48.

Oxygen desorption leads to systematic changes in the site-intensity ratio (I_n/I) of the various sites (n=A, B, C and D). These are displayed in Fig. 4b. Sites C and D extrapolate to near zero intensity at T_a=450°C when the samples become tetragonal ($\delta \sim 0.5$). Furthermore, the reader may note that oxygen-deficient $YBa_2(Cu_{1-x}Fe_x)_3O_{7-\delta}$ samples usually display an increased site intensity ratio (I_A/I) of site A, the site with the largest quadrupole splitting. Indeed the sharp variation of the Mössbauer site intensity ratio I_A/I as a function of oxygen content (see Fig. 4a for oxygen deficiency δ established by thermogravimetry) provides a simple, non-destructive, and microscopic calibration of oxygen content in these oxide superconductors.

Spectra of $YBa_2(Cu_{1-x}Fe_x)_3O_{7-\delta}$ as a function of Fe doping level 'x' taken from ref. [48] are displayed in Fig. 6. Several other groups have reported [59,61,64] similar results. Given the isomershifts and quadrupole splittings of the 4 sites (A, B, C and D) at an Fe doping level of x=0.015, lineshapes at other Fe doping levels in the range 0<x<0.06 were analyzed keeping the Mössbauer parameters of the sites fixed to values indicated in Table I. Fig. 6 displays the deconvoluted site-intensity ratios I_n/I as a function of x. We have also shown in Fig. 6c a plot of the product I_n/I and x as a

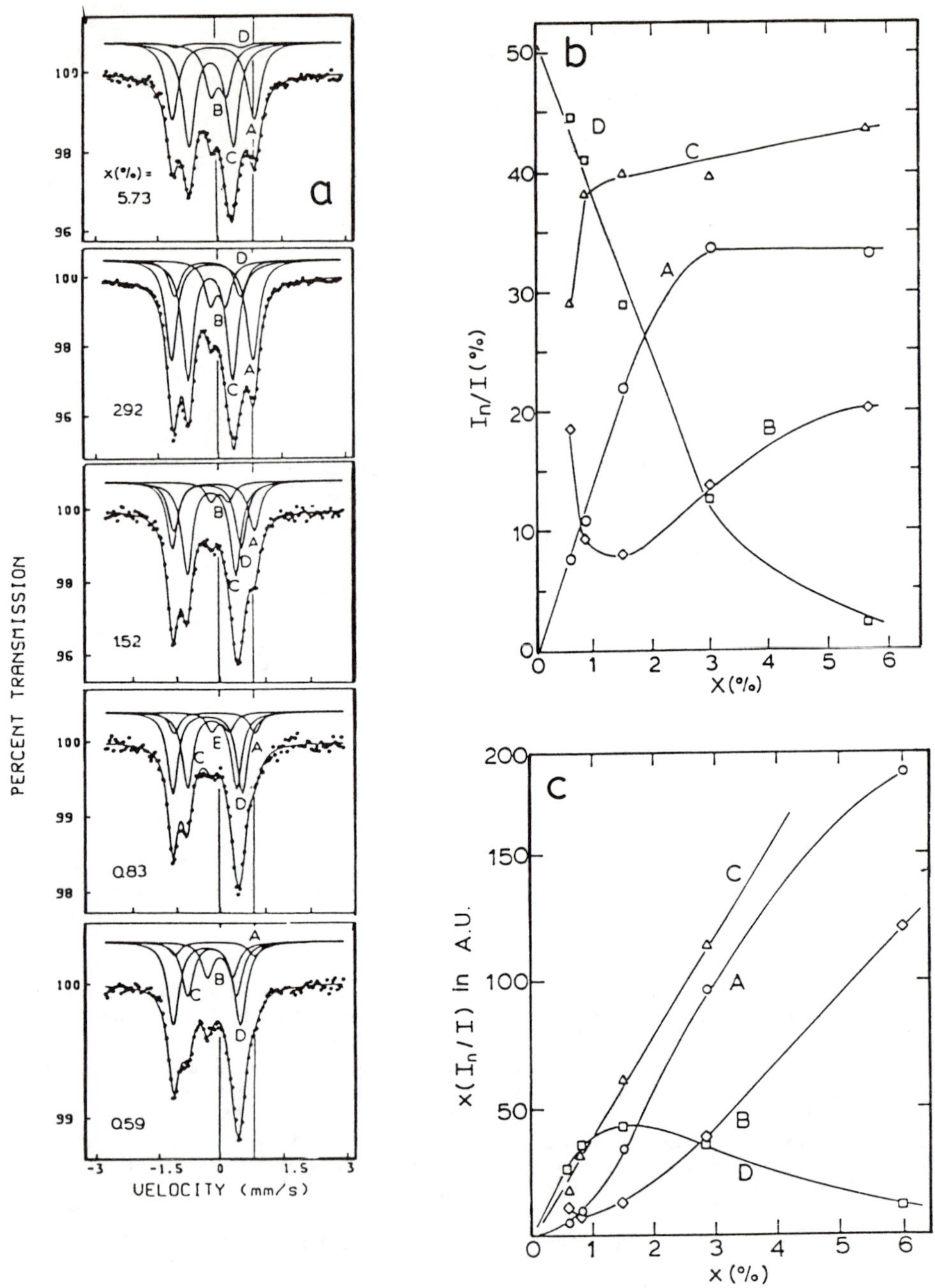

Fig. 6a (left panel) displays spectra of $YBa_2(Cu_{1-x}Fe_x)_3O_{7-\delta}$ samples at indicated Fe concentrations x in %. Fig. 6b (top right) shows the Mössbauer site intensity ratios I_n/I as a function of x. Fig. 6c (bottom right) shows the normalized Fe concentrations at various n=A,B,C and D sites affording a comparison at various x. Figure taken from ref. 48.

function of x. The ordinate on this plot is a measure of the relative Fe content present at various sites in these samples and permits an absolute comparison of site intensities for various x-values. Several features of the plot of Fig. 6c are worth noting. Fe dopant present in site D displays a maximum near x=0.015. Furthermore, Fe dopant present in sites C, A and B increases monotonically with x. A cursory glance at the $I_n/I(x)$ variation (of Fig. 6b) serves to demonstrate that at Fe doping levels x less than 0.01 or greater than 0.04, a 3-site fit to the spectra can be a good approximation to a 4-site fit largely because one of the 4 sites becomes weakly populated. Lineshape analysis in terms of 3-sites would, however, appear to be inadequate at doping levels in the range $0.01<x<0.04$ where all 4 sites appear to be populated. As noted by several groups [53,72], a 3-site fitting procedure then invariably leads to large quadrupolar doublet asymmetries in textureless samples and the width of the intermediate-sized quadrupole splitting (1.6 mm/s) doublet to be significantly larger than the widths of the other two doublets. These results are not peculiar to the Y-based 1-2-3 material but are a general feature of several RE-based 1-2-3 materials as well. Rich ^{57}Fe hyperfine structures are observed which can all be analyzed [79,80,81] in terms of 4 sites as shown in Fig. 7. We shall return to these observations later.

Before discussing the microscopic structure of the various sites A, B, C and D, we would like to present results of two other sets of experiments. One of these deals with spectra of $YBa_2(Cu_{0.985}Fe_{0.015})_3O_{7-\delta}$ samples taken as a function of oxygen loading ($\delta<0$) at elevated temperatures ($T_a \simeq 500°C$) and pressures ($P=10^7 Pa$) using a high pressure autoclave [82]. The results of this set of experiments appear in Fig. 8a and 8b as a function of loading time, at a fixed annealing temperature T_a and partial pressure of pure oxygen P_{O2}. X-ray θ-2θ scans of the loaded samples indicate an orthorhombic structure with steadily declining orthorhombicity as a function of loading time in the interval $0<t<48$ hours. At higher loading time intervals the integrity of the orthorhombic structure is lost and the crystal structure completely changes.

Perhaps the most significant result to emerge from the I_n/I trends of Fig. 8b is the complementary role of A- and C-sites: oxygen loading results in an increase of I_C/I at the expense of I_A/I, while oxygen desorption results in just the reverse behavior as noted earlier in Fig. 4b. These results rather conclusively demonstrate that one must assign a higher coordination number and charge state to C-site (Fe^{4+}) than to the A-site (Fe^{3+}), a point we shall return to in our discussion of the site microstructure. The second observation is that the ratio I_D/I displays a local maximum at an oxygen content of ~7. Increasing the oxygen content above 7 (Fig. 8) or decreasing the oxygen content below 7 (Fig. 4), in $YBa_2(Cu_{0.985}\ Fe_{0.015})_3\ O_{7-\delta}$ samples leads to a reduction in I_D/I. In fact the ratio I_D/I appears to be correlated to the degree of orthorhombicity (a-b)/(a+b) of the material and in particular I_D/I is qualitatively reduced once the material becomes tetragonal (either due to oxygen loading or due to oxygen depletion) leading to the natural suggestion that this site represents Fe impurity localized at twin-planes.

Mössbauer experiments on grain-aligned $YBa_2(Cu_{0.985}\ Fe_{0.015})_3O_{7-\delta}$ samples [71] have permitted establishing directly the sign, asymmetry parameter ($\eta = |V_{xx}-V_{yy}/V_{zz}|$) and orientation of the Electric Field Gradient (EFG) tensor at the 4 sites (A, B, C and D). The results of these experiments are particularly significant because these place severe restrictions over possible microstructures of the various sites. Details of the experiments appear in ref. 71 and in what follows we shall briefly outline the results.

Grain-alignment of $YBa_2(Cu_{0.985}Fe_{0.015})_3O_{7-\delta}$ powders in a magnetic field ($\vec{B}$) permits preparing samples with well defined texture. In this case it is known from x-ray measurements that the c-axis of the grains aligns parallel to the $\vec{B}$ field with the a and b axes oriented randomly in a plane perpendicular to $\vec{B}$. Mössbauer experiments

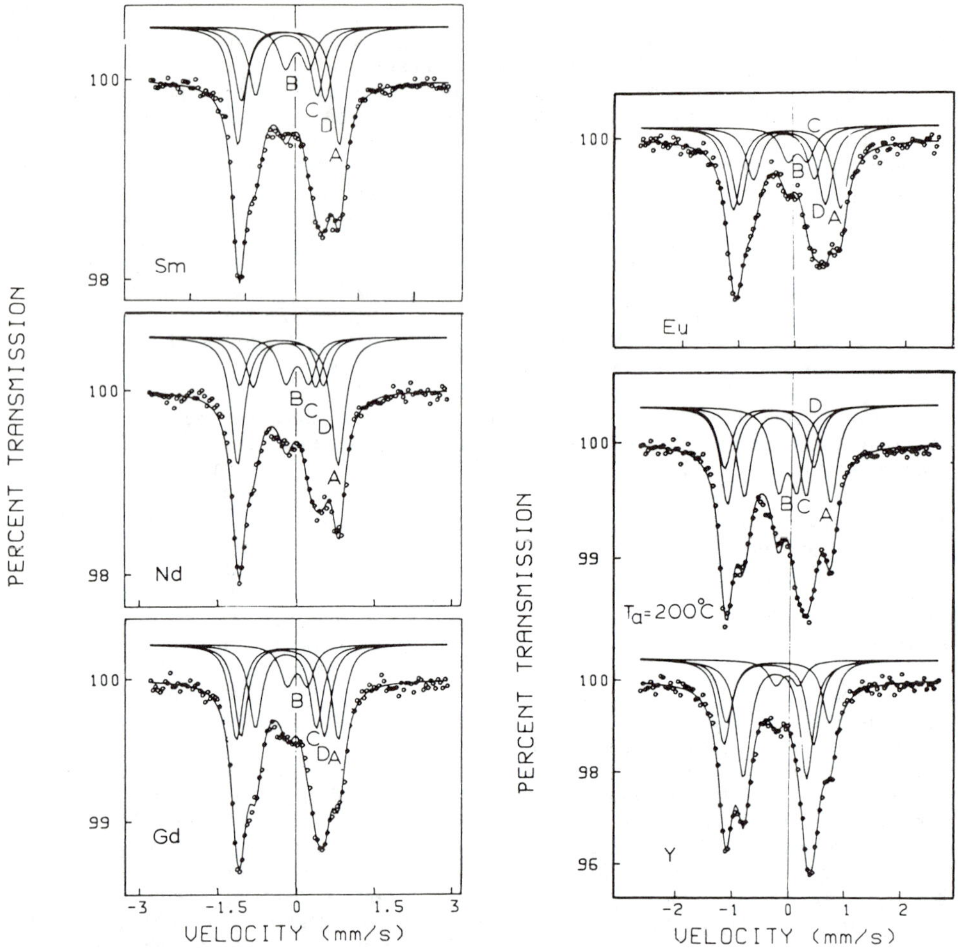

Fig. 7 Spectra of (Y or RE)$_1$Ba$_2$(Cu$_{0.985}$Fe$_{0.015}$)$_3$O$_{7-\delta}$ samples for indicated Rare-earths. In each case the lineshapes are analyzed in terms of 4 sites having values of the isomer shifts and quadrupole splittings which closely resemble those assigned to the Mössbauer spectrum of the Y-based compound. The oxygen content of the virgin samples, as indicated from the spectra, is Y>Gd>Eu>Sm>Nd. Except for the Nd-based sample, this is the same order as the values of b-a/b+a where a and b are the lattice constants reported by J.M. Tarascon, W.R. McKinnon, L.H. Green, G.W. Hull, and E.M. Vogel, Phys. Rev. B36 (1987) 226 (however the order of oxygen content of the samples as determined by Tarascon, et al., does not agree with the order indicated here). Note that the lineshape of a virgin Y-based sample that has been thermally annealed at 200°C, begins to have a resemblance to other RE-based virgin sample spectra. The unusual lineshape of the virgin Y-based sample is due to a strong population of C-site in such samples. Figure taken from ref. 81.

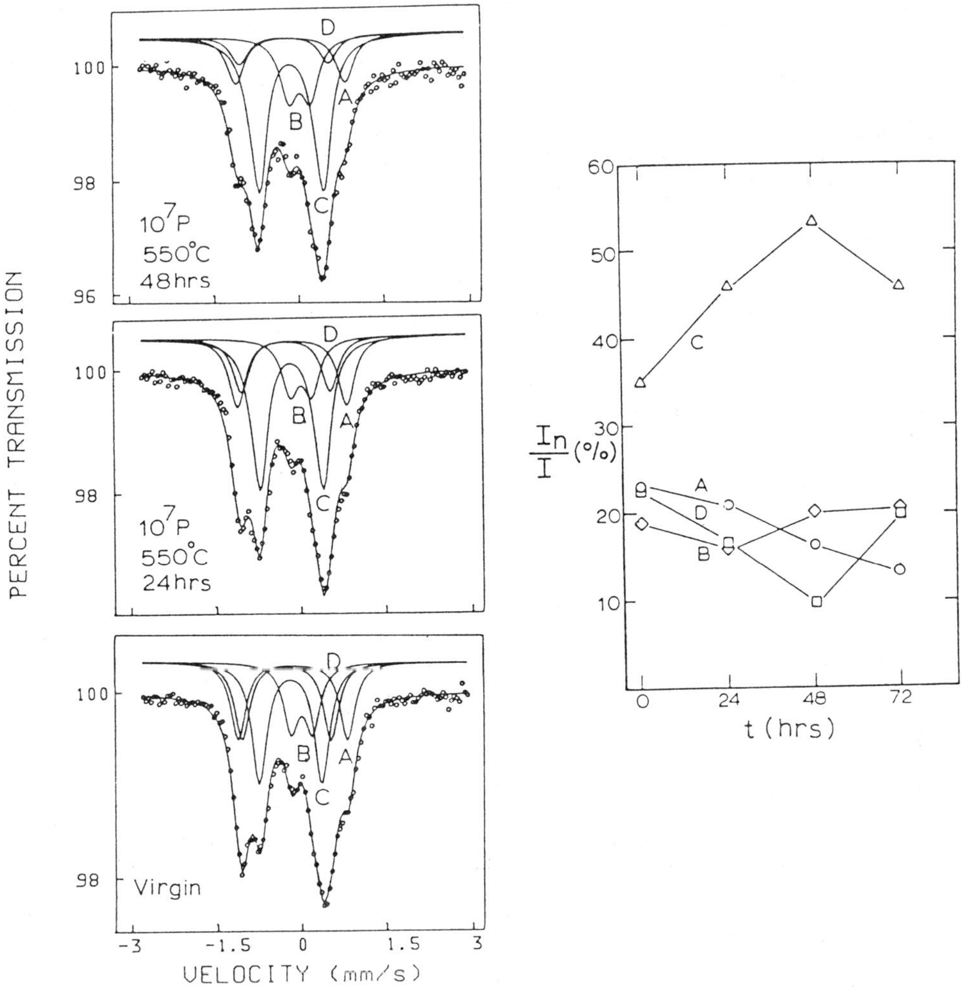

Fig. 8a (left panel) shows spectra of $YBa_2(Cu_{0.985}Fe_{0.015})_3O_{7-\delta}$ samples in the virgin state ($\delta \simeq 0$) and oxygen loaded state ($\delta < 0$). All oxygen loading was performed on the same sample at 10^7 Pascal and 550°C. After 24 hours the sample was removed from the autoclave, a Mössbauer spectrum obtained, and the sample returned for an additional loading period. Fig. 8b (right panel) shows Mössbauer site intensity ratio I_n/I as a function of loading time. After 72 hours of oxygen loading, the spectrum displayed a discontinuity caused by a structural change as indicated by x-ray powder diffraction. Figure taken from ref. 82.

performed systematically as a function of the tilt angle β between the transmitted γ-rays and the c-axis have permitted deconvoluting the quadrupolar doublet intensity asymmetry (I_+/I_-) of the various sites. Operationally, I_+/I_- is defined as the ratio of the integrated intensity of the positive to the negative velocity peak of a quadrupole doublet pair. Theoretically the ratio I_+/I_- is a manifestation of the radiation patterns of the two quadrupole components ($\pm 3/2 \xrightarrow{\pi} \pm 1/2$, $\pm 1/2 \xrightarrow{\sigma} \pm 1/2$) in the principal axis system (PAS) of the EFG tensor and is determined by the orientation of the PAS relative to the crystal set of axes and the nature of absorber texture [83]. In textureless samples, i.e. randomly oriented powders, I_+/I_- approaches 1, while in samples of specified texture, the ratio can be as large as 3 or as small as 0.6 depending on sample orientation and the EFG asymmetry parameter η.

Spectra at selected tilt angles β of samples in the virgin state (δ=0) and oxygen deficient state (δ=0.22) are reproduced in Fig. 9. It is quite apparent from these spectra that for site A, I_+/I_- is <1 at β=0 but becomes >1 at $\beta=\pi/2$. The β-dependence of I_+/I_- for the various sites, obtained from these spectra taken at several angles β in the range $0<\beta<\pi/2$, appear in Fig. 10. A cursory glance of the results in this figure clearly indicates that A, C and D sites display one type while B site displays another type of a behavior.

The smooth curves drawn in Fig. 10a are actually not a fit to the experimental results, but a plot of equation (1):

$$P_{\substack{\pi\\\sigma}}(\beta) = 1/2 \mp [(1-\eta)(3\cos^2\beta-1)]/[16(1+\eta^2/3)^{1/2}] \qquad (1)$$

This equation describes the angular distributions $P_{\substack{\pi\\\sigma}}(\beta)$ at a site having its EFG component V_{zz} in the a-b plane. We note that the overall change in the $P\pi/P\sigma$ ratio between the two extreme values of β declines with increasing value of the parameter η. In particular when η=1, the ratio $P\pi/P\sigma$ equals unity and becomes independent of β. It is quite clear that there exists a close correspondence of the observed intensity asymmetry $I_+/I_-(\beta)$ at sites A and C with the theoretical $P\pi/P\sigma(\beta)$ ratios deduced from equation (1) for η values of 0.2 and 0, respectively. These results imply that one must assign the positive velocity peak (I_+) to the π quadrupole component ($\pm 3/2 \rightarrow \pm 1/2$) and this leads unambiguously to a positive sign for the quadrupole interaction (e^2QV_{zz}) at sites A, C. Since the nuclear quadrupole moment of the $3/2^+$, 14.4 keV state of ^{57}Fe is known to be positive, it follows that the sign of V_{zz} at these sites is also positive.

The smooth curves drawn in Fig. 10b are plots of $P\pi/P\sigma(\beta)$ ratios deduced from equation (2) at several η values.

$$P_{\substack{\pi\\\sigma}}(\beta) = 1/2 \pm (3\cos^2\beta-1)/[8(1+\eta^2/3)^{1/2}] \qquad (2)$$

Equation (2) provides the angular distributions $P\pi_\sigma(\beta)$ for the circumstance when V_{zz} is directed parallel to the c-axis. In this case note that $P\pi/P\sigma(\beta=0) > P\pi/P_\sigma(\beta=\pi/2)$, a trend that is also displayed qualitatively by the ratio $I_+/I_-(\beta)$ at site B as shown in Fig. 10b. We consider these trends persuasive in suggesting that the B-site represents a site having its V_{zz} directed along the c-axis, as it would be if Fe substitutes Cu(2) sites. If such a site possesses a square pyramidal configuration, the local symmetry will require V_{zz} to be parallel to the c-axis. The lack of a quantitative agreement between $P_\pi/P_\sigma(\beta=0)$ and $I_+/I_-(\beta=0)$, in our opinion, is a reflection of the anisotropy of the Debye-Waller factor at such a site. One must remember that the anisotropy of atomic coordination at a Cu(2) site, requires the Debye-Waller factor along the c-axis (i.e., perpendicular to the buckled CuO_2 planes) to be lower than the one perpendicular to it. Since the $\pi(\sigma)$

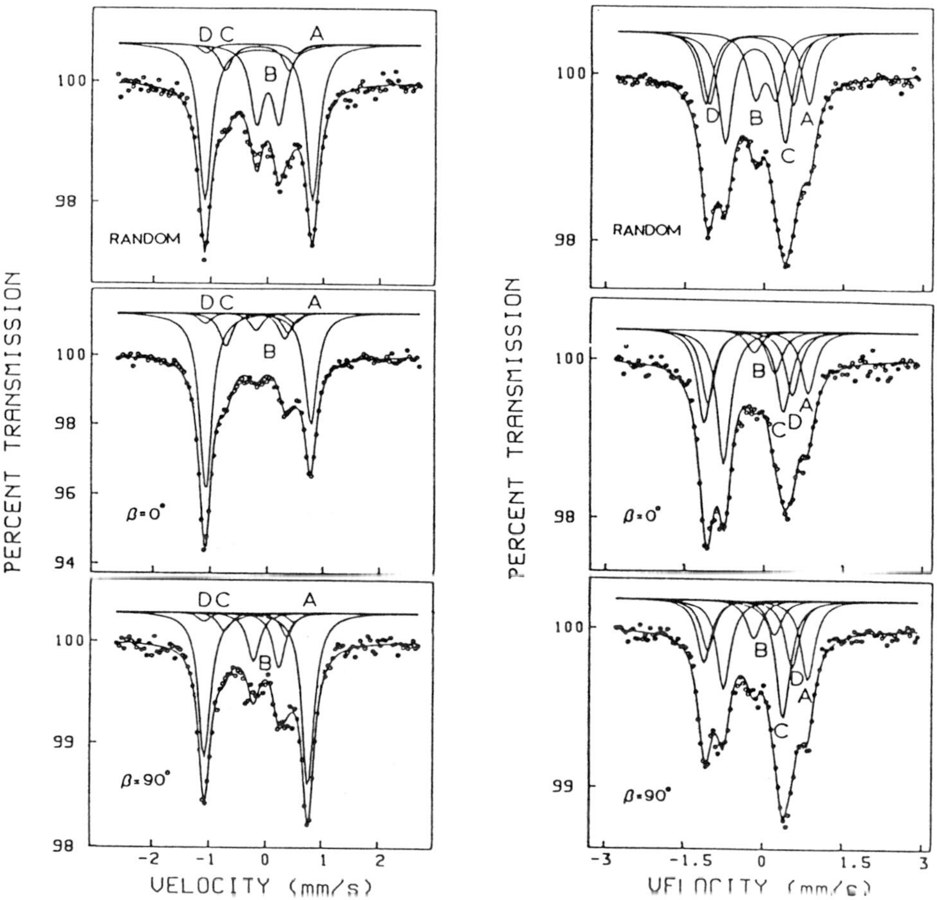

Fig. 9 Spectra of randomly oriented and grain-aligned $YBa_2(Cu_{0.985}Fe_{0.015})_3O_{7-\delta}$ samples in the virgin state (right panel) and oxygen desorbed state (left panel) at indicated tilt angles β. The tilt angle represents the angle between the γ-ray propagation direction and the c-axis of the grains. Note the quadrupolar doublet intensity asymmetries (I_+/I_-) emerging in the grain aligned samples. The I_+/I_- ratios for Fe in sites A, C, and D follow parallel trends as a function of tilt angle whereas the I_+/I_- value for the B site moves in the opposite direction. This gives an immediate qualitative indication that the highest symmetry axis for the B site is different from that for A, C or D sites. Figure taken from ref. 71.

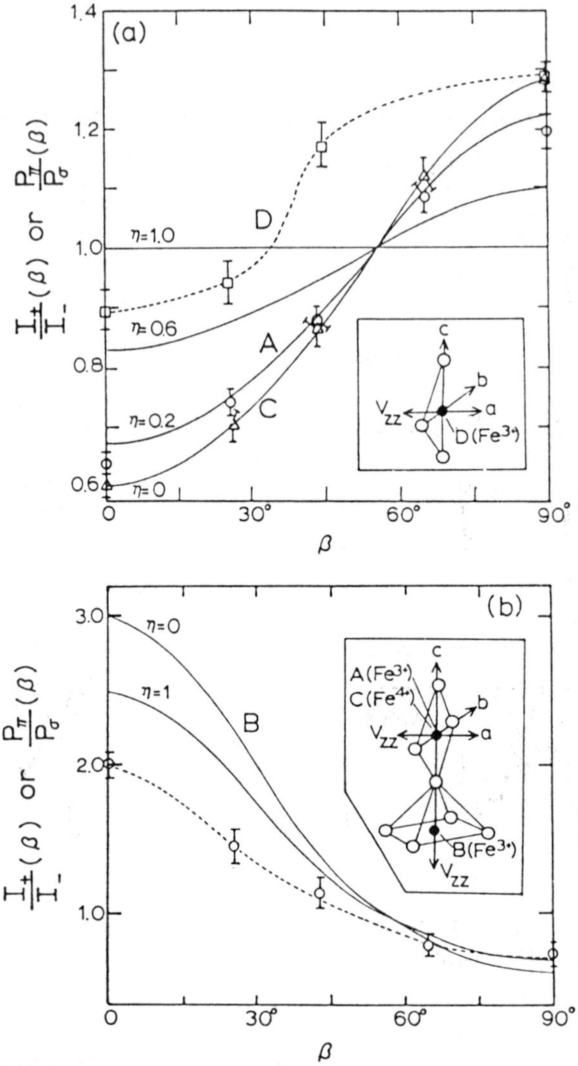

Fig. 10a (top) shows a plot of $P_\pi/P_\sigma(\beta,\eta)$ from equation (1) as a function of β for several η values. This represents the case when V_{zz} is directed in the a-b plane. Fig. 10b (bottom) shows a plot of P_π/P_σ (β,η) from equation (2), as a function of β for two η values and represents the case when V_{zz} is parallel to the c-axis. $I_+/I_-(\beta)$ for various sites is also plotted in the figure and show that while V_{zz} is ⊥ar to c-axis for sites A, C and D, it is along the c-axis for the B-site. Figure taken from ref. 71.

transitions preferentially probe the f-factors parallel (perpendicular) to the c-axis [83], one expects $f\pi/f\sigma < 1$. In general since the observed intensity ratio $I_\pi/I_\sigma(\beta)$ reflects the radiation patterns weighted by the anisotropy of the f-factor, one expects

$$I_\pi/I_\sigma(\beta) = P_\pi/P_\sigma(\beta)\ f_\pi/f_\sigma(\beta) \tag{3}$$

Note that if the ratio $f\pi/f\sigma(\beta=0) = 0.66$, equation (3) would be satisfied since $I_+/I_-(\beta=0)=2$ while $P\pi/P\sigma(\beta=0)=3$ for $\eta=0$.

As in the case of sites A, C and D, for site B as well, we must then identify the positive velocity component (I_+) with the π-component (Fig. 10b) and this leads to a positive sign for V_{zz} at this site as well. The oblate charge distribution due to the four oxygen near-neighbors in the basal plane primarily determines the sign of V_{zz} at those sites.

Microscopic structure of Fe-Doping sites in $YBa_2Cu_3O_{7-\delta}$. The foregoing body of experimental evidence, permits us now to discuss the site assignments. We will show that sites A, C, and D represent Fe present in CuO_3 chains while site B represent Fe present in buckled CuO_2 planes. This would suggest that a majority (or 90%) of Fe enters the 1-2-3 structure in chains and only a minority of ~ 10% in planes. If Fe dopant atoms were to randomly replace Cu cations in chains (Cu(1)) and planes (Cu(2)), one would expect the ratio of chain to plane sites to reflect the concentrations of the host cations Cu(1)/Cu(2) = 1/2. This is obviously not the case indicating that Fe dopant exhibits a strong preference to enter chains [51].

Site A. Quadrupole splitting: 1.90(5)mm/s, e^2qQ: + 3.8mm/s or 17.2 Mhz, η=0.20(5), V_{zz} positive and directed in the a-b plane. Isomer shift: + 0.05(2)mm/s relative to Fe metal at 300K. Saturation magnetic field B_{hf}: 21.5 Tesla.

This site is pronouncedly displayed in oxygen-deficient 1-2-3 samples. It is difficult to imagine on this basis that this site would be a high spin Fe^{4+} (S=2) as suggested elsewhere [54,68] to account for the smallness of the isomershift and B_{hf}. A more likely assignment is Fe^{3+} in the intermediate spin state (3/2) which could account for the B_{hf} in terms of the $-22\langle S_z\rangle$ Tesla rule.

The positive sign of V_{zz} directed in the a-b plane would clearly rule out an Fe site coordinated to two O(1) near-neighbors as found for the Cu(1) site in tetragonal $YBa_2Cu_3O_6$ and as proposed elsewhere [64] as the microstructure of site A in tetragonal $YBa_2(Cu_{1-x}Fe_x)_3O_{7-\delta}$ samples.

The suggested identification is Fe at Cu(1) site in a square planar configuration. This site will have a positive V_{zz} directed along the a-axis. ^{63}Cu NMR shows that the Cu(1) site in $YBa_2Cu_3O_{7-\delta}$ samples (δ~0) is characterized by a quadrupole coupling of 22Mhz with a large η~1, which has been ascribed [85-88] to the inequivalent roles of the O(1) and O(4) sites. The much smaller η~0.2 at Fe implies that the Fe-O(1) and the Fe-O(4) bond strengths nearly equalize; i.e., a local rearrangement of electronic charge on the pair of oxygen near-neighbors accompanies the change from $Cu^{2+}(d^9)$ to $Fe^{3+}(d^5)$ cation. This type of information would either need to be folded in calculations of the EFG parameters or would have to emerge from first-principles calculations of the EFG.

Site B. Quadrupole splitting: 0.41(1)mm/s e^2qQ: 0.82(2)mm/s, η unknown, V_{zz} parallel to c-axis, isomershift: 0.20(2)mm/s relative to Fe, Saturation Magnetic Field B_{hf}: 50(2) Tesla [51].

This is the least intense absorption in the spectra and its quadrupole coupling parameters e^2qQ and η are known rather imprecisely. In oxygen-deficient antiferromagnetic samples, this site displays a six-line magnetic hyperfine structure. Both the isomershift and saturation hyperfine field are characteristic of Fe^{3+} high-spin configuration.

The suggested identification is an Fe^{3+} high spin site replacing a Cu^{2+} site in the CuO_2 planes. Such a site will possess a positive V_{zz} directed along the c-axis. ^{63}Cu NMR experiments indicate an e^2qQ of 31.2Mhz and an η of 0.07(4) for the Cu(2) site [85-88]. The much higher quadrupole coupling at Cu in relation to Fe most likely is due to the presence of a d hole (d^9) at Cu but a half-filled d shell (d^5) at Fe which undoubtedly gives rise to a much larger covalent contribution to the EFG. We note that the underlying nuclear quadrupole moments, ^{57}Fe: $eQ(3/2^+)$ = 0.1897 barns, ^{63}Cu: $eQ(3/2)$=0.211 barns, are just about the same.

Site C. Quadrupole splitting: 1.14(1)mm/s e^2qQ: +2.28 mm/s or 10.3Mhz, η=0, V_{zz} directed in the a-b plane, isomershift: -0.01(1) mm/s relative to Fe metal at 300K, Saturation Magnetic Field B_{hf}: 24.3 Tesla [51].

The concentration at this site as observed in the spectra can be changed in a reversible manner in Y or RE based 1-2-3 samples by changing the oxygen content or the Fe content of such samples. Specifically increasing the oxygen content (7 or more) or Fe doping levels will result in an increased concentration at this site. Furthermore, the increase in site C concentration invariably appears at the expense of site A. This indicates that sites C and A are complementary to each other.

The suggested identification of site C is a quasi-octahedral chain site, possibly in Fe^{3+} intermediate spin or even Fe^{4+} state. Occupancy of the two O(5) sites at (a square planar) site A may be visualized as site C. It would also appear that the Fe-O(4) and Fe-O(1) bond strengths at such a site are equivalent, and their strength less than the Fe-O(5) bond strength. Such a coordination will give rise to a cylindrically symmetric (η=0) gradient having its symmetry axis directed along the a-axis; i.e., Fe-O(5) axis.

Site D. Quadrupole splitting: 1.60(1) mm/s e^2qQ: +3.20 mm/s, η~0.6, V_{zz} directed in the a-b plane, isomer-shift -0.11(2) mm/s relative to Fe at 300K, Saturation Magnetic field B_{hf}: unknown.

Experiments demonstrate that this particular site appears in the orthorhombic phase of Y or RE based 1-2-3 samples but is qualitatively suppressed in tetragonal ones, produced either by oxygen depletion, or oxygen loading or Fe doping. In $YBa_2(Cu_{1-x}Fe_x)_3O_{7-\delta}$ samples, our experiments demonstrate that the absolute concentration of this site exhibits a maximum at x~0.018 (see Fig. 6). Transmission electron microscopy (TEM) results on this system, recently reported by Y. Xu et al. [89]. demonstrate that the spacing between the (110) micro-twin planes exhibits a minimum of ~20nm at x=0.02. This indicates that the concentration of twin planes exhibits a maximum at this composition.

The correlation between Mössbauer spectroscopy and TEM is in harmony with the suggestion that site D represents an Fe dopant present at twin planes. Normally at a (110) twin plane one imagines a plane parallel to the c-axis across which the a-axis makes a 90° turn as illustrated in Fig. 11. The Cu(1) site coordination number at a twin-plane continues to be four as for the square planar site. The basic difference between the twin-plane associated Cu(1) site and the regular Cu(1) site however is one of geometry: a hop of an oxygen O(4) site to the vacant O(5) site will convert the square planar (regular Cu(1)) to a disphenoidal (twin plane Cu(1)) geometry. One may thus visualize the twin plane site as a defect chain-site. Fe replacement at a Cu(1) twin-plane site may yield a quasi-tetrahedral coordination if the Fe cation were positioned off-center. Neutron Bragg scattering experiments [90] do provide evidence for such a site. The poor local symmetry at such an Fe site would appear to be in harmony with the sizable η value suspected for this site.

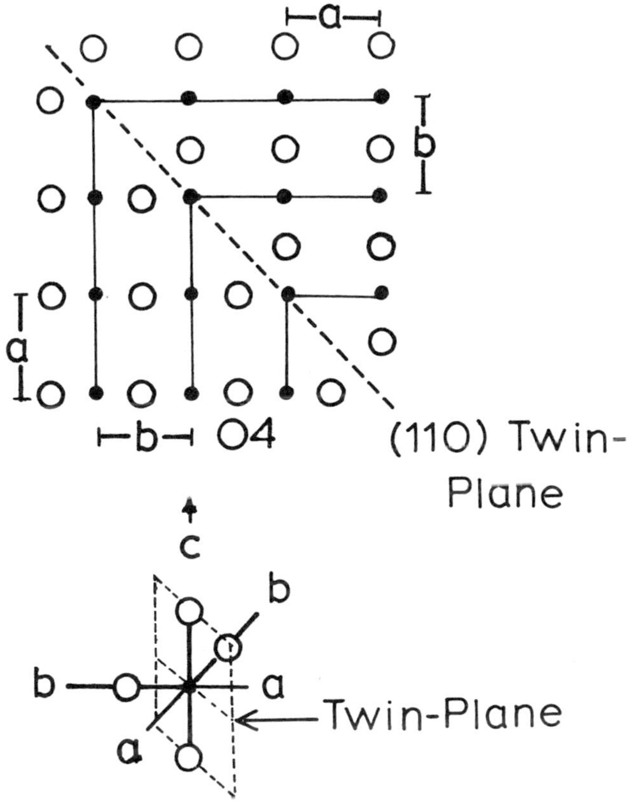

Fig. 11a (top) View along c-axis shows a (110) twin-plane. The CuO_3 chains abruptly turn by 90° at the dotted twin-plane giving rise to a Cu site having a trigonal pyramid coordination as shown in the bottom illustration (Fig. 11b). Fe replacing such a defect Cu(1) site can become quasi-tetrahedral by being offset from the center along the [110].

It is of interest to inquire how the microstructure evolves as a function of dopant concentration x, given the present Fe-site assignments. This is of interest because Bordet et al. [90] have examined in detail the microstructure of a tetragonal x=0.10 sample using both neutron and electron diffraction and have found evidence of strong intensity crosses at nodes located between reciprocal lattice layers. The latter is interpreted as evidence of linear clustering of Fe along [110] and [110] directions which are separated by domains of orthorhombic $YBa_2Cu_3O_{7-\delta}$.

Figure 6c provides the relative distribution of Fe-dopant amongst the various sites in the dopant concentration range $0<x<0.06$. These Mössbauer spectroscopy results may be used as a guide to extrapolate to $x \to 0.10$. Foremost, we note that $xI_D/I \to 0$ at $x \gtrsim 0.06$. The absence of site D, we have noted earlier, is compatible with the tetragonal symmetry prevailing in samples at these higher Fe concentrations where twin-planes are qualitatively absent. This would imply that tetrahedral Fe coordination (the twin-plane site) is absent at x=0.10.

We note that a smooth extrapolation of the trends shown in Fig. 6c from x=0.06 to $x \to 0.10$, suggests that the ratio $I_B : I_A : I_C \simeq 1:1:2$. Given that only the B-site is a planar-site, this would imply that about 75(10)% of the Fe dopant is present in chains, but not 100% as concluded by Bordet et al. [90] from a Rietveld refinement of their neutron scattering amplitudes. Here it would be pertinent to recall that Bordet et al. could not account for 20% of the Fe in their sample if they assigned all the dopant atoms to be present at Cu(1) chain-sites. In our view, this discrepancy in the neutron scattering results comes from not accounting for a small but finite amount of Fe present in the planes as suggested by the present Mössbauer results. The systematic lowering of T_c upon Fe-doping particularly once $x \gtrsim 0.015$, in our opinion, is the result of Fe systematically replacing Cu(2) sites in planes (growth in B site intensity in Fig. 6c). A cursive examination of the Fe-bearing linear chain-clusters proposed by Bordet et al. (Fig. 3 of ref. 90) indicates that one expects at least twice as many, if not more, of pyramidal (5-fold coordinated) Fe-sites than octahedral ones. From the Mössbauer results it is at present difficult to uniquely assign a coordination number to site C which could be 6 or 5.

It is, however, physically plausible that the growth in site C with x (see Fig. 6c) represents octahedral Fe segregating into [110] linear chains as suggested by Bordet et al. [90]. However to account for both the present Mössbauer results and Bordet's diffractions results, these Fe-containing linear-chains would have to be bordered by Cu-bearing and not Fe-bearing pyramidal units. Such a model would be chemically compatible with the bonding requirements of Fe choosing to be octahedral and Cu square-pyramidal. In such a model site A would then represent Fe replacing Cu(1) sites in $YBa_2Cu_3O_{7-\delta}$ grains residing between the Fe-rich linear-chains. One natural consequence of this linear-chain cluster model is co-existence of Fe-magnetic order in chains with superconductivity in CuO_2 planes at high $x(\gtrsim 0.05)$. Substantial experimental evidence to suggest this is available, as we shall discuss in Section II.

<u>Co dopant sites in $YBa_2(Cu_{1-x}Co_x)_3O_{7-\delta}$</u>. There is general recognition that Co-doping, much like Fe-doping in the Y-based 1-2-3 material leads to a crystallographic phase change from orthorhombic (at x=0) to tetragonal (at $x \simeq 0.033$) across which T_c remains continuous [45]. Mössbauer spectroscopy using ^{57}Co parent for the 14.4 keV gamma resonance in ^{57}Fe, provides an elegant method to probe Co dopant site occupancies in the cuprate superconductors even at rather miniscule concentrations ($x \simeq 10^{-5}$). Several groups have attempted these measurements by doping the tracer (^{57}Co) in solid state reacted $YBa_2Cu_3O_{7-\delta}$ samples. Since the carrier-free tracer is usually available as a $CoCl_2$ solution in dilute HCl, doping entails a conversion of the cobalt chloride to its oxide followed by a solid state reaction of the oxide with the cuprate superconductor. If ^{57}Co doping is carried out on the surface of a sintered pellet, the presence of texture on account of crystallite orientation will usually occur. The presence of texture will significantly alter the observed lineshapes as noted by Amar Nath et al. [70]. Analysis of the rich-lineshape in these circumstances becomes less unique, unless the degree of texture can be independently established by another technique. More unique fits and perhaps a better understanding of the spectra could result by using textureless samples obtained by merely crushing and randomizing crystallite orientation.

^{57}Co emission lineshape in a $YBa_2Cu_3O_{6.8}$ sample observed by Bottyan et al. [54] is rather similar to the one shown in Fig. 6a for a $YBa_2(Cu_{1-x}Fe_x)_3O_{7-\delta}$ sample at $x \simeq 0.008$. Broadly speaking, these results indicate that Fe and Co dopant site occupancies in this cuprate superconductor are rather similar. The unmistakable presence of the low quadrupole splitting site (B) is confirmed from the emission spectrum reported by Bottyan et al. This indicates that Co-dopant, much like the Fe-dopant site occupancy at Cu(2) planar-sites is small but finite ($I_B/I \sim 0.15$). The majority of the dopant in each case prefers occupancy at Cu(1) chain sites (~0.85).

Ion-implantation of ^{57}Co in these cuprate superconductors provides an attractive alternate doping route. Such a method overcomes difficulties associated with the chemical aspects of doping. Furthermore, if such implantations are performed at low enough temperatures so as to limit diffusion, site landing occupancies of the implanted ^{57}Co guest atoms should be random and at least for substitution at Cu sites reflect the ratio of Cu- cations in planes to chains of 2:1 of the host structure. Mössbauer experiments following ^{57}Co implantations could thus provide decisive tests of site-assignments suggested here.

II. MAGNETISM AND SUPERCONDUCTIVITY

Even before the discovery of the high-T_c superconductors, the rare earth ternary oxides were the subject of much theoretical and experimental activity concerning the possibility of the coexistence of magnetism and superconductivity [112]. Since the interaction of a magnetic center with a Cooper pair can destroy the pairing through a spin reversal of a pair member, the presence of atoms with magnetic moments generally destroys superconductivity. Accordingly, much excitement surrounds the occurence of magnetism and superconductivity in the cuprate superconductors where the underlying ordering temperatures (T_N and T_c) can be changed by two orders of magnitude by chemical doping. However, Cu-induced magnetism and superconductivity appear to be mutually exclusive. Both charge and spin degrees of freedom play a role in determining the ground states of the system. In this section we briefly review magnetism in these materials with emphasis on the results from Mössbauer studies.

$La_{2-2x}Ba_{2x}CuO_{4-\xi}$ samples display antiferromagnetism in the Ba composition range of $0<2x<0.016$, a spin glass state in the range $0.016<2x<0.05$, and superconductivity at $2x \gtrsim 0.05$ as illustrated in the magnetic phase diagram of Fig. 12a. The magnetism derives from the Cu^{2+} moments residing along the [110] direction which are antiferromagnetically coupled both in the a-c plane as well as along the b-axis as shown in Fig. 13.

The magnetic phase diagram of the prototypical Y-based 1-2-3 material is displayed in Fig. 12b. In this family of materials two distinct types of magnetic ordering have been observed, one associated with the Rare-earth (or Y) sublattice and the other with the Cu sublattice both Cu(1) as well as Cu(2)) sites. The latter is displayed in Fig. 12b where the two Neel temperatures observed in the oxygen-deficient (y>0.5) regime are ascribed to two distinct magnetic structures. The magnetic order at the higher temperature T_{N1} involves antiferromagnetic order of the Cu(2) moments in the buckled layers with no moment present on Cu(1) sites, while the one at the lower ordering temperature T_{N2} is ascribed to moments present on both Cu(2) and Cu(1) sites. These results have emerged from a wide variety of experiments including ^{139}La and ^{63}Cu NMR and NQR [91,93,94,95,100], μSR [98], magnetic susceptibility [92] and neutron scattering [96,97,99]. Among the interesting conclusions from the NQR studies are that for $YBa_2Cu_3O_7$ at room temperature all Cu atoms are in conducting environments, but for $YBa_2Cu_3O_{6.4}$ there are no conducting substructures [135]. For $YBa_2Cu_3O_{6.7}$ spin pairing occurs well above the T_c of 60 K [113]. The spin interactions of Cu(2) are mediated through the charge carriers.

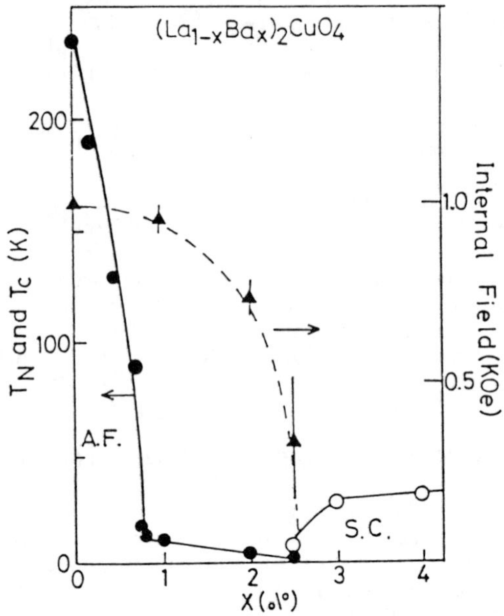

Fig. 12a shows magnetic phase diagram of $(La_{1-x}Ba_x)_2CuO_4$ as a function of 'x' in %

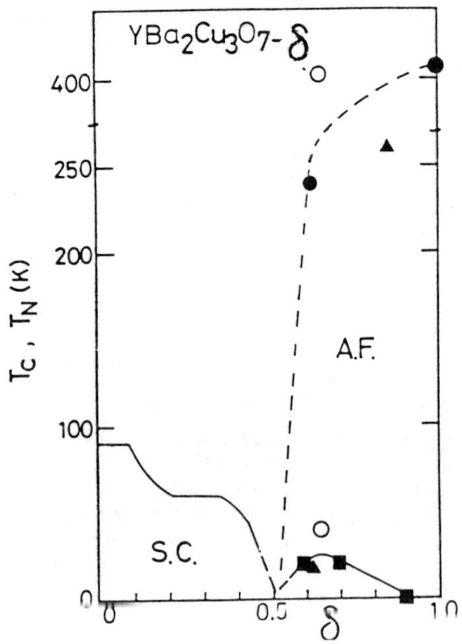

Fig. 12b shows magnetic phase diagram of $YBa_2Cu_3O_{7-\delta}$ as a function of oxygen vacancy concentration 'δ'. Note that in both cases, antiferromagnetism and superconductivity are mutually exclusive. Figures taken from Asayama et al. ref. 91.

Mössbauer spectroscopy has served as a useful probe of both Rare-earth and Cu-site magnetism, and this is a point we will discuss next with specific examples.

Rare-Earth Magnetism Co-exists with Superconductivity

Superconducting samples of the Gd-, Sm-, Er-, and Dy-based 1-2-3 materials display anomalies in calorimetric and susceptibilty measurements at a temperature T_N much below T_c. T_N is identified with onset of antiferromagnetism associated with the Rare-earth sublattice. Rare-earth magnetism thus appears to co-exist with superconductivity.

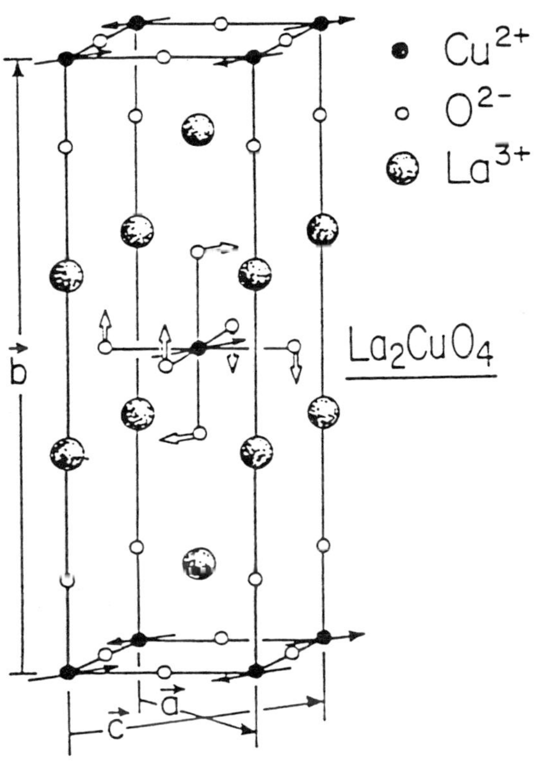

Fig. 13 Magnetic structure of La_2CuO_4 showing the Cu^{2+} moments along [110] antiferromagnetically coupled both in the basal plane (a-c plane) and along the b-axis. Figure taken from ref. 101

Temperature-dependent neutron scattering measurements [102] on $GdBa_2Cu_3O_{7-\delta}$ show that the scattering strength of the (1/2, 1/2, 1/2) Bragg peak grows precipiteously at 2.22(7)K as shown in Fig. 14. This is indicative of magnetic ordering. The temperature is identified as the Neel temperature of the Gd-sublattice with a moment of $7.4(6)\mu_B$ that lies along the c-axis. The (1/2, 1/2, 1/2) Bragg peak indicates that the magnetic unit cell is twice the chemical unit cell and the magnetic structure is illustrated in Fig. 15.

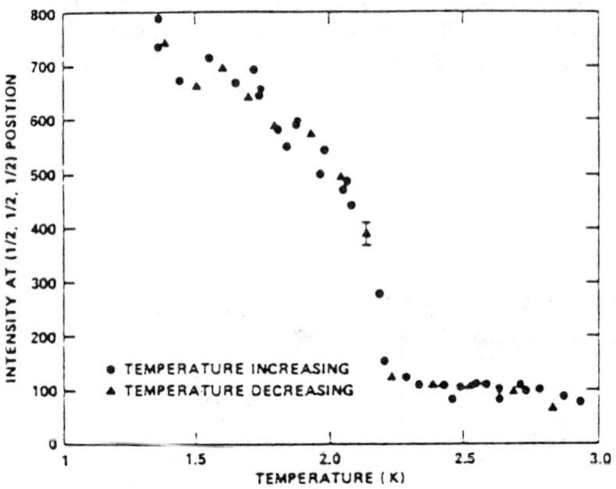

Fig. 14 Magnetic neutron scattering in $GdBa_2Cu_3O_{7-\delta}$ showing Bragg scattering of the (1/2, 1/2, 1/2) peak at $T=T_N=2.2K$. Figure is taken from ref. 102.

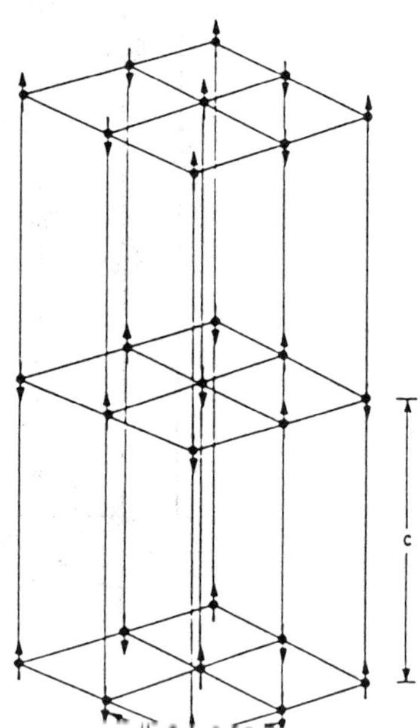

Fig. 15 Gd-sublattice magnetization in $Gd_2Ba_2Cu_3O_{7-\delta}$ deduced from neutron scattering measurements. Note that the magnetic unit cell is twice the size of the chemical unit cell. Figure taken from ref. 102.

^{155}Gd Mössbauer measurements display quadrupolar effects in the paramagnetic phase ($T>T_N$) but both quadrupolar and magnetic dipolar effects at $T<T_N$ as illustrated in Fig. 16a taken from the work of Bornemann et al. [103]. Several other groups [104-108] have confirmed these observations. Lineshape analysis of the spectra at $T<T_N$ shows that the angle θ between the EFG-V_{zz} and the magnetic field H_{int} is zero, as one would predict given that the Rare-earth moments lie along the c-axis, which is also the direction of V_{zz} imposed by the local site symmetry. The finite η value of the EFG is an indication of the lack of tetragonal symmetry about the c-axis.

The isomershift (0.5±0.03 mm/s relative to Gd^{3+} in $SmPd_3$ - the source matrix), the strength of the saturation (T=0K) magnetic field of 31.5 Tesla and the magnitude and sign of the quadrupole interaction unambiguously [104] assign a charge state of 3^+ for Gd. In the superconductor the realization of a magnetic field that is so close to the free ion (Gd^{3+}) value of 36 Teslas indicates that the interaction of the Rare-earth ion with conduction electrons is virtually absent [104]. The Gd^{3+} configuration leads to an $^8S_{7/2}$ ground state, indicating that the EFG has no valence contribution from the 4f electrons of Gd. The quadrupole interaction thus provides the crystalline Electric field parameters due to charges external to the Rare-earth ion, a quantity of general interest for other Rare-earth based 1-2-3 materials as well.

T-dependent μSR and ^{155}Gd Mössbauer spectroscopy experiments [108] in the antiferromagnetic phase (65mK<T<2.22K) of $GdBa_2Cu_3O_{7-\delta}$ have been performed to deduce the local magnetic field (B_{hf}). Surprisingly both probes yield a $B_{hf}(T)$ variation that is almost identical, which has led Dalmas de Réotier et al. to conclude that the low-field muon site observed in their experiments represents the muon localized interstitially either at (1/2, 1/2, 1/2) in the buckled layers or at (1/2, 0, 0) between the buckled layers (see Fig. 3). At these interstitial sites dipolar field vanish, and the muon probes the 4f magnetism of the magnetic Gd^{3+} ion at (1/2, 1/2, 1/2).

T-dependent ^{169}Tm Mössbauer spectra of $TmBa_2Cu_3O_{7-\delta}$ samples [109] at two specific oxygen stoichiometries (δ~0.1, $T_c\simeq$91K and δ ~ 0.4, T_c=58K) have been performed over the temperature range 0<T<600K by Gubbens et al. [109]. The results show that the magnitude of the quadrupole splitting exhibits a maximum at 91K for both samples. These authors conclude that the quadrupole splitting anomaly is crystal-field related and not superconductivity related.

^{170}Yb Mössbauer experiments on $YbBa_2Cu_3O_7$ have been performed in the paramagnetic and the antiferromagnetic phase. The results [110] shown in Fig. 16b clearly display the presence of a magnetic dipole interaction at $T<T_N$ which vanishes as $T>T_N$. Thus Mössbauer spectroscopy on the Rare-earth site can provide detailed microscopic information on the chemical and magnetic state of the cation in superconducting 1-2-3 materials. This information often complements results of macroscopic measurements - calorimetric and susceptibility [111]. The results demonstrate that the coexistence of RE-magnetism and superconductivity is due to the isolation of the 4f-valence electrons on the RE site from the d-like charge density localized on the buckled CuO_2 layers and CuO_3 linear chains.

The co-existence of RE magnetism and superconductivity independent of each other in these HTSCs is reminiscent of a parallel experience noted previously in two other classes of superconductors, the $RERh_4B_4$ [112] and $REMo_6S_8$ [113] systems. In each of these three instances, it appears that the magnetism derives from f-electron wavefunctions localized on the RE-ions while superconductivity results from the 'd' electron wavefunctions of Cu, or Rh or Mo. The weak overlap of 'f' and 'd' electron wave functions because of the physical remoteness of the RE-ions from the transition-metal-ions in respective crystalline structures is thought to be the reason why RE-magnetism and superconductivity can co-exist quite independently of each other in these materials.

Rare-earth sublattice magnetic ordering in the superconducting 1-2-3 materials do display variations, being 3-d in the Dy-based [114] and Gd-based materials [102] and both 3-d and 2-d in the Er-based materials [115]. Kistenmacher [116] has noted that Neel temperatures of various RE bearing 1-2-3 crystallites (Fig. 17) scale with the deGennes factor $(g_J-1)^2\ J(J+1)$ and changes in both these physical observables are correlated to the RE radii. Furthermore, since the RE sublattice magnetization is independent of electrical conductivity (i.e., whether the materials is a superconductor or semiconductor) it has been suggested that the coupling of RE moments is better understood as resulting from dipolar and superexchange via O(2) and O(3) sites rather than a Ruderman Kasuya Kittel Yosida (RKKY) exchange mediated via conduction electrons.

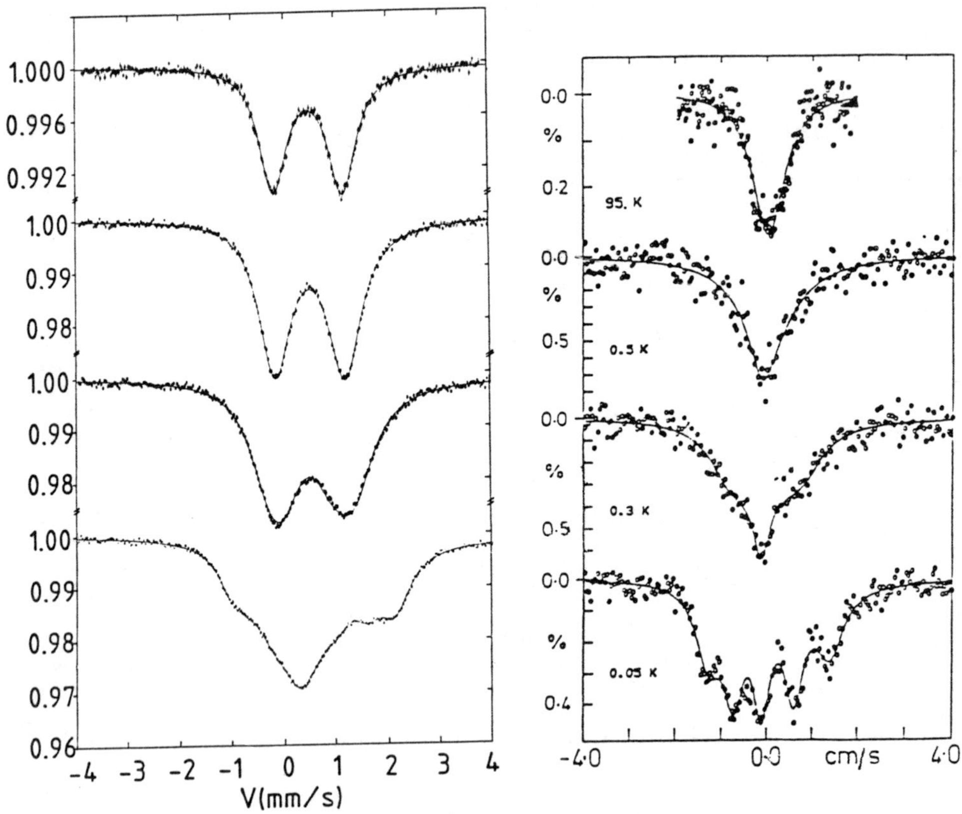

Fig. 16a (left) shows ^{155}Gd spectra of $GdBa_2Cu_3O_{7-\delta}$ taken above and below the Neel temperature T_N=2.22K, and indicate the presence of Zeeman splitting at T=1.5K<T_N as expected. The figure is taken from ref. 103. Fig. 16b (right) shows ^{170}Yb spectra of $YbBa_2Cu_3O_{7-\delta}$ taken above and below the Neel temperature T_N=0.4K and shows Zeeman splitting at T<T_N as expected. This figure is taken from ref. 110.

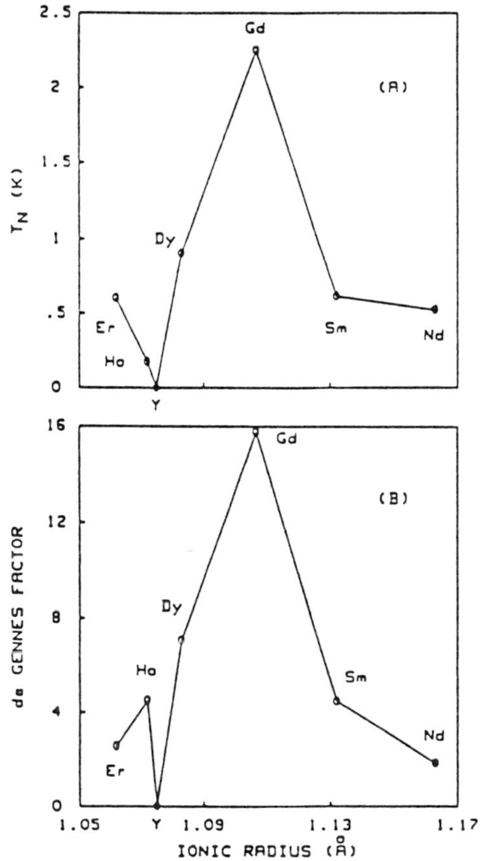

Fig. 17 The correlation between T_N and deGennes Factor as a function of Rare-earth radii in (Y or RE)$_1Ba_2Cu_3O_{7-\delta}$ superconductors. This figure is taken from ref. 116.

Cu-Magnetism in $La_{2-2x}Sr_{2x}CuO_{4-\xi}$

^{57}Fe Mossbauer spectroscopy results in the present family of superconductors has been used by Imbert, Jehanno and Hodges [76] to probe in detail the antiferromagnetic, spin glass and superconducting phase. An intrinsic appeal of the present system is the fact that the structure permits only one type of Cu site, that is, quasi-octahedral. Fe as a dopant can be expected to replace this particular site and display a quadrupole interaction because of the non-cubic local coordination noted earlier. Indeed spectra of

Fe-doped $La_2CuO_{4-\xi}$ samples taken at $T>T_N=220K$, Fig. 18 display a quadrupole doublet (1.60 mm/s). The isomershift of the doublet (~0.30 mm/s relative to Fe metal at RT) can be assigned to Fe^{3+} in a high spin state. The decisive evidence that the dopant replaces the Cu^{2+} site in the host material emerges from Mossbauer spectra taken at $T<T_N$ as we discuss next.

Spectra taken in the magnetically ordered phase (Fig. 18) display evidence of a magnetic dipole interaction in addition to the quadrupole interaction. Analysis of the spectral line shapes reveal that the angle θ between H_{hf} and the direction of the Electric Field Gradient V_{zz} is 90°. This is a significant result because it is signature that Fe guest mimics the local symmetry and magnetism of Cu in the tetragonal structure. The local 4mmm symmetry at Cu, requires the symmetry axis of the EFG to reside along the c-axis. Furthermore if the magnetic field aligns in the basal plane along the [110] direction, (i.e, the direction of the Cu^{2+} moments in the pristine material), the angle θ between V_{zz} and H_{hf} is expected to be 90°. Both the isomer-shift and magnetic field strength (~ 45 Tesla) of the dopant suggests that Fe enters as Fe^{3+} in a high-spin state in La_2CuO_4.

Two atom percent Sr-alloying leads to a pronounced reduction in the Neel temperature (Fig. 19) but not the magnetic field strength (H_{hf}) itself. Furthermore, the direction of the magnetic field continues to reside in the basal plane as the angle θ=90°. In the Sr concentration range of $0.02 < y < 0.07$, it is thought that a spin-glass state characterized by a rather low ordering temperature $T_m \simeq 10K$ persists in which the 2-d antiferromagnetic order in the CuO_2 planes is held intact, but the interplanar coupling giving rise to a 3-d order (that prevailed at $x \lesssim 0.02$) is apparently lost. The magnetic phase diagram of the $La_{2-y}Sr_yCuO_4$ system deduced from these Mössbauer experiments is shown in Fig. 19 and is analogous to the phase diagram of $La_{2-2x}Ba_{2x}CuO_4$ discussed earlier in Fig. 16a.

At higher alloying concentrations of Sr ($0.07 < y < 0.12$) the system becomes a superconductor and interestingly a magnetic hyperfine structure still persists at temperatures $T \lesssim T_m$ where $Tm \simeq 8K$. This result does not appear to be an Fe-impurity induced effect. It may suggest the presence of short-range antiferromagnetic correlations in the superconducting state. Indeed, it is possible that the finite H_{hf} detected in recent Muon spin rotation experiments on this superconductor [117] may be consistent with the existence of some short-range antiferromagnetic correlations.

Cu-Magnetism in $YBa_2(Cu_{1-x}Fe_x)_3O_{7-\delta}$

Neutron scattering [96,97], Cu^{63}-NMR [85-88,93,94] and μSR [98] experiments on single crystals and grain-aligned crystallites of the pristine (x=0) material clearly demonstrate that while antiferromagnetism occurs in oxygen-deficient samples ($\delta \gtrsim 0.5$) superconductivity occurs in oxygen-rich ($\delta \lesssim 0.5$) samples. Thus even though magnetism may not be the origin of pairing in these materials, the intriguing possibility that magnetic fluctuations play some role in the origin of superconductivity cannot be excluded.

Antiferromagnetic Ordering of Cu(2) sites at T_{N1}. There is broad recognition that oxygen-deficient (δ=1) 1-2-3 samples order antiferromagnetically with a Neel temperature T_N~400K. The magnetic structure revealed by neutron-scattering [118] experiments shows that Cu-sites in CuO_2 planes possess moments that are coupled antiferromagnetically in the a-b plane. The heavy twinning in such crystallites has precluded establishing the orientation of the moments, however. Figure 20 shows the

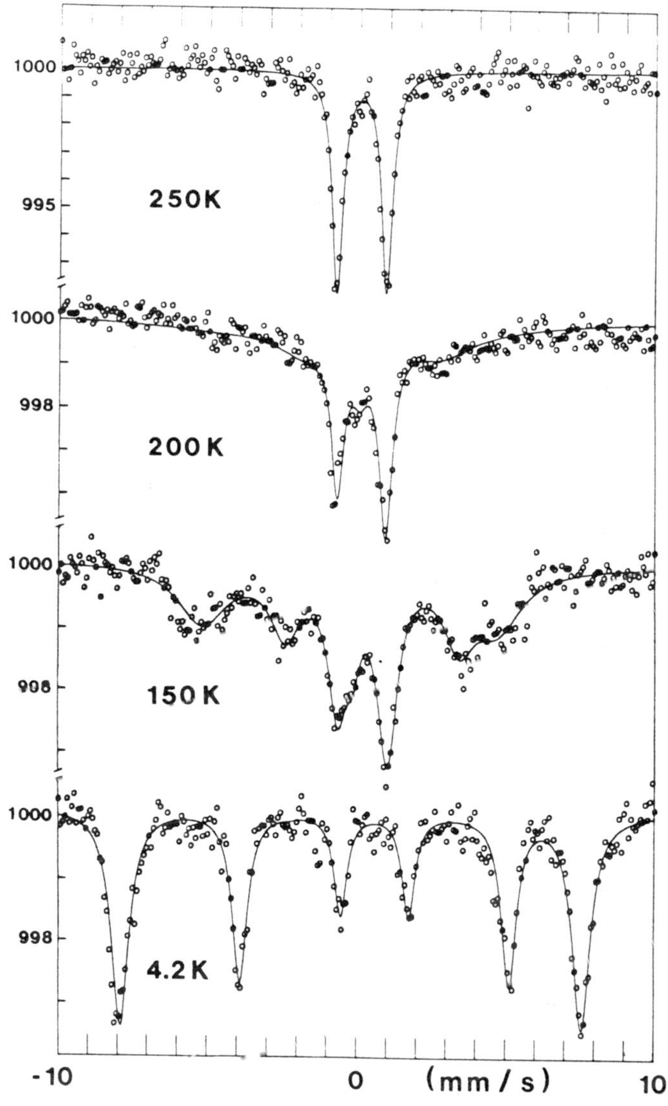

Fig. 18 ^{57}Fe spectra of $La_2Cu_{0.995}Fe_{0.005}O_4$ taken as a function of temperature and exhibit evidence of magnetic hyperfine splitting at $T \lesssim T_N$ = 220K. Figure is taken from Imbert et al. ref. 76.

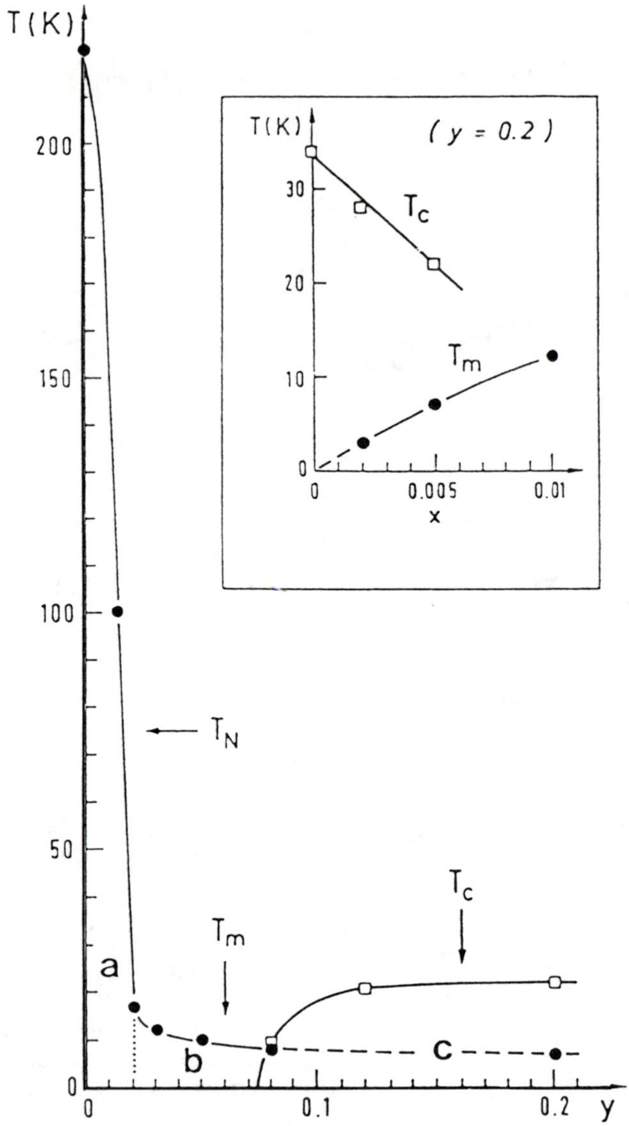

Fig. 19 Magnetic phase diagram of $La_{2-y}Sr_yCuO_4$ deduced from ^{57}Fe Mössbauer spectroscopy. The inset shows the effect of Fe dopant concentration x on T_c and T_m in $La_{1.8}Sr_{0.2}Cu_{1-x}Fe_xO_4$ samples. Figure is taken from Imbert et al. ref. 76.

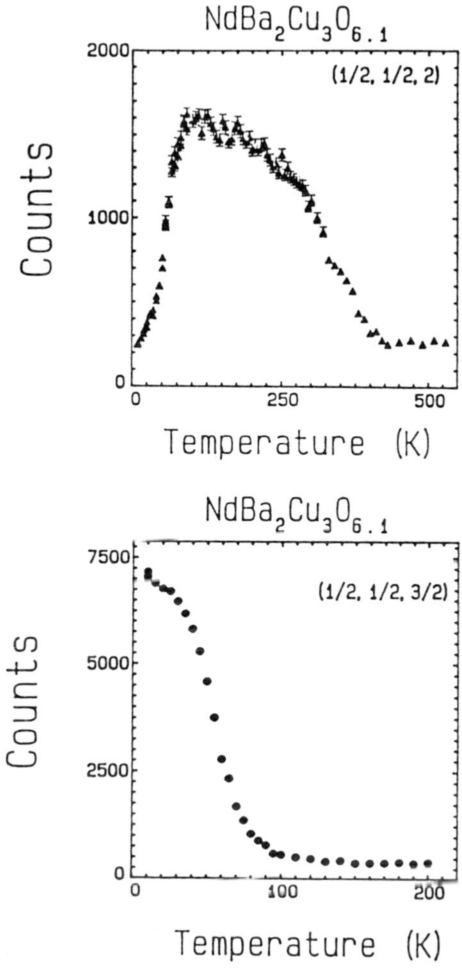

Fig. 20 shows Neutron scattering results on $NdBa_2Cu_3O_{6.1}$. Specifically, the strength of the (1/2, 1/2, 2) Bragg peak is found to increase (top) at $T \lesssim 430K = T_{N1}$ and to decrease at temperatures $T \lesssim 80K$. At $T \lesssim 80K$ concomitantly, the strength of the (1/2, 1/2, 3/2) peak increases. The lower temperature is identified as T_{N2}. The magnetic structures prevailing at T_{N1} and T_{N2} are illustrated in Figs. 21 and 24. The figure is taken from ref. 119.

T-dependence of the Magnetic Bragg scattering of the (1/2, 1/2, 2) reflection in $NdBa_2Cu_3O_{6.1}$ [119], at T_N=430(5)K. The corresponding magnetic structure is illustrated in Fig. 21.

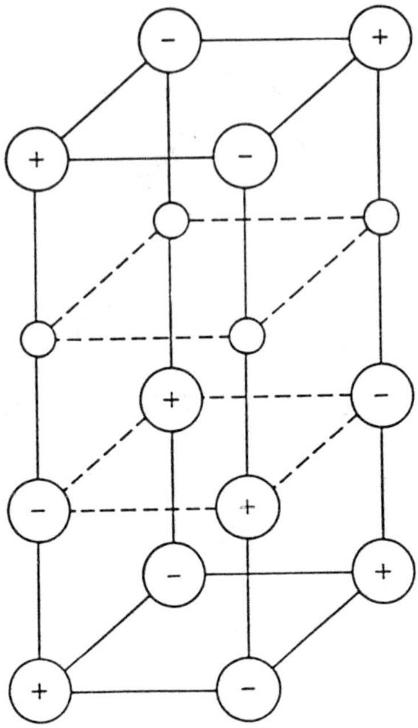

Fig. 21. Magnetic structure of $NdBa_2Cu_3O_{6.1}$ at T_{N1}=430K. Cu(1) sites indicated as small circles posses no ordered moments. Cu(2) planar sites couple antiferromagnetically in the a-b plane and along the c-axis. The magnetic unit cell length equals the crystallographic unit cell along the c-axis, but it is twice as large along the a- and b-axis. Figure taken from ref. 119.

Mössbauer Spectroscopy. Several groups [54,55,58,59,62,64,65,120,121] have intensively examined both the superconducting ($0<\delta<0.5$) as well as the magnetically ordered semiconducting ($0.5<\delta<1.0$) phase of Y-based 1-2-3 samples as a function of Fe concentration (0<x<0.15). Even though the sintered samples used have not always been characterized for oxygen and Fe content, lattice parameters, T_c, homogeneity and texture, a perusal of the literature, surprisingly, reveals a broad unanimity of the rather rich spectral lineshapes. Lineshape analysis of the spectra, interpretation of the results and particularly the underlying physical consequences remain in an evolutionary stage. The results appear to demonstrate that at low-doping concentrations ($x\lesssim0.04$), Fe mimics the magnetism of Cu sites in the 1-2-3 material as it does in the 2-1-4 material. At higher Fe doping levels ($x\gtrsim0.04$), the possibility that co-existence of superconductivity and magnetism may intrinsically be due to the inhomogeneous character of such samples is likely.

Spectra in the magnetically ordered phase (δ=1) taken at room temperature, i.e. $T<T_N$, display unambiguous evidence [55,62,65,73] of a six-line Zeeman split site characterized by a magnetic field H_{hf} (T=0) $\simeq$ 50 Teslas (Fig. 22). It is interesting to note that low-temperature $T_a \simeq 400°C$ oxygen anneal of such oxygen-deficient samples restores superconductivity as $\delta \to 0$, and concomitantly destroys antiferromagnetism. As noted elsewhere [62], in the superconducting phase, H_{hf} vanishes and the B-site doublet reappears unless the temperature of the sample is lowered below 1.3K or so, a point we shall return to later. These observations are readily understood in terms of oxygen O(4) chain-sites becoming vacant upon oxygen desorption and inducing antiferromagnetic ordering at Cu(2) planar-sites. Fe replacing Cu(2) sites in CuO_2 planes then experience a magnetic dipole interaction giving the magnetic hyperfine structure. Just the reverse sequence of events occur upon oxygen absorption. At low concentrations the guest dopant merely probes the local field at such a site. The magnetic field strength and isomer-shift suggest an Fe^{3+} high spin state for the B-site [51,62,64,68]. These results confirm the B-site assignment advanced earlier in section I from grain-aligned specimen (see Table I).

There exists a close analogy in the hyperfine structure of the B-site in $YBa_2(Cu_{1-x}Fe_x)_3O_{7-\delta}$ and the Fe dopant site in $La_{2-y}Sr_y CuO_4$. In both cases V_{zz} is positive and directed along the c-axis. In the antiferromagnetic phase, i.e. y<0.02 and δ>0.5, $H_{hf} \sim 50$ Tesla is observed which is characteristic of Fe^{3+} high spin with the angle θ between V_{zz} and H_{hf} of 90°. A perusal of the six-line spectrum of both these sites (see Fig. 18 for La_2CuO_4 and Fig. 22 for $YBa_2(Cu_{1-x}Fe_x)_3O_{7-\delta}$), shows that the spacing between the two right-most members(5-6) is smaller than the spacing between the two left-most members (1-2) of the 6-line spectrum. As noted earlier, θ=90° is in harmony with the local field directed in the basal plane as would be required by the known antiferromagnetic structures.

T-dependence of H_{hf} in the antiferromagnetic phase (δ=1) has been studied by several groups [55,65] and shows that $H_{hf}(T)$ at $T=T_N=423K$ vanishes as shown in Fig. 23. At low doping levels, the Fe-magnetism is not dopant induced but merely reflects the local Cu-magnetism of the host. Nasu et al. [65] have shown that a plot of $(H_{hf})^2$ versus $T^{3/2}$ displays a linear behavior which provides evidence for spin fluctuations in a weakly coupled antiferromagnet.

Antiferromagnetic ordering of both Cu(1) and Cu(2) sites at T_{N2}. Neutron scattering experiments reveal evidence of a new type of magnetic order in semiconducting ($\delta \gtrsim 0.5$) 1-2-3 materials at low temperatures $T \lesssim 80K$ [119]. This phase transition is associated with a new type of magnetic structure in which both Cu(1) chain sites and Cu(2) plane sites are antiferromagnetically ordered with the moments residing in the a-b plane. Proceeding along the c-axis, moments in neighboring Cu layers alternate their direction from layer to layer. Furthermore, since all three Cu-layers of the unit cell form part of the magnetic structure, the magnetic unit cell becomes twice the length of the chemical unit cell along the c-axis. Figure 24 displays the low-T magnetic structure. Thus the signature of ordering of both Cu(1) and Cu(2) sites at low temperatures and only Cu(2) sites at elevated temperature is the growth in the intensity of a half-integral (h/2, k/2, 1/2) reflection at the expense of an integral (h/2, k/2, 1) reflection as the temperature is lowered.

Since magnetic Bragg scattering is directly proportional to the sublattice magnetization, one can extract from the temperature-dependence of the scattering, not only the ordering temperatures but also the size of magnetic moments at Cu(1) and Cu(2) sites. For a $NdBa_2Cu_3O_{6.1}$ single crystal, neutron scattering results [119] are displayed in Figure 20. Note the growth in scattering of the (1/2, 1/2, 2) reflection at

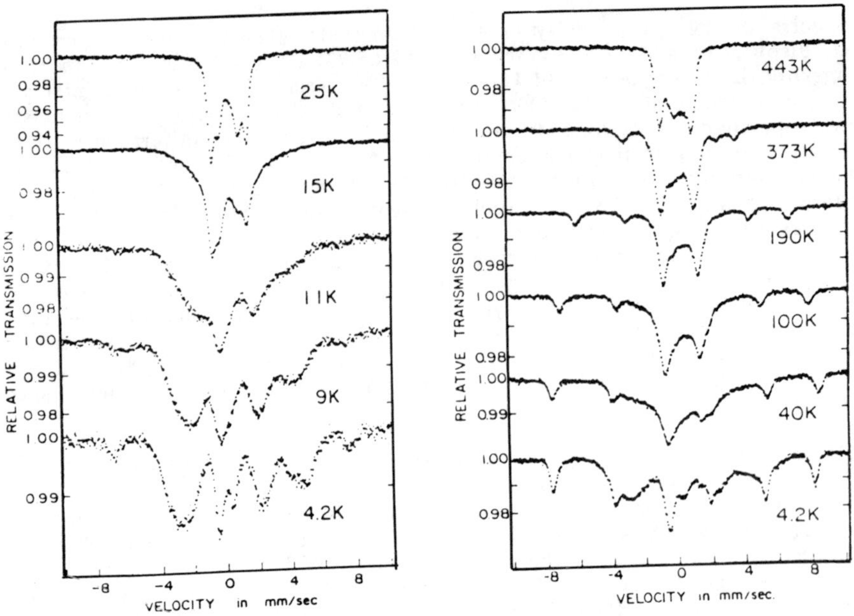

Fig. 22 Spectra of a $YBa_2(Cu_{0.92}Fe_{0.08})_3O_{7-\delta}$ sample taken at indicated temperatures in the superconducting (δ~0) state (left) and semiconducting (δ~1) state (right). Spectra are taken from ref. 62.

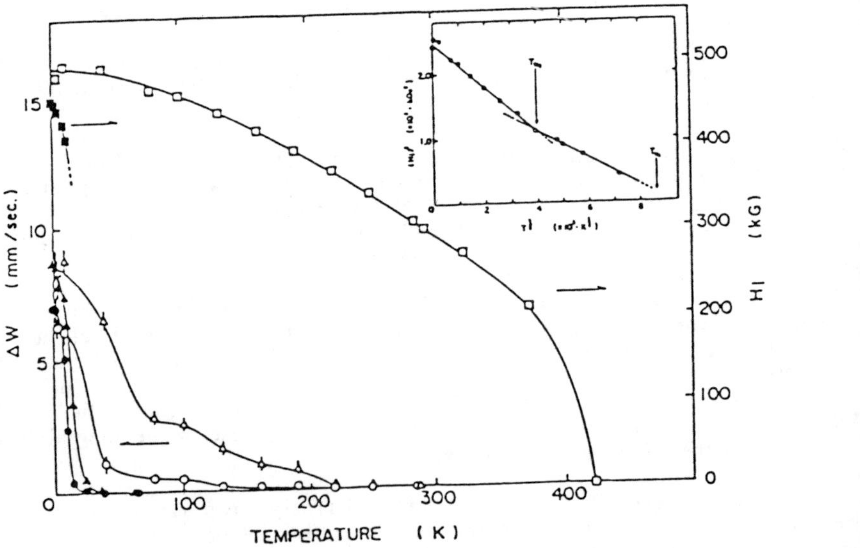

Fig. 23 T-dependence of hyperfine field deduced from $YBa_2(Cu_{0.92}Fe_{0.08})_3O_{7-\delta}$ samples in the superconducting phase (filled symbols) and semiconducting phase (open symbols). Figure taken from ref. 65.

T_{N2}=400K associated with ordering of the moments on the Cu(2) sites. The moments are found to have a value of about $0.6\mu_B$. As the temperature is lowered to $T=T_{N2}$=80K, intensity of the (1/2, 1/2, 3/2) Bragg peak begins to grow and concommitantly the intensity of the (1/2, 1/2, 2) peak begins to decrease. The values of the ordered moments at low temperatures are found to be Cu(1)=$0.46(6)\mu_B$ and Cu(2)=$0.97(a)\mu_B$. The ordering temperatures T_{N1} and T_{N2} are found to be oxygen stoichiometry sensitive as noted in the phase diagram of Fig. 12b. For a $YBa_2Cu_3O_{6.2}$ single crystal the corresponding ordering temperatures have been found [96,97,122] to be T_{N1}=400K and T_{N2}=40K.

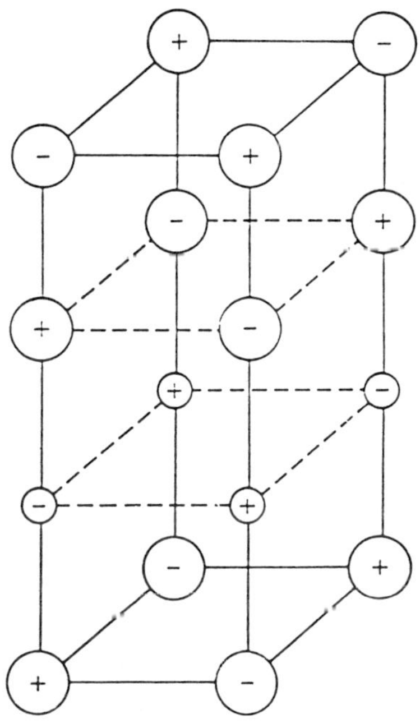

Fig. 24 Magnetic Structure of $YBa_2Cu_3O_6$ at T_{N2} showing antiferromagnetic ordering of both Cu(1) and Cu(2) sites. Figure is taken from ref. 119.

In the light of these neutron scattering results one can begin to understand the ^{57}Fe Mössbauer spectroscopy results in oxygen deficient $YBa_2Cu_3O_{7-\delta}$ samples. The results of Nasu et al (shown in Fig. 22) and those of several other groups display the onset of broadening or magnetic splitting (of spectral features in the velocity interval -2.0mm/s<v<2.0mm/s) as the temperature is lowered below 40K. We associate this behavior with Fe dopant present at Cu(1) chain sites displaying magnetic ordering at $T<T_{N2}$. As the temperature is lowered below T_{N2}, the Fe dopant (sites A, C and D)

responds to the local magnetism at chain sites and display a magnetic hyperfine structure. The magnetic structure is actually characteristic of a spin undergoing relaxation [51]. Furthermore, since Fe dopant displays a strong affinity to replace Cu(1) chain sites it is reasonable to expect the magnetic hyperfine structures associated with sites A, C and D to be dopant concentration dependent. The Mössbauer lineshapes observed at very low Fe-dopant concentrations $x \lesssim 0.01$, do indeed differ from those found at higher concentrations ($x \simeq 0.10$) conceivably because of clustering of Fe in the chains [48,90] at higher doping concentrations. For example, at $x \simeq 0.10$, one expects every 4th cation in the chains, on an average, to be an Fe atom - and consequently, the role of the dopant undergoes a change from being merely a probe of the chain magnetism to one that actually modifies the chain magnetism.

Spin fluctuations and superconductivity. In the superconducting phase of the 1-2-3 material, at low temperatures (T=4.2K) there is clear evidence of ^{57}Fe magnetic hyperfine structure, as illustrated by the results of Nasu et al in Fig. 22. One must remember, however, that these results were obtained with samples containing an Fe concentration x=0.08, and conceivably the magnetism observed is the consequence of Fe-dopant segregation. Indeed our attempts to observe a magnetic structure in superconducting samples ($T_c \gtrsim 90K$) that have a much lower Fe doping concentration x=0.015 have not been successful at 4.2K perhaps indicating that the ordering temperature is lower. Indeed there is a report by Pankhurst et al. [51] showing existence of magnetic hyperfine structure in a similar sample at T=1.8K. Suharan et al. [66] have reported evidence of magnetic hyperfine structure at 1.37K in a superconducting sample of $YBa_2(Cu_{0.99}Fe_{0.01})_3O_{7-\delta}$ possessing a T_c=92K. These results suggest that paramagnetic spin relaxation of Fe dopants at chain-sites coexists with superconductivity in CuO_2 planes. In this context, Nasu et al. have noted that magnetic ordering of Fe at Cu(2) sites in superconducting samples occurs once magnetic ordering of Fe at Cu(1) chain-sites has taken place. These observations are significant and demonstrate that Fe^{3+} spins in the planes may exhibit spin fluctuations that coexist with superconductivity.

III. MÖSSBAUER DEBYE-WALLER FACTORS AND VIBRATIONAL DENSITY OF STATES

The difficulty in accounting for the high-T_cs of cuprate superconductors in terms of the Bardeen-Fröhlich attractive electron-phonon interaction has been recognized generally. Several other avenues have been discussed to account for pairing of charge carriers and include (a) invoking a giant electron-phonon interaction mediated by fictive phonons [123] (b) or pairing mediated by polarization of an electron sub-system such as excitons [6] (c) or pairing induced by antiferromagnetic correlations between spins [124,125] (d) or pre-existing pairs at high temperatures condensing in the zero momentum state (Bose condensation) at some low temperature. It is difficult to say which one or combination of mechanisms indicated will appropriately describe results of experiments. If phonons play any role in pairing of charge carriers in cuprate superconductors, one expects to observe vibrational mode instabilities possibly leading to structural distortions. Inelastic neutron scattering, Raman scattering, thermal conductivity, specific heats, elastic constants, and Mössbauer Debye-Waller factors, all serve as important probes of lattice vibrations in crystalline solids and can play an important role in understanding these oxide superconductors as well.

Neutron and X-ray Bragg Scattering

Neutron Bragg scattering on powdered samples of $YBa_2Cu_3O_{7-\delta}$ has provided precise measurements of structure including lattice parameters, site coordination numbers, bond lengths, as well as mean-square displacements (msd) of various atoms including oxygen. These measurements performed as a function of temperature by Capponi et al. [21] have shown that there exists a significant anisotropy in the msd of the bridging O(4) chain sites even at the lowest temperature (5K) measured.

A more recent neutron diffraction study on a small (0.3x0.2x0.1 mm^3) single crystal of $YBa_2Cu_3O_{7-\delta}$ by You et al. [10] also displays some evidence of anisotropic msd (or Debye-Waller factor). Because of the heavy twining, these authors constrained the msd along the a-axis (U_{11}) and b-axis (U_{22}) to be identical and in general to be different from that along the c-axis (U_{33}), and found some evidence of anisotropy in the msd (U_{11}/U_{33} ~ 3.2(2.0)) of the Cu(1) site. As crystals of larger dimensions become available, conceivably these measurements will become more refined.

T-dependent x-ray diffraction on polycrystalline $YBa_2Cu_3O_{7-\delta}$ samples displays [126] evidence of increased orthorhombicity (2(b-a)/(b+a)) at low temperatures as illustrated in Fig. 25. We note that the orthorhombicity increases linearly as T is lowered in the range 300K>T>120K, to become superlinear at $T \lesssim 120K$. These results are quantitatively, if not qualitatively, different from those reported by A. Golovashkin et al. [127] who observe significant changes in lattice parameters of up to 0.02% near the same temperature (Fig. 26). The latter group has also noted that thermally cycling samples near T_c significantly reduces the lattice parameter anomaly and displays a hysterectic behavior. Such a result may be characteristic of a structural relaxation possibly involving movement of intergranular surfaces to relieve stress [123].

Inelastic Neutron Scattering

Inelastic neutron scattering from polycrystalline $La_{1.85}Sr_{0.15}CuO_4$ and $YBa_2Cu_3O_7$ at T=300K and at T=6K have been compared [128]. Surprisingly in both cuprate superconductors, the vibrational density of states (VDOS) at 6K, i.e. at $T<T_c$ display no obvious features, such as a vibrational mode strength change, or vibrational mode frequency shift relative to those obtained at T=300K. Only a small shift (1/2-1) meV in the energy range 10-20 meV has been suggested [129]. These results indicate that an immanent Peierls instability is not a characteristic of these cuprate superconductors.

A comparison of VDOS in superconducting $YBa_2Cu_3O_7$ (T=92K) and non-superconducting $YBa_2Cu_3O_6$ displays significant changes [128] on account of a change in oxygen stoichiometry. One observes (Fig. 27) a striking reduction in strength of the 19 meV mode, an overall reduction in vibrational strength in the 35-65 meV region, and the appearance of a narrow band at 80 meV, in the oxygen deficient $YBa_2Cu_3O_6$. Fortunately, Raman scattering results on these materials are now also available and provide additional insights on the microscopic origin of the possible vibrational modes, as we shall discuss next.

The difficulty of obtaining large single crystals of the 1-2-3 materials has restricted neutron scattering measurements to polycrystalline samples although some work on small single crystals has also been reported. Furthermore, with 13 atoms per unit cell, the VDOS are rather rich and complex. Nevertheless some progress in modelling the VDOS has been made [130] and permitted calculations of the T-dependence of specific heat

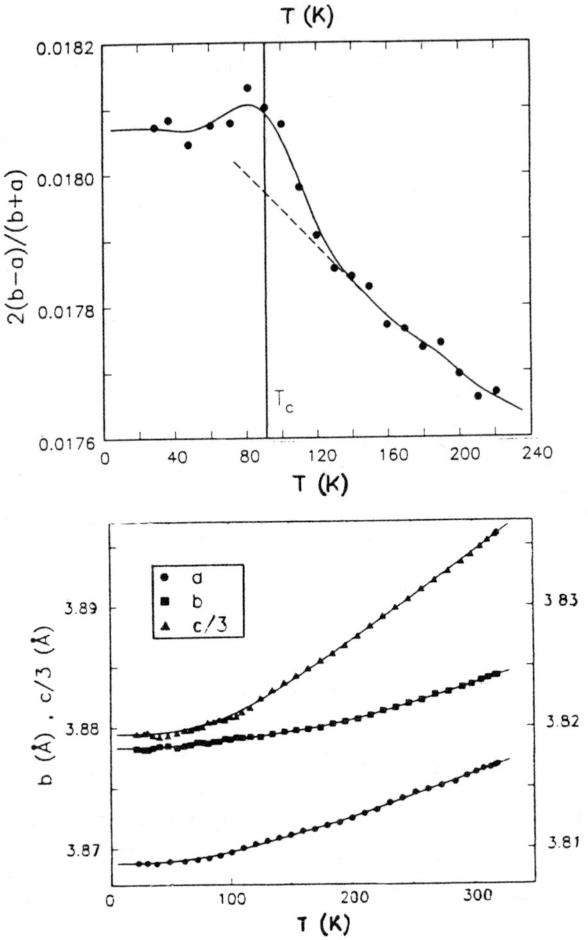

Fig. 25 T-dependence of lattice parameters a, b and c/3 of $YBa_2Cu_3O_{7-\delta}$ exhibiting a linear increase in orthorhombicity 2(b-a)(a+b) with decreasing temperature and a distinct anomaly near T_c. The figure is taken from ref. 126.

and Gd-site Debye-Waller factor. These calculations have been compared directly to the experimental results and the agreement appears by and large encouraging. With the availability of untwinned and large single crystals of the Bi- and Tl-based cuprate superconductors, it is likely that progress in understanding the DOVS will emerge in the foreseeable future.

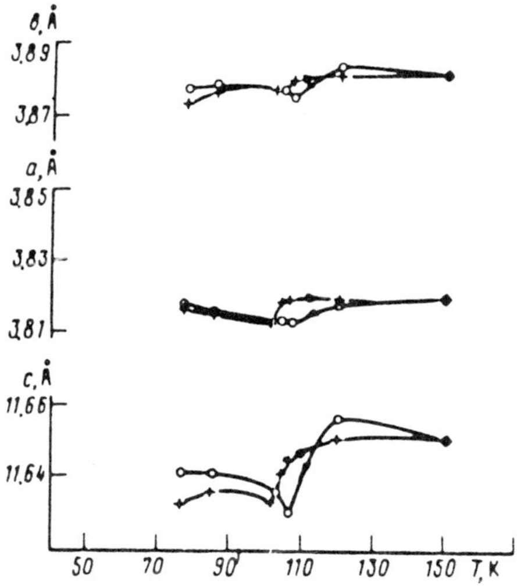

Fig. 26 Lattice parameters of $YBa_2Cu_3O_{7-\delta}$ sample displaying a hysterectic behavior near T_c upon thermal cyclic. Figure is taken from ref. 127.

Raman Scattering

Figure 27 compares the VDOS in $YBa_2Cu_3O_7$ and $YBa_2Cu_3O_6$ from two independent probes: inelastic neutron scattering [128] and Raman scattering [131]. There are similarities in the VDOS, and particularly in the optic mode region; i.e., at excitation energies E>30 meV as one might expect. In the superconducting sample, one notes that the 510 cm^{-1} and 336 cm^{-1} Raman modes have corresponding features at 70 meV and at 40 meV in the neutron DOVS. Furthermore, in $YBa_2Cu_3O_6$, the appearance of a high frequency phonon band at 600 cm^{-1} does indeed have a counterpart at 80 meV in the neutron DOVS. Of course one also recognizes that one may not push the analogy too far, since neutron DOVS include all excitations, while the Raman experiment selects a few crystal symmetry allowed zone center optic modes.

Cardona et al. [131] have given a summary of observed Raman and IR active phonons in superconducting (Y or RE), $Ba_2Cu_3O_{7-\delta}$ cuprates. There are sound reasons to suspect that the prominent 510 cm^{-1} phonon of A_{1g} symmetry represents a Cu(2)-O(1) stretch (see Fig. 27). The 336 cm^{-1} phonon is thought to be a z polarized Cu(2)-O(2) and Cu(2)-O(3) bond-bending mode of B_{1g} symmetry. Cooper et al. [132] were among the first to establish that this mode displays an anomalous temperature dependence. This phonon at first blue-shifts (300K>T>T_c) and then red-shifts

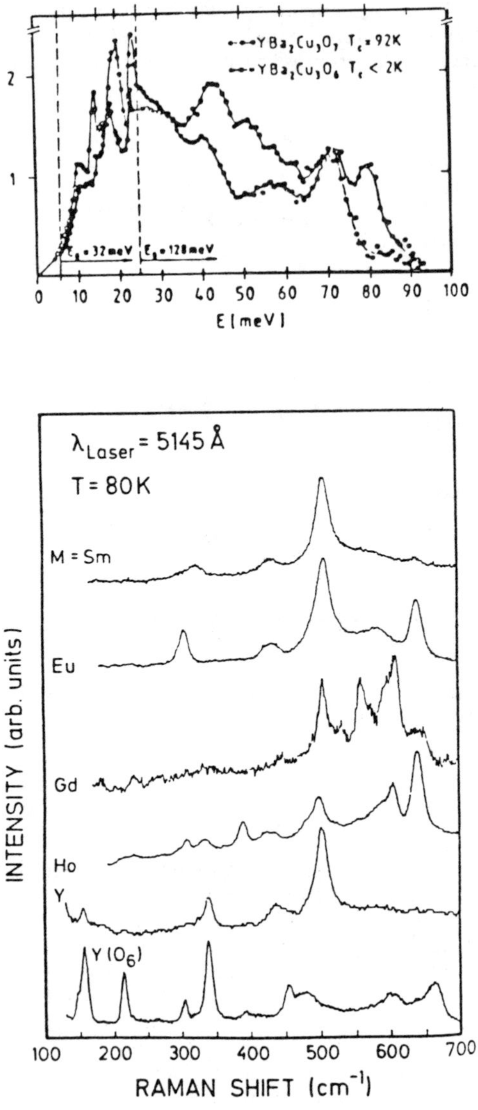

Fig. 27 DOVS from inelastic neutron scattering (top) and Raman scattering (bottom) in $YBa_2Cu_3O_7$ and $YBa_2Cu_3O_6$ compared. The top figure is taken from ref. 128 while the bottom one from ref. 131. The bottom figure also compares Raman spectra of several RE-based 1-2-3 materials with the corresponding Y-based material.

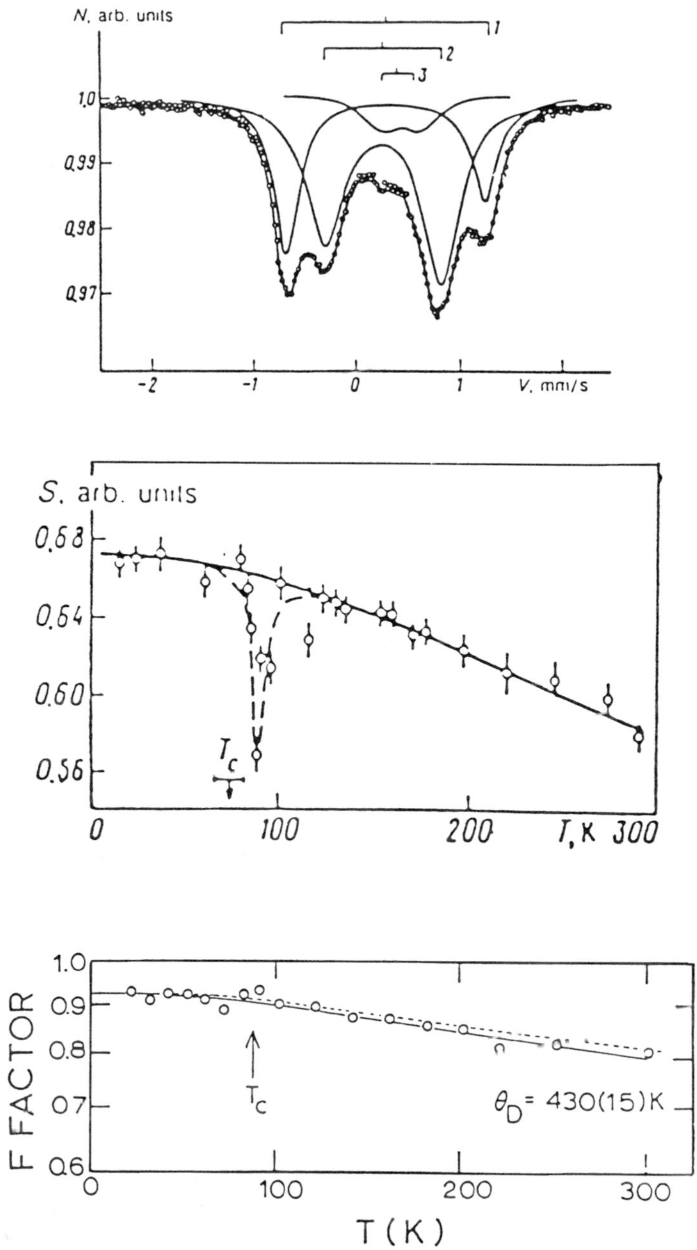

Fig. 28 ^{57}Fe spectrum of $YBa_2(Cu_{0.985}Fe_{0.017})_3O_7$ sample taken by Cherepanov et al. (ref. 72) deconvoluted in terms of 3 sites. Their T-dependence of the integrated area displays a giant anomaly above T_c as shown in (b). The lowest panel (c) shows T-dependence of the integrated area of the spectral lineshape for a $YBa_2(Cu_{0.985}Fe_{0.015})_3O_{7-\delta}$ sample studied by Blue et al. (ref. 136) and it displays no evidence of an anomaly near T_c.

(T_c>T>4K) with temperature, resulting in the phonon frequency having a maximum value near T_c. Such a threshold behavior in phonon frequency is an intriguing result and one would suspect be superconductivity related.

Burns et al. [133] have recently summarized observed Raman active phonons in several cuprate superconductors including those in Bi- and Tl-based materials. The 336 cm^{-1} B_{1g} mode of $YBa_2Cu_3O_{7-\delta}$ has counterparts in the Bi- and Tl-based materials at 282 cm^{-1} and 278 cm^{-1}. The 282 cm^{-1} mode in $Bi_2CaSr_2Cu_2O_8$ (T_c=85.K) is rather broad and apparently does not exhibit an anomalous behavior, although another mode at 464 cm^{-1}, an A_{1g} phonon associated with oxygen atoms above and below the Cu-O planes does exhibit [134] a similar anomalous behavior as the 336 cm^{-1} phonon in $YBa_2Cu_3O_{7-\delta}$.

Phillips [123] has taken the view that in these overcoordinated (or overconstrained) perovskites, internal stress may be partially relieved by static buckling of layers and partially by a new kinetic mechanism - phonon quenching into a few selected modes localized spatially on a few atoms or bonds in each unit cell. He has argued that the threshold behavior of the 336 cm^{-1} mode is evidence in support of a fictive phonon mechanism of high-temperature superconductors. The Fano lineshapes of Raman modes reported by Cooper et al. [132] are certainly in harmony with the presence of strong electron-phonon effects in these cuprates as advocated by Phillips.

Mössbauer Debye-Waller Factors

In the case of the 1-2-3 materials, the site selectively afforded by the Mössbauer effect permits probing independently the vibrational density of states at the Rare-earth or at the Cu-transition metal sites using dopants as probes. Such experiments have been performed and display noteworthy anomalies.

^{57}Fe spectroscopy. Cherepanov et al. [72] have examined the T-dependence of the size of the Mössbauer effect in an orthorhombic $YBa_2(Cu_{0.983}Fe_{0.017})_3O_{7-\delta}$ sample possessing lattice constants a=3.878Å, b=3.840Å and c=11.69Å and a transition temperature $T_c \simeq 75$K. The results appear in Fig. 28 and show that the integrated area under the complex lineshape displays an anomalous reduction of about ~15% at a temperature T~90K; i.e., about 15K above the onset of superconductivity at $T_c \simeq 75$K. The reduction in the integrated area implies an increase in the mean-square-displacement $\langle r^2 \rangle$; i.e., a softening of lattice vibrations.

A qualitatively different T-dependence of the size of the Mössbauer effect has been observed by Blue et al. [136] in an orthorhombic $YBa_2(Cu_{0.985}Fe_{0.015})_3O_{7-\delta}$ sample. The sample was characterized by lattice parameters a=3.822(5)Å, b=3.887(5)Å, c=11.682(5)Å and a transition temperature T_c=91.8K. The result displayed in lowest panel of Fig. 28 shows a normal behavior of the f-factor characterized by a θ_D=430(15)K. No anomaly is observed near, at or above T_c. It would be pertinent to remark that even though Cherepanov et al. deconvolute their lineshape in terms of 3 sites as shown in Fig. 27, a procedure that differs qualitatively from the one of Blue et al (who use deconvolution in terms of 4 sites, see Fig. 5a), the integrated area under the lineshape can be expected to be independent of the deconvolution procedure. Thus the difference between the results of Cherepanov et al. on one hand and those of Blue et al. on the other reflect intrinsic differences in sample processing. Here we note that even though the Fe dopant concentrations (x=0.017 and x=0.015) are very nearly the same for the two orthorhombic samples studied, neither the lattice parameters nor their transition temperatures are the same. The oxygen content of the sample used by Cherepanov et al. may be lower than the one used by Blue et al. and possibly this may have some connection to the reported Debye-Waller factor anomaly. This is a point that merits further study.

In an Fe-impurity stabilized tetragonal $YBa_2(Cu_{0.95}Fe_{0.05})_3O_{7.1}$ sample possessing a T_c=74K, Kimball et al. [60] have performed a comprehensive temperature dependence study of the resonant absorption fraction. These authors report observing a rather striking hardening of the vibrational behavior of Fe as the msd reduces in a small temperature window ΔT=20K centered around T_c (=74K) as illustrated in Fig. 29. Accompanying the hardening of the msd, is also a measurable change in the quadrupole splittings associated with the Fe-chain sites as shown in Fig. 30. These authors have also examined orthorhombic $YBa_2(Cu_{1-x}Fe_x)_3O_{7-\delta}$ samples at a lower Fe concentrations x=0.01 and x=0.02 and find no evidence of a vibrational anomaly near T_c in such samples, in accord with the results of C. Blue et al. [136] shown in Fig. 28c. According to Kimball et al., the anomaly observed in the tetragonal sample at x=0.05 is not superconductivity related but is the consequence of Fe dopant segregation in clusters at these higher dopant concentrations.

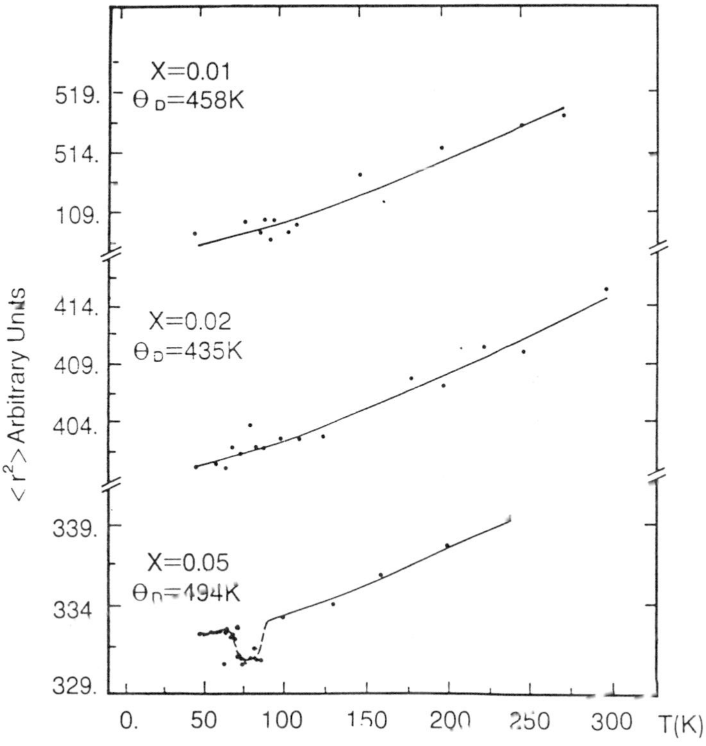

Fig. 29 T-dependence of ^{57}Fe mean-square displacement $<r^2>$ at indicated Fe-concentrations 'x' in $YBa_2(Cu_{1-x}Fe_x)_3O_{7-\delta}$ samples taken from ref. 60. The reduction in $<r^2>$ near T_c=74K observed at x=0.05 is ascribed to Fe-impurity induced clustering.

Rare-earth spectroscopy. The Debye-Waller factor of the rare-earth atom in Eu- and Gd-based 1-2-3 materials has been the subject of several ^{151}Eu [67,137-140] and ^{155}Gd [103] Mössbauer investigations. Fig. 31 displays result of a ^{151}Eu study in the T-range 10K<T<300K taken from ref. [137] and it is representative of several other reports. Tomy et al. [140] have examined the role of excess Eu in $EuBa_{2-x}Eu_xCu_3O_{7-\delta}$ samples and find T_cs to decline and approach zero as $x \rightarrow 0.40$. The charge state of Eu continues to be 3^+ at the Ba site. The T-dependence of the ^{155}Gd f-factor, taken from ref. [103] again displays a behavior similar to the one observed in the ^{151}Eu experiments. Bornemann et al. [103] have noted that to account for the f(T) variation in their ^{155}Gd experiments, the need to describe the vibrational density of states in terms of a superposition of a Debye-like and an Einstein-like DOVS is necessary. The latter term simulates an optical mode. Broadly speaking, these rare-earth site probes exhibit a normal vibrational behavior; i.e., a linear-temperature variation at high temperatures ($T \gtrsim \theta_D$) and a saturation at low-temperatures ($T \simeq 0$) on account of zero point motion. Specifically, no anomalies either in the msd, or in the isomer-shift or the quadrupole splitting are observed above, at or below T_c. This may be the result of a good measure of isolation of the rare-earth site from the superconductivity related CuO_2 planes in this class of materials.

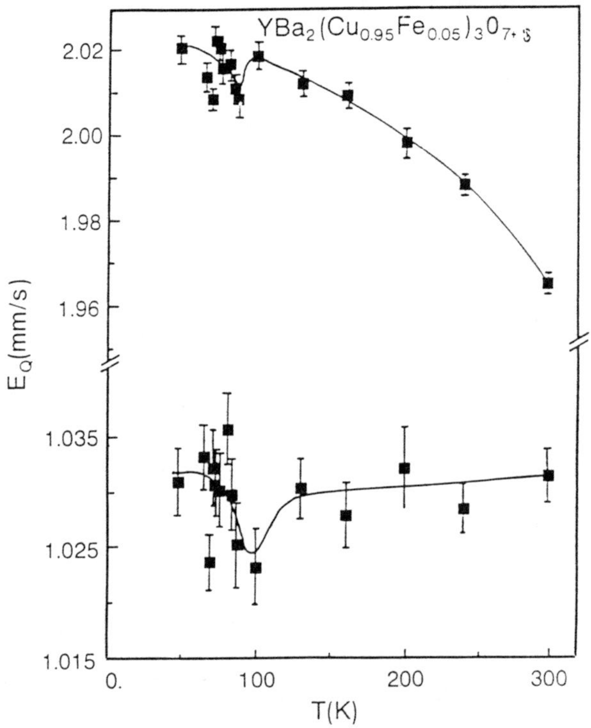

Fig. 30 Anomalous T-variation of the A-site and the C-site quadrupole splittings near $T=T_c$ in a tetragonal $YBa_2(Cu_{0.95})_3O_{7-\delta}$ sample reported by Kimball et al. [60].

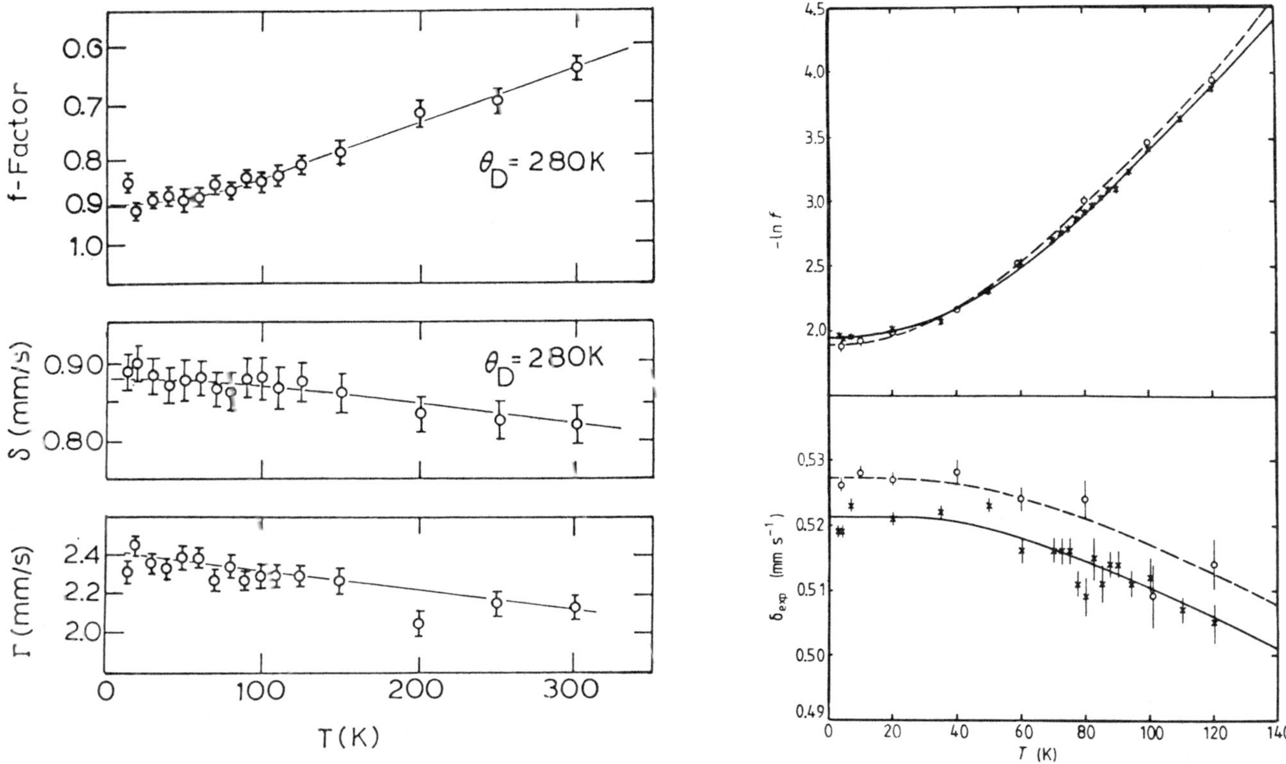

Fig. 31 (left) T-dependence of ^{151}Eu Mössbauer Debye-Waller factor (f-factor) isomershift (δ) and linewidth (Γ) in orthorhombic $EuBa_2Cu_3O_{7-\delta}$ samples taken from ref. 137. The panel on the right shows T-dependence of $-\ln_f$ (where f is the ^{155}Gd Mössbauer f-factor) and isomershift (δ_{expt}) in an orthorhombic $GdBa_2Cu_3O_{7-\delta}$ sample reported by Bornemann et al. ref. 103.

^{119}Sn spectroscopy. The case of Sn as a dopant in 1-2-3 materials is of interest because the radius of Sn^{4+}(=0.71Å) and Cu^{2+}(=0.72Å) are very nearly the same and, furthermore, these radii are significantly different from those of Y^{3+}(=0.89Å), O^{2-}(= 1.40Å) and Ba^{2+}(=1.34Å). Sn^{4+} cation in the perovskites (Ba or Ca)SnO_3, and Cu^{2+} cation in the perovskite La_2CuO_4, chemically and structurally may play a parallel role. Conceivably Sn as a dopant in the 1-2-3 materials could serve as a valuable probe of Cu sites. Driven by these considerations, we have probed [141] the vibrational and chemical behavior of Sn dopant in $EuBa_2(Cu_{1-x}Sn_x)_3O_{7-\delta}$ samples using the Mössbauer effect of the 23.8 keV gamma resonance in ^{119}Sn. Fig. 32 displays a semilog plot of f-factor as a function of temperature for a sample containing x=0.007, Sn dopant concentration. The sample was prepared the usual way by sintering the oxides at 950°C followed by either a slow-cool to room temperature in oxygen (to yield a T_c onset of 85K) or fast-cool to room temperature (to give a T_c onset of 60K). The lower panel of Fig. 32 represent results obtained on the slow-cooled sample while the upper panel of the figure shows results on the fast-cooled sample.

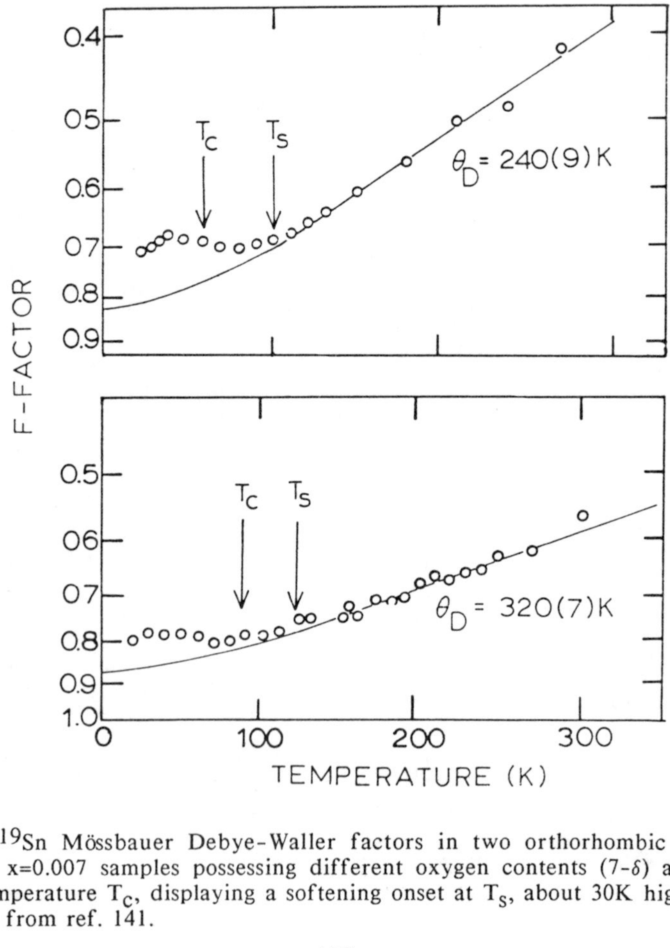

Fig. 32 ^{119}Sn Mössbauer Debye-Waller factors in two orthorhombic $EuBa_2(Cu_{1-x}Sn_x)_3O_{7-\delta}$, x=0.007 samples possessing different oxygen contents (7-δ) and therefore transition temperature T_c, displaying a softening onset at T_s, about 30K higher than T_c. Figure taken from ref. 141.

There are several noteworthy features of the results shown in Fig. 32. The fast-cooled sample having a reduced oxygen content compared to the slow-cooled one, possesses a lower T_c and a lower Debye-temperature. This is the expected result largely because the oxygen deficient sample is also the softer material [128]. But perhaps the more striking result displayed is the onset of a softening in the f-factor at a temperature T_s which is about 30K above T_c for both the slow-cooled and the fast-cooled samples. Thus lowering T_c by deliberately reducing the oxygen content of the samples, also lowers the softening temperature onset T_s of the f-factor. These results are persuasive in suggesting that the f-factor anomaly which tracks T_c, is superconductivity related. Yuen et al. [142] have independently examined the vibrational behavior of Sn in $YBa_2(Cu_{0.96}Sn_{0.04})_3O_{7-\delta}$ samples and report observing a softening of the f-factor near 120K as discussed above for the Eu-1-2-3 material. Remarkedly this softening temperature coincides with the onset temperature of orthorhombic distortion observed by P. Horn et al. [126] and ultrasound propagation reported by S. Bhattacharya et al. [143]. Anomalies in thermal conductivity are also observed [144] although these appear below T_c. Vibrational behavior of Sn dopant in $La_{1.85}Sr_{0.15}Cu_{1-x}Sn_xO_4$ cuprate superconductors at x=0.05, 0.10 and 0.15 has also been the subject of Mössbauer spectroscopy investigations [145]. Evidence of a softening in the Sn f-factor and anomaly in the isomer shift are reported at T~170K. We feel these results suggest that phonon instabilities related to superconductivity do play a role in these high-T_c cuprates. Unfortunately little is currently known about the Sn site location in these cuprates directly. An EXAFS measurement of this system would certainly aid in the structural interpretation of these rather fascinating Mössbauer effect results.

We conclude this section with a brief remark. Belogurov et al. [146] have reported a simple yet thought provoking Mössbauer experiments to show that charge carriers in high-T_c cuprates are coupled by phonons. The experiment consists of measuring the f-factor (f_{sn}) of a ^{57}Co in Cr source at T=78K; i.e., $T<T_c$ with a current (of 20 mAmps) flowing into the Cr-source from a HTSC ($YBa_2Cu_3O_7$; T_c=90K) and comparing it to the f-factor (f_{ns}) in the circumstance when the direction of the current flow is reversed. If charge carriers in the HTSC are coupled by phonons, one expects these phonons to be created and pairs to be broken as these charge carriers traverse from the HTSC to the normal metal (source matrix), thus decreasing f_{sn} (in relation to f_{ns}), when the current flows from the normal metal to the HTSC. The inequality $f_{sn}<f_{ns}$ requires that in the transmission Mössbauer experiment the count rate at resonance show the inequality: $N_{sn}>N_{ns}$. These workers measured the ratio $(N_{sn}-N_{ns})/N_{ns}$ and found it to be typically +(4.7±1.8) x 10^{-4} indicating that $f_{sn}<f_{ns}$. A critical check on the interpretation of this experiment would be to find $f_{sn}=f_{ns}$ as $T\gtrsim T_c$ and this remains to be demonstrated. Thus the experiment of Belogurov et al. and those of several others described in this section are strongly suggestive that phonons do play a role in the 1-2-3 family of HTSCs.

CONCLUDING REMARKS

In this review we have attempted to highlight the role of Mössbauer spectroscopy to understand the magnetic, microstructure and density of vibrational states in the two families of high-T_c cuprate superconductors -- La_2CuO_4 and $YBa_2Cu_3O_{7-\delta}$. Connection with other experiments including magnetic neutron scattering, Raman scattering, electron microscopy, neutron Debye-Waller factors, and density of vibrational states from inelastic neutron scattering measurements was made. Several new families of both cuprate and non-cuprate oxide superconductors have emerged in the past few months. Among these the Tℓ based materials that possess large unit cells and can be prepared in an (ordered) single phase will undoubtedly be the focus of Mössbauer work.

As single phase materials of the Bi- and Pb- and Ni-based materials are prepared, Mössbauer experiments on these materials will follow. Clearly the information presented here, we anticipate, will assist the interested reader in interpreting results on these new materials. Furthermore, as new processes are perfected to prepare better quality single phase materials, the present body of results on solid state reacted materials is likely to be refined.

Armed with the Mössbauer site signatures in pristine materials, the effects of chemical dopants, the effects of radiation damage and the role of Ag- alloying to form composites, are likely to be promising and perhaps rewarding avenues of investigations for the future. Such microscopic investigations on these large unit cell oxides are likely to provide critical complementary information to the more conventional macroscopic tools used to study superconductors. This type of information, we suspect, will eventually narrow down possible mechanisms responsible for superconductivity in these novel oxides.

ACKNOWLEDGEMENTS

First and foremost, we acknowledge with pleasure the active and enthusiastic participation of several graduate students on various HTSC projects who made possible several new results described in this review. The students include Chuck Blue, Khaled Elgaid, Ivan Zitkovsky, and Rezaul Siddique. We also wish to acknowledge collaborations with several faculty colleagues who have contributed to the present research effort on these materials and these include Professor David Farrell, Professor Warren Huff, Professor Attila Kilinc, Professor Bernie Goodman, Professor R. Coussement, Professor W.C.H. Joiner, Dr. J. Oostens and Dr. Douglas Ho. The rapid flow of results in this field has made it impossible for us to cite all relevant papers for which we apologize to appropriate authors. We are particularly grateful to Janie Runck for an excellent job in typing the manuscript and laying out this camera ready manuscript with its extensive illustrations. This work was supported in part by a grant from the U.S. National Science Foundation DMR-85-21005, and the University of Cincinnati through the office of the Dean for Graduate Research and Education.

REFERENCES

[1] J.G. Bednorz and K.A. Mueller, Z. Phys. B64 (1986) 189; H. Takagi, S. Uchida, K. Kitazawa and S. Tanaka, Jpn. J. Appl. Phys. 26, (1987) L123. Also see ibid, 26 (1987) L1.

[2] M.K. Wu, J.R. Ashburn, C.J. Torng, P.H. Hor, R.L. Meng, L. Gao, Z.J. Huang, Y.Q. Wang, C.W. Chu, Phy. Rev. Letters 58, (1987) 908.

[3] H. Maeda, Y. Tanaka, M. Fukutomi, and T. Asano, Japanese J. Appl. Phys. 27, (1988) L209.

[4] Z.Z. Sheng and A.M. Hermann, Nature 332, (1988), 138.

[5] V.J. Emery, Mat. Res. Soc. Bull. 14, (1989) 67.

[6] W.A. Little, Science 242, (1988) 1390.

[7] R.M. Hazen, L.W. Finger, R.J. Angel, C.T. Prewitt, N.L. Ross, H.K. Mao, C.G. Hadidiacos, P.H. Hor, R.L. Meng, C.W. Chu, Phys. Rev. B35, (1987) 7238.

[8] T. Siegrist, S. Sunshine, D.W. Murphy, R.J. Cava, S.M. Zahurak, Phys. Rev. B35, (1987) 7137.

[9] J.D. Jorgensen, H.B. Schuttler, D.G. Hinks, D.W. Capone II, K. Zhang, M.B. Brodsky, D.J. Scalapino, Phys. Rev. Lett. 58, (1987) 1024.

[10] Hoydoo You, R.K. McMullan, J.D. Axe, D.E. Cox, J.Z. Liu, G.W. Crabtree and D.J. Lam, Solid State Commun. 64, (1987) 739.

[11] R.J. Cava, B. Batlogg, C.H. Chen, E.A. Rietman, S.M. Zahurak and D. Werder, Nature 329, (1987) 423. Also see Phys. Rev. B37, (1988) 5912.

[12] R.J. Cava, B. Batlogg, A.P. Ramirez, D. Werder, C.H. Chen, E.A. Rietman, and S.M. Zahurak, Mat. Res. Soc. Symp. Proc. 99, (1988) 19.

[13] For a review see Proceedings of International Conference on High-Temperature Superconductors and Materials and Mechanism of Superconductivity, Interlaken, Switzerland, March 1988 published in Physica C volumes 153-155.

[14] Y. Tokura, H. Takagi, S. Uchida, Nature 337, (1989) 345. Also see V.J. Emery Nature 337, (1989) 306.

[15] J.B. Torrance, Y. Tokura, A.I. Nazzal, A. Bezinge, T.C. Huang and S.S.P. Parkin, Phys. Rev. Letters 61, (1988) 1127.

[16] A.W. Sleight, J.L. Gillson and P.E. Bierstedt, Solid State Commun. 17, (1975) 27. Also see A.W. Sleight, Science 242, (1988) 1519.

[17] R.J. Cava, B. Batlogg, J.J. Kajewski, R. Farrow, L.W. Rupp, Jr., A.E. White, K. Short, W.F. Peck and T. Kometani, Nature 332, (1988) 814.

[18] P. Villars and J.C. Phillips, Phys. Rev. B37, (1988) 2345.

[19] R.J. Cava, B. Batlogg, R.B. van Dover, D.W. Murphy, S. Sunshine, T. Siegrist, J.P. Remeika, E.A. Rietman, S. Zahurak, G.P. Espinosa Phys. Rev. Lett. 58, (1987) 1676.

[20] H. Steinfink, J.S. Swinnea, Z.T. Sui, H.M. Hsu and J.B. Goodenough, J. Am. Chem. Soc. 109, (1987) 3348.

[21] J.J. Capponi, C. Chaillout, A.W. Hewat, P. Lejay, M. Marezio, N. Nguyen, B. Raveau, J.L. Soubeyroux, J.L. Tholence, R. Tournier, Europhys. Lett. 3, (1987) 1301.

[22] L.M. Longo and P.M. Racah, J. Solid State Chem. 6, (1973) 526.

[23] D. Vaknin, S.K. Sinha, D.E. Moncton, D.C. Johnston, J.M. Newsam, C.R. Safinya, and H.E. King, Jr., Phys. Rev. Lett. 58, (1987) 2802.

[24] G. Shirane, Y. Endoh, R.J. Birgeneau, M.A. Kastner, Y. Hidaka, M. Oda, M. Suzuki, T. Murakami, Phys. Rev. Lett. 59, (1987) 1613.

[25] P.M. Grant, S.S.P. Parkin, V.Y. Lee, E.M. Engler, M.L. Ramirez, J.E. Vazquez, G. Lim, R.D. Jacowitz, R.L. Greene, Phys. Rev. Lett. 58, (1987) 2482.

[26] B.M. Tissue, K.M. Cirillo, J.C. Wright, M. Daeumling and D.C. Larbalestier, Solid State Commun. 65, (1988) 51.

[27] M. Francois, A. Junod, K. Yvon, A.H. Hewat, J.J. Capponi, P. Strobel, M. Marezio and P. Fischer, Solid State Commun. 66, (1988) 1117.

[28] J.D. Jorgenson, M.A. Beno, D.G. Hinks, L. Soderholm, K.J. Volin, R.L. Hitterman, J.D. Grace, Ivan K. Schuller, C.U. Segre, K. Zhang, and M.S. Kleefisch, Phys. Rev. B36, (1987) 3608; also see Physica C 153-155, (1988) 578.

[29] J.S. Swinnea, H. Steinfink, J. Mater. Res. 2, (1987) 424.

[30] Donglu Shi, Phys. Rev B (in press). Donglu Shi and D.W. Capone II, Appl. Phys. Letters, 53, (1988) 159.

[31] C. Chaillout, M.A. Alario-Franco, J.J. Capponi, J. Chenavas, P. Strobel, M. Marezio, Solid State Commun. 65, (1988) 283.

[32] K. Muraleedharan, C.V. Tomy, S.K. Malik, Ram Prasad, N.C. Soni, Solid State Commun. 68, (1988) 227.

[33] H. Küpfer, I. Apfelstedt, R. Flükiger, R. Meier-Hirmer, W. Schauer, T. Wolf and H. Wühl, Physica C 153-155, (1988) 367.

[34] A. Erle, S. Blumenröder, E. Zirngiebl, M. Veit, J. Langen, G. Güntherodt, Physica C 153-155, (1988) 296.

[35] N. Toghe, M. Tatsumisago, T. Minami, K. Okuyama, M. Adachi, Y. Kousaka, Jpn. J. Appl. Phys. 27, (1988) 292.

[36] P. Biswas, D. Zhou, I. Zitkovsky, C. Blue and P. Boolchand (unpublished).

[37] P.N. Peters, R.C. Sisk, E.W. Urban, C.Y. Huang and M.K. Wu, Appl. Phys. Letters 52, (1988) 2066.

[38] C.Y. Huang, Y. Shapira, E.J. McNiff, P.N. Peters, B.B. Schwartz, M.K. Wu, R.D. Shull and C.K. Chiang (private commun.).

[39] M. Tinkham, Phys. Rev. Lett. 61, (1988) 1658; Y. Yeshurun and A.P. Malozemoff, Phys. Rev. Lett. 60, (1988) 2202.

[40] P.L. Gammel, L.F. Schneemeyer, J.V. Waszczak, D.J. Bishop, Phys. Rev. Letts. 61, (1988) 1666.

[41] J.C. Barry, J. Elect. Micro. Tech 8, (1988) 325.

[42] I.N. Khlyustikov and A.I. Buzdin, Adv. in Physics 36, (1987) 272.

[43] F.M. Mueller, S.P. Chen, M.L. Prueitt, J.F. Smith, J.L. Smith and D. Wohlleben, Phys. Rev. B37, (1988) 5837.

[44] M.M. Fang, V.G. Kogan, D.K. Finnemore, J.R. Clem, L.S. Chumbley and D.E. Farrell, Phys. Rev. B37, (1988) 2334.

[45] J.M. Tarascon, P. Barboux, P.F. Miceli, L.H. Greene, G.W. Hull, M. Eibschutz and S.A. Sunshine, Phys. Rev. B37, (1988) 7458.

[46] Y. Maeno, T. Tomita, M. Kyogoku, S. Awaji, Y. Aoki, K. Hoshino, A. Minami and T. Fujita, Nature 328, (1987) 512.

[47] Gang Xiao, M.Z. Cieplak, A. Gavrin, F.H. Streitz, A. Bakhshai and C.L. Chien. Phys. Rev. Lett. 60 (1988) 1446. Also see Phys. Rev. B35, (1987) 8782.

[48] C. Blue, K. Elgaid, I. Zitkovsky, P. Boolchand, D. McDaniel, T. Davidson, W. Huff, T. Grigereit (unpublished); the Fe-concentration dependence of T_c found by Blue et al. is similar to the one reported by B. Ullman et al. in Physica C 153-155 (1988) 872; a region $0<x<0.023$ where T_c is constant at 92K, a region 0.023<x<0.;10 where T_c decreases linearly $dT_c/dx=5.4K/at$ % Fe; and a third region $0.10<x<0.16$ where T_c decreases more steeply $dT_c/dx=9K/at$ % Fe; $T_c \rightarrow O$ $x>0.16$.

[49] For a review of the method, the reader is referred to Mössbauer Spectroscopy by N.N. Greenwood and T.C. Gibb, Chapman and Hall Ltd., London 1971.

[50] C. Blue, K. Elgaid, I. Zitkovsky, P. Boolchand, Darl McDaniel, W.C.H. Joiner, Jean Oostens, Warren Huff, Phys. Rev. B37, (1988) 5905.

[51] Q.A. Pankhurst, A.H. Morrish and X.Z. Zhou, Phys. Lett. A127, (1988) 231. Also see J. Phys. C. Solid State Phys. 21, (1988) L7.

[52] E. Baggio-Saitovitch, I.S. Azevedo, R.B. Scorzelli, H. Saitovitch, S.F. da Cunha, A.P. Guimaraes, P.R. Silva and A.Y. Takeuchi, Phys. Rev. B37, (1988) 7967.

[53] E.R. Bauminger, M. Kowitt, I. Felner, I. Nowik, Solid State Commun. 65, (1988) 123.

[54] L. Bottyan, B. Molnar, D.L. Nagy, I.S. Szucs, J.Toth, J. Dengler, G. Ritter, J. Schober, Phys. Rev. B38, (1988) 11373.

[55] I. Nowik, M. Kowitt, I. Felner and E.R. Bauminger, Phys. Rev. B38, (1988) 6677.

[56] V. Sedykh, S. Nasu and F. Fujita, Solid State Commun. 67, (1988) 1063.

[57] M. Takano, Z. Hiroi, H. Mazaki, Y. Bando, Y. Takeda and R. Kanno, Physica B-C 153-155, (1988) 860.

[58] B.D. Dunlap and C.W. Kimball, in Proceedings of Nuclear Methods in Magnetism, Aug. 1988 Munich.

[59] C.W. Kimball, J.L. Matykiewicz, John Giapintzakis, Hsun Lee, B.D. Dunlap, M. Slaski, F.Y. Fradin C. Segre and J.D. Jorgensen, Mater. Research Society Symp. Proc. 99, (1988) 107.

[60] C.W. Kimball, J.L. Matykiewicz, Hsun Lee, John Giapintzakis, A.E. Dwight, B.D Dunlap, James D. Jorgensen, B.W. Veal and F.Y. Fradin (preprint).

[61] T. Tamaki, T. Komai, A. Ito, Y. Maeno and T. Fujita, Solid State Commun. 65, (1988) 43.

[62] S. Nasu, H. Kitagawa, T. Kohara, Y. Oda, T. Shinjo, K. Asayama and F.E. Fujita, Proc. of JIMIS-5, Suppl. to Trans. JIM 29, (1988) 543.

[63] A. Simopoulos and D. Niarchos, Phys. Rev. B38, (1988) 8931.

[64] M. Eibschutz, M.E. Lines, J.M. Tarascon and P. Barboux, Phys. Rev. B38, (1988) 2896. Also see ibid 38, (1988) 8858.

[65] S. Nasu, H. Kitagawa, T. Kohara, Y. Oda, T. Shinjo, K. Asayama, F.E. Fujita, Mat. Res. Symposium, (1988) Tokyo, Japan.

[66] S. Suharan, C.E. Johnson, D.H. Jones, M.F. Thomas, R. Driver, Solid State Commun. 67, (1988) 125.

[67] M. Takano and Y. Takeda, Jpn. J. Appl. Physics 26, (1987) L1862.

[68] M.W. Dirken, R.C. Thiel, H.H.A. Smit, H.W. Zandbergen, Physica C156, (1988) 303.

[69] H. Tang, Z.Q. Qiu, Y.W. Du and J.C. Walker, Hyperfine Interact. 42, (1988) 1243.

[70] A. Nath, S. Nagy, M.W. Barsoun, S.D. Tyagi and Y. Wei, Solid State Commun. 68, (1988) 181.

[71] P. Boolchand, C. Blue, K. Elgaid, Ivan Zitkovsky, Darl McDaniel, Warren Huff, B. Goodman, G. Lemon, D.E. Farrell and B.S. Chandrasekhar, Phys. Rev. B38, (1988) 11313.

[72] V.M. Cherepanov, M.A. Chuev, S.S. Yakimov, V.Y. Goncharov and S.A. Smirnov, JETP Lett. 47, (1988) 424.

[73] L.J. Swartzendruber, L.H. Bennett, M.Z. Harford, Mark Rubinstein, J. Superconductivity 1, (1988) 219.

[74] J. Bieg, J. Jing, H. Englemann, Y. Hsia, U. Gonser, P. Guetlich, R. Jacobi, Physica C 153-155, (1988) 952.

[75] B.D. Dunlap, J.D. Jorgensen, W.K. Kwok, C.W. Kimball, J.L. Matykiewicz, H. Lee, C.U. Segre, Physica C 153-155, (1988) 1100.

[76] P. Imbert, G. Jehanno and J.A. Hodges, Proc. Nuclear Methods in Magnetism, Aug. 1988, Munich to appear in Hyperfine Interactions; also see Nishihara et al. Jpn. J. Appl. Phys. 26, (1987) L1416. The corresponding ^{57}Co in $LaCuO_4$ source experiment has been reported by S. Jha et al. Physica C 153-155, (1988) 1555 who observe a six-line spectrum at 77K which is quite similar to the one shown in Fig. 18 at 4.2K.

[77] S.K. Malik, C.V. Tomy, D.T. Adroja, R. Nagarajan, R. Prasad and N.C. Soni, Solid State Commun. 66, (1988) 1097.

[78] V.A. Tsurin, N.P. Filippova, A.M. Sorkin, L. Ya, Kobelev, L.L. Nugaeva and A.P. Stepanov, JETP Lett. 46, (1987) 459.

[79] J.M.D. Coey, K. Donnelly, Z. Phys. B67, (1987) 513.

[80] K. Elgaid, C. Blue, I. Zitkovsky, P. Boolchand, Darl McDaniel, Warren Huff (unpublished).

[81] D. McDaniel, C. Blue, K. Elgaid, I. Zitkovsky, P. Boolchand, Warren Huff, W.C.H. Joiner (unpublished).

[82] C. Blue, K. Elgaid, Atilla Kilinc, Warren Huff, Darl McDaniel, P. Boolchand (unpublished).

[83] P. Boolchand, Ph.D. Thesis (unpublished) Case Western Reserve University 1969.

[84] R.R. Sharma, Phys. Rev. Lett. 26, (1971) 563.

[85] R.E. Walsted, W.W. Warren, Jr., R.F. Bell, G.F. Brennert, G.P. Espinosa, J.P. Remeika, R.J. Cava and E.A. Rietman, Phys. Rev. B36, (1987) 5727; also see W.W. Warren, Jr., R.E. Walstedt, G.F. Brennert, G.P. Espinosa, J.P. Remeika, Phys. Rev. Lett. 59, (1987) 1860.

[86] C.H. Pennington, D.J. Durand, D.B. Zax, C.P. Slichter, J.P. Rice and D.M. Ginzburg, Phys. Rev. B37, (1988) 7944.

[87] G. Guidi, C. Bucci, P. Carretta, R. De Renzi, R. Tedeschi, C. Vignali and F. Licci, Solid State Commun. 68, (1988) 759.

[88] M. Mali, D. Brinkmann, L. Pauli, J. Roos, H. Zimmermann and J. Hulliger, Phys. Lett. A124, (1987) 112.

[89] Youwen Xu, M. Suenaga, J. Tafto, R.L. Sabatini, A.R. Moodenbaugh and P. Zolliker (preprint). Phys. Rev. B (April, 1989).

[90] P. Bordet, J.L. Hodeau, P. Strobel, M. Marezio and A. Santoro, Solid State Commun. 66, (1988) 435.

[91] K. Asayama, Y. Kitaoka, and Y. Kohori, J. Magn. Mater. 76 (1988) 449; Y. Kitaoka, S. Hiramatsu, T. Kohara, K. Asayama, K. Oh-Ishi, M. Kikuchi, N. Kobayashi, Jpn. J. Appl. Physics 26, (1987) L397; J. Phys. Soc. Japan 56, (1987) 3024; ibid 57, (1988) 734.

[92] T. Fujita, Y. Aoki, Y. Maeno, J. Sakurai, H. Fukuba, H. Fujii, Jpn. J. Appl. Phys. 26, (1987) L368.

[93] I. Furo, A. Janossy, L. Mihaly, P. Banki, I. Pocsik, I. Bakonyi, I. Heinmaa, E. Joon and E. Lipmaa, Phys. Rev. B36 (1987) 5690.

[94] H. Lütgemeier and M.W. Pieper, Solid State Commun. 64, (1987) 267.

[95] Y. Kitaoka, S. Hiramatsu, K. Ishida, T. Kohara, Y. Oda, K. Amaya and K. Asayama, Physica 148B (1987) 298; also see J. Phys. Soc. Japan 57, (1988) 737.

[96] J.M. Tranquada, D.E. Cox, W. Kunnmann, H. Moudden, G. Shirane, M. Suenaga, P. Zolliker, D. Vaknin, S.K. Sinha, M.S. Alvarez, A.J. Jacobson and D.C. Johnston, Phys. Rev. Lett. 60, (1988) 156.

[97] H. Kadowaki, M. Nishi, Y. Yamada, H. Takeya, H. Takei, S.M. Shapiro and G Shirane, Phys. Rev. B37, (1988) 7932.

[98] N. Nishida, H. Miyatake, D. Shimada, S. Okuma, T. Yamazaki, Y. Watanabe, Y. Kuno, M. Ishikawa, T. Takabatake, K. Nagamine, K. Nishiyama, J.H. Brewer and S.R. Kreitzmann, Physica C 153-155, (1988) 761.

[99] P. Burlet, C. Vettier, M.J.G.M. Jurgens, J.Y. Henry, J. Rossat-Mignod, H. Noel, M. Potel, P. Gougeon, J.C. Levet, Physica 153-155, (1988) 1115.

[100] Y. Kitaoka et al. Proc. Int. Conf. Magnetism (1988) Paris.

[101] R.J. Birgeneau, D.R. Gabbe, H.P. Jenssen, M.A. Kastner, P.J. Picone, T.R. Thurston, G. Shirane, Y. Endoh, M. Sato, K. Yamada, Y. Hidaka, M. Oda, Y. Enomoto, M. Suzuki, and T. Murakami, Phys. Rev. B38, (1988) 6614.

[102] D. McK. Paul, H.A. Mook and A.W. Hewat, B.C. Sales, L.A. Boatner, J.R. Thompson, and Mark Mostoller, Phys. Rev. B37 (1988) 2341.

[103] H.J. Bornemann, G. Czjzek, D. Ewert, C. Meyer and B. Renker, J. Phys. F: Met. Phys. 17, (1987) L337.

[104] E.E. Alp, L. Soderholm, G.K. Shenoy, D.G. Hinks, D.W. Capone II, K. Zhang, B.D. Dunlap, Phys. Rev. B36, (1987) 8910.

[105] H.H.A. Smit, M.W. Dirken, R.C. Thiel, L.J. de Jongh, Solid State Commun. 64, (1987) 695.

[106] G. Wortmann, C.T. Simmons and G. Kaindl, Solid State Commun. 64, (1987) 1057; also see Physica C 153-155 (1988) 1547.

[107] P. Dalmas de Réotier, P. Vulliet, A. Yaouanc, P. Chaudouët, S. Garcon, J.P. Senateur, F. Weiss, L. Asch, G.M. Kalvius, Physica C 153-155, (1988) 1543. A related experiment has been performed by R.D. Taylor, J.O. Willis and Z. Fisk and reported in Hyperfine Interactions 42, (1988) 1257. This group has obtained the T-dependent of the ^{151}Eu hyperfine field in a $Eu_{0.1}Gd_{0.9}Ba_2Cu_3O_{7-\delta}$ sample and find $T_N \simeq 2.09(11)$K and a saturation field of 2.9(3) Tesla.

[108] P. Dalmas de Réotier, P. Vulliet, A. Yaouanc, O. Hartmann, E. Karlsson, R. Wappling, P. Chaudouët, S. Garcon, J.P. Sénateur, F. Weiss, T.O. Niinikoski, Physica C 153-155, (1988) 1541.

[109] P.C.M. Gubbens, J.J. van Loef, A.M. van der Kraan, D.M. deLeeuw, J. Mag. Magn. Mat. 76 and 77 (1988) 615.

[110] J.A. Hodges, P. Imbert, G. Jéhanno, Solid State Commun. 64, (1987) 1209.

[111] B.W. Lee, J.M. Ferreira, Y. Dalichaouch, M.S. Torikachvili, K.N. Yang and M.B. Maple, Phys. Rev. B37, (1988) 2368.

[112] M.B. Maple, H.C. Hamaker, L.D. Woolf in Superconductivity in Ternary Compounds II, Ed. by M.B. Maple and O. Fischer, Topics in Current Physics 34 (1982) 99, Springer Verlag. There are several chapters of interest in this book.

[113] W.W. Warren, Jr., R.E. Walstedt, G.F. Brennert, R.J. Cava, R. Tyeko, R.F. Bell, and G. Dabbagh, Phys. Rev. Lett. 62, (1989) 1193.

[114] A.I. Goldman, B.X. Yang, J. Tranquada, J.E. Crow, C.-S. Jee, Phys. Rev. B36 (1987) 7234.

[115] J.W. Lynn, W.-H. Li, Q. Li, H.C. Ku, H.D. Yang, R.N. Shelton, Phys. Rev. B36 (1987) 2374; also see T. Chattopadhyay et al. Europhysics Lett. 6, (1988) 363.

[116] Thomas J. Kistenmacher, J. de Physique, Colloque (1989).

[117] A. Weidinger, Ch. Niedermayer, A. Golnik, R. Simon, E. Recknagel, J.I. Budnick, B. Chamberland, C. Baines, Phys. Rev. Lett. 62, (1989) 102.

[118] M. Sato, S. Shamoto, J.M. Tranquada, G. Shirane, B. Keimar, Phys. Rev. Lett. 61 (1988) 1317.

[119] J.W. Lynn and W-H. Li, J. Appl. Phys. 64 (1988) 6065.

[120] X.Z. Zhou, M. Raudsepp, Q.A. Pankhurst, A.H. Morrish, Y.L. Luo, and I. Maartense, Phys. Rev. 36, (1987) 7230.

[121] H. Tang, Z.Q. Qiu, Y.-W. Du, Gang Xiao, C.L. Chien and J.C. Walker, Phys. Rev. B36, (1987) 4018.

[122] J.H. Brewer et al., Phys. Rev. Lett. 60, (1988) 1073.

[123] J.C. Phillips, Solid State Commun. 68, (1988) 769 and private commun.; J.R. Hardy and J.W. Flocken, Phys. Rev. Lett. 60, (1988) 2191.

[124] G. Baskaran, Z. Zou and P.W. Anderson, Solid State Commun. 63, (1987) 973.

[125] J.R. Schrieffer, X.-G. Wen and S.-C. Zhang, Phys. Rev. Lett. 60 (1988) 944.

[126] P.M. Horn, D.T. Keane, G.A. Held, J.L. Jordan-Sweet, D.L. Kaiser, F. Holtzberg, T.M. Rice Phys. Rev. Lett. 59, (1987) 2772.

[127] A.I. Golovashkin, O.M. Ivanenko, G.I. Leitus, K.V. Mitsen, O.G. Karpinski, and V.F. Shamrai, JETP Lett. 46, (1987) 410.

[128] B. Renker, F. Gompf, E. Gering, G. Roth, W. Reichardt, D. Ewert, H. Rietschel, Physica C 153-155, (1988) 272.

[129] J.J. Rhyne, D.A. Neumann, J.A. Gotaas, F. Beech, L. Toth, S. Lawrence, S. Wolf, M. Osofsky, D.U. Gubser, Phys. Rev. B36, (1987) 2294.

[130] W. Reichardt, L. Pintschovius, B. Hennion, and F. Collin, Supercond. Sci. Technol. 1, (1988) 173.

[131] M. Cardona, L. Genzel, R. Liu, A. Wittlin, Hj. Mattausch, F. Garcia-Alvarado and E. Garcia-Gonzalez, Solid State Commun. 64, (1987) 727.

[132] S.L. Cooper, M.V. Klein, B.G. Pazol, J.P. Rice and D.M. Ginzburg, Phys. Rev. B37, (1988) 5920.

[133] Gerald Burns, P. Strobel, G.V. Chandrashekhar, F.H. Dacol, F. Holtzberg, M.W. Shafer, Phys. Rev. B39 (1989) 2245.

[134] Gerald Burns, G.V. Chandrashekhar, F.H. Dacol, P. Strobel, Phys. Rev. B39, (1989) 775.

[135] A.J. Vega, W.E. Farneth, E.M. McCarron, and R.K. Bordia, Phys. Rev. B135, (1989) 2322.

[136] C. Blue, P. Boolchand (unpublished).

[137] P. Boolchand, R.N. Enzweiler, I. Zitkovsky, R.L. Meng, P.H. Hor, C.W. Chu and C.Y. Huang, Solid State Commun. 63, (1987) 521.

[138] M. Eibschutz, D.W. Murphy, S. Sunshine, L.G. Van Uitert, S.M. Zahurak, W.H. Grodkiewicz, Phys. Rev. B35, (1987) 8714.

[139] G. Wortmann, S. Blumenröder, A. Freimuth, D. Riegel, Phys. Lett. A126 (1988) 434.

[140] C.V. Tomy, R. Nagarajan, S.K. Malik, D.T. Adroja, N.C. Soni and Ram. Prasad, Solid State Commun. 68, (1988) 531.

[141] P. Boolchand, R.N. Enzweiler, I. Zitkovsky, Jack Wells, Wayne Bresser, Darl McDaniel, R.L. Meng, P.H. Hor, C.W. Chu and C.Y. Huang, Phys. Rev. B37, (1988) 3766.

[142] Tan Yuen, C.L. Lin, J.E. Crow, G.N. Myer, R.E. Salomon, P. Schlottmann, N. Bykovetz and Warren N. Herman, Phys. Rev. B37, (1988) 3770.

[143] S. Bhattacharya, M.J. Higgins, D.C. Johnston, A.J. Jacobson, J.P. Stokes, J.T. Lewandowski, D.P. Goshorn, Phys. Rev. B37, (1988) 5901.

[144] C. Uher and A.B. Kaiser, Phys. Rev. B36, (1987) 5680.

[145] John Giapintzakis, J.M. Matykiewicz, C.W. Kimball, A.E. Dwight, B.D. Dunlap, M. Slaski, F.Y. Fradin, Phys. Lett. A. 121, (1987) 307.

[146] V.N. Belogurov, A.N. Borzyak, N.N. Karklin'sh, JETP Lett. 48, (1988) 25.

RVB AND RELATED THEORIES OF HTSC

D. Schmeltzer
Bar-Ilan University, Isreal

INTRODUCTION

(a) Apology

Since the discovery of high-temperature superconductivity by Bednorz and Mueller (1) in 1986 a new theoretical effort has been launched. It seems that the understanding of the new superconductors will lead to a change in our basic understanding of the entire solid state physics.

The first step towards this possible revolution in solid state physics was started in the fall of 1986 by P.W. Anderson (2). Anderson made the bold suggestion that superconductivity in La_2CuO_4 and related materials might be caused by the occurrence in these materials of the "resonating valence bond" state, a magnetic spin liquid suggested by him in 1970 (3). Since 1986 the initial suggestion of Anderson has taken many directions.

In all the new directions the Fermi liquid principle fails in the high temperature superconductor.

This article will concentrate on a narrow aspect of the "High Temperature Superconductors". The article will deal with the formulation of a "Non-Fermi-Liquid Theory" and the implication of such a theory to superconductivity. Even this part of the Non-Fermi liquid theories have taken different directions and is difficult to give credit to all the contributors.

The natural solution will be to concentrate on those parts where I was personally involved and hope that other aspects will be covered by other authors.

(b) The CuO_2 materials

Stoichiometric La_2CuO_4 is a Mott insulator spin $\frac{1}{2}$ antiferromagnet (4). Doping the material by substituting Sr or La for about 3% destroys the magnetic order (5) leading to a spin liquid with a metal conductivity at zero temperature.

This behaviour is in many senses strange. The La_2CuO_2 oxide forms a Mott insulator (6). It has an even number of electrons per unit cell with a strong Coulombic repulsion. When we dope the oxide, we get a metallic behaviour. The situation is similar to semiconductors but because we are dealing with a Mott insulator (the existing gap is due to strong Coulombic repulsion) the situation is different than in a semiconductor.

Looking from the magnetic side we have difficulties in understanding in terms of classical percolation that only 3% of doping is enough to destroy the magnetic phase. The new disordered magnetic phase is called the quantum spin liquid phase. All these facts lead us to the conclusion that the Fermi liquid concept is not applicable in these materials. A Fermi liquid is by definition a system which has the low energy excitations similar to those of a noninteracting Fermi liquid. It seems that instead of a Fermi liquid theory we may have a situation where the quantum spin liquid is a chargeless Fermi liquid. [For the Heisenberg spin $\frac{1}{2}$ antiferromagnet in one dimension it is known that the excitation are neutral spin $\frac{1}{2}$ particles possessing a linear energy momentum relation. Those excitation form the quantum spin liquid or the neutral spin fermi liquid called spinon (7)].

The common thing of the oxide superconductors are the CuO layers. In those layers the Cu^{+2} spin $\frac{1}{2}$ form strong hybridization with the O^{-2}. Due to the Jahn-Teller effect and crystal splitting effect one has to consider only the hybridization between the $3d_{x^2-y^2}$ (Cu^{+2}) and $2p^6(O^{-2})$ orbitals. Including the effect of the Coulombic interactions we end up with a Mott-Hubbard Hamiltonian. Since the ratio between the Cu-O distances in the c direction to the distances in the plane (ab) is ~1.5 we have a strong anisotropy in the hopping term. In the lowest order approximation we may neglect the hopping integral in the c direction. We concentrate on a two dimensional model for the layers. This is indeed the situation in La-Ba(Sr)-Cu-O oxides. With doping we obtain a superconducting critical temperature T_c~40K°. The second class of materials are the y-Ba-Cu-O. The basic difference here is that we have two layers coupled together separted by Yttrium atoms. The critical temperature is $T_c \sim 90K°$ (in addition to the planes of CuO these materials have also chains of Cu-O).

The third class of materials are the Bi-Sr-Ca-Cu-O ($T_c > 115K°$) and $Tl_2Ba_2Ca_2Cu_3O_6$ ($T_c \sim 125K°$). The basic difference with respect to the other groups is the fact that here we have three layers of CuO coupled together. If indeed the numbers of layers play an important role in increasing T_c we may explain this by the fact that first superconductivity has to appear in the layers and then the coupling between the layers could reduce fluctuation leading to a higher T_c. For example assuming Josephson coupling between the superconductiving layers one has reached the conclusion that for $n \sim 3$-5 the maximum of T_c has to occur (8). Therefore the assumption is that pairs are formed in the 2D layers and they can tunnel between the layers. (The physical nature of this Josephson coupling can be the exchange interaction in the c direction).

An alternative mechanism for the increase of T_c with the number of stacked layers was suggested (9). This mechanism is based on a coupled phonon-plasmon mode with an oscillator strength which increases with the decrease of the distance between the layers.

In addition to the mentioned superconductors, we have additional materials with a low $T_c < 30K°$. We call these materials strange superconductors. Strange by the fact that they have no Cu planes or no hole conduction or do not form a spin $\frac{1}{2}$ system. $(Ba,K)BiO_3$ has no Copper (10), $Nd_{2-x}Ca_xCuO_{4-y}$ (11) and La-Ba-Ni-O has Ni instead Cu and form a S=1 system (12).

Here we will not consider those strange superconductors. We will consider explicitly only the CuO_2 layers.

(c) The Hubbard Hamiltonian - superconductivity in the copper oxides materials.

In order to find the microscopic model which describes the Copper oxides Mott insulator we use band structure calculations, experimental data such as photoemission and optical absorbtions (13).

From the band structure calculations we can estimate the overlapping integrals $t(t_{Cu-O})$ and t' (t_{O-O}). One finds $t \sim 1eV$ $t' \ll t$.

Photoemission and optical absorbtion are used for the determination of the on-site energies ϵ_d (for Copper in the state Cu^{+2}) and ϵ_p (oxygen O^{-2}) and Coulombic repulsion energies Ud (Hubbard U for copper), Up (Hubbard U for oxygen) and V_{pd} (the intrasite repulsion interaction between copper and oxygen).

The origin of t is the hybridization between the Cu^{+2} $3d_{x^2-y^2}$ orbital with $2p_x$, $2p_y$ oxygen orbitals. The direct overlapping between the $2p_x$ orbitals of oxygen gives rise to the t_{O-O}=t' overlapping. At the present time there is no agreement of the strength of this direct hopping integral. Most of the authors do not consider these orbitals. Some of the theories for superconductivity are based on the fact that t_{O-O} is the dominant hopping term (14,15). To my judgement this may not be the case. From photoemission data one finds that the Fermi level is strongly dominated by the O-2p and Cu-3d orbitals. The doped oxide photoemission experiment shows that the extra holes go into oxygen and form an oxygen band. The Hubbard model for the 2D square lattice with 3 atoms per unit cell (one copper and two oxygens) looks:

$$H = \sum_{i,\sigma=\uparrow,\downarrow} \epsilon_d d^+_\sigma(i) d_\sigma(i) + \sum_{\ell,\sigma=\uparrow,\downarrow} \epsilon_p P^+_\sigma(\ell) P_\sigma(\ell) + U_d \sum_i n_{d\uparrow}(i) n_{d\downarrow}(i) +$$
$$+ U_p \sum_\ell n_{p\uparrow}(\ell) n_{p\downarrow}(\ell) + V_{pd} \sum_{i,\ell} n_d(i) n_p(\ell) + t \sum_{<i,\ell>,\sigma=\uparrow,\downarrow} d^+_\sigma(i) P_\sigma(\ell) + H.C. +$$
$$+ t' \sum_{<\ell,\ell'>,\sigma=\uparrow,\downarrow} P^+_\sigma(\ell) P_\sigma(\ell') + H.C. \qquad (1)$$

where the parameters obtained from photoemssion experiments $\epsilon_p - \epsilon_d + V_{pd} \cong 5$-6eV, $V_{pd} \cong 0$-1eV, $U_d \sim 10$eV, $U_p \sim 4$-5eV, $t \sim 1$eV and $t' \lesssim 0.3$eV.

The Hamiltonian given in eq. 1 is still too complicated and can't be solved. Three types of models can be obtained for different ranges of parameters.

<u>The charge fluctuation model</u>. This model assumes that the nearest-neighbour repulsion V_{pd} between Cu and O plays an important role. For value $\epsilon_p - \epsilon_d = 0$, V_{pd}=4eV, U_d=10eV, U_p=8eV the resonanting charge transfer Cu^{+2}-O^{-2} $\Longleftrightarrow$ Cu^{+1}-O^{-1} gives rise to a plasmon which leads to pairing of holes and therefore to superconductivity (16-19).

<u>The two band model</u> initiated by Emery (20) and later extended by others (20-26). In this model the assumption is that there is no charge fluctuation on the Cu^{+2}. The copper atoms are frozen in the state Cu^{+2} forming a spin ½ system. Doping introduces holes on the oxygens leading to O^{-1}. Since O^{-1} has magnetic moment we have a magnetic coupling bewtween the Cu^{+2} and oxygen holes O^{-1} (20-26).

<u>The one band model</u> Following Anderson many authors have assumed that doping introduces holes into copper site and superconductivity appears as a double condensation, one of

singlets of spinon and the second as charge condensation (2, 27, 28). This direction has been followed by a large number of authors (29-45). For the range of parameters where $t' \approx 0$ and $\epsilon_p - \epsilon_d \simeq U_d$ it was shown that the two band model can be mapped into the one band model (46). When $\epsilon_d \sim \epsilon_p$, singlets are formed between Cu-O (47). An additional one band model which uses the small U/t limit is due to Schrieffer (48) which has introduced the spin bag model for superconductivity. The essence of this model is based on the existence of antiferromagnetic domains . Superconductivity occurs as a result of charge fluctuations in those domains. Similar ideas as the one introduced in ref. 48 have been used for an extended Hubbard model with the assumption that the charges move in an additional band (49).

THE TWO BAND MODEL

(a) The microscopic model

The starting point for this case is the Hamiltonian given in eq.(1). For the range of parameters $\epsilon_p - \epsilon_d < U_d, t, t' < U_d, U_p < Ud$ and $V_{pd} << U_p$, we assume that there is no charge fluctuation on copper. The copper sites are kept to be Cu^{+2} and form a spin $\frac{1}{2}$ system. Doping introduces holes into the oxygen bond leading to O^{-1} which has a magnetic moment. We form a model phase space which consists of the following electronic states: $|Cu^{+2}\rangle$, $|O^{-2}\rangle$ and $|O^{-1}\rangle$. The additional states $|Cu^{+3}\rangle$, $|Cu^{+1}\rangle$ and $|O^{0}\rangle$ correspond to high energy configuration which plays no roles for the ground state properties. Using the projections technique (50,51), I together with A. Muramatsu and R. Zeyher have derived an effective Hamiltonian for copper and oxygen (see ref. 21).

The basic assumption is that the $Cu3d_{x^2-y^2}$ bonding band remains half filled in the presence of a dopant, while the additional holes go into the oxygen band O^{-1}. This leads to a square lattice of spins with spin $\frac{1}{2}$ antiferromagnetic Heisenberg interaction with a coupling constant J between the Cu^{+2} spins and to an exchange interaction with a coupling constant K (K>0 for the oxygen P_σ orbitals and K<0 for oxygen P_π orbitals) between the Cu^{+2} and O^{-1}. We consider explicitly the case with K>0.

$$H = H_o + H_M \tag{2a}$$

$$H_o = \int d^2r \sum_{\alpha=\uparrow,\downarrow} P^{+}_{\alpha}(r)\hat{\epsilon}[\nabla^2]P_{\alpha}(r) \tag{2b}$$

$$H_M = J\sum_{n,n'} \vec{S}(n)\cdot\vec{S}(n') + K\sum_{n}\vec{S}(n)\cdot\vec{R}(n) \tag{2c}$$

$\vec{S}(n)$ represents the spin $\frac{1}{2}$ variables of Cu^{+2} on a square lattice, $\vec{R}(n)$ represents the magnetic moment of the O^{-1} holes, $\epsilon(\nabla^2)$ represents the kinetic energy operator of the holes and the p holes at the copper sites have $d_{x^2-y^2}$ symmetry, $P^+_\alpha(n) = \Sigma f(K) e^{iKn} P^+_\alpha(K)$ where $f(K)$ is the symmetry function $f(K)=[1-\frac{1}{2}(\cos K_x+\cos K_y)]^{1/2}$.

The exchange coupling constant $J=J(t,\epsilon_p,\epsilon_d,U_d,U_p,\delta)$ and $K=K(t,\epsilon_p,\epsilon_t,U_d,U_p)$ depend explicitly on the microscopic parameters given in eq. 2. J is calculated to fourth order in t and K to second order in t. We find $K \approx 1eV$, $J=0.1-0.4\delta$ ("δ" is the hole concentration). We find that with doping the exchange Cu-Cu is reduced and finally vanishes for $\delta \simeq 0.22$. Those parameters explain why the antiferromagnetic order is destroyed and why superconductivity is destroyed. Experimentally one finds that the antiferromagnetic order disappears for $\delta \simeq 0.05$. This means that $J(\delta=0.05)$ has to be below the critical values J_c for the antiferromagnetic long range order (the Neel state). Superconductivity is destroyed for $\delta \sim 0.25$ and indeed this is in agreement with our value $\delta=0.22$ for which J vanishes.

(b) Derivation of the effective hole-hole potential and the root towards superconductivity

In order to derive the effective interaction between the holes we will use eq. 2. Two approaches will be presented:

I - The adiabatic method (52)
II - The gauge field method (53)

I. We first consider the adiabatic method using the coherent state representation for the Cu^{+2}. We integrate the Cu^{+2} magnetic fluctuations. The Cu^{+2} variables form an antiferromagnetic Heisenberg model.

The result of the adiabatic integration of the Cu^{+2} variables is given by:

$$H_{eff} = \int d^2r \sum_{\alpha=\downarrow,\uparrow} P^+_\alpha(r)\hat{\epsilon}[\nabla^2]P_\alpha(r)$$

$$+ 4J \sum_{n\neq n'} \frac{\vec{R}(n)\cdot\vec{R}(n')}{[\langle R^2(n)\rangle\langle R^2(n')\rangle]^{\frac{1}{2}}} - \frac{1}{4}|K|\sum_n \frac{\vec{R}^2(n)}{\langle\vec{R}^2(n)\rangle} \tag{3a}$$

where $\vec{R}(n)=P^+_\alpha(n)\vec{\sigma}_{\alpha,\beta}P_\beta(n)$ is the magnetic moment of the O^{-1}. The result given in equation 3a was obtained in the limit $J/K \ll 1$. The concentration of holes $\delta=K_F/2\pi$ and $\langle\vec{R}^2\rangle=(3/4)\delta$.

The effective interaction given in eq. 3a can be understood in the following way: Due to the strong exchange interaction Cu-O all the holes belonging to a given copper site interact effectively ferromagnetically, in contrast to holes belonging to different copper sites interact antiferromagnetically.

Expressing the magnetic interaction with the aid of the hole fermion variables we obtain an effective BCS interaction.

$$V(q) = \frac{3}{8}\frac{K}{<R^2>^{\frac{1}{2}}} - \frac{4J}{<R^2>}\,[\cos q_x + \cos q_y] \tag{3b}$$

Using this potential we find a "d" wave solution with a critical superconducting temperature T_c

$$T_c = tK_F^2 \exp(-\frac{1}{\lambda}) \tag{3c}$$

where $\lambda=(8/7)(J/\pi t\langle\vec{R}^2\rangle$, t is the overlapping integral ($t\langle 1eV$) and δ the hole concentration ($\delta=K_F^2/2\pi$, $\langle R^2\rangle \simeq 3/4\delta$).

We emphasize again that the results obtained in eq. 3 were obtained for $J/K \ll 1$, the integration of the Cu^{+2} variables was possible by assuming a slow time variation of those variables with respect to the motion of the oxygen holes.

II. Instead of making use of the adiabatic method we use the gauge field method (53).

The exchange interaction Cu-O can be expressed with the aid of a Su(2) matrix g. If $(-1)^n\vec{M}(n)$ represents the Cu^{+2} magnetic order parameter we can write for the exchange interaction:

$$\sum_n (-1)^n P_\alpha^+(n)\,[\vec{M}(n)\cdot\vec{\sigma}]_{\alpha,\beta}\, P_\beta(n) = \sum_n (-1)^n P_\alpha^+(n)\,[g^{-1}(n)\cdot\sigma_3 g(n)]_{\alpha,\beta}\, P_\beta(n) \tag{4a}$$

where $g\in SU(2)$ and $\vec{M}\cdot\vec{\sigma}=g^{-1}\sigma_3 g$. We perform a local gauge transformation and introduce the spinor $\Psi^+(n)=P^+(n)g^{-1}(n)$; $\Psi(n)=g(n)P(n)$. Due to this transformation the vector potential is changed to $A_\mu \longrightarrow A_\mu+(\partial_\mu g)g^{-1}$.

The gauge transformation separates the AF interaction into two parts: a constant AF magnetic background field and a gauge field $(\partial_\mu g)g^{-1}$. The effect of the constant AF background field will be to transform the two component spinor $\Psi(n)$ into one component spinless fermion with the spin determined by the position in space (odd sites spin up and even sites spin down or vice versa). We use the CP^1 representation for the Heisenberg AF: $\vec{M}=Z^{+}\sigma Z$, $Z^{+}=(z_1^{+},z_2^{+})$, $|Z_1^{+}|^2+|Z_2|^2=1$ with the gauge field $A_\mu=(-i/2)[z^{+}\partial_\mu Z-(\partial_\mu z^{+})z]$ and the lagrange muiltiple λ. (An explicit description for the CP^1 representation is given in ref. 54).

Using the CP^1 representation we obtain for the hamiltonian given in eq. 2 the following effective action:

$$S = \int_0^\beta d\tau\int d^2r\{\psi^+(r,\tau)[\partial_\tau+(\partial_\tau g)g^{-1}]\psi(r,\tau)+\psi^+(r,\tau)\hat{\epsilon}[(\nabla+eA_\mu^{E.M.}+(\partial_r g)g^{-1})^2]\psi(r,\tau)$$

$$+\tfrac{1}{2}K\cos(Q\cdot r)\psi^+(r,\tau)\sigma_3\psi(r,\tau)+\int_0^{\beta c}d\tau\int d^dr\{\frac{1}{2\gamma}(\partial_\mu+iA_\mu)Z^+(\partial_\mu-iA_\mu)Z+ \tag{4b}$$

$$\lambda(r,\tau)[|Z|^2-1]\}$$

where $\mu=(\vec{r},\tau)$, $A_\mu=-(\partial_\mu g)g^{-1}{}_{1,1}$, $A_\mu=(\partial_\mu g)g^{-1}{}_{2,2}$. The wave vector $\vec{Q}=(\pi/a,\pi/a)$ reproduces the AF modulation $(-1)^n$, $\gamma\sim J^{-1}$ and c is the spin wave velocity. In equation 4b we have used the fact that hopping of holes occurs in the CuO_2 planes and the Cu-Cu exchange interaction is three dimensional ($J_z=\eta J$, $\eta<1$, where J_z is the exchange coupling between Cu-Cu in the "z" direction).

The assumption is that after doping the exchange coupling is reduced such that $J<J_c$ and the Neel state is suppressed.

In order to investigate the effect of the AF field on the holes we diagonalize the oxygen hole hamiltonian in the absence of the gauge field. After this diagonalization we obtain

$$S = \int_0^\beta d\tau\int d^2r\{\tilde{\psi}^+(r,\tau)[(\partial_\tau+iA_\tau\sigma_3)+$$
$$\epsilon(\nabla+eA_r^{E.M.}+iA_r\sigma_3)^2]\tilde{\psi}(r,\tau)+\Gamma(A_\mu,\lambda)\} \tag{4c}$$

where $\Gamma(A_\mu,\lambda)$ is the effective action obtained after the integration of the z_1,z_2 variables:

$$\Gamma(A_\mu,\lambda) = \int_0^\beta d\tau\int d^dr\{NT_r Ln[-(\partial_\mu+iA_\mu)+\lambda]-\frac{\lambda}{2\gamma}\} \tag{4d}$$

We expand $\Gamma(A_\mu,\lambda)$ and use the fact that the Heisenberg AF is in the disordered phase.

The gauge field $A_\mu=(A_r,A_\tau)$ has the temporal component A_τ which is analogous to the longitudinal scalar field in electrodynamics. We eliminate the temporal component and find the following effective Hamiltonian.

$$H = \int d^2r\tilde{\psi}^+(r)\hat{\epsilon}[(\nabla_r+eA^{E.M.}+ie^*\sigma_3A_r)]\psi(r) +$$
$$+\int d^dr\int d^dr'[\tilde{\psi}^+(r)\sigma_3\tilde{\psi}(r)]\frac{e^{*2}}{\nabla^2}(r,r')[\tilde{\psi}^+(r')\sigma_3\tilde{\psi}(r')]+$$
$$+\tfrac{1}{2}\int d^dr[(\partial_\tau A_r)^2+(\vec{\nabla}\times\vec{A}_r)^2] \tag{4e}$$

We have .an attractive Coulombic potential between the holes with opposite spin. Those holes gain an additional charge e^* (positive and negative) in addition to the electromagnetic charge e. The origin of the additional charge is due to the fact that in the disorder AF phase the spin variable for the copper is composed from two massive charged bosons (z_1, z_2). Therefore when separation of those bosons occurs we have new charges $e^* \sim \pm M/2\sqrt{2}$ with M being the inverse correlation length of the massive bosons z_1, z_2 $(M(J \geq J_c)=0,\ M(J<J_c) \neq 0)$. The Hamiltonian given in eq. 4e allows for a bound state solution of two holes $|\emptyset\rangle = \sum \chi(r_1 - r_2) b^+(r_1, r_2) |0\rangle$ where $b^+(r_1, r_2) = \sum \sigma \tilde{\varphi}^+_\sigma(r_1) \varphi^+_{-\sigma}(r_2)$. Performing the variation $\langle\emptyset|H|\emptyset\rangle$ with the condition $\langle\emptyset|\emptyset\rangle = 1$ gives the following eigenvalue equation in the center of mass of the two holes.

$$[-2\tilde{t}\nabla^2_r + v(r)]\chi(r) = E\chi(r) \tag{4f}$$

where t is the effective overlapping integral and v(r) is the attractive hole-hole potential. Considering the anisotropy in the exchange interaction we have $v(r) = -e^{*2}/(r^2+d^2)^{1/2}$. Here r is the distance between the holes in the plane and d is the distance between the planes.

As a result the binding occurs between holes in adjacent planes. Such a potential has a bound state solution.

The existence of a bound state solution for two holes leads to a picture similar to the Wanier excitons. First one obtains a bound state for two holes and assumes that the pairs of holes interact via the residual interaction between the pairs. This picture is valid for the dilute limit of holes. We consider the case where the binding radius ξ is smaller than the average distance L between the holes $(\xi < 1/\sqrt{\delta} = L)$.

We have here a boson model with a charge 2e and a hard core repulsive interaction. According to the general theory, the 2-D boson system with hard core repulsive interaction has superfluid transition. The $\delta/2$ pairs of bosons have the critical superfluid transition, $T_c \sim \tilde{t}\delta/Ln(Ln(\Lambda^2/\delta))$ ($\Lambda \sim 1/a$ where a is the lattice spacing).

SUPERCONDUCTIVITY FOR THE NON-FERMI LIQUID CASE

(a) The one band model

For the range of parameters $\epsilon_p - \epsilon_d > U_d \longrightarrow \infty$, $t \sim 1eV$ and $t'_{o-o} \longrightarrow 0$ it is possible to map the problem into an effective one band model (55). For the non-half filled band case we obtain the following effective model

$$H = + \sum_{\sigma=\uparrow,\downarrow} C^{+}_{i\sigma} C_{J\sigma} + H.C. + J_1 \sum_{n.n} \vec{S}_i \cdot \vec{S}_J + J_2 \sum_{nnn} \vec{S}_i \cdot \vec{S}_J \tag{5a}$$

$\vec{S}$ is the spin $\frac{1}{2}$ quantum operator and $n_i = \sum_{\sigma=\uparrow,\downarrow} C^{+}_{i\sigma} C_{i\sigma}$ is a constrained fermion operator, $n_i = 0,1$ ($n_i \neq 2$, double site occupation is excluded). Exactly at the half-filled band $n_i = 1$ ($J_2 = 0$ and the kinetic term is missing). The presence of holes induce a next nearest neighbour term J_2 (53).

(b) The magnetic hamiltonian - spin liquid and chiral spin liquid

We consider the magnetic part of the Hamiltonian given in eq. 5a.

$$H_M = J_1 \sum_{n.n} \vec{S}_i \cdot \vec{S}_J + J_2 \sum_{n.n.n} \vec{S}_i \cdot \vec{S}_J \tag{5b}$$

when $J_1 \neq 0$, $J_2 = 0$ and $J_1 > J_c$ one has an AF Neel state (56). It was shown (56) that the quantum Heisenberg Hamiltonian in d=2 dimensions can be mapped into the classical one in D=d+1 dimension. This mapping is not exact, non adiabatic corrections are missing. The effect of nonadiabatic corrections is included by a Berry phase γ (57). Suppose that $|\psi(t)\rangle$ is the adiabatic wave function of the Hamiltonian 5b $|\tilde{\psi}(t)\rangle = \exp[-i\int_0^t E_{AF}(\vec{S}(t'))dt'] \; |\psi(o)\rangle$ with $E_{AF}(\vec{S}(t))$ the adiabatic energy. The complete wave function $|\psi(t)\rangle$ which takes into consideration nonadiabatic corrections is given by:

$$|\psi(t)\rangle = \exp[i\gamma(t)]\exp[-i\int_o^t E_{A.F.}(\vec{S}(t')]\,|\psi(o)\rangle \tag{5c}$$

where

$$i\gamma(t) = \int_o^t \langle\tilde{\psi}(t')|\,i\partial_{t'}|\tilde{\psi}(t')\rangle dt' \tag{5d}$$

Using the coherent state representation (58) it was shown that for $J_2 = 0$, we have $\gamma = 0$ (59) and the results are in agreement with the one given in ref. 56. On the other hand when $J_1 \neq 0$, $J_2 \neq 0$ and $1 > J_2/J_1 > 0.5$ it was shown that $\gamma \neq 0$ (52). When $\gamma \neq 0$ we have a chiral spin liquid phase (such a phase with $\gamma \neq 0$ breaks parity and time reversal symmetry).

An alternative method for investigating the Heisenberg spin ½ AF is the method of the many body wave function. A possible representation of the wave function is given by the S_z representation $|\uparrow\rangle$, $|\downarrow\rangle$.

$|\psi_{AF}\rangle = \Sigma\, a[S_1 \ldots S_n]\, |S_1, S_2, \ldots S_n\rangle$, where $|S_i\rangle = |\uparrow\rangle, |\downarrow\rangle$. Using a variational procedure one determines the ground state energy E_g and the coefficients $a[S_1 \ldots S_n]$. $|\psi_{AF}\rangle$ represent the wave function for a magnetic phase.

An alternative wave function which corresponds to a complex order parameter is the Resonant Valence Bond (RVB) wave function (60). For the spin ½ AF on a square lattice the RVB wave function is given as a superposition of singlets.

$$|\psi\rangle = \sum_{\substack{i_\alpha \in A \\ J_\beta \in B}} h(i_1,J_1) \ldots h(i_n,J_n)(i_1,J_1) \ldots (i_n,J_n) \qquad (6a)$$

where h stands for any positive function and can be interpreted as a weight factor for a bond as a function of its length. Also any two lattice sites i and J, $(i,J) = |\uparrow,\downarrow\rangle - |\uparrow,\downarrow\rangle$ represent a singlet bond connecting term.

The advantage of the wave function given in eq. 6a is that as a function of the weight h we can have different ground states. In the limit when the weight function h is taken to be independent of the bond length, we get the projection of the Neel state on the singlet subspace, if h decreases with distance we get a smooth way to transform a Neel state into a disordered state. In the intermediate stages of this process, we find a featureless liquidlike singlet RVB state as proposed by Anderson (27,28). When h is limited to nearest neighbour bonds we obtain the short range RVB (29). Using the short range RVB it was shown that the spin ½ Heisenberg AF is equivalent to the classical statistical problem of a dimer like gas (61).

A variational wave function built from singlets was suggested for the spin liquid phase (the disorder Heisenberg AF). This variational function is similar to the BCS wave function.

$$|\psi_{R.V.B}\rangle = P_G[\sum_{(i,J)} h(i,J) b^+(i,J)]^{n/2}|0\rangle \qquad (6b)$$

where $b^+(i,J) = C^+_{i\uparrow}C^+_{J\downarrow} - C^+_{i\uparrow}C_{J\downarrow}$ and $P_G = 1 - n_\uparrow(i)n_\downarrow(i)$

In "K" space $|\psi_{RVB}\rangle$ looks identical to the BCS wave function with the difference that there is no charge!

$|\psi_{RVB}\rangle = P_G\pi(U_K+V_K C_{K\uparrow} C_{-K\downarrow})|0\rangle$ where $|U_K|^2+|V_K|^2=1$ and $h(K)=U_K/V_K$ ($h(K)=F_{Transform}h(i,J)$). $h(K)$ is determined by the variation method. One can find different types of solutions:

$$h_s(K) = h_o(\cos K_x+\cos K_y) \quad , \quad \text{s-wave;}$$
$$h_d(K) = h_o(\cos K_x-\cos K_y) \quad , \quad \text{d-wave;}$$
$$h_{S+id}(K) = h_s(K)+ih_d(K) \quad , \quad \text{S+id-wave.}$$

(c) The lattice Gauge field method

The spin order parameter $\langle\vec{S}\rangle$ is represented in quantum mechanics as a combination of two operators. According to Schwinger the angular momentum operator can be expressed as a combination of two harmonic oscillators (the two harmonic oscillators can be bosons or fermionic Grassman variables).

The analogue to the classical operator $\langle\vec{S}\rangle$ is $\langle\vec{S}\rangle=\langle C^+_\alpha\vec{\sigma}_{\alpha\beta}C_\beta\rangle$ ($\vec{\sigma}_{\alpha,\beta}$ are the Pauli matrices).

An alternative order parameter can be constructed - a bond order parameter. We can have $X(i,J)=\langle \sum_{\alpha=\uparrow,\downarrow} C^+_\alpha(i)C_\alpha(J)\rangle$ or $\Delta(i,J)=\langle b^+(i,J)\rangle$. The last two order parameters have no classical analogues, they involve the breaking of the U(1) symmetry. In cases where χ, Δ are the order parameters (and not $\vec{S}$) we have a disordered spin liquid. This phase involves in some sense a process of deconfining of the two charged harmonic oscillators which compose the S order parameters and describe the paramagnetic disordered phase.

Using the Hubbard-Stratonovici transformation we introduce the complex order parameter in the Heisenberg Hamiltonian

$$H = J_1\Sigma\vec{S}_i\cdot\vec{S}_J = \sum_{\alpha=\uparrow,\downarrow} \chi(i,J)C^+_\alpha(i)C_\alpha(J)+\text{H.C.}+\frac{|\chi(i,J)|^2}{J_1} \tag{6c}$$

Since the number of particles is constrained, $n_i=1$ we introduce a lagrange field A_τ. Eq. 6c resembles

$$H = \sum_{i,J,\alpha=\uparrow,\downarrow} \chi(i,J)C^+_\alpha(i)C_\alpha(J)+\text{H.C.}+|\chi(i,J)|^2/J_1 + +i\sum_i A_\tau(i)\left[\sum_{\alpha=\uparrow,\downarrow} C^+_\alpha(i)C_\alpha(i)-1\right] \tag{6d}$$

The order parameter $\chi(i,J)$ can be expressed as $\chi(i,J)=\rho_{i,J}\exp[i\int_i^J\vec{A}\cdot d\vec{r}]$

The fields $(\vec{A},A_\tau)$ form the U(1) representation of ordinary electrodynamics. ($\vec{A}$ is the vector potential and A_τ is the longitudinal scalar potential in electrodynamics) . When $\oint\vec{A}\cdot d\vec{r}=\gamma\neq0$ we have a chiral spin liquid!

(d) The spinon and holons in the slave-boson representation

Suppose that the ground state is composed of superposition of singlets, when a bond is broken we have a free spin. If the bond is "moving" the free spin will appear at a different site. Exactly at the half filled band case n=1 the excitations form a spin liquid (spinon): $f^+_\sigma(i)|\psi_{RVB}\rangle = |\sigma(i)\rangle$, $|\sigma\rangle = |\uparrow\rangle, |\downarrow\rangle$. f^+_σ, f_σ are the spinon Grassman variables. When we dope the system the <u>destruction of a bond</u> and the <u>removing of an electron</u> gives rise to a <u>holon</u>, a charged hole with no <u>spin</u> (all the particles participate in the singlet configuration, empty sites correspond to holons): $e^+(i)|\psi_{RVB}\rangle = |+e\rangle$. An alternative approach which was introduced in references 27, 30, 33, 44 is based on: $\sum_{\sigma=\uparrow,\downarrow} C^+_\sigma(i)C_\sigma(i)=0,1$, $C^+_\sigma = f^+_\sigma e$, $C_\sigma = e^+ f_\sigma$, $e^+e + \sum f^+_\sigma f_\sigma = 1$ (f^+_σ, f_σ are fermion and e, e^+ are boson, it is also possible to choose the f bosons and the e fermions). In this language an electron is a bound state of a spinon and holon, and the superconducting order parameter appears as a combination of four operators $\langle C^+_\sigma(i)C^+_{-\sigma}(J)\rangle = \langle f^+_\sigma(i)e(i)f^+_{-\sigma}(J)e(J)\rangle$. Non conventional superconductivity will involve separation of charge and spin.

$$\langle C^+_\sigma(i)C^+_{-\sigma}(J)\rangle \simeq \langle f^+_\sigma(i)f^+_{-\sigma}(J)\rangle\langle e(i)e(J)\rangle$$

(e) The U(1) excitations in the chiral spin liquid phase - Derivation of the Chern Simon term.

As a result of doping the charged particles (holons) move in the presence of the spinon fluctuations. Those fluctuations are equivalent to the U(1) electrodynamics.

Since the holons form a bose fluid of charge 1e superconductivity may appear as a result of a superfluid transition (44). When $1 > J_2/J_1 \geq 0.5$ the frustration effects become important. The presence of $J_2 \neq 0$ gives rise to a new plaquet of the form $\langle \chi(1,2)\chi(2,3)\chi(3,1)\rangle = \exp i\gamma$ where $\gamma = \oint \vec{A}\cdot dr \neq 0$, with $\gamma \neq 0$ we have a chiral spin liquid (the presnce of triangular plaquets is due to $J_2 \neq 0$). When the minimum of $\chi(i,J)$ occurs such that $\gamma = \pm\pi/2$ the holons moving around closed path accumulate a phase $\pi/2$ like in the presence of a magnetic field $\vec{B} = \vec{\nabla}\times\vec{A}$. Since this phase depends on the frustrations and there fore on the holons density one finds $e^+(i)e(i) = (1/\pi)(\eta/|\eta|)\vec{B}(i)$ ($\eta \sim J_2$). Therefore the holon boson flux $\pi/2$ is equal to a half fermion!

In the remaining part of this section we will derive the effective action for the charged holons in the spin liquid phase. I will follow the derivations given in references 30,40 and 62.

$$H = -\sum_{i,J,\sigma=\uparrow,\downarrow} t_{i,J} C^+_{i\sigma} C_{J\sigma} + H.C. + J_1 \sum_{n.n} \vec{S}_i\cdot\vec{S}_J + \qquad (\,/a)$$

$$+ J_2 \sum_{n.n.n.} \vec{S}_i\cdot\vec{S}_J$$

If we want to include an external magnetic field we substitute $t_{i,J} \longrightarrow t_{i,J}\exp[i\int_i^J \vec{A}_{ext}.d\vec{r}]$.

We use the slave boson formalism (30,40), $C^+_\sigma(i) = f^+_\sigma(i)e(i)$, $C_\sigma(i)=f_\sigma(i)e^+(i)$. Instead of eq. 7a we obtain

$$H = -\sum_{(i,J)} t_{i,J}e_i^+e_J\chi(i,J)+\sum_i f_\sigma^+(i)f_\sigma(J)\chi(i,J)+H.C.+\sum_{n.n}\frac{|\chi(i,J)|^2}{J_1}+$$

$$+\sum_{n.n.n}\frac{|\chi(i,J)|^2}{J_2}+v\Sigma e_i^+e_ie_J^+e_J+A_\tau(i)[e_i^+e_i+f_\uparrow^+(i)f_\uparrow(i)+f_\downarrow^+(i)f_\downarrow(i)-1] \quad (7b)$$

$\chi(i,J)$ is a collective field variable such that: $\chi(i,J)=\langle J\sum f^+_\sigma(i)f_\sigma(J) + t_{i,J}e_i^+e_J\rangle$ and $v=4t^2/J$. $\chi(i,J)$ can be represented by:

$$\chi(i,J) = \rho_{i,J}\exp[i\int_i^J\vec{A}\cdot d\vec{r}]$$

$(\vec{A},A_\tau)$ form the representation of the U(1) electrodynamic. We find the configuration which minimizes the spinon free energy This happens for $\chi_{n.n}=\rho e^{i\pi/4}$ and $\chi_{n.n.n}=\pm\eta$ when $\eta\neq 0$ we have $\langle\chi(1,2)\chi(2,3)\chi(3,1)\rangle=\rho^2\eta\, e^{\pm i\pi/2}$

We find that when a gap of the order J_2 is opened in the spinon spectrum we have $\gamma=\oint A=\pm\pi/2$ (62). (This will happen for $J_2/J_1\geq 0.5$).

The effective action which is obtained for the holons in the chiral spin liquid phase is:

$$H = \int d^2r[e^+(r)[\frac{1}{2m}(-i\nabla-\vec{A}_{ext}-\vec{A})^2]e(r)+v_r\sum_{r,r'}e^+(r)e(r)e^+(r')e(r') +$$

$$+\sum_r A_\tau(r)[e^+(r)e(r)-\frac{1}{\pi}\frac{\eta}{|\eta|}(\vec{\nabla}\times\vec{A})]+\tfrac{1}{2}[\dot{A}^2+(\vec{\nabla}\times\vec{A})^2] \quad (8)$$

The details of this derivation are given in reference 62.

The essential physics of this effective hamiltonian is that the holon hamiltonian is identical to the problem of interacting electrons in a strong magnetic field (63).

(f) Half integer holons in the chiral spin phase and binding of two holons due to gauge fields

The fact that the density of holons $e^+(r)e(r) = (1/\pi)(\eta/|\eta|)B(r)$ is proportional to the magnetic field allows us to use a fictitious magnetic field which produces a phase $\pi/2$. The relation $B(r)=\pi e^+(r)e(r)$ is replaced in a mean field approximation by $B=\pi\delta$. The result is that the holons move in an effective magnetic field. The presence of this effective magnetic field gives rise that two holons bind via the magnetic field. The result is that two holons which repel each other move in the presence of a magnetic field as they were attracted!! Therefore two holons which behave as two half fermions form a boson with a charge 2e! Solving the relative motion of two holons in the presence of the magnetic field $B=\pi\delta$, we find that in the relative coordinate system we have a harmonic oscillator. The lowest energy corresponds to the zero point energy $=\frac{1}{2}h\omega_c$. (ω_c is the cylotron frequency $\omega_c=hB/m$). The energy $E=(\hbar^2 B/2m)$ corresponds to the two holon binding energy. Therefore the critical temperature for destroying the pair holons is $T \sim \omega_c \sim \delta/m$

(g) Superconductivity

When $J_2/J_1 \geq 0.5$ a gap is opened in the spinon spectrum. The presence of this gap breaks the parity and the time-reversal symmetry leading to an effective magnetic field. Therefore the spin liquid phase form the chiral spin liquid phase ($\gamma \neq 0$). In this phase the two holons (half fermions) give rise to a boson bound state with a charge of 2e!!

Superconductivity will appear as a superfluid transition of the boson system with the charge 2e! (62).

In the limit $J_2/J_1 \ll 0.5$ we have the spin liquid phase, $\gamma=0$. In this phase the holon boson form a superfluid state with a charge of 1e. The critical temperature is estimated to be $T_c \sim t\delta$.

EXPERIMENTAL IMPLICATIONS

It seems that in order to prove the existence of the non-Fermi liquid for the new superconductors the question of spin and charge separation is crucial. More directly it will be interesting to investigate the properties of a chiral spin phase. Here we suggest that if such a phase exists a gap in the spinon spectrum should be opened. Experimentally this has to be seen by a gap in the spin-wave spectrum! An additional effect which can prove the existence of the chiral spin phase is the Farady effect.

Since the charge density is proportional to the induced magnetic field, Maxwell's equations will be different (62)

$$(\partial_z^2 - 1/c^2 \partial_t^2)\vec{A} = \beta\sigma_1\partial_t\vec{A} \tag{9}$$

where $\vec{A}-\binom{A_k}{A_y}$, σ_1 is the Pauli matrix and $\beta=(1/\mu)(\hbar/|n|)$.

The left side of equation 9 describes the propagation of the vector potential $\vec{A}=\binom{A_x}{A_y}$ in the z direction (the propagation vector K is in the z direction perpendicular to the CuO layers). The effect of the frustration $J_z \neq 0$ is to induce the polarization corrections. The right side of equation 9 is similar to the Faradey effect, we have; $K_{\pm}^{\pm}=\sqrt{(w/c)^2 \pm \beta^2}$. This means two propagation vectors $K_{\pm}^{+}$ for left hand and right hand polarizations. Measuring the phase difference between right and left polarization as a function of the width in the Z direction will allow to estimate the vlaue of β, the time reversal breaking coefficient.

REFERENCES

[1] J.G. Bednorz and K.A. Mueller, Z. Phys. 64, 198 (1986)

[2] P.W. Anderson, Science 235 (1987)

[3] P.W. Anderson, Matter Res. Bull 8, 153 (1973)

[4] D. Vaknin et al., Phys. Rev. Lett. 58, 2802 (1987); Y.J. Uemura et al,ibid. 59, 1045 (1987); J.M. Tronquada et al., ibid. 60, 156 (1988)

[5] G. Shirane et al., ibid 59, 1613 (1987); Y. Endoh et al., Phys. Rev. B 37, 7443 (1988)

[6] N.F. Mott, Proc. R. Soc. London Ser. A62,414 (1949); B.H. Brandow, Adv. Phys. 26, 651 (1977); I.Yu. Pomeranchuk, J. Phys. (Moscow) 4, 356 (1941)

[7] H.A. Bethe, Z. Phys. 71, 205 (1931); L.D. Fadeev and L.A. Takhtajan, Phys. Lett. 85A, 375 (1981)

[8] J.L. Birman and J.P. Lu, Modern Physics Letters B, vol. 2, Nos. 11, 12, 97 (1988)

[9] A. Bishop, private communication.

[10] R.J. Cava et al., Nature, 332, 814 (1988)

[11] H. Takagi, S. Uchida and Y.L Tokura, Phys. Rev. Lett. 62, 1197 (1989)

[12] A. Bishop, private communication

[13] F. Mila - Miniworkshop on "Mechanisms for high temperature superconductivity" 20 June - 29 July 1988, Trieste

[14] Y. Gao, J.M. Longlois and W.A. Goddard III 239, 896 (1988)

[15] R.J. Birgenau, M.A. Kastner and A. Aharony, Z. Phys. B 71, 57 (1988)

[16] C.M. Varma, S. Schmidt-Rink and Elihu Abrahams, Solid State Commun. 62, 681 (1987)

[17] J.E. Hirsh and J. Tang, E. Loh, Jr. and D.J. Scalapino, Phys. Rev. Lett. 60, 1668 (1988)

[18] C.A. Balseiro, A.G. Rojo, E.R. Gagliono and B. Alacsio, Phys. Rev. B38, 9315 (1988)

[19] I. Batistic, A.R. Bishop, R.C. Martin and Z. Tesanovic, preprint (1989) and A. Bishop - Miniworkshop on "Mechanisms for high temperature superconductivity" 20 June - 29 July 1988, Trieste. In this reference the authors consider the charge transfer mechanism between Cu-O layers.

[20] V.J. Emery, Phys. Rev. Lett. 58, 2794 (1987)
[21] A. Muramatsu, R. Zeyher and D. Schmeltzer, Europhysics Letter 7 (5), 473 (1988)
[22] D. Schmeltzer, Modern Physics Letter, vol. 1, No. 5, 803-810 (1988)
[23] P. Prelovsek, Physics Lett. A 126, 282 (1988)
[24] E.Y. Loh, T. Martin, P. Prelovsek and D.K. Campbell, Phys. Rev. B38, 2494 (1988)
[25] C. Castellani, C. Dicastro and M. Grill, Physica C 153-155, 1659 (1988)
[26] D. Schmeltzer, J. Phys. Condens. Matters, Letter to the Editor (1989)
[27] P.W. Anderson, G. Baskaran, Z. Zou and T. Hsu, Phys. Rev. Lett. 58, 2790 (1987)
[28] Z. Zou and P.W. Anderson, Phys. Rev. B37, 627 (1988)
[29] S. Kivelson, D. Rokshar and J. Sethna, Phys. Rev. B35, 885 (1987)
[30] D. Schmeltzer, J. Phys. C21, L139 (1988)
[31] D. Schmeltzer, Physica C 153-155, 1291-1292 (1988)
[32] G. Kottliar, Phys. Rev. B37, 3664 (1988)
[33] A.E. Ruckenstein, P.J. Hirshfield and J. Appel, Phys. Rev. B36, 857 (1987)
[34] F.C. Zhang , C. Gross, T.M. Rice, H. Shiba and R. Joynt Physica C 153-155, 1251-1252 (1988)
[35] D. Poilblanc, Physica C 153-155, 1259-1260 (1988)
[36] E. Mueller-Hartman, M. Drazga, A. Kampf and H.A. Wishmann, Physica C153-155, 1261-1262 (1988)
[37] J. Spalek, Physica C 153-155, 1265-1268 (1988)
[38] M. Lavagna, Physica C 153-155, 1271-1277 (1988)
[39] M. Gabay, S. Doniach and M. Inui, Physica C 153-155, 1281-1282 (1988)
[40] J. Carmelo, Physica C 153-155, 1281-1282 (1988)
[41] P. Fazekas , Physica C 153-155, 1283-1284 (1988)
[42] P. Horsh and W. Linden, Physica C 153-155, 1285-1286 (1988)
[43] Atsushi Nakamura and Tetsuo Matsui, Physica C 153-155, 1301-1302 (1988)
[44] D. Schmeltzer, Phys. Rev. B38, 893 (1988)
[45] D. Schmeltzer, Modern Physics Letter, vol.1, No.5, 631-645 (1988)
[46] F.C. Zhang and T.M. Rice, Phys. Rev. B 37, 3759 (1988)
[47] D. Schmeltzer, J. Phys C 21, L139 (1988)
[48] J.R. Schrieffer, X.G. Wen and S. C. Zhang, Phys. Rev. Lett. 60, 944 (1988)
[49] D. Schmeltzer, Physica C 153-155, 1673-1674 (1988)
[50] J.E. Hirsch, Phys. Rev. Lett. 54, 1317 (1985)
[51] Z. Lingred, J. Phys. B7, 2441 (1974)
[52] D. Schmeltzer, J. Phys.: Condens. Matter 1 (1989) 3295-3300
[53] D. Schmeltzer, Binding of holes induced by gauge fields preprint (January 1989)

[54] A.M. Polyakov, Gauge fields and Strings, Harwood Academic (1988)
[55] F.C. Zhang, Phys. Rev. B 39, 7375 (1989)
[56] S. Chakravarty, B.I. Halperin and D. Nelson, Phys. Rev. Lett. 60, 1057 (1988)
[57] M.V. Berry, Proc. R. Soc. A393, 45 (1984)
[58] A. Perelomov, Springer Verlag (1986)
[59] F.D. Haldane, Phys. Rev. Lett. 61, 1029 (1988)
[60] S. Liang, B. Doucot and P.W. Anderson, Phys. Rev. Lett. 61, 365 (1988)
[61] B. Sutherland, Phys. Rev. B 37, 3786 (1988)
[62] D. Schmeltzer (Binding of holons in the chiral spin liquid phase) submitted for publication (1989) January
[63] R.B. Laughlin, Phys. Rev. Lett. 50, 1396 (1983)
[64] A.S. Goldhaber, Phys. Rev. Lett. 36 1122 (1976) and 49, 905 (1982)

FABRICATION OF 107K SUPERCONDUCTING FILMS AND THEIR NORMAL STATE TRANSPORT PROPERTIES

S. Miyazawa, M. Mukaida, K. Kuroda and Y. Tazoh
NTT LSI Laboratories, Kanagawa, Japan

1. INTRODUCTION

Since a new perovskite oxide $(La,Ba)_2CuO_4$ with K_2NiF_4 structure was reported to be superconductive below ~ 30K [1], extensive efforts to search for high Tc superconducting oxide compounds have been carried out worldwide. As a result, $(La,Sr)_2CuO_4$ with a Tc of ~ 40K [2] and $YBa_2Cu_3O_x$ with a Tc of ~ 90K [3] were found and their normal state electronic properties have been well studied. On the other hand, a number of new oxide compounds not consisting of rare-earth elements have been also synthesized. The first compound was $Bi_2Sr_2CuO_{7+x}$ with a Tc of ~ 22K [4], and the other was Bi-Sr-Cu-O compound with a Tc below 8K [5].

More recently, a new superconducting Bi-Sr-Ca-Cu-O system exhibiting a Tc of ~ 110K has been found [6], and extensive examinations on synthesis of this high Tc superconductor are now being carryied out. It has been verified that the Bi-Sr-Ca-Cu-O system includes at least three kinds of superconducting phases satisfying the general formula $Bi_2Sr_2Ca_{n-1}Cu_nO_x$(n=1,2, and 3) ; $Bi_2Sr_2CuO_x$(Tc=22K, abbreviated as 2201 phase, hereafter), $Bi_2Sr_2CaCu_2O_x$(Tc=75K, 2212 phase or low-Tc phase), and $Bi_2Sr_2Ca_2Cu_3O_x$ (Tc=107K, 2223 phase or high-Tc phase). Among them, the 2223 phase is very attractive in applications for microelectronics because of a higher Tc than that of $YBa_2Cu_3O_x$ and being relatively insensitive against oxygen deficient atmosphere. Many techniques for preparing thin films of this 2223 phase are now being performed ; sputtering [7,8], co-evaporation [9,10], laser ablation[11], reactive ion beam deposition

[12], liquid-phase epitaxy(LPE)[13], and sequential evaporation [14,15]. Most of the results, however, show the difficulty of film formation of this high-Tc 2223 single phase, because the 2212 phase is the most stable and formed readily over the temperature range of 850 ~ 880° C in the case of ceramics bulk formation.

While a Pb-substituted Bi-Sr-Ca-Cu-O system was later reported to be easier for obtaining single phase, exhibiting a Tc of ~107K [16], a film exhibiting a Tc of ~107K has not yet been reported. On the other hand, a single crystal of the 2212 phase was successfully grown and electronic transport properties were measured using the single crystal [17], but the properties of the 2223 phase were not clear because of the difficulty of growing single crystals.

In this Chapter, the synthesis of thin films exhibiting a Tc of ~107K is described, and the evaluation of the films is also reported. Although many techniques for preparing superconducting films have been carried out, a sequential deposition technique using a single electron beam gun is simple and very suitable for varying composition in order to search conditions appropriate for obtaining films of Tc ~ 110K in the Bi-Sr-Ca-Cu-O system. The aim of this study is to know electronic transport phenomena of the high-Tc 2223 phase in the Bi-Sr-Ca-Cu-O system, especially Hall coefficient of this phase, which has not yet been reported.

2. PREPARATION OF FILMS

A sequential deposition of four constituents was performed for preparing thin films of Bi-Sr-Ca-Cu-O polytypoids [14]. As raw materials, Cu, Bi, CaF_2 and SrF_2 were used. The fluorides of Sr and Ca can be easily evaporated by a conventional evaporation system and they stabilize the as-deposited films, while films consisting of metallic Sr and Ca are easily corroded in air. A 10-KW single electron gun and four-hearth vacuum system were used. Background pressure was kept below $5x10^{-6}$ Torr. The raw materials were evaporated in the order of Cu, Bi, CaF_2 and SrF_2 on a substrate that was maintained at an ambient temperature. A modification of deposition sequence did not show any remarkable differences. As for substrates, (100) MgO, (100) $SrTiO_3$, Al_2O_3, Y-stabilized ZrO_2(YSZ), and thermally oxidized (100) Si were examined. However, only MgO and $SrTiO_3$ were finally found to be suitable for obtaining superconducting films, and only the films on MgO substrates showed a zero-resistance transition above 100K. Thus, the following experiments were carried out using the MgO substrates.

Only a single deposition cycle, i.e., Cu → Bi → CaF_2 → SrF_2 , was performed for each film. The composition was varied by adjusting the thickness of each constituent layer. A typical total thickness was about 2000A. The film composition was determined by X-ray Fluorescence Analysis(XFA) calibrated for as-deposited films by the Inductive Coupled Plasma(ICP) method. The as-deposited films were amorphous and showed mirror-like. The appearance of the film surface was found to change to some extent after lengthy storage in an ambient atmosphere, which may be caused by oxidation of Cu and/or Bi. But, the as-deposited films were rather insensitive to the ambient atmosphere, as compared to the films deposited from metallic Ca and Sr as source materials.

The as-deposited films were annealed in oxygen flow at 250° C for 3 hours and subsequently at 850 ~ 900° C for 1 hour to obtain superconducting films. The heating and cooling rates were fixed about 300° C/hr. This first low temperature annealing process seemed to be neccessary to oxidize Bi efficiently because of the melting temperature of metallic Bi, 271° C. Figure 1 shows typical X-ray diffraction(XRD) patterns of the films (as-deposited nominal composition was 2223) annealed at 510°, 640°, and 770° C. Below 510° C, no crystal structure was detected. Above ~ 640° C, a crystallization occured. The pattern indicates that the crystallized film seems to have a preferred orientation, but the peaks could not be indexed with known substances. Above 770° C, XRD showed strong peaks, indicating the formation of superconducting phases in the film. Assuming 2θ =5.8° to be the reflection from (002),

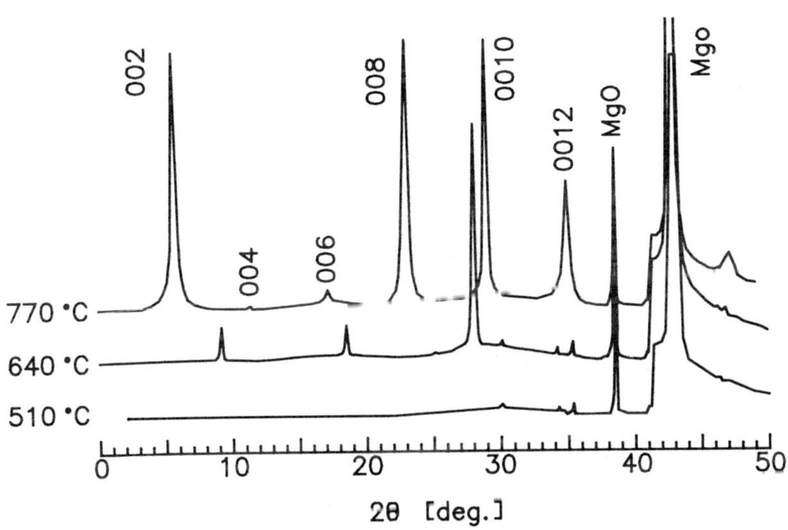

Fig. 1 Successive change of X-ray diffraction pattern as a function of the annealing temperature for the film of 2223.

the lattice constant c was estimated to be 30.6A, which agrees well with that of the low-Tc phase 2212 of the Bi-Sr-Ca-Cu-O system. This implies that the low-Tc 2212 phase readily forms in stable.

The optimum annealing temperature for obtaining a Tc more than 100K was determined eventually to be around 875° C [18]. Annealing at 5° C higher or lower than 875° C did not result in such high transition temperature to zero-resistance. The films annealed higher than 900° C were recognized to be melted partially under an optical microscope. Any remarkable compositional change due to annealing was not detected, and also fluorine was not found in the annealed films by the examinations using an Ultra Light Element Detector(ULED) of Electron Probe Microscope Analysis(EPMA) within the detection limits.

3. Film Thickness Dependence on ρ - T Characteristics

The zero-resistance transition above 110K was found to be as sensitive to film thickness as it was to annealing temperature. Hereafter, the film thickness is defined to the as-deposited one, because the thickness of the annealed film was not measurable due to irregular surface morphology.

Figure 2 shows the resistance vs. temperature (ρ - T) curves for

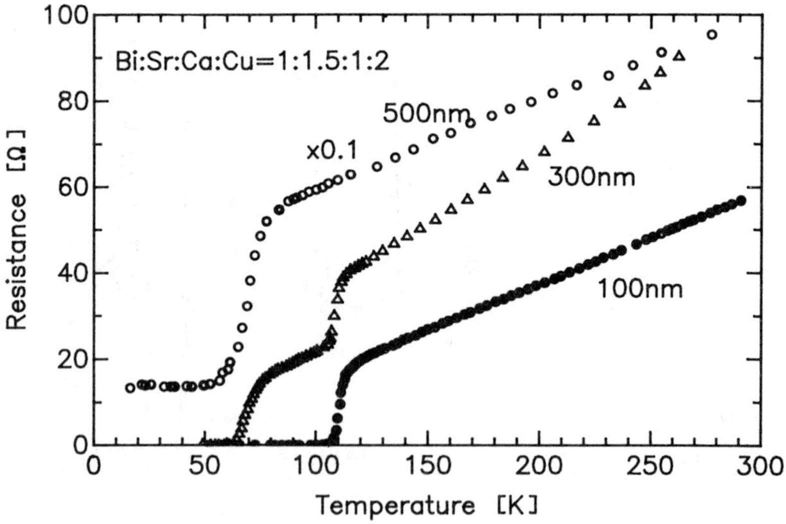

Fig. 2 Temeperature dependence of resistance of the films with different thickness. The as-deposited nominal composition was 1:1.5:1:2 in the order Bi-Sr-Ca-Cu.

three different thickness films [19]. The as-deposited nominal film composition was $BiSr_{1.5}CaCu_2O_x$. The electrical resistance was measured by a conventional four-probe pressure contact method with a probe spacing of 1mm. A zero-resistance was defined as a resistance of less than 10^{-5} of the ambient temperature resistance. The sensitivity was 1 μV with 1 mA for 100 Ω. The films of less than 2000A in thickness exhibited a full resistance drop at about 110K(Tc-on) and a zero-resistance at 107K(Tc-zero), while the 3000A thick film showed a two-step multi-phase superconducting characteristic(110K and 75K) as was reported first for the 1112 bulk sample [6]. The 5000A thick film showed only a transition of the low-Tc phase at around 75K, but did not exhibit a zero-resistance in the measured temperature range. It was also obtained that thinner films had lower resistance at the normal conductive temperature range.

This ρ - T characteristic depending on the film thickness can be explained regarding the surface morphology. In order to study the change in surface morphology, an in-situ microscopic observation during the annealing was carried out under an optical microscope installed with a hot stage. The sample was placed on a SiC coated susceptor that can be heated up to 1000° C by an infrared light. Careful observation revealed that the as-deposited film surface started to melt partially around 820° C, and several different features appeared as a result.

Figure 3(a) shows a Scanning Electron Microscope(SEM) image of typical surface morphology of the sample annealed at ~ 875° C, where very thin sheets(A) adhered to the substrate surface, platelet-like(B), needle-shaped(C) and bulky granular(D) crystallites. The as-deposited nominal film thickness was about 2000A in this case. The composition of crystallites were examined using EPMA. The needle-shaped crystallites were measured to be very rich in Ca and Cu (Ca/Bi=7.9 and Cu/Bi=5.4), and the platelets and thin sheets were a little Ca-poor (Ca/Bi=0.64) and Cu-rich (Cu/Bi=1.4), respectively. On the other hand, the bulky granular crystallites were very rich in Sr. The needel-shaped crystallites were observed in the films annealed above 875° C and platelet-like crystallites were often found in low-Tc films, while the thin sheets were always found in the high-Tc films. Figure 3(b) shows an SEM micrograph of a cleaved surface including the interface between the substrate and the annealed film. The thickness of the thin sheets adhered on the substrate was estimated to be roughly 300 A, while the size of the bulky crystallites was more than 1 μm. The thin sheets looked like epitaxially grown thin films. It is possible that current passes in the thin sheets, because crystallites are isolated from each other.

It was found that the needle-shaped and bulky granular crystallites spreaded over the surface with the increase of the nominal thickness. This possibly resulted in the poor percolation or connection of the

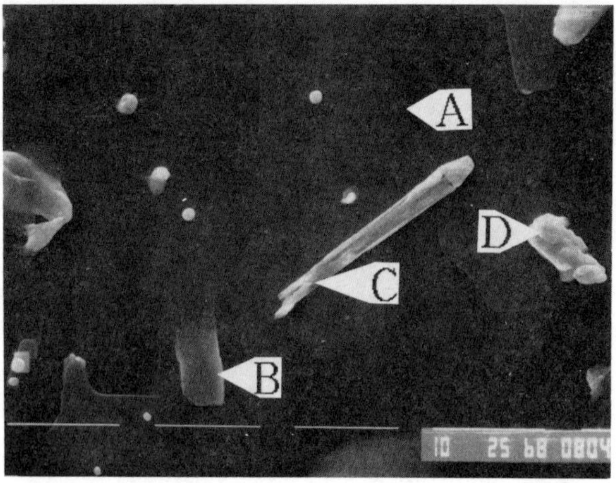

(a)

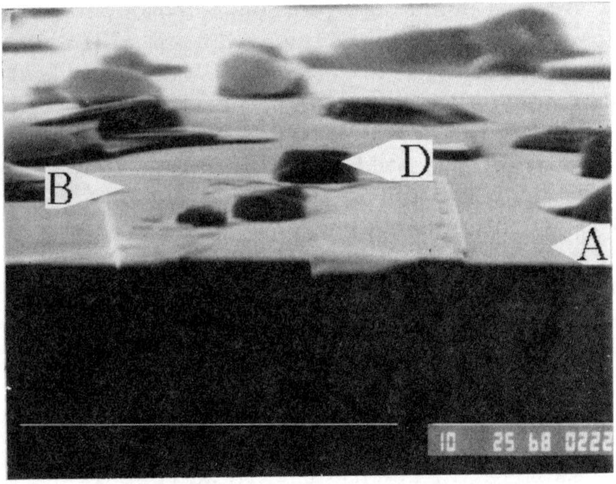

(b)

Fig. 3 Surface morphologies of the as-annealed films observed by an SEM. (a) is a plane-view, and (b) is a cross-sectional view. The bar indicates 10 μ m.

high-Tc phase. The optimum thickness of the as-deposited film was determined to be less than 2500A for obtaining a 107K of a zero-resistance temperature.

4. Composition dependence on ρ - T characteristics

The composition dependence on ρ - T characteristics was investigated based on the annealing conditions and film thickness dependence on the appearance of the high-Tc transition described above. The annealing conditions were fixed at 875° C for 1 hour, and the nominal film thickness was 2000A. It was recognized that the films with a zero resistance above 100K can be obtained by a relatively wide range of Sr concentration, i.e., the as-deposited nominal composition range from 1112 to 1312.

Figure 4 shows typical ρ - T curves for films with various compositions. When the nominal film composition was fixed to 2223, which is well established as the composition exhibiting a zero-resistance of 107K, the ρ - T curve exhibited a two-step resistance drop around 110K and 80K, and zero-resistance was 75K. The linear dependence of the resistance above Tc's was clearly observed for each film.

In this preliminary experiment, it was found that the film with a

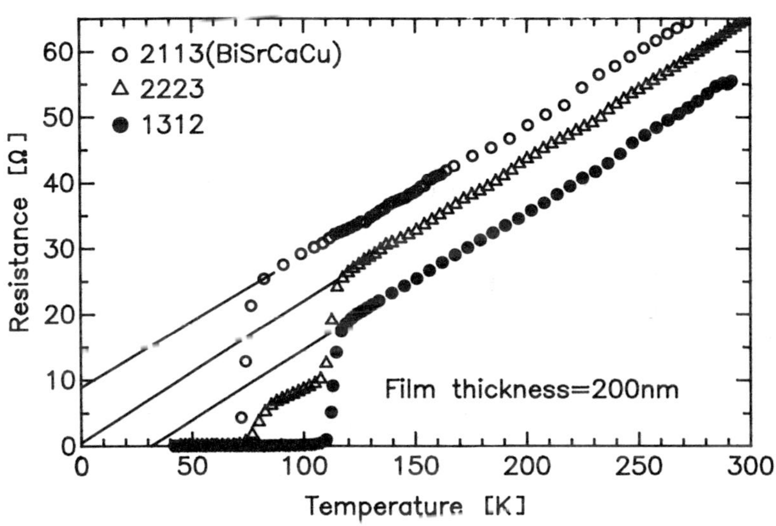

Fig. 4 Temperature dependence of resistance of the films with different compositions in Bi/Sr/Ca/Cu. The lines are the extrapolation of normal state resistance.

nominal composition of 1312 resulted in a zero-resistance transition at 107K. The temperature dependence of the normal resistance normalized as d(R(T))/d(R(300K)) is 3.6×10^{-3}/K and 3.3×10^{-3}/K for the high-Tc and low-Tc films, respectively. Here, we noted [18] that the extrapolation of the normal resistance to zero degrees was always negative for the high Tc film (●), but usually positive for the low-Tc films (○). In the case of the two-step ρ - T film, 2223 for example, the intersected temperature on the temperature axis depended on the volume ratio of high-Tc to low-Tc phases. The larger the volume fraction, the higher the intersected temperature from zero on the temperature axis. This characteristic aspect will be discussed later.

A composition mapping with respect to the zero-resistance transition temperature was readily obtained as shown in Figs.5(a) and (b), where Bi+Cu, Sr, and Ca are used as the three end components in (a) and Bi, Sr+Ca, and Cu are used in (b) [19]. In the figures, closed circles, triangles, open circles and X's represent the film exhibited a zero-resistance temperature above 100K (high-Tc), a two-step mixed transition (70 ~ 80K and ~ 100K), around 70K, and insulative, respectively. We notice here that the film composition that exhibited a Tc of 107K lies along the (Bi+Cu):Ca = 3:1 line in (a) and Bi: Cu= 1:2 line in (b), and composition area outside these lines resulted in multi-phase resistance behavior, i.e., the two-step ρ - T curve. These results strongly suggest that Ca ion plays an important role in high-Tc phase formation. In fact, the Tc of Ca-free $Bi_2Sr_2CuO_x$ was reported to be as low as 20K [4].

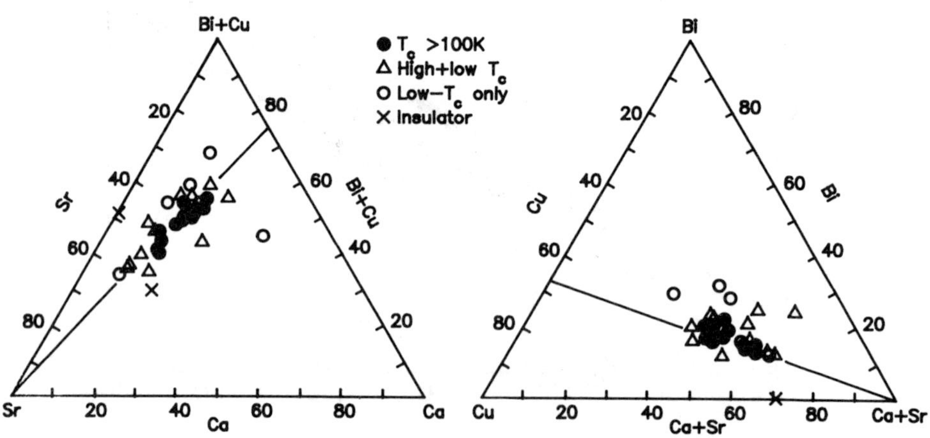

(a) (b)

Fig. 5 Composition mapping with respect to the zero-resistance temperature.

The reason is not clear yet why the high Tc film is obtained over a relatively wide range of compositions, especially Sr. In the case of preparing the Bi-Sr-Ca-Cu-O bulk material, the most stable 2212 phase is formed readily over the temperature interval of 850 ~ 900° C. As was well established elsewhere, the 2223 high-Tc phase begins to form within the 2212 phase in a very narrow temperature range close to the melting point (~ 880° C). By careful setting of the sintering temperature and controlling the cooling process, quite a large amount of the 2212 phase can be converted to the 2223 phase [20]. Regarding that the optimum annealing temperature determined experimentally is close to the melting temperature, ~ 880° C, of the Bi-Sr-Ca-Cu-O compound, and assuming that the formation of the high-Tc phase is kinetic rather than equilibrium, the formation of the high-Tc is strongly suggested to be caused by a peritectic or monotectic reaction among the constituents over a narrow temperature range. Excess constituent of Sr might lower the partial melting temperature and play a role as flux for the growth of the high-Tc phase, possibly similar to the addition of Pb, in which case a single phase of the high-Tc can be obtained at a reaction temperature lower than that for Pb-free compound [16]. From the in-situ surface observation during the annealing, it was clearly recognized that the granular crystallites were precipitated in front of a partially melted portion. The fact that the granular crystallites on the annealed film were very Sr-rich may support this assumption.

We recall here the extrapolation of the normal resistance to zero degrees on the ρ T curve. Figure 6 shows ρ - T curves for the films with different Sr concentrations in nominal that exhibited a zero-resistance transition above 100K. Here, it is also recognized that the temperature dependence of resistance was quite linear. The zero-resistance temperature is almost the same, around 107K, and the temperature dependence of resistance for these films are very similar; the normalized ratio of resistance at 300K and 150K, R(300)/R(150), was estimated to be close to 2.1, independent of the Sr concentration. This means that the ρ - T characteristic measured is possibly peculiar to the high-Tc phase in the films examined, that is, the thin sheets adhered on the substrate surface shown in Figs.3(a) and (b) are responsible for this electric behavior, independent of the Sr concetration. The linear ρ - T relation has attracted much interest and has been well discussed to be linked to the two-dimensional character for electron transport, indicating a characteristic signature of the current carrying in the CuO_2 planes.

It should be noted that all the extrapolated lines of normal resistance are not directed toward zero but intersect the temperature axis at 20 ~ 25K. This "characteristic temperature" was observed only for the high-Tc films, and seemed to be unique for the high-Tc phase in the Bi-Sr-Ca-Cu-O system, because a carefully prepared Pb-substituted bulk of (Bi:Pb):Sr:Ca:Cu = (0.8:0.2):0.8:1:2 [16] exhibited the "characteristic

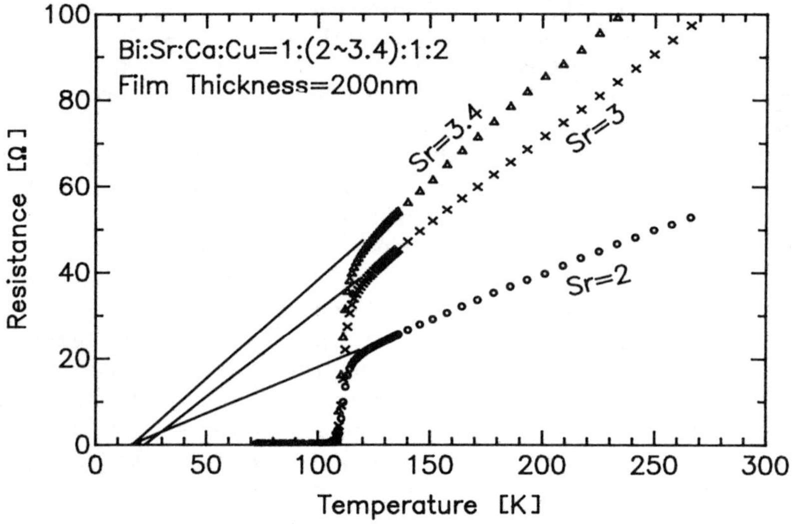

Fig. 6 Temperature dependence of the resistance of the films with different Sr concentration in nominal. The lines are the extrapolation of the normal state resistance.

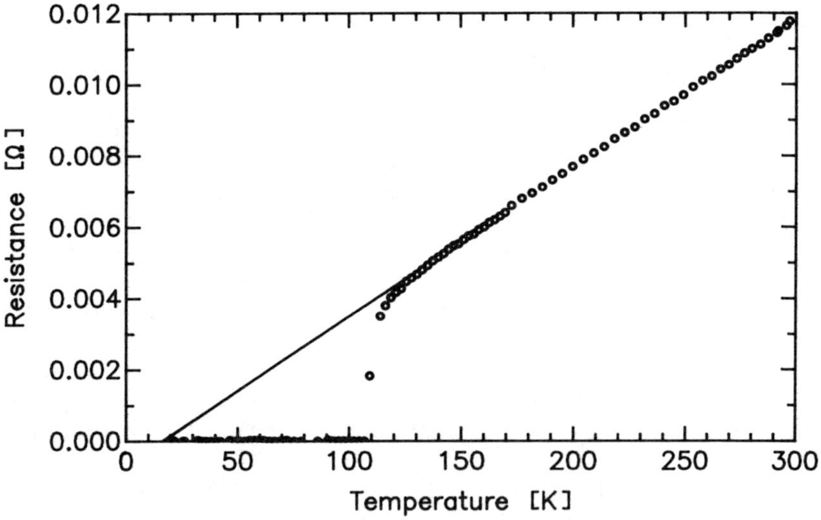

Fig. 7 Temperature dependence of the resistance of the high-Tc bulk of Pb-substituted Bi-Sr-Ca-Cu-O compound. The extrapolated line from the normal state resitance crosses at about 16K.

temperature" of about 16K, as shown in Fig.7. This type of "characteristic temperature" was also observed for highly oriented films and single crystals of the $YBa_2Cu_3O_x$ system [21,22]. Very recently, it has been reported that a new class $Y_2Ba_4Cu_8O_x$, different from the $YBa_2Cu_3O_x$ structure, showed a similar result [23]. The $Y_2Ba_4Cu_8O_x$ structure has a strong two-dimensional lattice framework. Therefore, it can be said that the zero or positive "characteristic temperature" may be the characteristic common to two-dimensional perovskite oxide superconductors, but the physical meaning has not yet been clarified.

5. X-ray Diffraction Analysis

When the film of nominal composition 1112 was annealed at 870° C, there was a full drop of resistance at around 110K, as shown in Fig.8, but a zero-resistance temperature was recognized to be as low as 60K, as is enlarged in the figure. Figure 9 shows the X-ray diffraction pattern taken for this film, where only (00L) reflections are observed indicating that the major part of the film is the low-Tc phase and the film is highly c-axis oriented. Assuming the peak at $2\theta = 5.74°$ to be a reflection from (002) and using different strong peaks indexed as (00L), the lattice constant c was calculated to be 30.60(5)A, which is in good agreement with that measured for the low-Tc bulk material [19].

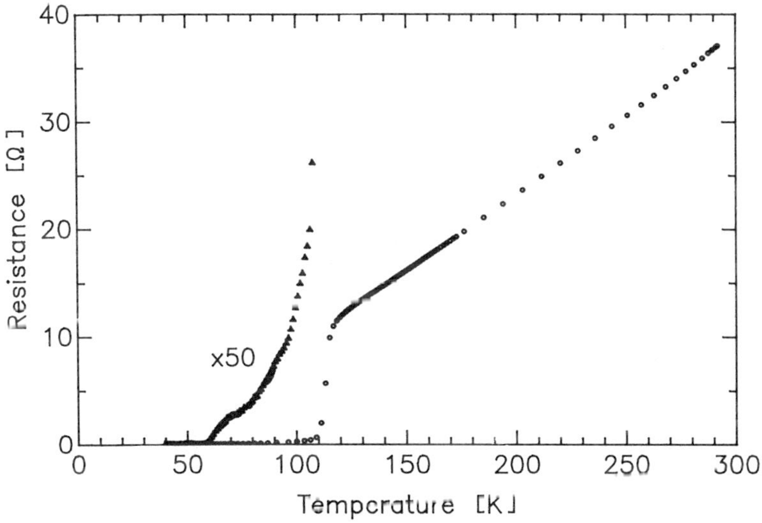

Fig. 8 Temperature dependence of resistance of the film with 1112 composition annealed at 870° C. A zero-resistance temperature reached about 60K.

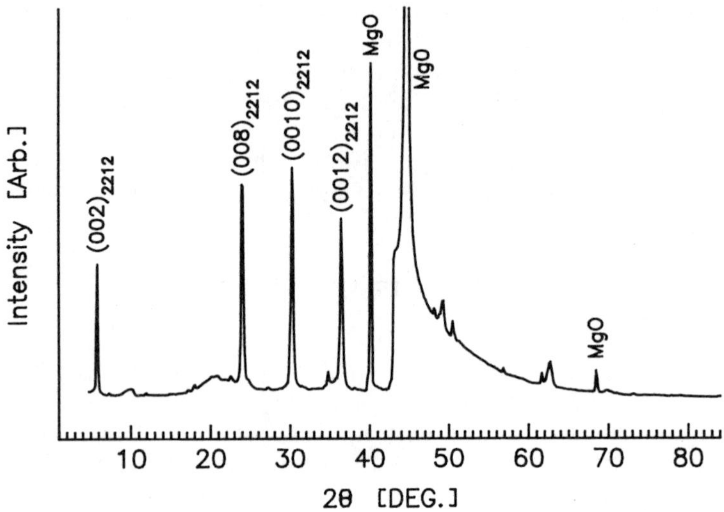

Fig. 9 X-ray diffraction pattern of the as-annealed film with 1112 composition.

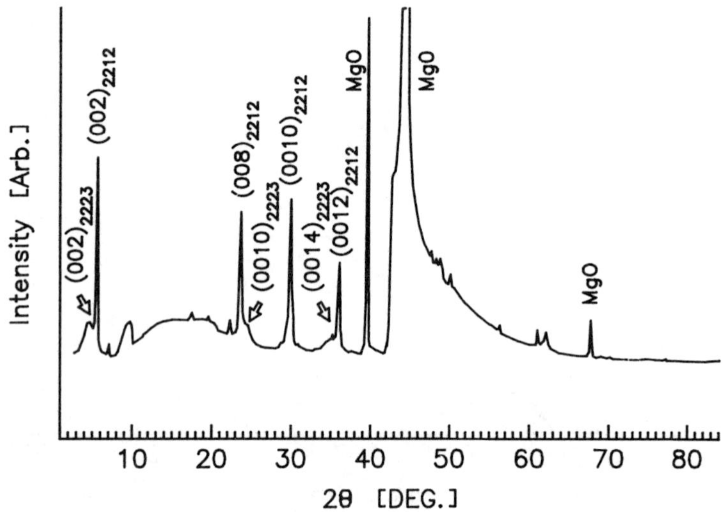

Fig. 10 X-ray diffraction pattern of the film exhibited a zero-resistance at 107K. The arrows indicate an existance of the 2223 high-Tc phase.

To determine whether the films exhibited a zero-resistance transition at 107K, includes the high-Tc phase or not, the X-ray diffraction pattern was taken carefully. Figure 10 shows a typical result, where it can be recognized that all the strong peaks can be indexed as (00L) corresponding to the lattice constant c=30.6A of the low-Tc phase. But, a few additional weak peaks around 2θ =4.8°, 23.8°, and 34.3° can be recognized, as indicated by the arrows. It is well known that the peak at 2θ =4.8° corresponds to the (002) reflection of the high-Tc 2223 phase. The lattice constant c of the high-Tc phase in the film was estimated to be about 36.8A, which is also in good agreement with that of the high-Tc phase. Both the high-Tc and low-Tc phases oriented preferentially along c-axes of each compound perpendicular to the substrate surface. This is not surprising in that, generally speaking, materials with strong two- dimensional lattice framework possibly crystallize in a thin plate form according to its reciprocal lattice constants. In addition, a weak peak at 2θ =7.2° can be observed, corresponding to the 2201 phase. It has been reported that the formation of the high-Tc phase increases or decreases with a reaction between 2201 and 2212 phases, depending on annealing conditions [24]. This strongly suggests that the high-Tc phase is formed kinetically.

From the X-ray diffraction experiments, it can be said that the volume fraction of the high-Tc phase in the films is small, even though zero resistance is 107K. It is considered that, assuming very thin sheets adhered over the whole area of the substrate to be the high-Tc phase, as described previously, the estimated thickness of about 300A possibly provides a small yield for X-ray diffraction efficiency, while bulky granular and plate-like crystallites yielded more. A relatively small volume fraction of the high-Tc phase in the film exhibiting a 107K transition was recognized by measuring the magnetic susceptibility.

6. Magnetic Susceptibility Measurement

The magnetic susceptibility of a 107K film, a low-Tc film, and high Tc phase bulk was measured as a function of temperature from 20K to 150K, as shown in Fig.11 [25]. The high-Tc bulk was successfully prepared according to the reported by Endo et al. [16] The open circles give the temperature dependence for the low-Tc film, showing the on-set of about 75K, although the transition is not sharp. This possibly implies that the film is poor in crystallinity. The open triangles give one for the high-Tc phase bulk indicated a sharp transition at 105K. On the other hand, the 107K high-Tc film exhibited a small change in the susceptibility at around 110K (closed circles), but large at around 70K.

From the result, the volume fraction of the high-Tc phase in the 107K film was estimated to be as small as 10%, as was supposed from the X-ray

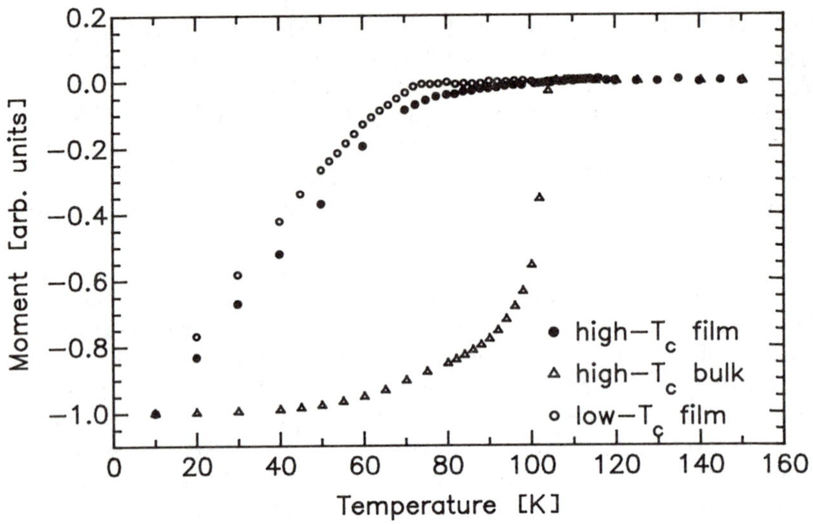

Fig. 11 Temperature dependence of magnetic susceptibility of a 107K film, a 75K film, and high-Tc Pb-substituted bulk compound.

diffraction experiments. However, the SEM observation revealed that very thin sheets adhered on the substrate, that is thought to be a high-Tc phase; approximately 300A thick. This thickness is comparable to the penetration depth of the magnetic field, and, in addition, the H_{c1} of the high-Tc phase was very small, only 60 Oe [26], in comparison to the applied magnetic field of 100Oe used in the present measurement. Therefore, the volume fraction of the high-Tc phase was probably underestimated.

In order to increase the volume fraction of the high-Tc phase in the 107K films, we reexamined annealing conditions, deposition processes, and so on, but all trials failed. Therefore, it is stressed that the sequential deposition of metal elements by means of electron beam does not seem to be suitable for this purpose, because of the necessity of post-annealing at relatively high temperatures and of the kinetic growth of the high-Tc phase.

7. HALL MEASUREMENT

The Hall effect was measured as a function of temperature for the 107K high-Tc film, the 75K low-Tc film, and the high-Tc bulk. The measurement was carried out using a conventional van der Pauw technique under an applied magnetic field of 5000 Oe. For the high-Tc bulk sample, the applied magnetic field was 5 T. In the measurements, the current

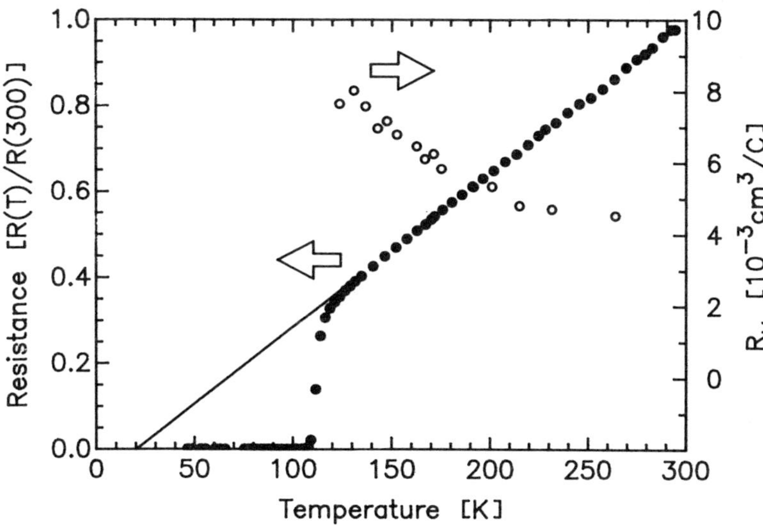

(a)

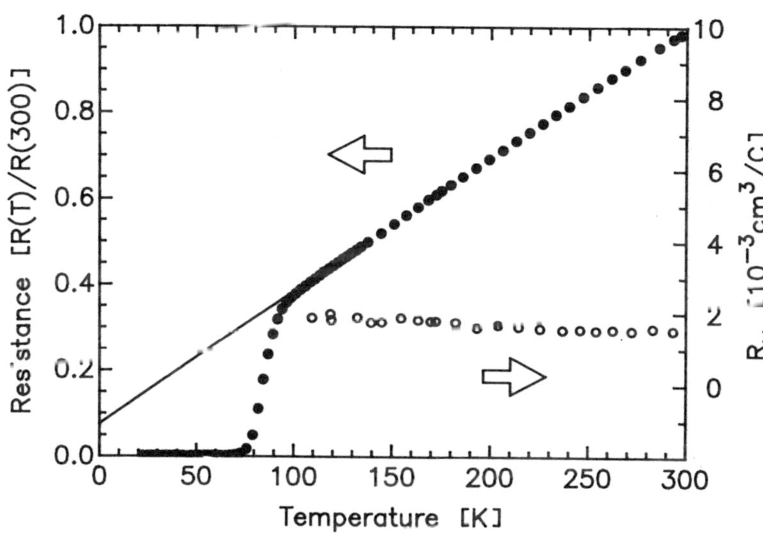

(b)

Fig. 12 Tempetarurc dependence of resistance and Hall coefficient of (a) the high-Tc and (b) low-Tc phase films examined.

direction and magnetic field direction were altered to cancel the thermoelectric power and non-symmetry effect. This procedure was necessary to measure the Hall effect of high resistive oxide superconductors [25]. Ag was used as ohmic contact electrodes, and ohmic I-V characteristics were recognized for each contact.

Figures 12(a) and (b) show the temperature dependences of resistance (ρ) and Hall coefficient(R_H) for the high-Tc and low-Tc films used for the Hall effect measurements, respectively. Here, the resistance R(T) normalized by the resistance at 300K, R(300), was plotted against the temperature. Normal resistance varied linearly with temperature. The Hall measurement showed a hole-type Hall effect for both films, common to the $(La,Sr)_2CuO_4$ [27] and $YBa_2Cu_3O_x$ [28] systems. It is evident that the 107K high-Tc film has characteristics very different from that of the low-Tc film.

Figure 13 summarizes temperature dependences of the Hall coefficient(R_H) of the 107K high-Tc film(●), the low-Tc film(○), and the high-Tc bulk(△) [25]. In the figure, the data for the low-Tc phase $Bi_2Sr_2CaCu_2O_x$ single crystal reported by Takagi et al. [17] is provided for comparison (▲). As can be seen in the figure, the temperature dependences of R_H of the high-Tc and low-Tc films are in good agreement with those of the high-Tc bulk and low-Tc single crystal, respectively.The R_H value of the low-Tc film at 300K was 1.6×10^{-3} cm^3/C, which is close

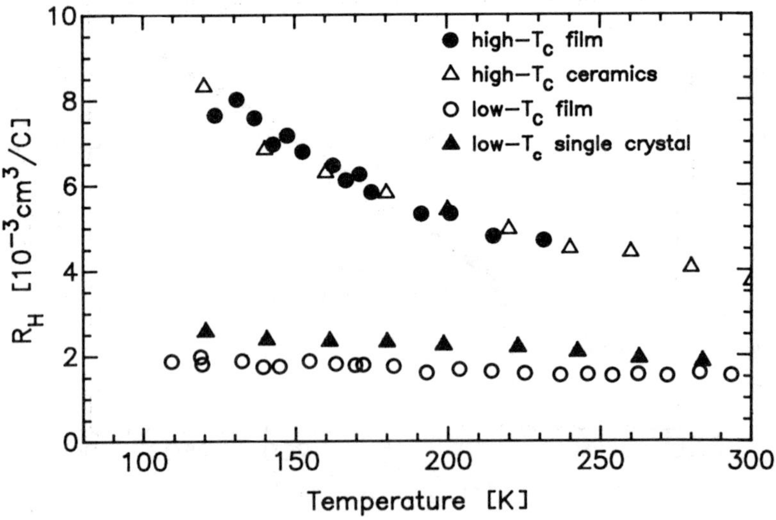

Fig. 13 Summarized temperature dependences of Hall coefficient R_H of the 107K film, 107K bulk, 75K film, and low-Tc phase single crystal reported in Ref.[2].

to that of the low-Tc phase single crystal. The small difference in R_H between the film and single crystal may be due to a poor crystallinity of the low-Tc film, as was revealed by the magnetic susceptibility measurement (see Fig.11). The R_H value of the high-Tc bulk at 300K is estimated to be about 3.8×10^{-3} cm^3/C. Assuming that the thickness of the 107K high Tc film was 300A as described before, the R_H of the 107K film was calculated to be 4.0×10^{-3} cm^3/C, which agrees with that of the bulk. From these estimations, it can be deduced that the thin sheets appeared inevitably in the high-Tc films must be of the high-Tc 2223 phase. It is also stressed that the Hall coefficient of the high-Tc 2223 phase is 2～2.5 times larger than that of the low-Tc phase.

The measured Hall coefficient for the high-Tc phase film was thought to be one of the very thin sheets adhered on the substrate as is seen in Fig.3. This can be explained as follows; if the film is composed of two phases, one of them is very thin but covers a whole area, and the other is bulky and isolated. The volume of the bulky part is greater than that of the thin part. The measured Hall voltage is the sum of the voltage from both. As is well known, the voltage depends on the applied magnetic field, Hall coefficient of materials to be measured, the current density in the materials, and the width of the materials through which the current passes, independent of the thickness of the material to be measured. Therefore, it can be concluded that the measured Hall coefficient mainly reflects the part covering the substrate surface.

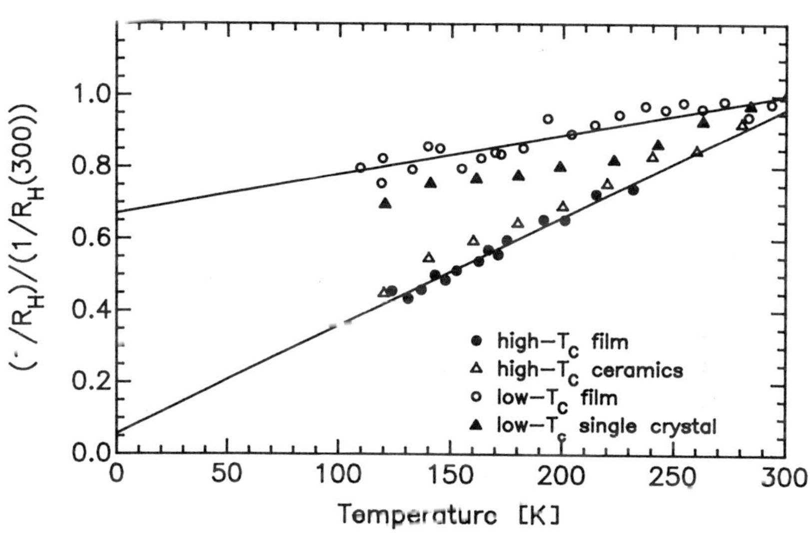

Fig. 14 Inverse Hall coefficient as a function of temperature of both high-Tc and low-Tc phases. The lines are given by the least squares method for high- and low-Tc films.

A very interesting and important result was deduced, as shown in Fig.14, from the temperature dependence of the inverse Hall coefficient. The inverse R_H of the low-Tc film and single crystal was not proportional to the temperature, while that of the high-Tc film and bulk was almost proportional to the temperature. The inverse R_H of the $YBa_2Cu_3O_x$ system was reported to be proportional to the temperature [23]. The temperature dependence of the inverse R_H, $d(R_H(300)/R_H(T))/dT$, is about 3.30×10^{-3}/K, while that of the low-Tc film and single crystal is about 1.10×10^{-3}/K, approximately one-third that of the high-Tc compound. It is also noted that the temperature dependence of the inverse Hall coefficient was linear and the extrapolated line passes almost to the origin. The physical meaning of this zero-passing is still unclear, but this passing to zero seems to be a characteristic feature of the high-Tc phase in the Bi-Sr-Ca-Cu-O system.

The Hall effect, together with the ubiquitous linear ρ - T behavior, may be the most striking anomalies in the normal state of two-dimensional oxide superconductors. According to the theoretical analysis by Micnas et al.[29], the relation $\rho \propto T/\sqrt{n}$ for $k_BT > 2\hbar p_F s$ is deduced, where n denotes the carrier concentration, p_F the Fermi momentum, and s the sound velocity. Based on the conventional single-band model, the carrier concentration can be determined by $1/R_H \cdot e$. Then, we plotted the resistivity ρ against the $T\sqrt{R_H}$, and found the linear relation between them, as shown in Fig.15. A fit of R_H to a 1/T behavior is problematical

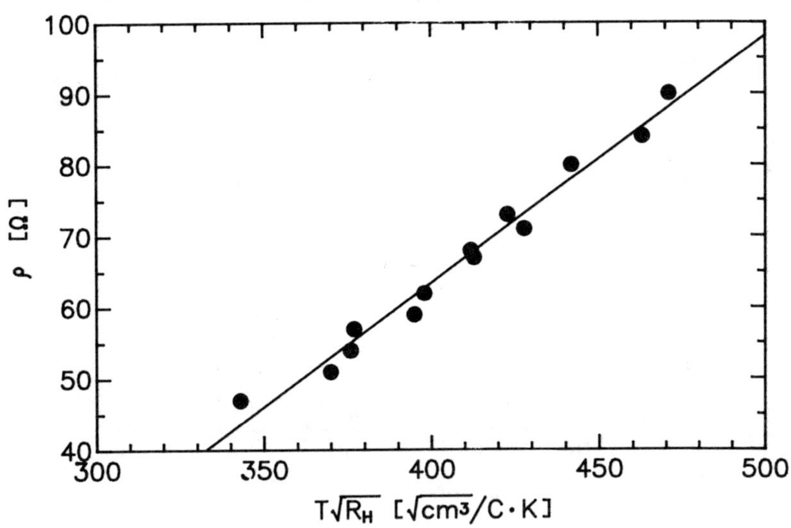

Fig. 15 A linear relationship between ρ and $T\sqrt{R_H}$ for the high 107K film. $\rho \propto T/\sqrt{n}$ is deduced theoretically by Micnas et al. [29]

in a single-band model, and a multi-band model and/or another model is strictly required to explain the temperature dependence of Hall coefficient and resistivity. Further investigations are required for understanding this relation and the mechanism of superconducting behaviors in oxide superconductors with two-dimensional perovskite structure.

8. SUMMARY

Thin films with a zero-resistance transition of 107K in the Bi-Sr-Ca-Cu-O system were successfully synthesized by the sequential deposition of elemental metals using SrF_2, CaF_2, Bi, and Cu, and the succeeding annealing treatment around 875° C for 1 hour. The annealing conditions were found to be very critical. The film composition necessary for obtaining a Tc of 107K was in the region between 1112 and 1312 for the order Bi-Sr-Ca-Cu, and it was found that the ratio of (Bi+Cu):Ca is restricted to 3:1, where the ratio of Bi:Cu was 1:2.

Microscopic observations revealed that the films exhibited 107K of the zero- resistance always contained about 300A thickness thin sheets which are adhered on the substrate surface. The thin sheets were observed independent of the Sr concentration. It was determined that the thin sheets were attributed to the high-Tc 2223 phase, although X-ray diffraction experiments revealed the main phase in the film to be the low-Tc 2212 phase.

It was noted that the extrapolation of the normal resistance to zero degrees always showed negative resistance intersect only for the films exhibited a Tc of 107K.

The Hall coefficients of the high-Tc film, low-Tc film, and also high-Tc bulk were successfully measured as a function of the temperature, and it was verified that the 107K high-Tc phase of the Bi-Sr-Ca-Cu-O system has a stronger temperature dependence than that of the low-Tc phase ; approximately three times larger. In addition, the temperature dependence of the inverse Hall coefficient of the high-Tc phase was proportional to the temperature, passing close to the origin.

The linear-in T of the resistance and the inverse Hall coefficient of the high-Tc 2223 phase in the Bi-Sr-Ca-Cu-O system seems to be peculiar to the two-dimensionality of perovskite layer structure oxide high-Tc superconductors. More detailed investigations, either theoretically or experimentally, are strictly required for understanding the mechanism of superconducting behaviors.

REFERENCES

[1] J.G.Bednorz and K.A.Müller, Z.Phys., B64 (1986) 189.

[2] H.Takagi, S.Uchida, K.Kitazawa, and S.Tanaka, Jpn.J.Appl.Phys., 26 (1987) L123.

[3] M.K.Wu, J.R.Ashburn, C.J.Torng, P.H.Hor, R.L.Meng, L.Gao, Z.J.Huang, Y.Q.Wang, and C.W.Chu, Phys.Rev.Lett., 58 (1987) 908.

[4] C.Michel, M.Herveu, M.M.Borel, A.Grandin, F.Deslandes, J.Provot, and B.Raveau, Z.Phys., B68 (1987) 421.

[5] J.Akimitsu, A.Yamazaki, H.Sawa, and H.Fujiki, Jpn.J.Appl.Phys., 26 (1987) L2080.

[6] H.Maeda, Y.Tanaka, M.Fujimota, and T.Asano, Jpn.J.Appl.Phys., 27 (1988) L209.

[7] M.Nakano, H.Kuwahara, R.Yuasa, H.Mukaida, and A.Mizukami, Jpn.J.Appl. Phys., 27 (1988) L378.

[8] K.Setsune, K.Hirochi, H.Adachi, Y.Ichikawa, and K.Wasa, Appl.Phys. Lett., 53 (1988) 600.

[9] C.E.Rice, A.F.J.Levi, R.M.Fleming, P.Marsh, K.W.Baldwin, M.Anzlowar, A.E.White, K.T.Short, S.Nakahara, and H.L.Stormer, Appl.Phys.Lett., 52 (1988) 1828.

[10] T.Yoshitake, T.Satoh, Y.Kubo, and H.Igarashi, Jpn.J.Appl.Phys., 27 (1988) L1262.

[11] B.F.Kim, J.Bohandy, T.E.Phillips, W.J.Green, E.Agostinelli, F.J.Adrian, and K.Moorjani, Appl.Phys.Lett., 53 (1988) 321.

[12] A.B.Harker, P.H.Kobrin, P.E.D.Morgan, J.F.DeNatale, J.J.Ratto, I.S.Gergis, and D.G.Howitt, Appl.Phys.Lett., 52 (1988) 2186.

[13] R.S.Liu, Y.T.Huang, P.T.Wu, and J.J.Chu, Jpn.J.Appl.Phys., 27 (1988) L1470.

[14] K.Kuroda, M.Mukaida, M.Yamamoto, and S.Miyazawa, Jpn.J.Appl.Phys., 27 (1988) L625.

[15] A.Mogro-Campero, L.G.Turner, M.F.Garbauskas, and R.W.Green, Appl. Phys.Lett., 53 (1988) 327.

[16] U.Endo, S.Koyama, and T.Kawai, Jpn.J.Appl.Phys., 27 (1988) L1476.

[17] H.Takagi, H.Eisaki, S.Uchida, A.Maeda, S.Tajima, K.Uchinokawa, and S.Tanaka, Nature, 332 (1988) 334.

[18] K.Kuroda, M.Mukaida, and S.Miyazawa, Jpn.J.Appl.Phys., 27 (1988) L2230.

[19] K.Kuroda, M.Mukaida, and S.Miyazawa, in Extend Abstract of 20th Conf. Solid State Devices and Materials (1988, Tokyo) 431.

[20] J.M.Tarascon, W.R.McKinnon, P.Barboux, D.M.Hwang, B.G.Bagley, L.H.Green, G.Hull, Y.LePage, N.Stoffel, and M.Giroud, Phys.Rev., B8 (1988) 8885.

[21] M.Naito, R.H.Hammond, B.Oh, M.R.Han, J.W.P.Hsu, P.Rosental, A.F.Marshall, M.R.Beasley, T.H.Geballe, and A.Kaptulnik, J.Mater.Res., 2 (1987) 713.

[22] S.W.Tozer, A.W.Kleinsasser, T.Penney, D.Kaiser, and F.Holtzberg, Phys.Rev.Lett., 59 (1987) 1768.

[23] K.Char, M.Lee, R.W.Barton, A.F.Marchall, I.Borovic, R.H.Hammond, M.R.Beasley, T.H.Geballe, A.Kapitulnik, and S.S.Landrman, Phys.Rev.Rapid Commun. 38 (1988) 834.
[24] H.Nobumasa, K.Shimizu, Y.Kitano, and T.Kawai, Jpn.J.Appl.Phys., 27 (1988) L846.
[25] M.Mukaida, K.Kuroda, and S.Miyazawa, Appl.Phys.Lett.,(in press, Sept. 1989).
[26] N.Murayama, E.Sudo, M.Awano, K.Kani, and Y.Torii, Jpn.J.Appl.Phys., 27 (1988) L1629.
[27] M.Suzuki and T.Murakami, Jpn.J.Appl.Phys., 27 (1987) L524.
[28] S.-W.Cheong, S.E.Brown, Z.Fish, R.S.Kwok, J.D.Thompson, E.Zirngiebl, G.Gruner, D.E.Peterson, G.L.Wells, R.B.Schwarz, and J.R.Cooper, Phys.Rev., B36 (1987) 3913.
[29] R.Micnas, J.Ranninger, and S.Robaszkiewicz, Phys.Rev., B36 (1987) 4051.

ION BEAM ANALYSIS OF HIGH Tc SUPERCONDUCTORS

J. Keinonen, J. Raisanen and E. Rauhala
University of Helsinki, Helsinki, Finland

INTRODUCTION

The discovery of high critical temperature (T_c) superconductivity in the Cu-O-based compounds $LaBa_2Cu_3O_{7-x}$ [1] and $YBa_2Cu_3O_{7-x}$ [2,3] has led to numerous efforts to prepare these new materials with ever increasing T_c's. Since the first reports, a variety of copper oxides having high T_c's has been reported, e.g. $La_{1.8}Sr_{0.2}CuO_4$ [4], $Bi_2Sr_2Cu_2O_{7-x}$ [5], $Bi_2Sr_2CaCu_2O_{8+x}$ [6], $Tl_2Ba_2Ca_2Cu_3O_{10+x}$ [7]. From the viewpoints of the physical mechanism underlying the superconductivity and its obvious technological potential, various modified perovskite structures present new, major challenges in the synthesis of high-T_c superconductors (HTSCs).

The preparation of bulk samples, using conventional ceramic techniques (grinding, firing and pelletizing), has been well established. The technological and scientific importance of these samples is, however, very limited due to the facts that they are quite porous and brittle and show only very moderate critical current densities. A decisive improvement of the critical current density has been achieved with thin films [8]. For electronic applications as well as for the basic understanding of this new class of materials, it is essential to prepare HTSC samples in the form of thin films on suitable substrates. Unlike the bulk material, it is difficult to prepare high-T_c superconducting thin films since the complex composition as well as the appropriate crystal structure of the material have to be reproduced in a planar geometry. Various methods, such as sputtering [9,10], thermal and electron beam evaporation [11,12,13], molecular beam deposition [14], pulsed laser evaporation [15], multilayer deposition [16] and sol-gel [17], have been utilized. The major problems encountered in preparing high quality HTSC films are the need to achieve strict stoichiometry in order to avoid second phases as well as the interdiffusion and reaction between the substrate and the film. Due to the high temperature treatments, required for the formation of the perovskite structure (near 900°C) and the adjustment of the correct oxygen content (above 500°C in oxygen atmosphere), superconducting thin films cannot yet be made on technologically important substrates like silicon. In order to reduce the interdiffusion and interfacial reactions of the HTSC constituents with substrates, different substrates such as Al_2O_3, ZrO_2, MgO, $SrTiO_3$, BaF_2, CaF_2, SrF_2 [12,18], have been studied. So far the best films have been obtained on $SrTiO_3$ substrates.

A very important requirement for the study of HTSC compounds therefore lies in the precise analysis of their composition. Ion beam techniques such as Rutherford backscattering spectrometry (RBS) [19] have long been used to study the elemental stoichiometry of materials including HTSCs. The content of the metal constituents has been measured frequently by RBS using around 2 to 4 MeV ^4He ions, e.g. Refs. [12,15,18]. The oxygen content has been analyzed by non-Rutherford scattering of around 3 and 9 MeV ^4He ions [20-25] and around 2 MeV protons [21,26]. In addition, 25 MeV oxygen ions in the RBS analysis of metals [27] and 0.9 and 2 MeV deuterons in the nuclear reaction analysis of oxygen [28,29] have been reported in the literature. 2 MeV protons have also been utilized in the analysis of metallic constituents by particle induced X-ray emission (PIXE) [28,30]. In this article we shall consider the additional possibilities to use ^7Li and ^{12}C ion beams in the RBS analysis.

The present article reviews the use of ion beam techniques in the analysis of HTSC films. The procedures can also be applied to bulk samples if only the near-surface (outer few μm) stoichiometry is desired. The HTSC materials most frequently studied in the literature, i.e. $LaBa_2Cu_3O_{7-x}$, $YBa_2Cu_3O_{7-x}$, $La_{1.8}Sr_{0.2}CuO_4$, $Bi_2Sr_2Cu_2O_{7-x}$, $Bi_2Sr_2CaCu_2O_{8+x}$ and $Tl_2Ba_2Ca_2Cu_3O_{10+x}$, will be used as examples to illustrate the ion beam analysis. We first discuss the stopping power of HTSCs for different MeV-energy ions and the energy deposited due to their slowing down. In this connection we describe some effects which ion beams have on HTSCs. The techniques to be considered are not strictly nondestructive. The first method to be discussed is Rutherford backscattering spectrometry (RBS). It is generally considered a good method for depth profiling in thin films. In the axial channeling mode, RBS is sensitive to long-range crystalline order and provides a depth-dependent measure of crystal defects. The role of particle induced X-ray emission (PIXE) for compositional analysis will be discussed. The third technique to be presented comprises the use of different nuclear reactions. Emphasis will be laid on the validity of the proton induced γ-ray emission (PIGE) method developed in our laboratory, for the analysis of HTSCs. For a comparison of the different methods, the bulk sample of superconducting $Y_{1.0}Ba_{2.2}Cu_{3.8}O_{7.3}$, which we have analysed previously by 2 MeV ^4He RBS and non-Rutherford scattering of 2.3 MeV protons [26], has been further studied in this article using the PIXE and PIGE techniques. One can thus look for any impurities which can destroy the superconducting phase. A mention will be made of the secondary ion mass spectrometry (SIMS) and the elastic recoil detection (ERD) methods in the analysis of HTSCs. A comprehensive list of references relevant to illustrate the topics considered will be provided.

Electronic Stopping Power

Quantitative analysis of HTSC samples by MeV-energy ion beams requires reliable and accurate stopping powers of HTSCs for the probing ion beams. The stopping powers used in the literature have been based on the assumption of Bragg's rule [31] of additivity of stopping cross sections and on the elemental stopping cross sections according to the consistent parametrization by Ziegler et al. [32]. The semiempirical electronic stopping cross sections [32] are illustrated in Fig. 1 for ^{1}H, ^{4}He, ^{7}Li, and ^{12}C ion beams in $YBa_2Cu_3O_7$ and $Bi_2Sr_2CaCu_2O_8$.

In the case of ^{4}He ions in $YBa_2Cu_3O_7$ compounds, the semiempirical electronic stopping data have been found to be accurate to within 5% for Y and Cu and 7% for Ba [33]. In the cases of other ions and HTSCs, larger deviations from the semiempirical stopping power could be expected. In many cases the tabulated stopping data are based on interpolation between rather sparse data for neighbouring elements. In addition, correction of the stopping data for the contribution of oxygen in solids is taken as an average over the individual perturbations of the electronic systems in various oxides. This correction has been estimated to be at most about 3%

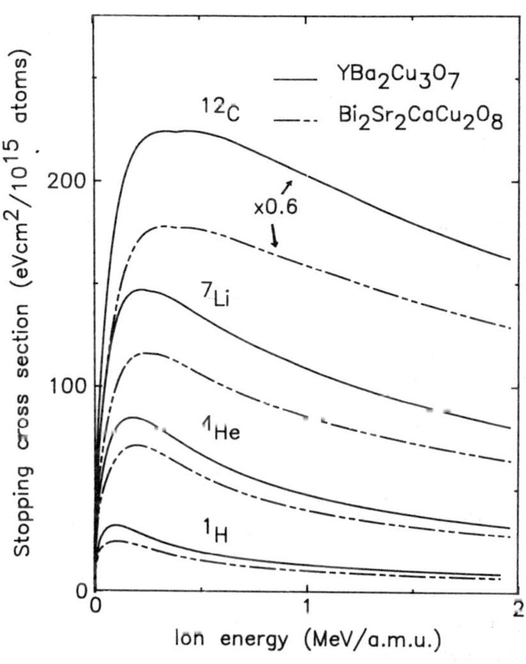

Fig. 1. Electronic stopping powers of $YBa_2Cu_3O_7$ and $Bi_2Sr_2CaCu_2O_8$ for ^{1}H, ^{4}He, ^{7}Li, and ^{12}C ions.

for 3 MeV ^{4}He ions in $YBa_2Cu_3O_7$ [33]. Further, larger deviations (>7%) from the semiempirical stopping data have been obtained for some ions and elemental solids [34]. We therefore think that without the validitation by experimental data, it will not be realistic to assume that the

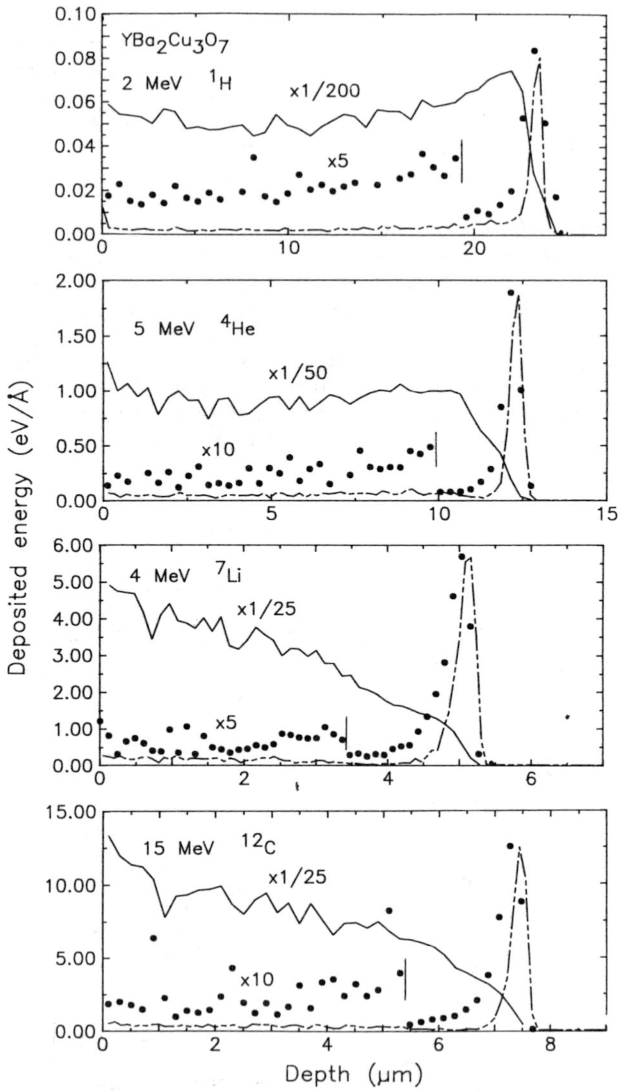

Fig. 2. Range and deposited energy distributions of 2 MeV protons, 5 MeV ^{4}He ions, 4 MeV ^{7}Li ions, and 15 MeV ^{12}C ions in $YBa_2Cu_3O_7$. Dark points, deposited energy due to the nuclear stopping power; full curve, deposited energy due to the electronic stopping power; chain dotted curve, range distribution. The density of $YBa_2Cu_3O_7$ has been taken as 6.4 g/cm^3.

tabulated stopping cross sections are accurate to better than ±7%. Excluding the work by Børgesen and Lilienfeld [33], we have not seen any studies on the electronic stopping power of HTSCs.

Ion Beam Induced Damage

The use of MeV-energy ion beams in the analysis of different materials, is generally thought of as a nondestructive method. However, in the analysis of HTSCs this technique is not strictly nondestructive. The superconducting phases of $YBa_2Cu_3O_{7-x}$ [35-39] and $La_{1.8}Sr_{0.2}CuO_4$ [40] have been observed to be sensitive to the defects created by ion bombardment. The composition has been found to be insensitive to ion damage produced during the analysis [20,21,26,36] and the defects due to the ion beam probing have been observed to disappear in annealing the $YBa_2Cu_3O_{7-x}$ samples at room temperature [36].

The irradiation by MeV-energy ions can produce electrical and structural changes in HTSCs. In the analysis of HTSC films the ion ranges exceed the conventional film thicknesses (100-1000 nm). The damage is approximately uniform throughout the film and does not include complications due to chemistry of dopants or end-of-range damage due to elastic collisions between the ions and the HTSC atoms. In spite of the fact that the dominant energy loss mechanism is the electronic stopping power, the damage produced in thin films of $YBa_2Cu_3O_7$ by different MeV-energy ion beams (63 MeV H, 65 MeV He, 3.5 MeV Be, 6.4 MeV B, 0.8 and 9 MeV N, 0.5 MeV O, 1.0 MeV Ne, and 2.0 MeV Ar) has been shown to be dominated by nuclear collisions [36,37].

It has been shown that bulk superconductivity disappears at fluences significantly below those required to disorder the crystal [36]. The effect of ion beams has been interpreted to result in a progressive decoupling of superconducting grains and to destroy the phase coherence in the sample [35-37]. The mechanism whereby radiation damage affects the grain boundaries is not yet known. The observation that oxygen defects diffuse to the grain boundaries of an HTSC film and stimulate the growth of an amorphous layer [35], is not believed to be the damage mechanism behind the loss of superconductivity [36,37]. In the GeV-energy xenon ion irradiation, removal of oxygen atoms as a result of high energy-density concentration due to electronic energy loss effects has been seen to lead to a crystalline-to-amorphous transformations and rapid decrease in T_c [39].

Calculated deposited energies in elastic and electronic collisions along with the simulated ranges are illustrated in Fig. 2 for $YBa_2Cu_3O_7$ using the same set of ions as considered in Fig. 1. The simulated distributions were obtained through computer programs [34,41,42]. In the obtained of the nuclear stopping power, the scattering angles of the ions were obtained directly from the classical scattering integral [43] and the inter-atomic interaction was derived from the Thomas-Fermi potential [44]. The electronic stopping power was deduced from the semiempirical data [32] using the Bragg's rule [31]. The nuclear energy losses in the MeV energy ion beam studies of damage in HTSCs have varied between 0.2 to 20 eV/Å

[36,37]. The typical fluences ranged between 10^{12} to 10^{16} ions/cm^2 [36,37] and thus overlap those required in ion beam analysis of HTSCs.

RUTHERFORD BACKSCATTERING SPECTROMETRY

General

Low-energy Rutherford backscattering spectrometry (RBS) is generally rapid, quantitative and capable of depth analysis. In the axial channeling mode, it is sensitive to long range crystalline order and provides a depth-dependent measure of crystal defects. It is non-selective, all elements contribute to the spectrum. No standards are needed and usually an overall statistical accuracy of better than 5% is easily achieved for compositional and depth analysis. The main drawbacks of the conventional ^{4}He backscattering method at 2 MeV are the poor detection sensitivity of light elements and limited mass resolution for the analysis of heavy elements. To overcome these shortcomings, either the enhancement of the scattering cross section at high energy non-Rutherford backscattering or the better mass resolution of heavy ion RBS at higher energies may be utilized.

Table 1. Kinematic factors and scattering cross sections of elements in the most frequently studied HTSCs for selected analysing beams in RBS. Scattering angle of θ=170° with respect to the incoming beam and energies of 0.5 (^{1}H), 2 (^{4}He), 4 (^{7}Li) and 15 MeV (^{12}C) are assumed. It has been reported by several groups that replacing Y with rare earth elements (except Ce, Pr and Tb) in $YBa_2Cu_3O_{7-x}$ had little effect on the superconductivity [49-52]. From the RBS point of view these cases can be approximately considered by interpolating between the values given for Ba and Tl.

Z	Element	Kinematic factor				Scattering cross section (mb/sr)			
		^{1}H	^{4}He	^{7}Li	^{12}C	^{1}H	^{4}He	^{7}Li	^{12}C
8	O	0.779	0.362	0.154	-	0.334	0.074	0.031	0.003
20	Ca	0.905	0.672	0.496	0.293	2.103	0.516	0.278	0.070
29	Cu	0.939	0.778$^{a)}$	0.644$^{a)}$	0.468$^{a)}$	4.424	1.098	0.608	0.165
38	Sr	0.955	0.834	0.727	0.579	7.597	1.892	1.056	0.293
39	Y	0.956	0.836	0.731	0.583	8.003	1.993	1.112	0.309
56	Ba	0.971	0.891	0.816	0.7096	16.502	4.119	2.309	0.650
57	La	0.972	0.892	0.818	0.709	17.097	4.268	2.392	0.674
81	Tl	0.981	0.952	0.873	0.792	34.527	8.627	4.844	1.372
83	Bi	0.981	0.927	0.875	0.796	36.253	9.057	5.087	1.441

a) Different isotopes with significant natural abundances, whose separation exceeds the detector resolution at 0.5 MeV/amu.

An extensive amount of publications where ion backscattering spectrometry has been employed in the characterization of HTSC films can be found in the literature. To illustrate the applications, some examples of recent studies are listed next. Rutherford backscattering has been conventionally used to characterize the various metallic elements in the HTSC materials, e.g. Refs. [12,15,18,23]. Diffusion and its temperature dependence at the interface between the superconductor and the substrate have been investigated with respect to the choice of the substrate material [12,45]. RBS channeling has been employed to examine the surface crystallinity, amorphization and annealing behavior of $YCa_2Cu_3O_{7-x}$ single crystal [46] as well as to study the radiation effects and ion beam induced destruction of its superconducting properties [36]. For the analysis of oxygen, the use of non-Rutherford backscattering with 2.5 MeV ^1H beams [21,26,28] and 3 and 9 MeV ^4He beams have been reported [20-25]. The RUMP program package [23,47] has been used for data reduction of backscattering spectra in a number of studies on HTSCs at different laboratories.

In the following we shall briefly discuss analytical properties of RBS with respect to the studies of HTSC films. In the simulations of the RBS spectra, we utilize the elastic scattering data analysis program ELSA developed in our laboratory [48].

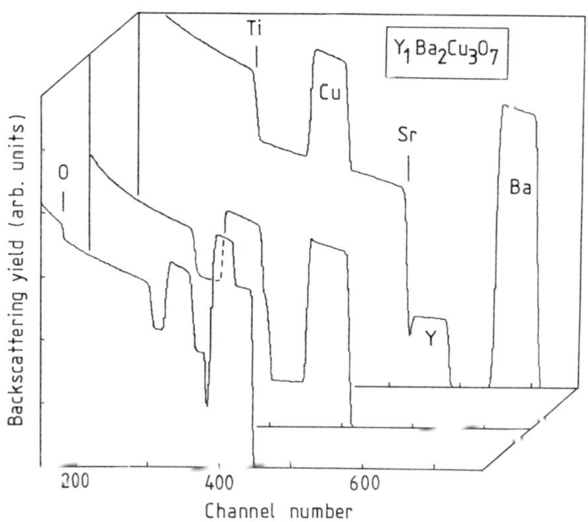

Fig. 3. Simulated backscattering spectra for 2 MeV ^4He (front), 4 MeV ^7Li and 15 MeV ^{12}C ions (back) incident on a 2×10^{18} atoms/cm^2 $YBa_2Cu_3O_7$ film on a $SrTiO_3$ substrate. Particle energies per channel were chosen as 4 keV/ch, 6.4 keV/ch, and 11.6 keV/ch, respectively, to reproduce equal signal widths of the heaviest element (Ba). Scattering angle of $\theta=170^\circ$, system resolutions of 13, 30 and 80 keV, and Bohr straggling [53] were assumed. The yields for ^7Li and ^{12}C ions have been multiplied by 1.8 and 9.7, respectively, with respect to the yield of ^4He ions.

Mass Resolution

The ability to resolve elements having different masses depends on the scattering kinematics. Table 1 presents the kinematic factors of the elements in the most frequently studied HTSCs for ^{1}H, ^{4}He, ^{7}Li and ^{12}C ions. The kinematic factor multiplied by incident ion energy, gives the energies of the leading edges for the elements at the HTSC sample surface. As evident from the table, the kinematic mass resolution decreases with increasing difference between the masses of the colliding particles. The separation of the front edges of two selected elements is dependent on the system's energy resolution. Typical values for the surface barrier detector are 12-15, 15-20, 25-30 and 60-80 keV for one MeV ^{1}H and ^{4}He, 2-4 MeV ^{7}Li and 10-15 MeV ^{12}C ions, respectively. One finds for example that La and Ba are not separated by backscattering of any of these ions. On the other hand, the front edges of the other elements in HTSCs quoted in this article are well separated by using ^{7}Li or ^{12}C ions. To study the metallic elements ^{4}He or heavier ions should be chosen instead of ^{1}H. ^{4}He ions with energies above 2.5 MeV or heavier ions are the preferred choices to distinguish Tl and Ba.

For elements beneath the surface layer the probing ion energy loss has to be considered. The overlapping of the signals of different elements depends on the choice of the ion and its energy, the mass of the target element and the thickness of the layer. Figs. 3-5 illustrate simulated RBS spectra for the 2×10^{18} atoms/cm^2 of $YBa_2Cu_3O_7$, $Bi_2Sr_2CaCuO_8$ and $Tl_2Ba_2Ca_2Cu_3O_{10}$ films on $SrTiO_3$ substrates assuming 2 MeV ^{4}He, 4 MeV ^{7}Li and 15 MeV ^{12}C as incident ions. The increasing signal separation with ion mass and energy is evident. The use of heavy ions makes it possible to investigate compositional changes within an HTSC film like interdiffusion and mixing of heavy elements without any complications due to the overlapping signals from different elements.

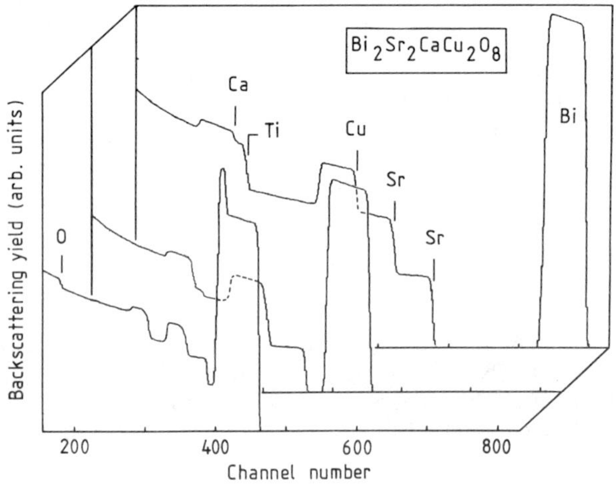

Fig. 4. As for Fig. 3 but for a $Bi_2Sr_2CaCu_2O_8$ film.

Sensitivity

A typical feature of RBS is the poor detection sensitivity with good mass resolution and vice versa. Fig. 6 illustrates the detection sensitivity of some relevant elements in HTSCs probed by 2 MeV ^{4}He and 15 MeV ^{12}C ions. Due to the possibility that a HTSC film may not be thin enough for peaks from different elements to be totally separated, a bulk sample is assumed. Further, a signal is assumed to be detectable if its height is three times the statistical error of the underlying background from the HTSC constituents. Rutherford scattering cross sections for selected ions and energies are given in Table 1 for comparison. The yield from Rutherford scattering is greater for high-Z elements (e.g. Ba) as compared to low-Z elements (e.g. O), and the amount of backscattering decreases as the incident ion energy increases. In the laboratory frame of reference, the cross section is proportional to $(Z/E)^2$, where Z is the atomic number of the target element and E the incident ion energy. The inverse proportionality of the cross section on the square of ion energy facilitates the scaling of cross sections to other energies. In RBS the signal area is directly proportional to the areal density (atoms/cm^2) and the cross section of the element. However, the height of the signal is inversely proportional to the stopping cross section factor [19]. Due to the low cross section, oxygen is not reliably detectable by RBS in bulk and thick film HTSC samples as the oxygen peak rests on a high background due to heavy metallic constituents. To obtain good statistical accuracy in RBS with ions heavier than ^{4}He, the lower scattering cross sections and higher stopping powers (Fig. 1) have to be compensated by higher incident ion doses. Using heavier ions the rapid increase of the deposited energy (Fig. 2) and radiation damage have to be taken into account. Analysed samples may not be usable for superconducting studies any more.

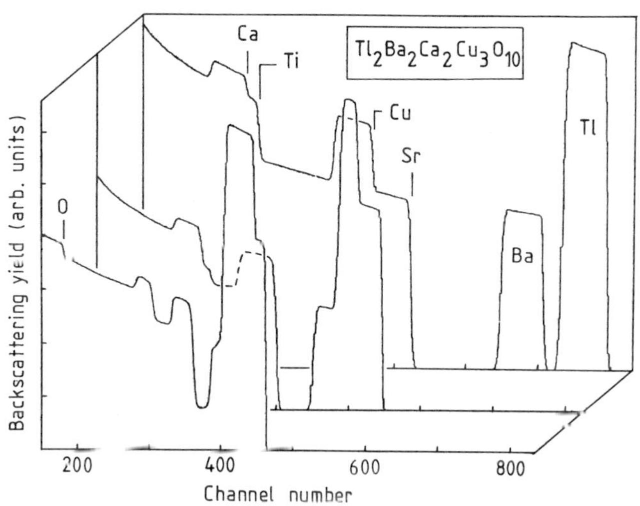

Fig. 5. As for Fig. 3 but for a $Tl_2Ba_2Ca_2Cu_3O_{10}$ film.

Accessible Depth

To reach greater depths in RBS, either lighter ions or higher energies are needed. Table 2 shows the accessible depth in the case of $YBa_2Cu_3O_7$ for the ions of Table 1 at selected energies. A criterion $E_1 \geq (kE_o)/4$ has been adopted, where E_1 and E_o are the detected and incident ion energies respectively, and k the kinematic factor of the element in question. It may be noted that the analysis range for 2 MeV ^4He and 4 MeV ^7Li ions is quite similar. This range is somewhat exceeded by 15 MeV ^{12}C ions and even more by 5 MeV ^4He. In the case of 2 MeV ^1H ions, depths of about 10 μm may be reached.

Depth Resolution

Depth resolution is the ability to sense composition changes or variations in impurity distributions with depth. It is defined as $\delta x = \delta E_1/[S]$, where δE_1 is the energy resolution at depth x and [S] the energy loss factor [19]. Both the system resolution and the straggling contribute to δE_1. Table 3 gives the depth resolutions at the $YBa_2Cu_3O_7$ sample surface and at 1 μm depth for the ions and energies of Table 2. Bohr straggling [53], scattering angle of $\theta = 170^o$, system resolutions of 12, 15, 30 and 80 keV for ^1H, ^4He, ^7Li and ^{12}C ion beams, respectively, have been assumed. The average atomic mass of $YBa_2Cu_3O_7$ is taken as 50 amu and the specific gravity as 6.4 g/cm^3. Similar depth resolution behavior of ^4He, ^7Li and ^{12}C ion beams at the energies used is evident.

The resolving power of RBS has facilitated the measurements of diffusivities for different metallic constituents of thin $YBa_2Cu_3O_7$ films on MgO, sapphire, quartz, and Si substrates by 2.7 MeV ^4He ions [45]. The film thicknesses were chosen so that different constituents could be identified through separate peaks. We observe from tables 1-3 and Figs. 3-5 that the use of ^7Li or ^{12}C RBS would be very suited for such studies on considerably thicker HTSC films.

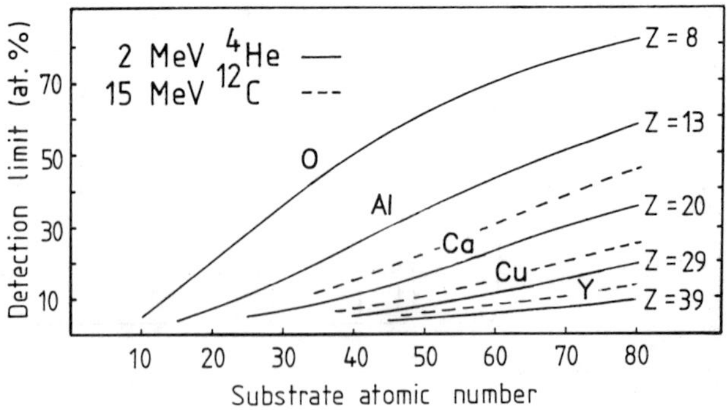

Fig. 6. Detection limits of some relevant elements in bulk HTSC materials estimated for 2 MeV ^4He and 15 MeV ^{12}C ion backscattering at $\theta = 170^o$.

Table 2. Accessible depths for analysis of the elements of $YBa_2Cu_3O_7$ with ^1H, ^4He, ^7Li and ^{12}C ions.

Ion	Energy (MeV)	Accessible depth (μm)			
		Ba	Y	Cu	O
^1H	2	10.7	10.5	10.3	8.8
^4He	2	1.4	1.3	1.3	0.8
	5	4.8	4.5	4.3	2.1
^7Li	4	1.6	1.5	1.5	0.9
^{12}C	15	2.2	2.0	1.8	-

Channeling

It has been demonstrated that RBS channeling will contribute significantly to an understanding of the thermal and chemical stability of interfaces between the HTSCs and substrates, overlayers and ambients with which HTSCs can come in contact [46]. The bulk channeling results have been shown to agree well with expectations for high-quality crystals based on the crystal structure and lattice dynamics of $YBa_2Cu_3O_{7-x}$ derived from neutron-diffraction studies [46].

The dynamics of the Cu-O layers and the effects of oxygen vacancies on high temperature superconductivity have attracted considerable theoretical and experimental attention. The structure of oxygen sublattice is clearly a central issue in HTSCs [54]. Many diffraction experiments have been conducted to determine the structural parameters of oxygen atoms. RBS channeling is potentially very sensitive to the dynamic and static displacements of oxygen from their ideal lattice sites, since out-of-row atoms get emphasized in a channeling spectrum.

The orientation of HTSC films with respect to substrates have been observed by RBS channeling techniques. $Bi_2Sr_2CaCu_2O_{8+x}$ films have been observed to be epitaxial on (100) and (110) $SrTiO_3$ substrates [55] and

Table 3. Depth resolutions at the heaviest element (Ba) signal in $YBa_2Cu_3O_7$ for ^1H, 2 ^4He, ^7Li and ^{12}C ions.

Ion	Energy (MeV)	Depth resolution (nm)	
		surface	at 1 μm depth
^1H	2	100	230 (500 at 5 μm)
^4He	2	16	50
	5	25	85 (120 at 3 μm)
^7Li	4	20	50
^{12}C	15	25	45

$YBa_2Cu_3O_{7-x}$ films on (100) $SrTiO_3$ substrates, whereas only a small degree of orientation has been observed on (110) $SrTiO_3$ [56]. A considerable amount of interface reaction has been seen to take place during annealing, which puts a lower limit on the thickness of the HTSC films. The RBS channeling provides a good method to study the minimum thicknesses of HTSC films with single crystal surface structures as illustrated by the case of epitaxial GaAs films on single crystal Si [57].

Non-Rutherford Backscattering

The structural, electric and superconducting properties of HTSCs are extremely sensitive to oxygen content [58-60]. There are a number of techniques known to be useful in oxygen detection such as electron probe microanalysis [61], thermogravimetry [62], inert gas fusion nondispersive infrared method [63] and neutron diffraction [64]. However, most of the methods are limited when absolute measurement of small oxygen concentrations is required. This has led to a search for more sensitive methods of measuring oxygen concentrations. Backscattering experiments have been performed in the non-Rutherford energy regime, where the oxygen scattering cross section is greatly enhanced relative to pure Rutherford scattering. Either resonant or nuclear potential scattering is then utilized. In the former case the cross section vs. energy exhibits sharp resonances, e.g. at 3.05 MeV for ^{4}He ions, while in the latter case it smoothly enhances with ion energy. The scattering cross sections for ^{1}H

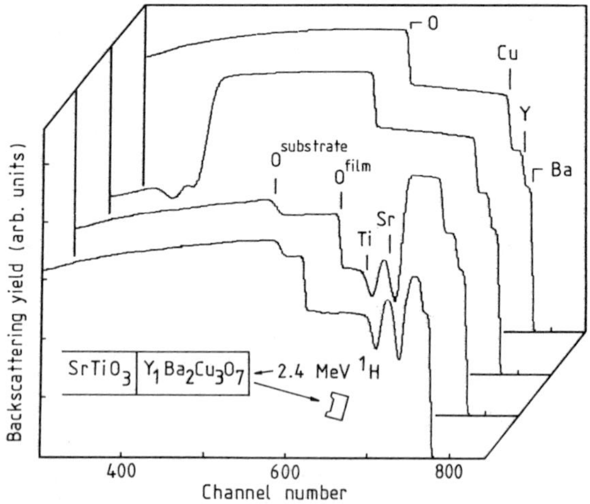

Fig. 7. Simulated non-Rutherford backscattering spectra for 2.4 MeV ^{1}H ions incident on $YBa_2Cu_3O_7$ films of different thickness on a $SrTiO_3$ substrate. The film thicknesses were chosen as 1 μm (front), 2.5 μm, 10 μm and >15 μm (back). Non-Rutherford cross sections of Ref. [67] for O and of Ref. [69] for Ti were assumed. Other parameters are as in Fig. 3.

have been evaluated in Refs. [26,65,66,67] and for ^{4}He in Refs. [22,23,25,68]. In the smooth region at 8.7 MeV the cross section of oxygen for ^{4}He ions is 22 times higher than the Rutherford cross section [20,21]. For ^{1}H the smooth cross section region extends up to 2.5 MeV where the enhancement ratio is 5 [67].

Fig. 7 presents simulated 2.4 MeV ^{1}H ion backscattering spectra from the $YBa_2Cu_3O_7$ superconductor films of different thickness on a $SrTiO_3$ substrate. The simulations have been carried out by the elastic scattering data analysis program ELSA [48]. The program handles any ions and accounts for non-Rutherford scattering for any multielemental target material when experimental cross sections as a function of ion energy are provided. In addition to simulations, spectrum analysis may also be accomplished. Non-Rutherford cross sections for O [67] and for Ti [69] were assumed in the simulations. The Ti cross sections deviate only slightly from the Rutherford cross sections at 2.4 MeV and these deviations are not significant, so far as oxygen analysis in the HTSC films is concerned. From the figure it may be observed that the leading edge of the oxygen signal is uniquely discernable for ready analysis regardless of the film thickness. This is also the case for other HTSC films on a $SrTiO_3$ substrate. With thinner films the oxygen signal from the substrate will, however, impede unambiguous oxygen analysis.

PARTICLE INDUCED X-RAY EMISSION

The main features of particle induced X-ray emission (PIXE) analyses have been known from the early 70's [70]. Its use has increased considerably since and it is now a well established method. To our knowledge, the PIXE method has been reported in the literature for the analysis of HTSC materials only in two cases; 2.5 MeV protons have been utilized to check the chemical composition of a La-Cu oxide [30] and 2 MeV protons have been employed in deducing the relative concentrations of Cu, Y and Ba in $YBa_2Cu_3O_7$ superconductors [28]. The PIXE milliprobe was also used to check the lateral homogeneity of the bulk samples [28]. No PIXE spectra have however been displayed in the literature.

In the present article emphasis is put on the practical problems related to the analysis of the most frequently studied HTSC materials. Since the use of protons as bombarding particles is most common, only they will be considered.

The usability of the PIXE method can be characterized by the magnitudes of the cross sections for X-ray production, i.e. $\sigma_p=\sigma_{ion}\cdot\omega$, where σ_{ion} is the ionization cross section and ω the fluorescence yield. Table 4 presents the selected X-ray production cross sections for constituents of HTSCs. The observation that the replacement of Y with rare earth elements (except Ce, Pr and Tb) in $YBa_2Cu_3O_{7-x}$ had little effect on the superconductivity [49-52] has been taken into account by listing the cross sections for all the rare earth elements. For the rare earth elements heavier than Ba the cross sections for the K X-ray production are insignificant. The L X-rays, which have higher production cross sections, should be used instead. Although the L X-ray production cross sections for the elements lighter than Ba are significantly higher than the corresponding K X-ray production cross sections, the analytical use of L X-rays is limited by the absorption of the soft X-rays in the sample material, the medium between the sample and the detector and in the detector window material. The energies of X-ray lines are given in Table 5 for the HTSC elements considered in Table 4. Due to the limited resolving power of the X-ray detectors some overlapping of X-ray peaks may affect the analysis of HTSC samples. The CuKα peak has about the same energy as the TmLβ lines and the CuKβ peak coincides with the LuLβ lines. The BaKβ peak can not be resolved from the NdKα line and the BaLβ peak from the CeLα and LaLα lines. In an analysis of these elements other peaks may be used instead. The peaks for Ca, Sr, Y, Tl and Bi do not have overlapping energies with transitions presented in the table.

When considering a PIXE analysis of thin HTSC films, the substrate material also should be taken into account. Light constituents (O, F, Mg, Al, Si) of the substrate materials are not relevant in this respect. The $SrTiO_3$ and ZrO_2 substrates can cause overlapping of X-ray peaks since the TiK lines have about the same energy as the BaL lines. The HTSCs $La_{1.8}Sr_{0.2}CuO_4$, $Bi_2Sr_2Cu_2O_{7-x}$ and $Bi_2Sr_2CaCu_2O_{8+x}$ on the ZrO_2 substrate will result in an overlapping of the ZrKα line with the SrKβ line.

Table 4. X-ray production cross sections ($\sigma_{ion} \cdot \omega$) in barns for 2.5 MeV protons. The ionization cross sections and K-fluorescence yields have been taken from the tables of reference [70]. The average L-fluorescence yields have been taken from Ref. [71].

Element	Z	K X-ray σ_p	L X-ray σ_p
Ca	20	411	
Cu	29	68	
Sr	38	9.4	
Y	39	7.6	
Ba	56	0.27	345
La	57	0.23	318
Ce	58	0.19	291
Pr	59	0.16	269
Nd	60	0.14	246
Sm	62		206
Eu	63		189
Gd	64		173
Tb	65		161
Dy	66		148
Ho	67		136
Er	68		125
Tm	69		114
Yb	70		105
Lu	71		96
Tl	81		41
Bi	83		34

The overlapping of X-ray peaks may not be avoidable but the proper choice of the detector may in some cases be helpful. The PIXE spectra obtained from the same $Y_{1.0}Ba_{2.2}Cu_{3.8}O_{7.3}$ bulk sample [26] with 2.5 MeV protons are compared in Fig. 8, one employing an intrinsic Ge detector (50 mm^2 × 6 mm) and the other using Si(Li) detector (80 mm^2 × 5 mm). The important difference between the spectra is the presence of escape peaks in the spectrum measured with the intrinsic Ge detector. The cases where escape peaks of both types of detectors may disturb the analysis of HTSC materials, are listed in Table 6. Only the escape peaks for the low energy X-rays which are significant for the Si(Li) detector are considered in Table 6.

Due to the high counting rates from Cu and Y X-rays, a special pin hole absorber was used to detect the low energy L X-rays of Ba. The absorber consists of a 150 μm thick Al plate which has a 0.2 mm diameter pin hole. The soft X-rays may then reach the detector only through this small hole. By varying the hole size, i.e. the solid angle of the detector for the soft X-rays, the relative peak intensities may be substantially influenced to obtain a more "balanced" spectra. Also a conventional 1500 μm thick Kapton absorber was tried but the observation of the BaL lines was not as clear as with the pinhole Al absorber. The presence of large amounts of

Table 5. K and L X-ray energies for the elements of HTSCs.

Element	X-ray energy (keV)					
	Kα		Kβ		$\overline{L\alpha}$	$\overline{L\beta}$
Ca	3.69		4.01			
Cu	8.04		8.91			
Sr	14.14		15.85		1.81	1.87
Y	14.93		16.75		1.92	2.00
Ba	31.82	32.19	36.38	37.26	4.47	4.83
La	33.03	33.44	37.80	38.73	4.65	5.04
Ce	34.28	34.72	39.26	40.23	4.84	5.26
Pr	35.55	36.02	40.75	41.77	5.03	5.49
Nd	36.85	37.36	42.27	43.30	5.23	5.72
Sm	39.52	40.12	45.40	46.55	5.63	6.20
Eu	40.90	41.54	47.03	48.24	5.84	6.45
Gd	42.31	43.00	48.72	49.96	6.05	6.71
Tb	43.74	44.48	50.39	51.74	6.27	6.97
Dy	45.21	46.00	52.18	53.49	6.49	7.25
Ho	46.70	47.55	53.93	55.29	6.73	7.53
Er	48.22	49.13	55.69	57.09	6.95	7.81
Tm	49.77	50.74	57.58	58.95	7.18	8.10
Yb	51.35	52.39	59.35	60.96	7.42	8.40
Lu	52.96	54.07	61.28	62.95	7.66	8.71
Tl	70.83	72.87	82.56	84.90	10.27	12.21
Bi	74.81	77.11	87.34	89.83	10.84	13.02

heavy elements results in high counting rates in the detector and makes the analysis of light elements difficult even with use of ordinary X-ray absorbers.

Another important point to consider is the choice of the experimental system. In our laboratory we have concentrated on the use of external beams. Protons are extracted into the atmosphere through a thin exit foil [72]. Due to the effective sample cooling with an external beam system, possible elemental losses during irradiation are minimized.

The advantages of the PIXE method are that it is fast, sensitive, multielemental (all elements except oxygen could be determined in a single

Table 6. Escape peaks of practical importance which may disturb the PIXE analysis of HTSC materials.

Intrinsic Ge-detector		Si(Li)-detector	
origin of escape peak[a)]	escape peak energy (keV)	origin of escape peak[b)]	escape peak energy (keV)
$SrK\alpha$	4.26 3.16	$TiK\alpha$[c)]	2.8
$SrK\beta$	5.97 4.87	$TiK\beta$	3.2
$YK\alpha$	5.05 3.95	$CuK\alpha$	6.3
$YK\beta$	6.87 5.77	$CuK\beta$	7.2
$ZrK\alpha$[c)]	5.87 4.77	$Ba\overline{L}\alpha$	2.7
$ZrK\beta$	7.81 6.71	$Ba\overline{L}\beta$	3.2
$BaK\alpha$ $BaK\beta$	~21-22 ~25-27	$La\overline{L}\alpha$ $La\overline{L}\beta$	2.9 3.3
$LaK\alpha$ $LaK\beta$	~22-23 ~27-29		
$Bi\overline{L\beta}$	3.14 2.04		
$Tl\overline{L\beta}$	2.33 1.23		

a) The magnitude of the escape peaks from the rare earth elements is decreasing with increasing atomic number becoming rapidly insignificant.

b) For the rare earth elements the escape peak energies are from 3.1 keV to 7.0 keV.

c) Substrate material.

run) and only a small amount of sample material is needed. PIXE may be well used for contamination and impurity control purposes and to check possible inhomogeneities of HTSC materials. As may be seen from the spectra of Fig. 8, the $Y_{1.0}Ba_{2.2}Cu_{3.8}O_{7.3}$ bulk sample [26] contains a small amount of Sr as an impurity. As the sensitivity of the method depends upon the experimental setup used and the nature of the sample, no general sensitivity values are given. The detection limit in the present case for example for Sr is of the order of few ppm (weight). The method is however not as accurate and straightforward for obtaining the elemental concentration ratios as are the RBS and proton induced γ-ray emission (PIGE) methods.

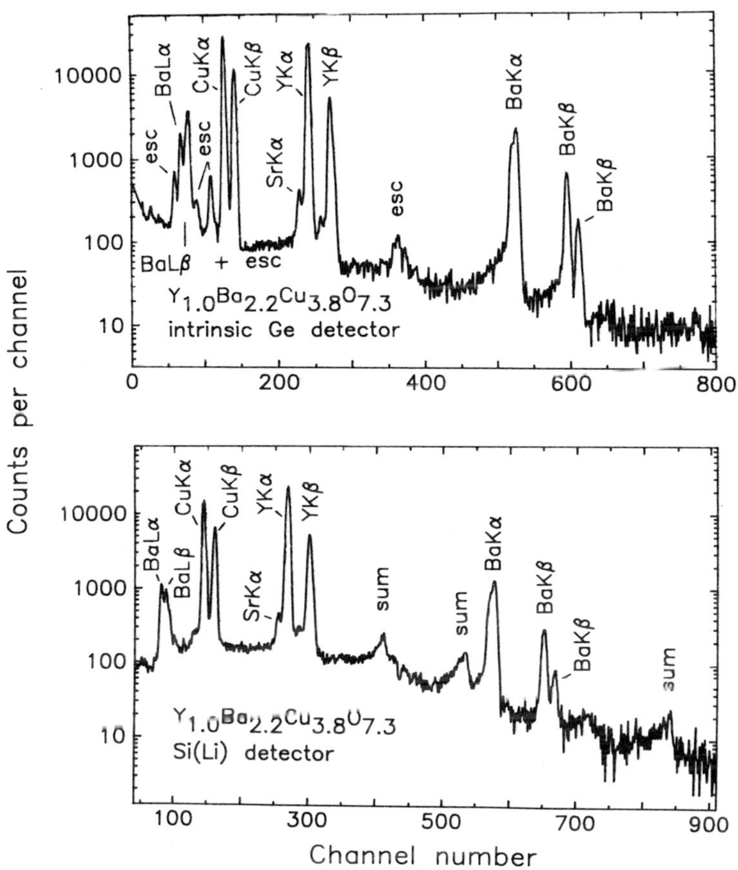

Fig. 8. Typical PIXE spectra obtained from the $Y_{1.0}Ba_{2.2}Cu_{3.8}O_{7.3}$ bulk sample [26] in employing 2.5 MeV protons. The upper spectrum was obtained in an intrinsic Ge detector and the lower spectrum in a Si(Li) detector. A 150 μm thick Al pinhole (0.2 mm ø) absorber was used. The beam currents were ~1 nA on the sample and the measuring times typically 20 minutes.

NUCLEAR REACTION ANALYSIS

To our knowledge, oxygen is the only main constituent of HTSCs which has been analyzed in the literature by employing nuclear reactions. The utilization of the reactions $^{16}O(\alpha,\alpha)^{16}O$ and $^{16}O(p,p)^{16}O$ has been mentioned above in the context of non-Rutherford scattering. The reaction $^{16}O(d,p)^{17}O$ has been used to compare oxygen contents amongst different samples but not to deduce absolute amounts or to obtain directly the ratios between oxygen and metallic atom contents [28,29]. For the deduction of the atomic ratios a standard reference sample has to be used [29].

Due to the high Coulomb barrier of the metallic nuclei for all bombarding particles, non resonance reactions are possible. In their absence the nuclear reaction technique is not sufficiently sensitive for the depth profiling of the constituents in HTSCs. For the analysis of the HTSC film compositions, different nuclear reactions could be used. One of the largest cross sections for the oxygen analysis is provided by the 30 MeV O-beam induced reaction $^{16}O(^{16}O,np)^{30}P$ [73]. A portion of a γ-ray spectrum measured with the $Y_{1.0}Ba_{2.2}Cu_{3.8}O_{7.3}$ bulk sample [26] is shown in Fig. 9. Near the Coulomb barrier the cross section decreases very rapidly with the decreasing O-beam energy [73]. By using 24 MeV O-beam, there results an insignificant contribution of oxygen from the substrate in HTSC films of about 2 μm thickness. In the case of metallic atoms, the heavy ion reactions are not feasible for the analysis.

Proton induced γ-ray emission (PIGE) technique with 9 MeV protons, provides large cross sections for nuclear reaction in most of the HTSC

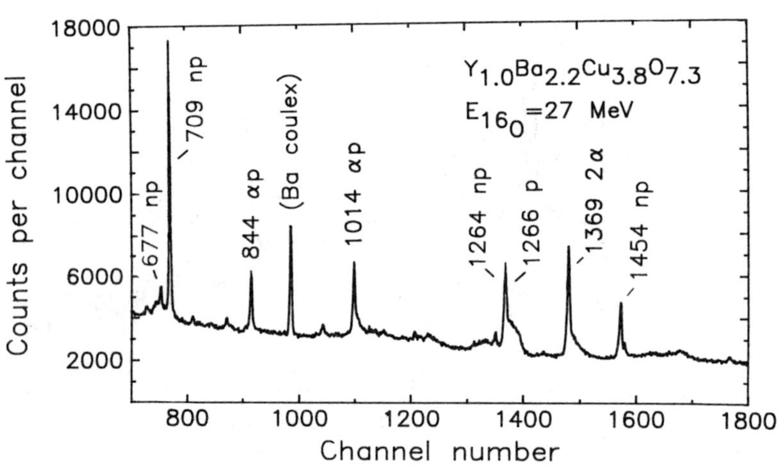

Fig. 9. Portion of a γ-ray spectrum measured with the 24 MeV ^{16}O beam on the $Y_{1.0}Ba_{2.2}Cu_{3.8}O_{7.3}$ bulk sample [26]. The γ-ray energies are given in keV. The strong peak at about 910 keV has been interpreted to be due to the Coulomb excitation of Ba isotopes.

constituents [74]. We have very recently described a method where 9 MeV protons can be used to obtain the O/Cu ratios in HTSCs to a few percent accuracy [75]. Simultaneously, the relative amounts of some other metallic atoms can also be obtained. Portions of a γ-ray spectrum measured with the $Y_{1.0}Ba_{2.2}Cu_{3.8}O_{7.3}$ sample [26] is shown in Fig. 10. The atomic ratios O/Cu = 1.98±0.09, O/Y = 7.6±0.4, and Y/Cu = 0.259±0.012 are in a good agreement with the ratios 1.92±0.12, 7.3±0.4, and 0.263±0.010, rescpectively, obtained in the backscattering studies of the same sample [26]. Barium content could not be measured with PIGE [75]. The main drawback for this case, like in all the cases where nuclear reactions are used, is the fact that the cross sections are not known to an accuracy which would make the absolute measurements or direct determination of atomic ratios meaningful. In our method the use of pure Cu and other metal samples as standards helped to obtain reliable atomic ratios. The rather slow decrease of the cross sections with the decreasing energy [74] along with the low stopping power for high energy protons (Fig. 1) make it difficult to measure composition of HTSC films which have thicknesses less than 4 μm.

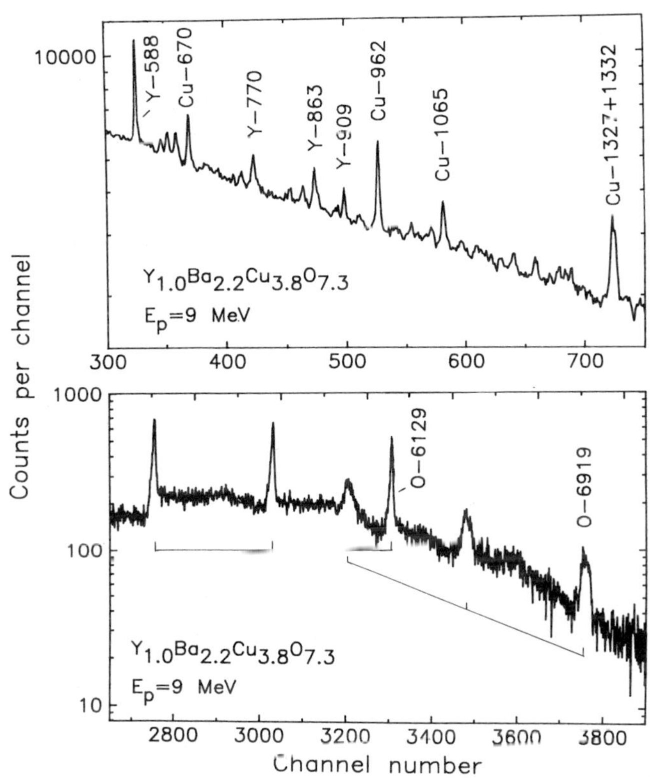

Fig. 10. Portions of a γ-ray spectrum measured with the 9 MeV proton beam on the $Y_{1.0}Ba_{2.2}Cu_{3.8}O_{7.3}$ sample [26]. The γ-ray energies are given in keV.

Hydrogen has been shown to influence the superconducting behavior of $YBa_2Cu_3O_{7-x}$ [76-81]. Up to a concentration of about 1.5 at.% (z=0.2) hydrogen will dissolve in the oxide of $H_zYBa_2Cu_3O_{7-x}$ and slightly increase the T_c [76,77,80]. At higher H contents hydride phases precipitate and the superconductivity depends on the formed hydride systems [77,80]. The migration of hydrogen into the defect regions due to the granularity of HTSCs as well as H implantation have observed to reduce T_c [78,79].

In most cases, the uptake of H by an HTSC sample has been deduced indirectly from the pressure change in H atmosphere. This method is not applicable for HTSC films. In a H implantation of an HTSC film, the content can be evaluated from the fluence and the non-uniform distribution deduced from the range calculations. Experimentally, the H concentration distribution can be obtained accurately by the nuclear resonance broadening (NRB) technique using a ^{15}N ion beam in conjunction with the 6.385 MeV resonance of the $^{1}H(^{15}N,\alpha\gamma)^{12}C$ reaction [82]. Concentration distributions at the level of z=0.1 can easily be measured to a depth of about 2 μm. To our knowledge, this very potential technique has been utilized in only one study of H in HTSCs [81]. The good resolving power of this technique is specially suitable for studies of H at the interfaces of HTSC films with their substrates.

The increase of T_c due to the introduction of fluorine into $YBa_2Cu_3O_{7-x}$ and $EuBa_2Cu_3O_{7-x}$ has been studied recently [83-85]. No evidence for a F-containing superconducting phase [83] has been found in later studies [84,85]. By the utilization of the depth resolving power of the resonance reaction $^{19}F(p,\alpha\gamma)^{16}O$, it was observed that fluorine diffuses out of the HTSC samples at the near surfaces [84]. It is detrimental to superconductivity and that 123 oxides are sensitive to surface effects [85].

OTHER ION BEAM TECHNIQUES

The secondary ion mass spectrometry (SIMS) has been utilized, in addition to RBS, to characterize $La_{1.8}Sr_{0.2}CuO_4$ and $YBa_2Cu_3O_{7-x}$ films on different substrates and to study interdiffusion [18]. Another technique for depth profiling is the use of sputtering combined with Auger electron spectroscopy (AES). This method has been also used to study interdiffusion of $YBa_2Cu_3O_{7-x}$ films on $SrTiO_3$ substrates [56]. Both the techniques are very sensitive in detecting constituents and their distributions in HTSC films. A known common problem of the sputter-based analytical techniques is preferential sputtering. It is especially important in materials containing both low and high masses. To quantify such measurements the use of a standard is required and composition of the latter must be as close as possible to that of the material under measurement. The chemical binding may also effect the results in the AES technique. Both techniques are however very potential for the analysis of very small impurity contents.

Very recently, the use of the elastic recoil detection (ERD) method has been demonstrated as a new technique in the analysis of HTSCs [86]. Results for ^{127}I incident ions at energies up to 240 MeV show that the elements of $YBa_2Cu_3O_7$ and $Bi_2Sr_2CaCu_2O_8$ can be determined simultaneously in about 1 μm thick layer with a depth resolution of a few hundreds Angstrom and a detection limit of about 10^{14} atoms/cm^2.

REFERENCES

[1] J.G. Bednorz and K.A. Müller, Z. Phys. B64, (1986) 189.

[2] C.W. Chu, P.H. Hor, R.L. Meng, L. Gao, Z.J. Huang, and Y.Q. Wang, Phys. Rev. Lett. 58, (1987) 405.

[3] M.K. Wu, J.R. Ashburn, C.J. Torng, P.H. Hor, R.L. Meng, L. Gao, Z.J. Huang, Y.Q. Wang, and C.W. Chu, Phys. Rev. Lett. 58, (1987) 908.

[4] B. Reihl, T. Riesterer, J.G. Bednorz, and K.A. Müller, Phys. Rev. B35, (1987) 8804.

[5] C. Michel, M. Hervieu, M.M. Borel, A. Grandin, F. Deslandes, J. Provost, and B. Raveau, Z. Phys. B68, (1987) 412.

[6] H. Maeda, A. Koizumi, N. Bamba, E. Takayama-Muromachi, F. Izumi, M. Onoda, Y. Kuroda, H. Maruyama, Y. Yoshikawa, T. Ishii, M. Hida, and H. Yamazaki, Jpn. J. Appl. Phys. 27, (1988) L807.

[7] Z.Z. Sheng and A.M. Hermann, Nature 332, (1988) 138.

[8] P. Chaudhari, R.H. Koch, R.B. Laibowitz, T.R. McGuire, and R.J. Gambino, Phys. Rev. Lett. 58, (1987) 2684.

[9] Y. Enomoto, T. Murakami, M. Suzuki, and K. Moriwaki, Jpn. J. Appl. Phys. 26, (1987) L1248.

[10] M. Scheuermann, C.C. Chi, C.C. Tsuei, D.S. Yee, J.J. Cuomo, R.B. Laibowitz, R.H. Koch, B. Braren, R. Srinivasan, and M.M. Plechaty, Appl. Phys. Lett. 51, (1987) 1951.

[11] R.B. Laibowitz, R.H. Koch, P. Chaudhari, and R.J. Gambino, Phys. Rev. B35, (1987) 8821.

[12] M. Naito, R.H. Hammond, B. Oh, M.R. Hahn, J.W.P. Hsu, P. Rosenthal, A.F. Marshall, M.R. Beasley, T.H. Geballe, and A. Kapitulnik, J. Mater. Res. 2, (1987) 713.

[13] D.K. Lathrop, S.E. Russek, and R.A. Buhrman, Appl. Phys. Lett. 51, (1987) 1554.

[14] J. Kwo, T.C. Hsieh, R.H. Fleming, M. Hong, S.H. Liou, B.A. Davidson, and L.C. Feldman, Phys. Rev. B36, (1987) 4039.

[15] D. Dijkkamp, T. Venkatesan, X.D. Wu, S.A. Shaheen, N. Jisrawi, Y.H. Min-Lee, W.L. McLean, and M. Croft, Appl. Phys. Lett. 51, (1987) 619.

[16] Z.L. Bao, F.R. Wang, Q.D. Jiang, S.Z. Wang, Z.Y. Ye, K. Wu, C.Y. Li, and D.L. Yin, Appl. Phys. Lett. 51, (1987) 946.

[17] C.E. Rice, R.B. van Dover, and G.J. Fisanick, Appl. Phys. Lett. 51, (1987) 1842.

[18] P. Madakson, J.J. Cuomo, D.S. Yee, R.A. Roy, and G. Scilla, J. Appl. Phys. 63, (1988) 2046.

[19] W.K. Chu, J.W. Mayer, and M.-A. Nicolet, Backscattering Spectrometry, Academic Press, N.Y., (1978).

[20] J.A. Martin, M. Nastasi, J.R. Tesmer, and C.J. Maggiore, Appl. Phys. Lett. 52, (1988) 2177.

[21] J.C. Barbour, B.L. Doyle, and S.M. Myers, Phys. Rev. B38, (1988) 7005.

[22] P. Berning and R.E. Benenson, Nucl. Instrum. Methods B36, (1989) 335.

[23] B. Blanpain, P. Revesz, L.R. Doolittle, K.H. Purser, and J.W. Mayer,

Nucl. Instrum. Methods B34, (1988) 459.
[24] K.G. Prasad, M.B. Kurup, P. Singh, A.K. Grover, and G.V. Subba Rao, Nucl. Instrum. Methods B36, (1989) 485.
[25] J.A. Leavitt and L.C. McIntyre, Jr., Nucl. Instrum. Methods, B40/41, (1989) 797.
[26] E. Rauhala, J. Keinonen, and R. Järvinen, Appl. Phys. Lett. 52, (1988) 1520.
[27] O. Eryu, K. Murakami, K. Takita, K. Masuda, H. Uwe, H. Kudo, and T. Sakudo, Jpn. J. of Appl. Phys. 27, (1988) L628.
[28] G. Demortier, F. Bodart, G. Deconninck, G. Terwagne, Z. Gabelica, and E.G. Derouane, Nucl. Instrum. Methods B30, (1988) 491.
[29] Z.Y. Hua, H.S. Cheng, C.L. Jia, A.R. Jiang, and B.C. Miao, Int. J. of Mod. Phys. 1, (1987) 1035.
[30] Gy. Hutiray, A. Jánossy, G. Kriza, L. Mihály, S. Pekker, P. Ségransan, Z. Szökefalvi-Nagy, and E. Zsoldos, Solid State Comm. 63, (1987) 907.
[31] W.H. Bragg and R. Kleeman, Philos. Mag. 10, (1905) S318.
[32] J.F. Ziegler, J.P. Biersack, and U. Littmark, The Stopping and Ranges of Ions in Matter, Vol. 1, Plenum, N.Y., (1985).
[33] P. Børgesen and D.A. Lilienfeld, Nucl. Instrum. Methods, B36, (1989) 1.
[34] A. Kuronen, J. Räisänen, J. Keinonen, P. Tikkanen, and E. Rauhala, Nucl. Instrum. Methods B35, (1988) 1.
[35] G.J. Clark, A.D. Marwick, R.H. Koch, and R.B. Laibowitz, Appl. Phys. Lett. 51, (1987) 139.
[36] A.E. White, K.T. Short, D.C. Jacobson, J.M. Poate, R.C. Dynes, P.M. Mankiewich, W.J. Skocpol, R.E. Howard, M. Anzlowar, K.W. Baldwin, A.F.J. Levi, J.R. Kwo, T. Hsieh, and M. Hong, Phys. Rev. B37, (1988) 3755.
[37] D.B. Chrisey, W.G. Maisch, G.P. Summers, A.R. Knudson, and E.A. Burke, IEEE Transactions on Nucl. Sci., 35, (1988) 1456.
[38] S.V. Antonenko, I.Yu. Besotosniy, A.I. Golovashkin, A.I. Grigoriev, N.N. Degtyarenko, V.V. Evstigneev, V.F. Elesin, I.A. Esin, V.E. Zhuchrov, I.V. Zacharchenko, S.I. Krasnosvobodtsev, A.S. Molchanov, E.V. Pechen, I.A. Rudnev, and S.V. Shavkin, Physica C153, (1988) 336.
[39] D. Bourgault, M. Hervieu, S. Bouffard, D. Groult, and B. Raveau, Nucl. Instrum. Methods, B42, (1989) 61.
[40] J. Geerk, G. Linker, O. Meyer, C. Politis, F. Ratzel, R. Smithey, B. Strehlau, and G.C. Xiong, Z. Phys. B67, (1987) 507.
[41] J. Keinonen, Capture γ-ray Spectroscopy and Related Topics-1984, (Ed. S. Raman), A.I.P., N.Y., (1985) p. 557.
[42] P. Tikkanen, J. Keinonen, V. Karttunen, and A. Kuronen, Nucl. Phys. A456, (1986) 337.
[43] N. Goldstein, Classical Mechanics Addison-Wesley, Massachusets, (1980) p. 74.
[44] M. Bister, M. Hautala, and M. Jäntti, Radiat. Eff. 42, (1979) 201.
[45] H. Nakajima, S. Yamaguchi, K. Iwasaki, H. Morita, H. Fujimori, and Y. Fujino, Appl. Phys. Lett. 53, (1988) 1437.
[46] N.G. Stoffel, P.A. Morris, W.A. Bonner, and B.J. Wilkens, Phys. Rev. B37, (1988) 2297.
[47] L.R. Doolittle, Nucl. Instrum. Methods B9, (1985) 344.
[48] E. Rauhala, J. Appl. Phys. 56, (1984) 3324.

[49] P.H. Hor, R.L. Meng, Y.Q. Wang, L. Gao, Z.J. Huang, J. Bechtold, K. Forster, and C.W. Chu, Phys. Rev. Lett. 58, (1987) 1981.
[50] D.W. Murphy, S. Sunshine, R.B. van Dover, R.J. Cava, B. Bartlogg, S.M. Sahurak and L.F. Schneemayer, Phys. Rev. Lett. 58, (1987) 1888.
[51] E.M. Engler, Y.Y. Lee, A.I. Nazzal, R.B. Beyers, G. Lim, P.M. Grant, S.S. Parkin, M.L. Ramirez, J.E. Vazquez, and R.J. Savoy, J. Am. Chem. Soc. 109, (1987) 2848.
[52] X.D. Wu, T. Venkatesan, A. Inam, E.W. Chase, C.C. Chang, Y. Jeon, M. Croft, C. Magee, R.W. Odom, and F. Radicati, Proceedings of the SPIE, Vol. 948 (Ed. R.B. van Dover) SPIE, Bellingham, WA, (1988) p.50.
[53] N.Bohr, K. Dan. Vidensk. Selsk. Mat. Fys. Medd. 18, (1948) no. 8.
[54] Progress in High Temperature Superconductivity, Vol. 14 (Eds. S. Lundqvist, E. Tosatti, M.P. Tosi, and Yu Lu), World Scientific, Singapore, (1988).
[55] C.E. Rice, A.F.J. Levi, R.M. Fleming, P. Marsh, K.W. Baldwin, M. Anzlowar, A.E. White, K.T. Short, S. Nakahara, and H.L. Stormer, Appl. Phys. Lett. 52, (1988) 1828.
[56] X.D. Wu, D. Dijkkamp, S.B. Ogale, A. Inam, E.W. Chase, P.F. Miceli, C.C. Chang, J.M. Tarascon, and T. Venkatesan, Appl. Phys. Lett. 51, (1987) 861.
[57] J. Varrio, H. Asonen, A. Salokatve, M. Pessa, E. Rauhala, and J. Keinonen, Appl. Phys. Lett. 51, (1987) 1801.
[58] J.M. Tarascon, W.R. McKinnon, L.H. Greene, G.W. Hull, and E.M. Vogel, Phys. Rev. B36, (1987) 226.
[59] R.J. Cava, B. Batlogg, C.H. Chen, E.A. Rietmann, S.M. Zahurak, and D. Werder, Phys. Rev. B36, (1987) 5719.
[60] A. Santoro, S. Miraglia, F. Beech, S.A. Sunshine, D.W. Murphy, L.F. Schneemeyer, and J.V. Waszczak, Mater. Res. Bull. 22, (1987) 1007.
[61] M.P.A. Viegers, D.M. de Leeuw, C.A.H.A. Mutsaers, H.A.M. van Hal, H.C.A. Smoorenburg, J.H.T. Hengst, J.C.W. de Vries, and P.C. Zalm, J. Mater, Res. 2, (1987) 747.
[62] J.F. Marucco, C. Noguera, P. Garoche, and G. Collin, J. Mater. Res. 2, (1987) 757.
[63] H. Oyanagi, H. Ihara, T. Matsubara, T. Matsushita, M. Hirabayashi, M. Tokumoto, K. Murata, N. Terada, K. Senzaki, T. Yao, H. Iwasaki, and Y. Kimura, Jpn. J. Appl. Phys. 26, (1987) L1233.
[64] J.D. Jorgensen, M.A. Beno, D.G. Hinks, L. Soderholm, K.J. Volin, R.L. Hitterman, J.D. Grace, I.K. Schuller, C.U. Serge, K.Zhang, and M.S. Kleefisch, Phys. Rev. B36, (1987) 3608.
[65] M. Braun and T. Fried, Z. Phys. 311, (1983) 173.
[66] R.A. Laubenstein, M.J.W. Laubenstein, L.J. Koestler, and R.C. Mobley, Phys. Rev. 84, (1951) 12.
[67] M. Luomajärvi, E. Rauhala, and M. Hautala, Nucl. Instrum. Methods, B9, (1985) 255.
[68] J.A. Leavitt, P. Stoss, D.B. Cooper, J.L. Seerveld, L.C. McIntyre,Jr., R.E. Davis, S. Gutierrez, and T.M. Reith, Nucl. Instrum. Methods, B15, (1986) 296.
[69] E. Rauhala, Nucl. Instrum. Methods, B40/41, (1989) 790.
[70] S.A.E. Johansson and J.L. Campbell, PIXE: A Novel Technique for

Elemental Analysis (Wiley, Chichester, 1988).
[71] D.D. Cohen, Nucl. Instr. and Meth. B22, (1987) 55.
[72] J. Räisänen and A. Anttila, Nucl. Instrum. Methods, 196, (1982) 489.
[73] A. Kuronen, J. Keinonen and P. Tikkanen, Phys. Rev. C35, (1987) 591.
[74] J. Räisänen, T. Witting, and J. Keinonen, Nucl. Instrum. Methods, B28, (1987) 199.
[75] J. Keinonen and J. Räisänen, to be published.
[76] J.J. Reilly, M. Suenaga, J.R. Johnson, P. Thompson, and A.R. Moodenbaugh, Phys. Rev. B36, (1987) 5694.
[77] C.Y. Yang, X.-Q. Yang, S.M. Heald, J.J. Reilly, T. Skotheim, A.R. Moodenbaugh, and M. Suenaga, Phys. Rev. B36, (1987) 8798.
[78] T. Kato, K. Usami, J. Kuniya, and S.-P. Matsuda, Jpn. J. Appl. Phys. 27, (1988) L1104.
[79] M. Nicolas, J.N. Daou, I. Vedel, P. Vajda, and J.P. Burger, Solid State Comm. 66, (1988) 1157.
[80] H. Fujii, H. Kawanaka, W.Ye, S. Orimo, and H. Fukuba, Jpn. J. Appl. Phys. 27, (1988) L525.
[81] P. Berning, J. Collazo, A. Mashayekhi, and R.E. Benenson, Materials Research Society Symposium Proceedings, Vol. 99 (Eds. M.B. Brodsky, R.C. Dynes, K. Kitazawa, and H.L. Tuller), Materials Research Society, Pennsylvania, (1988) p. 939.
[82] H.J. Whitlow, J. Keinonen, M. Hautala, and A. Hautojärvi, Nucl. Instrum. Methods B5, (1984) 505.
[83] H. Xian-Ren, R. Yan-Ru, L. Ming-Zhu, T. Quing-Yun, L. Zhen-Jin, S. Li-Hua, D. Wei-Qing, F. Min-Hua, M. Qing-Yun, L. Chang-Jiang, L. Xiu-Rai, Q. Guan-Liang, and C. Mou-Yuan, International Symposium on High T_c Superconductors, Beijing, People's Republic of China (1987) unpublished.
[84] J.R. Tesmer, C.J. Maggiore, M. Nastasi, S.-W. Cheong, and C.-M. Dick, Materials Research Society Symposium Proceedings, Vol. 99 (Eds. M.B. Brodsky, R.C. Dynes, K. Kitazawa, and H.L. Tuller), Materials Research Society, Pennsylvania, (1988) p. 643.
[85] N.N. Sauer, E. Garcia, J.A. Martin, R.R. Ryan, P.G. Eller, J.R. Tesmer, and C.J. Maggiore, J. Mater. Res. 3, (1988) 813.
[86] J.P. Stoquert, G. Guillaume, M. Hage-Ali, J.J. Grob, C. Ganter, and P. Siffert, private communication, and Nucl. Instrum. Methods B, to be published.

THERMAL CONDUCTIVITY OF HIGH TEMPERATURE SUPERCONDUCTORS

A. Jezowski and J. Klamut
Polish Academy of Sciences, Wroclaw, Poland

1. INTRODUCTION

Processes of energy transport have great significance in solid state physics. Therefore it is hard to imagine a monograph devoted to description of solids in which — to greater or smaller extent — the problem of thermal conductivity would not find place. We present results and discussion of thermal conductivity of high-temperature superconductors. The discovery of superconductivity in the new materials is very recent and the share of thermal conductivity investigations in the otherwise great number of scientific reports is rather moderate. This is connected not only with necessity of conducting long-term, precise measurements in very large ranges of temperature, but also with really poor definition of the investigated material, both because of non-satisfactory technology and because of lack of detailed knowledge about physical constitution (degree of crystallization, porosity, anisotropy, *etc.*). Anyway, experimental results hitherto obtained turned out to be particularly interesting, evoking in many cases new view on properties of high-temperature superconductors. Thermal conductivity investigations not only are significant for explanation of physical processes occurring in these materials but also are really necessary from the point of view of their future applications. In the recent two years above 35 works, done in 17 most important research centers, and dealing with measurements of thermal conductivity were published. The present review, basing on the available literature, sums up the results of investigations of La-CuO, La-Sr-CuO, Y-Ba-CuO, RE-Ba-CuO and presents result of the sole measurement of Tl-Ca-Ba-CuO.

On the phase transition of a substance from normal to superconducting state its physical properties change fundamentally. The electromagnetic, thermal and kinetic properties of superconductors differ from properties of metals in normal state. The basic difference follows mainly the presence of the condensate of electron (charges) pairs, which motion is not connected with entropy transport. Thus the kinetic phenomena in superconductors have particular character, related to the dependence of number of electron excitations on temperature (below T_c their number tends to zero while $T \to 0$). The differences can also involve the mechanisms leading to superconductivity itself, the strength of interaction between elementary excitations or the appearance of singularities connected with phase transitions. These and other (presently not known) requisites can cause anomalous thermal conductivity of high-temperature superconductors, showing themselves up also in anomalies of, *e.g.*, specific heat, electrical resistivity and thermal expansion. Extensive interpretation will be thus possible after gathering more data, involving not only thermal conductivity.

1.1. Basic Mechanisms of Thermal Conductivity

Thermal conductivity is one of most complicated kinetic phenomena occurring in solid matter. In the transport of thermal energy takes part directly the electron- and phonon- subsystems. Assuming fulfillment of the Mathiessen's Rule one can adopt that the total thermal conductivity κ is equal to the sum of two components: electron thermal conductivity and lattice thermal conductivity.

$$\kappa = \kappa_e + \kappa_{ph} = [(\kappa_e^{d})^{-1} + (\kappa_e^{ph})^{-1} + (\kappa_e^{e})^{-1}]^{-1} + \\ + [(\kappa_{ph}^{e})^{-1} + (\kappa_{ph}^{ph})^{-1} + (\kappa_{ph}^{d})^{-1}]^{-1} \quad (1)$$

Thermal conductivity differs from electrical conductivity not only in twice the number of relaxation mechanisms (characteristic time τ) but also in different weight of their contribution to conductivity. Summing up, we may distinguish six mechanisms of relaxation: electron scattering on admixtures (κ_e^{d}), on phonons (κ_e^{ph}) and on electrons (κ_e^{e}), and also phonon interactions with electrons (κ_{ph}^{e}), with phonons (κ_{ph}^{ph}) and scattering of phonons on admixtures, defects and borders (κ_{ph}^{d}) - (in brackets there are designations of the thermal conductivity coefficients relating to the respective processes). The thermal resistivity (κ^{-1}) connected with the particular type of carriers can be, in the first approximation, determined as sum of the thermal resistivities corresponding to the respective mechanisms (1). Each of them is characterized by its temtemperature dependence, hence the whole picture is unusually complicated,

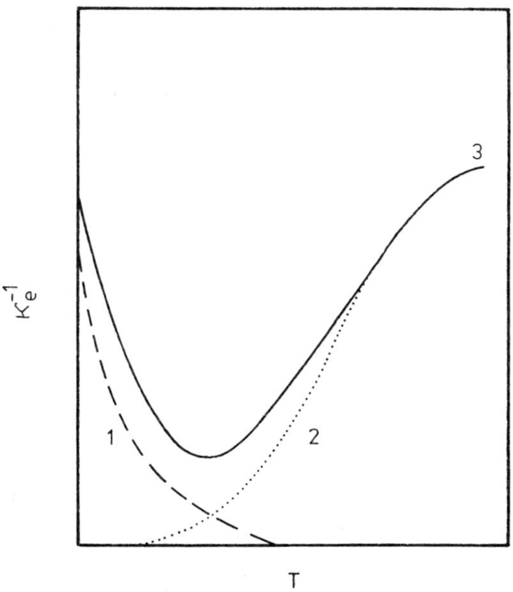

Fig. 1 *a* Temperature dependence of electronic thermal resistivity for metals: 1 — electron scattering on physical defects, 2 — electron scattering on thermal vibrations of crystal lattice, 3 — total electronic thermal resististivity.

making the interpretation of experimental data difficult. However, in various conditions, determined, *e.g.*, by the structure and the state of a substance and temperature, one can expect that dominating role will be played by one or few mechanisms of relaxation. The general character of temperature dependence of thermal conductivity relies upon these mechanisms and is connected with the temperature dependence of mean free path of the respective heat carriers.

1.2. Electron Thermal Conductivity

The electron component of thermal conductivity of metals, semimetals and semiconductors is approximately described by the Wiedemann-Franz-Lorenz Rule:

$$\kappa_e = L \sigma T = L T/\rho , \tag{2}$$

where $L = \pi^2/3(k_B/e)^2$ — Lorenz number, ρ — electrical specific resistivity. For strongly defected metals L in the whole range of temperature adopts constant value: $L = L_0 = 2.44 \times 10^{-8}$ [V^2K^{-2}]. For metals of intermediate purity, for $T < \Theta$ (where Θ — Debye temperature) the value of L changes from zero to a constant value L_0, depending on the concentration of admixtures. For clean metals L is constant above Θ, and in low temperatures Lorenz number decreases and tends to zero at $T \rightarrow 0$.

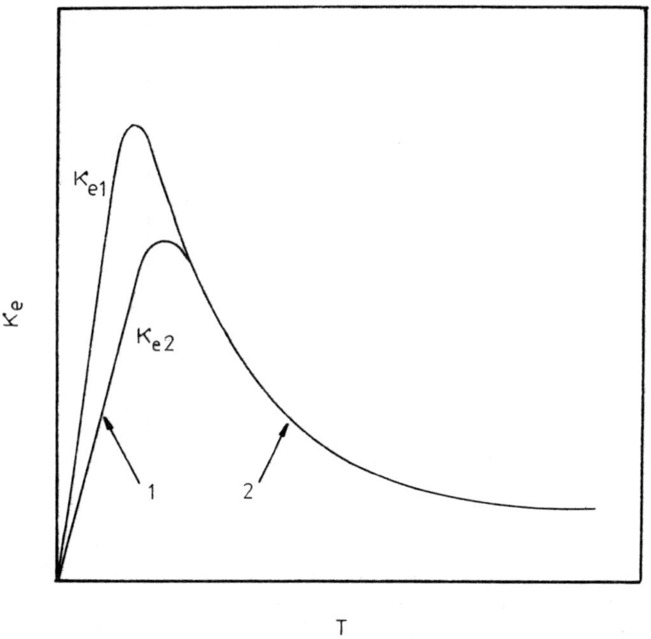

Fig. 1 *b* Temperature dependence of electronic thermal conductivity for metals with dominating electron scattering on: 1 - defects, 2 - phonons; the upper curve corresponds to a sample of greater purity.

The temperature dependence of electronic thermal conductivity of metals κ_e (Fig. 1) in the range of low temperatures is described by the expression:

$$\kappa_e^{-1} = \alpha\ T^2 + \beta/T\ , \tag{3}$$

where the coefficient $\beta \equiv \rho_0/L_0$ depends on the amount of admixtures (ρ_0 is the electrical residual resistivity) and α depends on the number of free electrons per unit volume. The first term of the equation (3) is connected with interaction of electrons with phonons while the second one — with the scattering of electrons on admixtures, defects and crystallite boundaries in polycrystalline samples. In an ideal crystal thermal conductivity would tend to infinity by $T \rightarrow 0$. In reality, in sufficiently low temperatures, the scattering of electrons on defects and admixtures dominates, what causes linear decrease of thermal conductivity by $T \rightarrow 0$. It also means that at particular temperature (the value depends on the sample purity) the maximum of thermal conductivity is observed. The increase of the number of admixtures causes lowering of the maximum and its shifting towards higher temperatures (Fig. 1 *b*). In normal metals decisive role is played by the electron component of thermal conductivity.

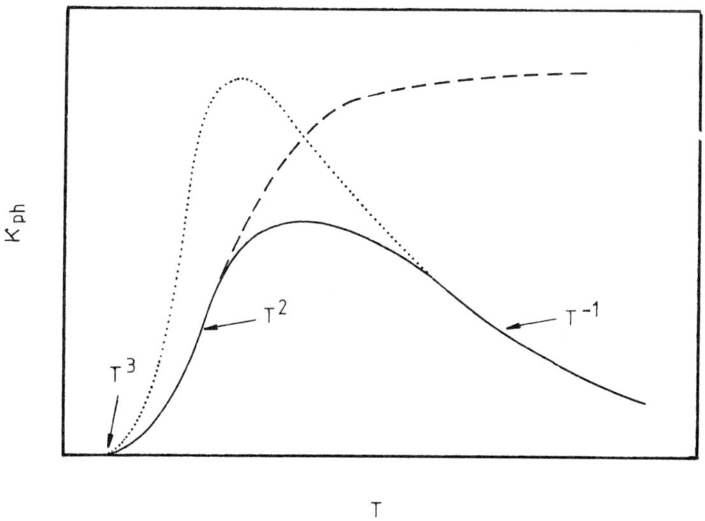

Fig.2 Schematic course of $\kappa_{ph}(T)$ dependence for metals and dielectrics: dotted curve — dielectrics, solid line — metals, dashed line — $\kappa_{ph}^{e}(T)$ dependence for phonon scattering on conductivity electrons only.

1.3. Phonon Thermal Conductivity

The effect dominating in low temperatures is scattering of phonons on crystal walls or crystallite borders. This scattering, playing fundamental role, is responsible for the thermal conductivity coefficient tending to zero when $T \to 0$. If the free path is constant, as it is in case of scattering of phonons on borders, κ_{ph} is proportional to T^3. In metals and semimetals when temperature increases (in the range $T \ll \Theta$), scattering on conductivity electrons joins up ($\kappa_{ph} \propto T^2$). Near the maximum the scattering of phonons on defects prevails, and in temperatures above the maximum thermal conductivity the interaction phonon-phonon becomes significant and the fundamental contribution is supplied by anharmonic oscillations ($\kappa_{ph} \propto T^{-1}$) (Fig.2). In the middle range of temperature both the scattering of phonons on electrons and phonon-phonon scattering should be taken into consideration. It should be noted that at $T > \Theta/3$ defects and admixtures cause independent of temperature change of lattice thermal conductivity.

1.4. Thermal Conductivity of Superconductors

Below the superconducting transition temperature the conducting electrons in the metal can be divided into two parts, normal and superconducting. The normal one preserves its properties (as in normal state) and can absorb and transport thermal energy. Thus it contributes to thermal conductivity and diminishes phonon conductivity due to interactions (scattering). The superconducting part neither absorbs nor transports energy and does not interact with phonons. Therefore, transition into superconducting state, due to decrease of number of conducting heat electrons leads to diminishing of electron thermal conductivity.

Additionally, the phonon thermal conductivity connected with the electron-phonon interaction is a quantity increasing when temperature falls, contrary to the case in normal metals. This peculiarity is connected with the increase of mean free path ($l_{ph} \propto (n_e\sigma)^{-1}$ where n_e is the number of electron excitations, σ — their cross-section for interaction with phonons) due to exponential fall of the excitations number below T_c ($n_e \propto \exp[-T_c/T]$).

In clean superconductors, as well as in samples with small amount of admixtures ($\kappa_e \gg \kappa_{ph}$) thermal conductivity in superconducting state changes in the same way as the electron component of thermal conductivity, falling down with temperature decrease (Fig.3, curve 1). This fall, slow at first, becomes more steep according to the character of changes of electron excitations. For clean superconductors the electron-phonon scattering is dominating, while for superconductors with small amounts of admixtures the scattering phonon — admixture becomes most important.

In low temperatures, for $T \to 0$ K, significantly increases the role of phonon component, and the scattering of phonons on admixtures, defects and crystallite borders becomes decisive for thermal conductivity ($\kappa_{ph} \to 0$ for $T \to 0$ K); one can observe that the character of heat transfer becomes similar to that of dielectrics. Such course of κ (Fig.3, curve 1) was observed experimentally in, *e.g.*, Sn [1], Pb, Al, Zn [2].

Heavy doping of superconductor diminishes the mean free path of electrons and in the same way the electron contribution to thermal conductivity ($\kappa_{ph} \gg \kappa_e$). In temperature range above $0.3T_c$ — $0.5T_c$ and for $T \to T_c$ responsible for thermal conductivity are scatterings of phonons on electrons. In superconducting state, in temperature range where thermal conductivity connected with these scattering (κ_{ph}^e) becomes close to the thermal conductivity connected with phonons scattering on defects (κ_{ph}^d), one can observe the maximum of thermal conductivity (Fig.3, curve 2). Below the temperature corresponding to maximum, similarly as in the first case, for the course of κ the character of scattering of phonons on defects and crystal boundaries is responsible. That kind of thermal conductivity was observed in superconducting alloy Pb + 10% at Bi [3].

For superconductors belonging to the middle category (moderately dirty) the characteristics $\kappa(T)$ is composition of the first and second type (Fig.3, curve 3). Below T_c thermal conductivity falls according to the course of κ_e. In sufficiently low temperatures κ_{ph} becomes greater than κ_e and thermal conductivity increases. This dependence has been observed, *e.g.*, for In-Pb alloy [4], doped Sn [5] and doped Hg [6].

Basing on the actual microscopic theory of superconductivity one can describe the thermal conductivity phenomena to sufficient extent. Thus, for classical (not high-temperature) superconductors the processes of transport may be thought as closed part of physics of superconducting state.

Obviously, the presented above description of mechanisms of thermal conductivity is simplified and suits only calling attention to the basic problems of heat transport. More about this subject one can find in some monographs, *e.g.*, [7-12], whereas more on theory of thermal conductivity of superconductors derived from BCS theory can be found in works of: Bardeen, Rickayzen and Tewordt (BRT) [12], Geilikman and Kresin [13], or Geilikman [14].

Concluding, we may state that in classical superconductors the electron component of thermal conductivity has decisive meaning by $T \to T_c$, while the part of lattice component increases considerably for doped superconductors and for pure metals in temperatures well below T_c.

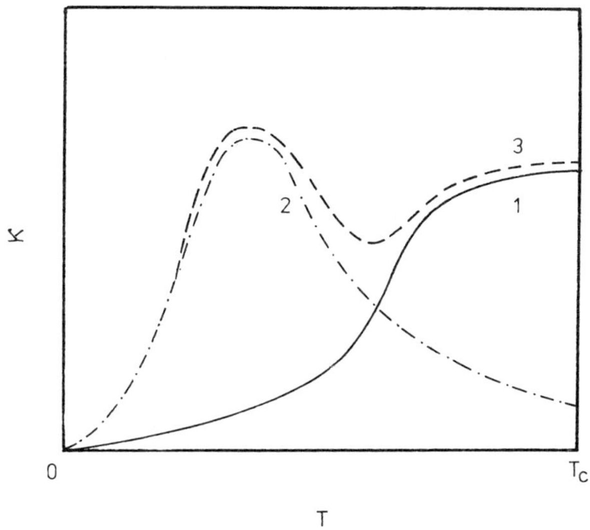

Fig.3 Basic temperature dependences of thermal conductivity of superconductors: 1 — pure; 2 — heavily doped; 3 — intermediate case.

In the case of high-temperature superconductors it turned out that the electron part (obtained from equation (2) with assumption that the carriers are scattered elastically) is inconsiderable and the decisive contribution to thermal conductivity, both above and below T_c comes from the lattice part. This conclusion, drawn from the obtained experimental data, was first formulated in [15]. Since the basic properties of superconducting state are connected with reconstruction of electron subsystem, thermal conductivity in temperature range below T_c has similar character as in dielectrics.

Extremely useful phenomenon for investigation of thermal properties of superconductors is the possibility of evoking transition from superconducting to normal state by application of magnetic field. Due to this, comparison of thermal conductivity in normal and superconducting state becomes important tool for decomposition of the phonon and electron parts, allows to investigate their mutual interactions, interaction with defects as well as allows to determine properties and character of defects. Unfortunately, for high-temperature superconductors this opportunity still does not exist, since the increase of slope of critical field characteristics as a function of temperature is very steep, amounting $dH_{c2}/dT \geq 5$ T/K [16], what means that just at temperature slightly below the superconducting transition, magnetic field would be needed of heights presently unattainable in any physical laboratory. The above fact considerably limits possibility of investigations of these materials.

2. THERMAL CONDUCTIVITY OF HIGH-TEMPERATURE SUPERCONDUCTORS

It is known that many of physical properties of high-temperature superconductors do not show similarities to properties of classical superconductors. Investigations discovered several such "anomalous" properties, as, *e.g.*, linear electrical resistivity dependence above T_c , linear term in thermal conductivity, small or lack of isotope effect. One of the physical quantities which differ prominently is thermal conductivity. First of all, it considers the existence of a T^n ($n = 1, 2, 3$) type dependence of κ below 10 K, appearance of a strong anomaly in the vicinity of T_c, essential dependence of the value and course of $\kappa(T)$ on the oxygen content, as well as presence of hysteresis phenomenon in normal state. Among the transport properties (electrical resistance ρ, thermal conductivity κ and thermoelectric force S) thermal and electrical conductances appear to bring most surprising and simultaneously ambiguous results. Exploiting data about these two properties in both normal and superconducting state one can value the strength of the electron-phonon interaction.

It is worth to note that considerations of conductivity of high-temperature superconductors led to diametrally opposite results, namely — — to adopt weak strength, what may lead to the conclusion that physical mechanism causing superconductivity lies not in phonon coupling of electrons into BCS-type pairs — and, on the other hand, to adopt strong coupling, which existence may be an argument for application of BCS theory.

The obtained till now temperature dependencies of thermal conductivity coefficient in high-temperature superconducting ceramics are all qualitatively similar (Fig.4) and confirm the first results obtained in [15]. Thermal conductivity of the samples is, first of all, phonon conductivity (the contribution of electron part changes, depending on a sample, from 1 to 20%) with distinctly marked maximum in the temperature range of 40 - 85 K, depending on the substance. The value of the thermal conductivity coefficient at maximum varies from 3 to 8 $Wm^{-1}K^{-1}$. The transition to superconducting state leads to the step change of the slope of temperature dependence of κ. The above description concerns the compounds RE-Ba-CuO and Tl-Ba-CuO; in the samples of La-(Sr,Ba)-CuO there have been no essential change in $\kappa(T)$ course in the vicinity of T_c found. In low temperatures thermal conductivity is qualitatively similar in all mentioned above compounds.

The complexity of interpretation of the obtained results lies not only in too small amount of data and in the lack of adequate theory, but also in — impossible on the basis of published results of investigation — necessary elimination of influence of effects that may be connected with thermodynamical instability of states realized in the whole samples or in their parts, non-determined stresses and defects introduced in the processes of pressing and sintering of the samples. Therefore necessary is suitable amount of data in the purpose of finding general, *i.e.* independent of the state of samples, properties and comparing them with data obtained with other than TC methods. The particular significance have X-ray and neutron diffraction investigations, determination of phonon spectrum, measurements of acoustic, transport, elastic and thermodynamical properties (specific heat, thermal expansion). Unfortunately, not all of the results of these investigations allow co-interpretation with the observed anomalies in thermal conductivity, however some general regularities are separable.

2.1. Low Temperature Thermal Conductivity

The most interesting for thermal conductivity region in low temperatures turned out to be the region below 10 K, where κ may be described by simple T^n dependence, $n = 1,2$ or 3, in various temperature ranges. For $T < 0.3$ K, κ is proportional to T, for $0.5 < T < 1$ K, $\kappa \propto T^3$ and for $1 < T$ $T < 10$ K, $\kappa \propto T^2$. Below 1 K the thermal conductivity measurements were done for polycrystalline samples, both superconducting and non-superconducting, of Y-Ba-CuO [17-20], La_2CuO_4 [22], La-Sr-CuO [22], and for

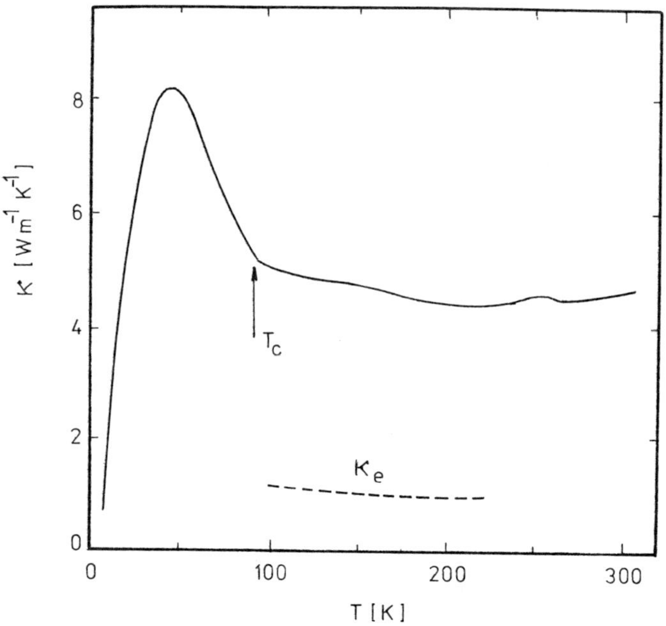

Fig.4 Thermal conductivity of $YBa_2Cu_3O_{6.81}$ sample; the dashed line denotes maximum contribution of electronic thermal conductivity [15].

single crystals Y-Ba-CuO and Ho-Ba-CuO [17]. It should be stressed that there is reasonably good qualitative agreement between the results, evidencing that they are characteristic properties of the investigated materials. Essential differences are in the interpretation of the obtained results and in the absolute values of thermal conductivity coefficient. The differences between published values happened to be twofold.

In low temperatures thermal conductivity of superconductors is comparable with thermal conductivity of amorphous materials, *e.g.*, SiO_2. The differences in the value of thermal conductivity can be explained, to some extent, by differences in materials density. Measurements on other porous materials [25] indicate that below 1 K phonons can be scattered not only on grain borders, but also on the pores. This scattering may both limit the phonons mean free path and exclude heat flux from some volume of the sample. In the first case, for Y-Ba-CuO, from the dependence $\kappa = 2.72 \times 10^{10}\ RT^3/v^2r$ [18], where v — sound velocity ($v_l = 7 \times 10^3$ m/s from extrapolation down to T = 0 K [26]), R — diameter of a pore, r — material porosity, one managed to evaluate $R \cong 2\ \mu$m. The obtained value is consistent with electron microscopy observations of the investigated samples. Substituting the value R to the above expression for κ one obtains that about 30% density change causes more than twice change of the thermal conductivity coefficient. Therefore important, by comparisons of κ values, are measurements on polycrystalline material of high, possibly close to the theoretical value, density (6.31 g/cm³) [27]. This is particularly relevant to low temperature region, as in higher temperatures the sample porosity should influence thermal conductivity much lesser [28].

In low temperatures measurements of thermal conductivity of samples with various values of X-ray-density (70 - 90%) were conducted by Sparn *et al.* [20]. It turned out that the course of $\kappa(T)$ (existence of the term bT^3 and the linear term), in spite of different values of the thermal conductivity coefficient, do not depend on the sample density. Bernasconi *et al.* [22] measured $\kappa(T)$ in samples of different grain size (150 - 20 μm). Thermal conductivity of the sample of greater grains is less than of samples of smaller grains, but, like in the case of samples of different density, the course of $\kappa(T)$ in low temperatures is independent of grain dimensions, which influence shows up only in insignificant shift — on the scale of temperature — of the origin of the portion with T^3 dependence. It should be pointed out that the measurements of thermal conductivity of single crystal samples of Y-Ba-CuO [17] gave the value of the thermal conductivity coefficient one order of magnitude greater than those of polycrystalline samples. The possible cause of such difference is, probably, besides the higher density effect, anisotropy of thermal conductivity, in result of which κ in the directions of copper-oxygen planes is greater than in the direction perpendicular to them [22].

Since the observed in low temperatures dependence, $\kappa \propto T^3$ is typical not only for porous but also for polycrystalline dielectric materials, where the phonon mean free path is determined by scattering on grain borders, one has to choose which of the mechanisms is responsible for this dependence. One can do this estimating the value of the phonon mean free path and comparing to the dimensions of the grains and pores.

In casė of RE-Ba-CuO, when $T \ll T_c$, temperature dependence of κ follows, first of all, processes of excitations relaxation of phonon system. The lattice component influences decisively also the character of changes of specific heat. It allows to estimate the effective mean free path of phonons l_{ph} on the basis of data on thermal conductivity and specific heat. In case of heat transport by phonons, which are considered as elementary excitations of the system, thermal conductivity may be described by the expression [7]:

$$\kappa(T) = \tfrac{1}{3} \sum_{i} \int_{0}^{\omega_D} C_i(\omega)\, v_i(\omega)\, l_i(\omega)\, d\omega, \tag{4}$$

where: v_i and l_i are phonons velocity and their mean free path, $C(\omega)$ is the lattice component of specific heat for phonons of frequency ω, and ω_D denotes maximum Debye frequency. The description of heat transport by phonons may be considerably simplified if one exploits the approximation for dominating phonons:

$$\kappa = \tfrac{1}{3}\, Cvl = (2\pi^2 k_B{}^4 l / 15\hbar^3 v^2) T^3 \; . \tag{5}$$

In spite of the fact that all the estimations of the value of l_{ph} obtained on the basis of data of specific heat, sound velocity and thermal conductivity vary in the cited works within almost one range of magnitude, it could be positively stated that it is only when temperature falls down to 1 K the mean free path becomes comparable with the characteristic for the sample grain sizes, estimated usually to be 5 to 40 μm. Thus one may adopt that the scattering of phonons on grains as well as on pores is the reason of observed dependence of $\kappa(T)$ below 1 K.

The dependence of the type $\kappa_{exp} = aT + bT^3$ in ceramic superconductor samples of Y-Ba-CuO in low temperature range has been observed by: Graebner *et al.* [17], Gottwick *et al.* [19], Sparn *et al.* [20], Salce *et al.* [21], Bernasconi *et al.* [22], Cohn *et al.* [23*a*], also, in the range 0.5 - 1 K, $\kappa \propto T^3$ has been observed by Efimov *et al.* [24]. On the other hand, Freeman *et al.* [18] found in their samples the dependence $\kappa \propto T^2$

for $T < 0.3$ K while above 0.3 K — $\kappa \propto T^3$. The linear term aT have appeared in the cited works below temperature 0.3 - 0.5 K while the exponent of the term bT^n varies in the limits 2.6 - 3 and the dependence $\kappa \propto T^3$ appears usually below the range $0.8 < T < 1$ K. In some works [18,21] existence of that dependence was observed notably above 1 K ($T < 7$ - 10 K). In non-superconducting samples of Y-Ba-CuO thermal conductivity is proportional to T^3 [20,21,23*a*].

Interpretation of the existence of the term bT^3 do not raise any doubts and is obviously connected with manifesting the phonon size effect. On the other hand, the appearance of linear dependence in low temperatures is interpreted in various ways. According to Graebner *et al.* [17] the distinct change of the dependence κ(T) in ceramic samples is caused by a resonance scattering of phonons around 0.1 K (Rayleigh scattering), when the phonon wave length becomes comparable with the mean grain dimensions. However, there is lack of such anomaly close the same temperature ($T = 0.1$ K) in non-superconducting samples of the same grain dimensions and the same bT^3 dependence above that temperature, what rather excludes possibility of that contribution to phonon scattering mechanism. Sparn *et al.* [20] on the basis of comparison of the dependence $\kappa(T)$ in some superconducting and non-superconducting Y-Ba-CuO samples conclude that, independently of phonons, some other low energy excitations must give contribution to low temperature thermal conductivity (*e.g.*, normal electron carriers of small effective mass). Also Gottwick *et al.* [19] tied the existence of linear $\kappa(T)$ dependence to the presence of normal electrons. On the other hand, Salce *et al.* [21] on the basis of the similar as in [20] comparison suppose that the nature of linear term is not connected with the presence of charge carriers, but it lies in phonon scattering on structure defects, like the grain boundaries, pores and particularly walls between orthorhombic domains. The last are being preferred in the discussion, since the respective authors believe that twinning is present only in superconducting samples.

This type of interpretation excludes possibility of searching for common causes for low-temperature linear term γT characteristic for specific heat [*e.g.* 29,30], and for linear term αT observed also in low temperature thermal expansion [*e.g.* 31,32]. The low-temperature dependence of thermal conductivity, specific heat and thermal expansion on temperature might be explained by the existence in the considered samples a glassy state. It should be however noted that the observed dependences for specific heat and thermal expansion can also be connected with the presence of admixtures in the material.

In the above context very interesting and dependable appear the results of Cohn *et al.* [23*a*] obtained in the temperature range 0.1 - 1 K. They have conducted measurements of a series of Y-Ba-CuO samples, which structural and therefore superconducting properties (transition from orthorhombic to tetragonal structure) were changed by suitable thermal processing of the same sample. Simultaneously a scanning electron micrography (SEM) has been performed, what allow to exclude any changes connected with grain size and porosity. It turned out that in superconducting samples thermal conductivity at temperature 0.1 K (Fig.5, see also Fig.19 *b*) is one order of magnitude higher than in non-superconducting ones. Below 0.5 K, for superconducting sample, a change of character of the dependence occurs, from $\kappa \propto T^3$ to $\kappa \propto T$. This result is rather unexpected one, since in such low temperatures one may expect heat transport by phonons only and identical κ course in both samples, independently of their other physical properties. Since the effect cannot be tied to differences in grain sizes and porosity or with greater defect number of oxygen superstructure, which can be the cause of additional scattering, the anomalous behavior of the superconductor can only be explained by

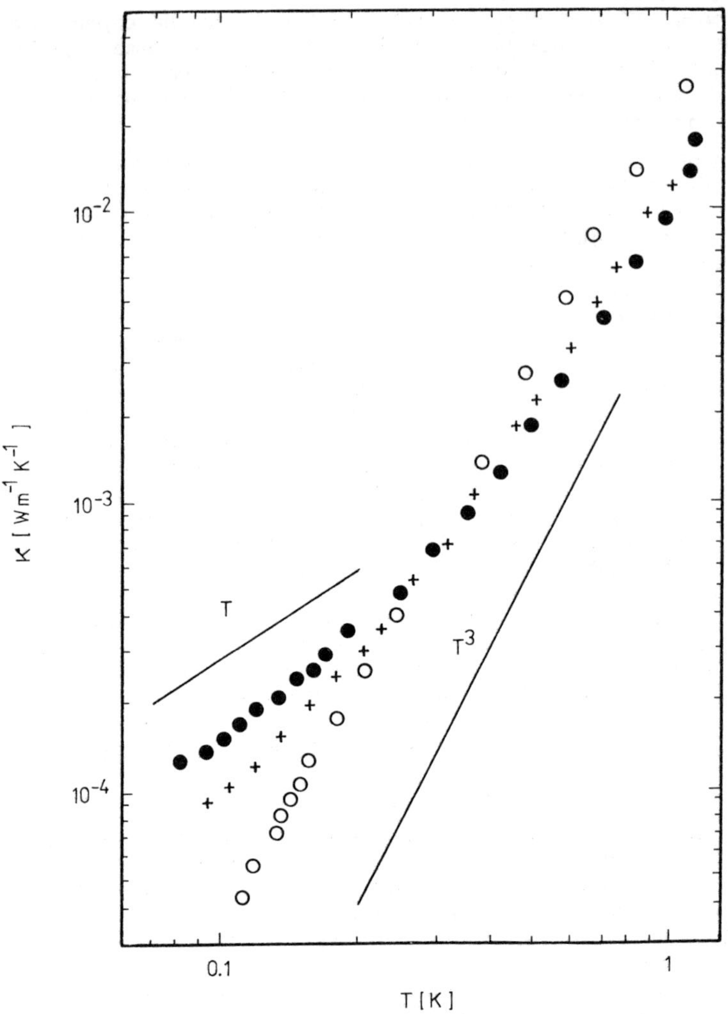

Fig.5 Thermal conductivity of a $YBa_2Cu_3O_{7-x}$ sample subjected to heat treatment: • — virgin state, o — heated in vacuum, + — reheated in oxygen [23*a*].

the presence of a non-phonon component of thermal conductivity.

In very low temperatures the phonon component falls down rapidly and even small number of free carriers may dominate the thermal transport. In that range of temperatures, where contribution of electron scattering on admixtures and defects should be dominating, one may expect that thermal conductivity should change linearly with temperature, what indeed is observed in experiments. Thus the hypothetical presence of electron component in thermal conductivity points out that below T_c some number of electrons stay uncondensed, their number estimated to be about 5% of carriers existing in normal state. The estimation of the number of normal electrons in superconducting state differs from that, amounting

15% [19]. It should be pointed out here that the recent measurements of the samples $La_{1-x}Sr_xCuO_4$ in low temperature have showed also identical qualitative dependence as in $YBa_2Cu_3O_{7-x}$ [23*b*]. In the light of the above, it is surprising that thermal conductivity of La_2CuO_4 in low temperatures can be described also by the expression of the type $aT + bT^3$ [22]. It is also surprising, that there exists the dependence $\kappa \propto T^2$ below 0.5 K instead of $\kappa \propto T$ for superconducting Y-Ba-CuO presented in [18]. The presence of normal electrons can also explain strong linear term γT in low temperature specific heat of high-temperature superconductors [29,30].

It is intriguing that the thermal conductivity of Bi-Sr-Ca-Cu-O below 3 K follows an approximatelely quadratic temperature variation with no indication of the limiting T-linear dependence [33]. It should be underlined that specific heat measurements of single crystal samples did not show any significant difference between γ values of superconducting and non-superconducting Y-Ba-CuO samples [34,35], contrary to situation of linear term in thermal conductivity of ceramic (polycrystalline) samples. Measurements on single crystal samples of $YBa_2Cu_3O_7$ and $HoBa_2Cu_3O_7$ in temperature range $0.03 < T < 5$ K [17] did not exhibit similar to the abovementioned dependence and in the whole investigated temperature range κ changes like $T^{1.8}$ or $T^{1.9}$ (Fig. 6). Appearance of the dependence of type $\kappa \propto T^2$ is explained by the authors of work [17] as the result of scattering of phonons on tunneling system (*cf.* further considerations below). It looks like the appearance of another type of dependence in single crystals does not negate totally a result indicating the presence of normal electrons below T_c. Phonons in polycrystalline samples are less sensitive to eventual scattering on tunnel system than in single crystals, and in low temperatures for the course of κ the size effect is decisive. Therefore the anomaly connected with the presence of addition al heat transport by electrons is easier to detect.

It was found, in the temperature range $1 < T < 10$ K, that thermal conductivity changes proportionally to T^2. For the first time this dependence was reported in works of Uher and Kaiser for Y-Ba-CuO [36], Nùñez-Requiero *et al.*for La_2CuO_4 and La-Sr-CuO (proportionality to $T^{1.8}$ and to $T^{1.5}$) [37], Er-Ba-CuO [38], and in works of Jeżowski *et al.* for RE-Ba-CuO (where RE = Y,Gd,Sm) [39,40], Y-Ba-CuO of tetragonal structure [41], Tl-Sr-Ba-CuO [42] and La-CuO with oxygen surplus superconducting ($La_{1.85}CuO_{4+x}$), and non-superconducting ($La_{2.10}CuO_{4+x}$) [43]. As it was noted previously, the dependence of the type of $\kappa \propto T^2$ is exhibited also by single crystal samples (Fig. 6). In the Fig.7 fulfilling of the T^2 relation in case of Y-Ba-CuO with different oxygen content was demonstrated. However, one should not overlook works, which authors either do not mention the existence of that (*i.e.* T^2) law for $T < 10$ K [19,20], or definitely did not found the presence of other than $\kappa \propto T^3$ dependence below that temperature [18,21,24]. Also there is lack of univocal opinion on the reasons for the $\kappa \propto T^2$ dependence. For example, Freeman *et al.* [18] remark that the like dependence is visible in many other sintered ceramic materials being insulators and the reasons for it are not clear. Scattering of phonons on electrons causes also the same T^2 dependence in the same temperature range [8] but it seems that very small number of free electrons should not make such strong scattering medium. It is worth remembering that the same dependence is observed also in samples of tetragonal structure (Fig.7). Merisow *et al.* [45] suggest as a mechanism responsible for the considered phenomenon a "surface" phonons, which contribution to the total thermal conductivity, due to great internal surface (grains of sintered material) may be considerable in low temperatures. It seems that the presence of the T^2 dependence in single crystals strongly shakes the last hypothesis.

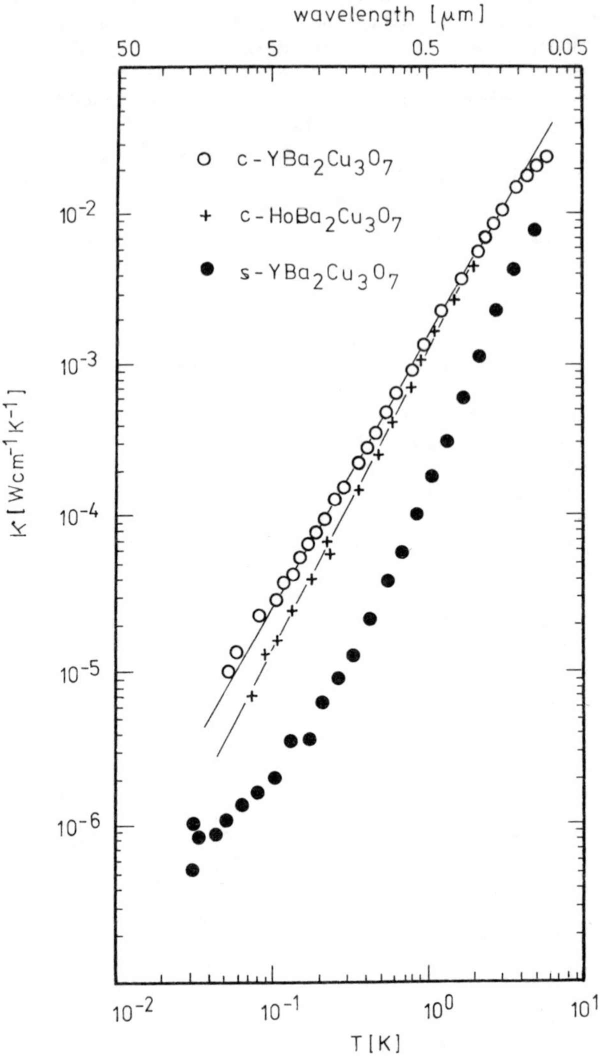

Fig.6 Thermal conductivity of single crystals of $YBa_2Cu_3O_7$ and $HoBa_2Cu_3O_7$ and also of a polycrystalline sample of $YBa_2Cu_3O_7$. Solid lines correspond to dependences: $\kappa \propto T^{1.8}$ and $\kappa \propto T^{1.9}$. The upper scale in the Figure is related to wave length of thermal phonons [17].

In connection with the above considerations, the interpretation proposed by Núñez-Regueiro *et al.* [37] and Graebner *et al.* [17] deserves attention. It ties the considered effect to resonance scattering of phonons on tunneling system (TS) - phenomenon known, among others, from investigations of properties of amorphous materials. Standard tunneling model is widely applied to description of low temperature properties of these materials [46,47]. Similar value of thermal conductitivity of compounds based on La and Y, particularly in low temperatures and the presence of $\kappa \propto T^2$ dependence, characteristic for glass allow for ana-

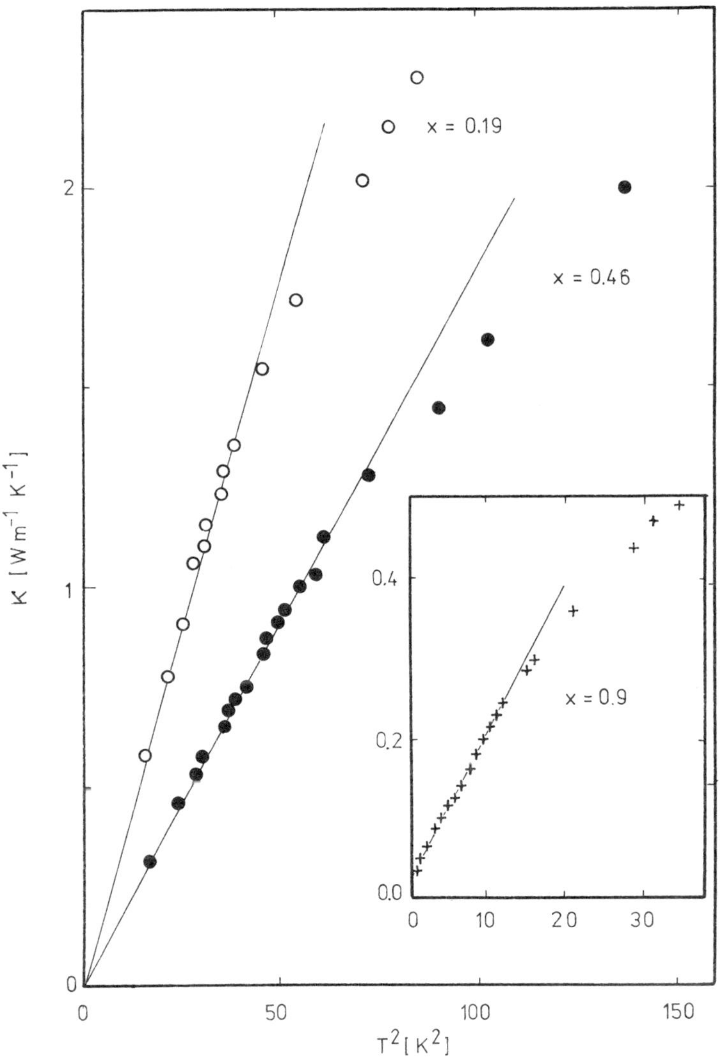

Fig.7 Thermal conductivity of $YBa_2Cu_3O_{7-x}$ dependence on T^2 for samples of different oxygen content [41].

logical approach to high-temperature superconductors, the more that recently different measurements of other than thermal conductivity physical properties indicate the possibility for existence in superconducting ceramics a localized tunneling system with widely dispersed relaxation times and energetic excitations. Namely, phonons induced in acoustic measurements are also relaxatively scattered on tunneling system what is demonstrated in temperature dependence (*e.g.*, plateau in sound attenuation). Measurements of the velocity of sound of low frequency [38,48] show logarithmical temperature dependence below 1 K, characteristic both for amorphous dielectric and for metals [49]. For higher frequencies or lower temperatures a maximum of sound velocity is observed, in accordance with tunneling model. This phenomenon was confirmed experimentally also in Eu-Ba-CuO [38] and Y-Ba-CuO [48] indicating the presence of resonance and relaxative interaction of phonons with tunneling system. The measurements of sound attenuation in La-Sr-CuO [50] also indicate the presence of a plateau, typical for amorphous materials, in temperature range 4 — 40 K. The dependence of internal friction Q^{-1} on temperature [38] in superconductors in low temperatures can again be explained within the tunneling model. Attention should be called here to the works [32,37] where the possibility of connecting anomalous linear thermal expansion or specific heat with the presence of tunneling system was indicated.

Let us come back again to the thermal conductivity measurements. Exploiting the proposed by Freeman and Anderson [51] description of thermal conductivity depending on reduced variables, being parameters of material, the authors [52,53] observed that in La_2CuO_4 and Eu-Ba-CuO the value of the thermal conductivity efficient as well as its temperature dependence show similarity to these of other amorphous and unordered systems (Fig. 8). It allow them to attempt explanation of the observed dependence of thermal conductivity ($\kappa \propto T^2$) on the ground of tunneling model.

In case of disordered materials (glass) the mean free path of phonons of frequency ω changes as $l(\omega) \propto \omega^{-1}$ with respect to resonance scattering by a tunneling system, *i.e.* atoms or groups of atoms occupying a double potential well in a configuration space. The expression for the mean free path in a two- level tunneling model has the form [47]:

$$l^{-1} = \pi\omega/\rho v^3 \; \bar{P}\gamma^2 \; \tanh(\beta\hbar\omega/2) \tag{6}$$

where: ρ — mass density, v — sound velocity, $\overline{P}$ — mean density of tunneling states, γ — phonon-TS coupling constant, $\beta \equiv (k_B\, T)^{-1}$. Using the expressions (4) and (6) one can obtained the relation for thermal conductivity,

$$\kappa = (\rho k_B{}^3 v \;/6\pi \;\hbar^2\bar{P}\gamma^2)T^2. \tag{7}$$

As it is seen, the T^2 dependence is characteristic for glassy state and has been confirmed experimentally, also for superconductors. An important parameter in tunneling model is $C = P\gamma^2/\rho v^2$. Its value for various disordered and amorphous solids is approximately the same, $C \doteq 1\times10^{-4}$. It turned out that similar value was obtained also for high-temperature superconductors [52,53]. Therefore, there is basis for assuming the presence of TS in these materials.

However, the microscopic interpretation of tunneling states presents a serious problem. There are premisses indicating that great concentration of oxygen vacancies causes increase in the density of tunneling states [53]. This leads to a hypothesis that appearance of TS in considered compounds is connected with oxygen or respective vacancies movement. This interpretation can be founded not on the presence of oxygen vacancies but on the recognized fact that in high-temperature superconductors oxygen ions in some crystallographic positions are in potential with the double minimum. As a simple example one can present RE-Ba-CuO superconductors, where oxygen ions situated along the copper-oxygen chains (between Ba-Ba planes) are configurated in zig-zag. It means that they can be distributed with equal probability on both sides of the chain direction. Adopting this interpretation leads to the conclusion that increasing of oxygen content (in some limits) should increase the number of TS. However, the correlation between oxygen vacancies number (and hence oxygen content) and C parameter is distinctly visible only in La-Sr-CuO. On the other hand, in case of RE-Ba-CuO it is not so evi-

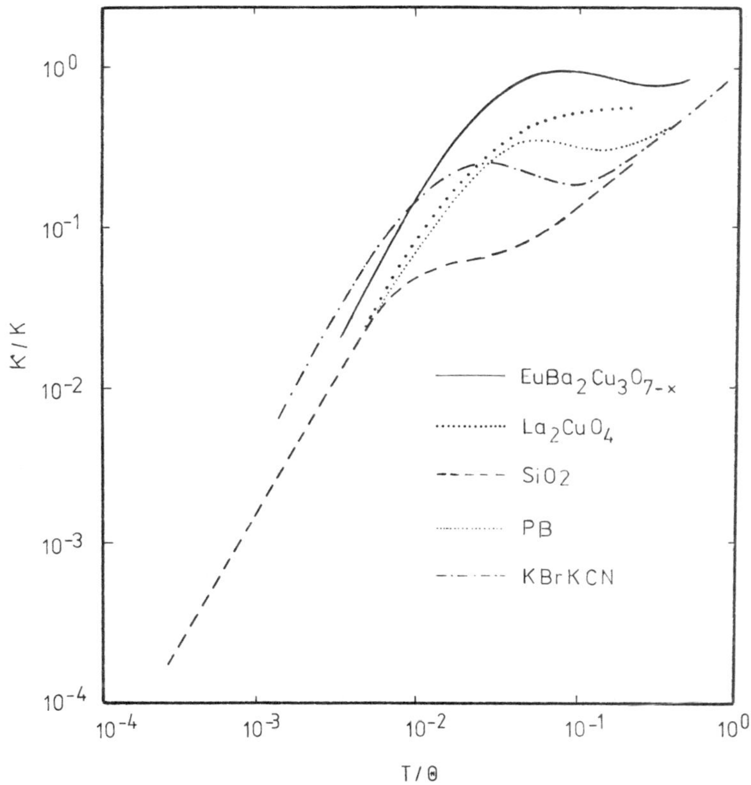

Fig.8 Comparison of thermal conductivity of $EuBa_2Cu_3O_{7-x}$ and La_2CuO_4 and thermal conductivity of other amorphous materials [52,53], $K = f(\Theta)$.

dent [53]. Also the linear term in Y-Ba-CuO specific heat is practically independent of the vacancies concentration [34,35]. It means that for these materials another reason for eventual presence of tunneling states should be searched. It seems also that the presence of TS is not connected with superconductivity [53] since La-Sr CuO samples richer in oxygen vacancies and thus having higher density of TS are non-superconducting.

2.2. Influence of Magnetic Phase Transitions and Magnetic Field on Thermal Properties of Superconductors

In magnetic materials below the magnetic ordering temperature T_N (Neel temperature) additional component of thermal conductivity may appear as the result of heat transport by the system of ordered magnetic moments. The ordered magnetic system excitation leads to heat transport by magnons (κ_{mag}) and to additional scattering of phonons on magnons ($\kappa_{ph}{}^{mag}$). In low temperatures κ_{mag} may be of the same order of magnitude as κ_{ph}. As temperature approaches T_N the magnon conductivity diminishes [54]. In the critical region of magnetic phase transition strong phonon scattering on spin system (fluctuations) can occur, what may reflect itself in the temperature dependence of lattice thermal conductivity [55].

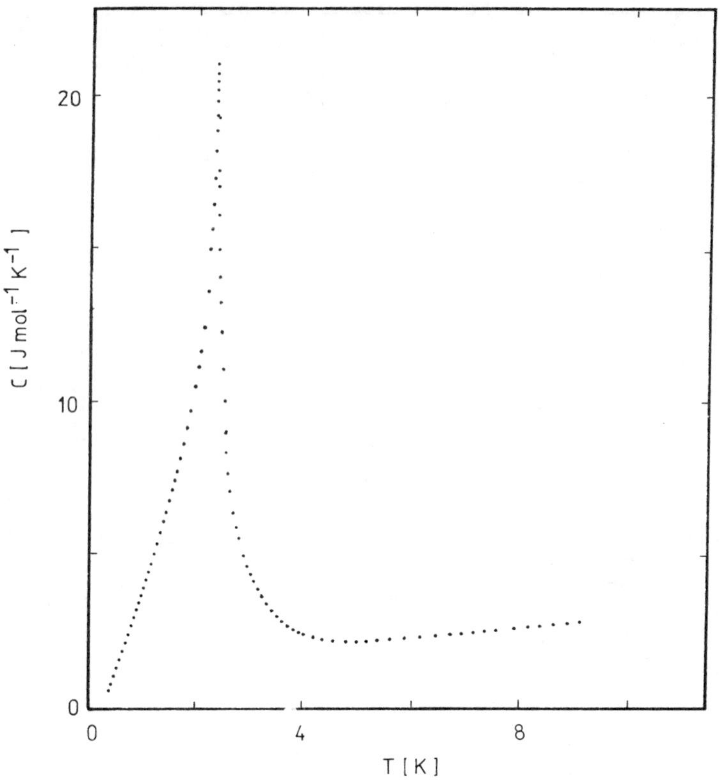

Fig.9 Low temperature specific heat of $GdBa_2Cu_3O_{7-x}$ [59].

Coexistence of superconductivity and magnetic order is one of fascinating properties of the new superconductors. When Y is substituted by trivalent paramagnetic ion of rare earth, Gd, Dy, Sm or Er, one can observe the antiferromagnetic ordering in temperature range 2.3 - 0.6 K (for Gd compounds - 2.2 K, Dy - 0.9 K, Sm - 0.6 K, Er - 0.6 K) without any change in superconducting properties. Specific heat measurements in these compounds show existence of anomalies of the *lambda* type, characteristic for a magnetic phase transition [56-58]. In the Fig.9 the temperature dependence of specific heat of Gd-Ba-CuO in the range of temperature 1.4 - 6 K [59] is presented. The results show distinctly an anomaly of λ type at 2.22 K, *i.e.*, at the antiferromagnetic ordering temperature. Appearance of a sharp peak at the same temperature was discovered also in measurements of thermal expansion of Gd-Ba-CuO [60]. Below 1 K, in Ho-Ba-CuO an abrupt change of specific heat temperature dependence has been discovered with an anomaly around 0.17 K, which also has been connected with a magnetic phase transition [61].

In temperature range, in which the magnetic phase transitions occur, thermal conductivity of the following samples has been investigated: single crystal $HoBa_2Cu_3O_7$ (Graebner *et al.* [17]) and polycrystalline Gd-Ba-CuO (Mori *et al.* [62], Jeżowski *et al.* [63]). Measurements of Ho-Ba-CuO have not revealed any changes of $\kappa(T)$ (Fig.6), what may evidence very weak coupling between phonons and magnetic structure. On the

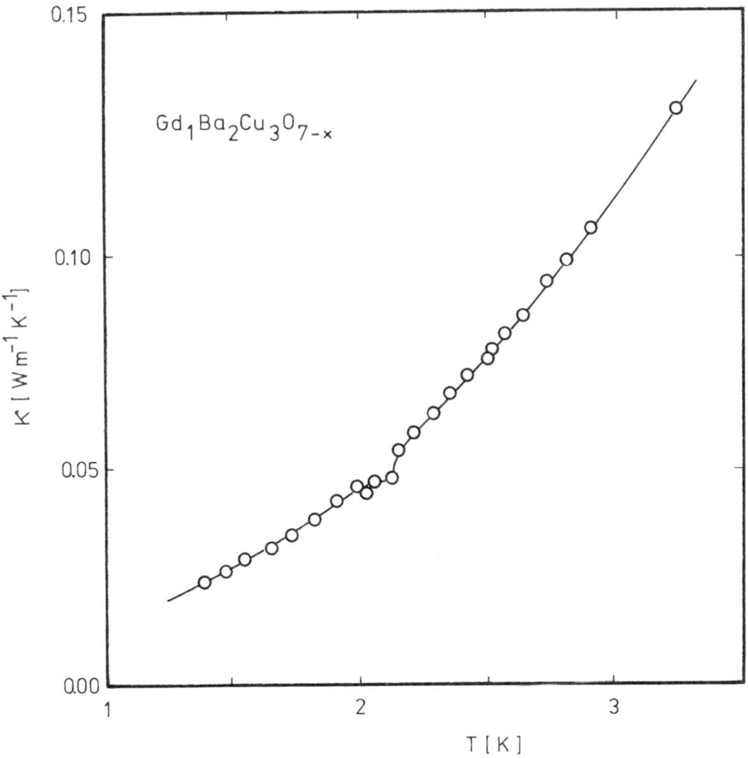

Fig. 10 Low temperature thermal conductivity of $GdBa_2Cu_3O_{7-x}$ [63].

other hand, the results obtained for Gd-Ba-CuO turned out to be more interesting - see Fig. 10. The distinct change of the $\kappa(T)$ course occurs in the vicinity of T_N. This indicates that below T_N character of phonon scattering changes; in low temperatures, $T < T_N$, κ is proportional to T, while in temperatures above T_N $\kappa \propto T^2$ [63]. The observed jump of κ at T_N is within the range of experimental error, hence no particular meaning should be tied to this observation; it may however indicate that a change of scattering character connected with spin orientation takes place in very narrow temperature range.

As it follows from the available literature, measurements of thermal conductivity in magnetic field were done for Gd-Ba-CuO [62] and Y-Ba-CuO [21], and two contradictory results were obtained. In [62] the results of measurements conducted at 4.41 K, for limited range of field change, 0 - 2.5 kOe. The observed change of κ for that magnetic field thermal conductivity is connected by the authors with appearance of scattering on normal regions or on normal-superconducting phase borders. This kind of dependence is similar to dependence usually found in type-II superconductors. On the other hand, Salce *et al.* [21] conducted measurements of $\kappa(T)$ in Y-Ba-CuO sample without magnetic field and in a stationary field in the temperature range 150 mK — 10 K, observing no particulars. They found neither changes of temperature dependence character of $\kappa(T)$ ($aT + bT^3$), nor changes of the value of κ (Fig. 12).

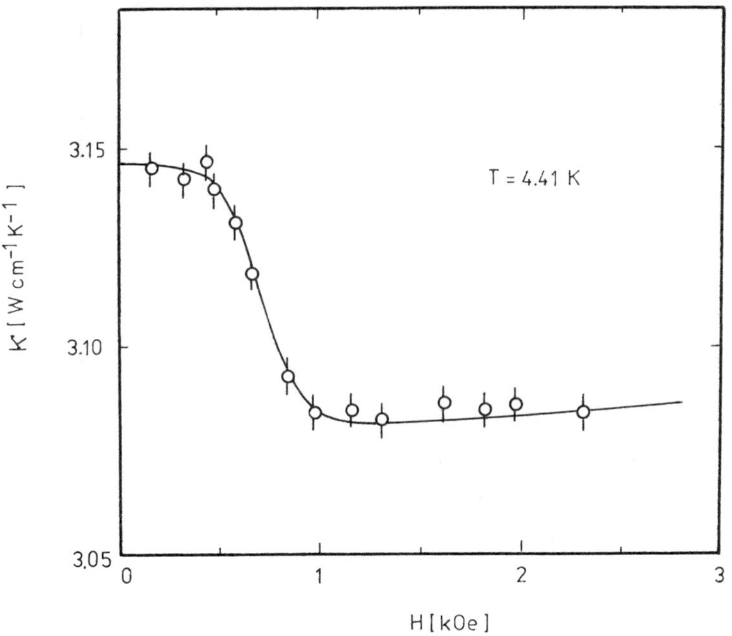

Fig.11 Dependence of thermal conductivity of Gd-Ba-CuO on applied magnetic field [62].

2.3. Thermal Conductivity Near T_c

Unsuspected and surprising effect has been discovered by Jeżowski *et al.* [15] in thermal conductivity of Y-Ba-CuO, namely abrupt increase (jump in the first derivative) of κ below the superconducting transition temperature (Fig.4). Similarly as in many other investigations of high-temtemperature superconductors, almost simultaneously the same was reported in the works of: Bayot *et al.* [64], Gotwick *et al.* [19], Uher and Kaiser [36] and Morelli *et al.* [65]. Superconducting La-Sr-CuO have exhibited some similarity in the course of κ(T), but the effect turned out to be very small [36,66] (Fig. 13) and looks like rather as insignificant convexity of the κ(T) curve than the abrupt increase as it is the case in Y-Ba-CuO. It should be also noted that other authors [37, 67] did not stated any changes in thermal conductivity of La-Sr-CuO. Also the measurements on oxygen-surplus superconducting sample of La_2CuO_{4+x} did not did not exhibit any changes of the course of κ(T) in the vicinity of T_c [43]. Possibly, the reason for this lies in the small amount of the superconducting phase in the investigated samples or in the accuracy of the measurement, very important in the case of such small effect. Thermal conductivity of themultiphase sample of Y-Ba-CuO was measured by Merisov *et al.* [68]. In the vicinity of superconducting transition they observed a few jumps in the derivative of κ(T) with respect to temperature. This result is however hard to interpret because of undefined phase composition of the sample.

Although thermal conductivity of classical superconductors usually falls down below T_c (electron pairs do not transport heat), the observed

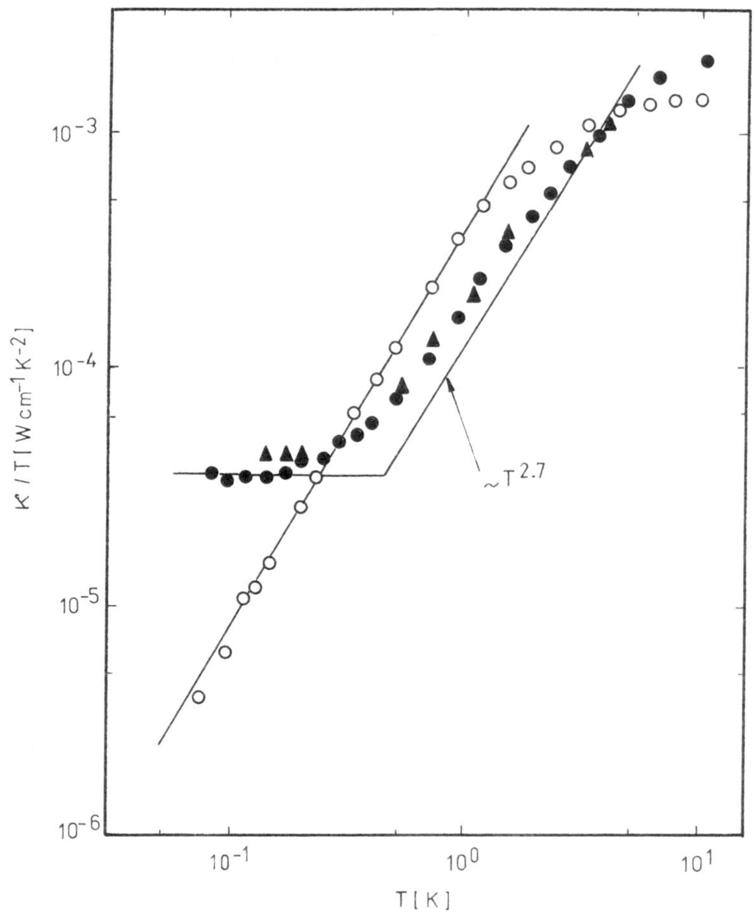

Fig.12 Thermal conductivity temperature dependence for $YBa_2Cu_3O_{7-x}$ [21]: • — with no field applied; ▲ — in magnetic field of 7 T; o — for $YBa_2Cu_3O_{6.1}$.

increase of κ in ceramic superconductors can be easy to understand if the lattice component of thermal conductivity dominates over the electron component. Since the paired electrons (or charges) do not scatter thermal vibrations, phonon-governed thermal conductivity should increase in the superconducting state. This effect should prevail over the decrease of thermal conductivity resulting from diminishing the electron component κ_e. This conclusion is acceptable only in case $\kappa_e \ll \kappa_{ph}$. The calculations of the electron component from the Wiedemann-Franz Law (2) have indicated that the contribution of the electron component to the total thermal conductivity is within the limits 1 to 20%, it is hence relatively small. Moreover, the highest value have been attained in Y-Ba-CuO samples with oxygen index close to 7 [69].

The dependence of the electrical resistivity on temperature for $T > T_c$ is close to linear, having hence metallic character In metal alloys of high specific electric resistivity the Wiedemann-Franz Law is

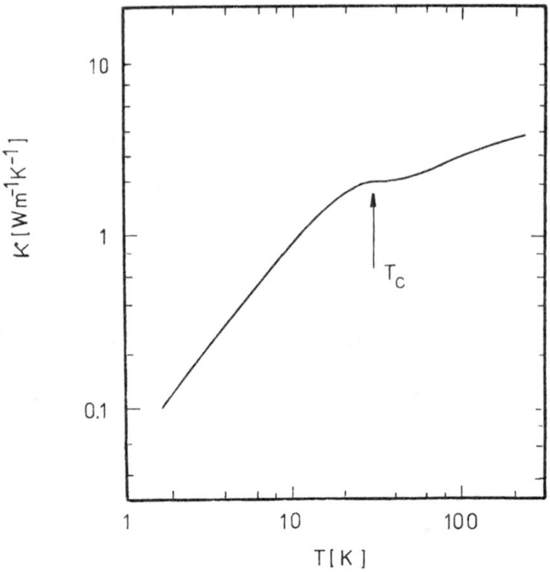

Fig. 13 *a* Thermal conductivity of $La_{1.8}Sr_{0.2}CuO_4$ sample [36].

fulfilled in wide range of temperatures, and deviations from it are connected with peculiarities of electron-phonon scattering (inelastic scattering), may lead to decrease of Lorenz number with respect to L_0 (§ 1.2). Therefore the values of κ_e determined from the Wiedemann-Franz Law reflect an overestimated electron component. The estimated in [70] value of the effective Lorenz number for the investigated sample of Eu-Ba-CuO at $T \cong 100$ K amounts $L_{eff} \cong 1.4 \times 10^{-8}$ V^2K^{-2}.

The sole earlier observation of κ increase below T_c concerned disordered alloys, where $\kappa_e \ll \kappa_{ph}$. In new superconductors, which are also materials of some internal degree of disorder, small electron component κ_e is, one may think, due to another cause: — small density of carriers (it follows from, among others, measurements of electrical resistivity and Hall effect).

Since there was problem with finding sound mechanism responsible for such abrupt change of κ near T_c and connected solely with behaviour of either electrons or phonons, some investigators ascribed this fact to a strong electron-phonon interaction. It seems that now, with increasing collection of experimental data, there are premisses for searching also other reasons [83,84] responsible for such intriguing behaviour of thermal conductivity near the superconductive transition. There was also a hypothesis formulated, based on the results of investigations of Raman and infrared spectra temperature dependence, about the possibility of appearing of additional collective lattice vibrations (librons). Including thus these new thermal vibrations may lead to appearance of new mechanism of thermal transport, hence to increase of thermal conductivity coefficient [84].

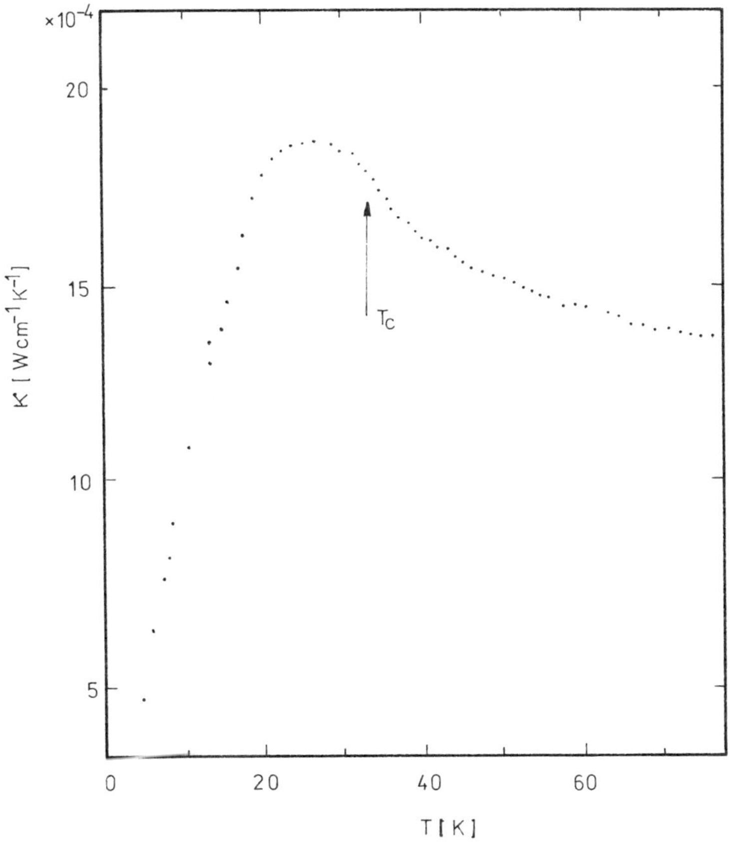

Fig. 13 *b* Thermal conductivity of $La_{1.83}Sr_{0.17}CuO_4$ sample [66].

Using data of thermal conductivity one cannot quantitatively estimate the value of electron-phonon interaction strength. It is however possible to determine electrons density exploiting the model usually applied to metals, as it has been done by Bayot *et al.* [64]. In cases when the phonon heat transport is limited by scattering on electrons, the ratio of the thermal resistivity $(\kappa_{ph}{}^{e})^{-1}$ to the ideal electrical resistivity is described by the expression [7]

$$(\kappa_{ph}{}^{e}\rho_i)^{-1} \propto n_a{}^2(e/k_B)^2\ (C_{max}/C)^2, \tag{8}$$

where: e — electron charge n_a — number of free electrons per atom, $C_{max} = 3NK$ — the maximum value of specific heat C, N — atom density. As it follows from the above discussion, the lattice thermal resistivity connected with scattering on electrons $(\kappa_{ph}{}^{e})^{-1}$ should take considerable part in the total lattice thermal resistivity $(\kappa_{ph})^{-1}$, therefore the density of free carriers would be a quantity of the following order:

$$n_a \leq (k_B/e)(C/\ C_{max})(\rho_i\kappa_{ph})^{-1/2}\ 1^{1/2} \tag{9}$$

Taking the value $\kappa \simeq 3 - 5$ W/cmK and $C/C_{max} \cong 0.45$ (both values for

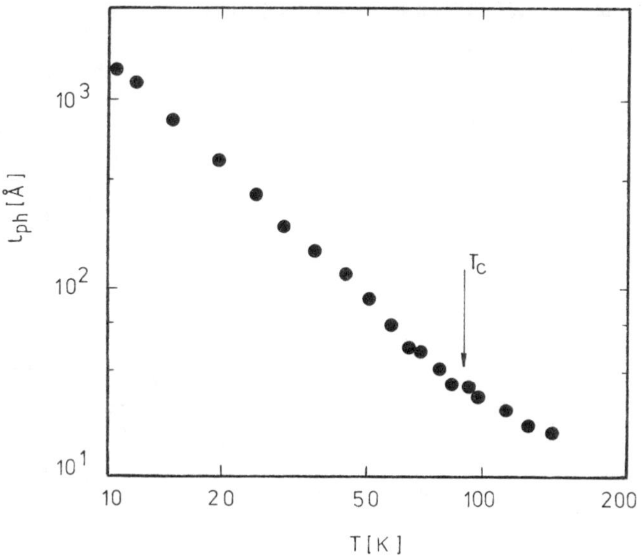

Fig.14 Temperature dependence of phonon mean free path for $YBa_2Cu_3O_{6.9}$ [24].

temperature about 100 K), and ρ_1 (residual resistance) $\cong$ 100 $\mu\Omega$cm, one obtains the density of carriers $n \leq 10^{22}$cm^{-3}. This appears to be a value of right order, what confirms the possibility of exploiting ideas proper for metallic conduction.

The values of thermal conductivity of Y-Ba-CuO and La-Sr-CuO above T_c fall in between that for typical good crystals and for amorphous materials. Similarly as we did in case of low temperatures we can now determine the phonon mean free path from the relation (5), exploiting, of course, the fact of domination of phonon conductivity. Taking the values for Y-Ba-CuO: $\kappa \cong 3$ - 5 W/cmK, C = 1.4×10^6 J/m^{3}K [71] and the sound velocity $v \cong 3$ - 4×10^3 m/s [72] we obtain for the mean free path value at 100 K equal $l_{ph} \cong 25$ - 40 Å. It turns out that near T_c the value of l_{ph} is of the order of magnitude of the lattice constant along c axis in these substances. In the Fig.14 the dependence of l_{ph} on temperature is depicted [24]; it can be seen that the mean free path increases with lowering of temperature and $l_{ph} \propto T^{-2}$ being 10^{-5} at 10 K (when the grains diameter $d \cong 10^{-4}$cm). It follows from this considerations, and those concerning the mean free path in low temperatures, that above 1 K the scattering of phonons within the grain volume prevails: on microscopic defects (vacancies, dislocations, twin boundaries), on phonons and on normal electron excitations. The value of mean free path and its temperature dependence is similar to that observed in disordered dielectrics. Similar is the situation in La-Sr-CuO samples.

The authors of the work [70] attempted to decompose the components of thermal conductivity κ_e^{ph}, κ_{ph}^{e} and κ_{ph}^{d} using fundamental dependencies applied to this purpose for isotropic metals. Basing on the calculated values of κ_{ph} in normal and in superconducting states they determine the energy gap as the function of temperature using the relation from BRT theory [12]. The obtained value of the gap (Fig.15), $2\Delta(0) = (3.3 \pm 0.7) k_B T_c$ agrees with some data from other experiments, being typical

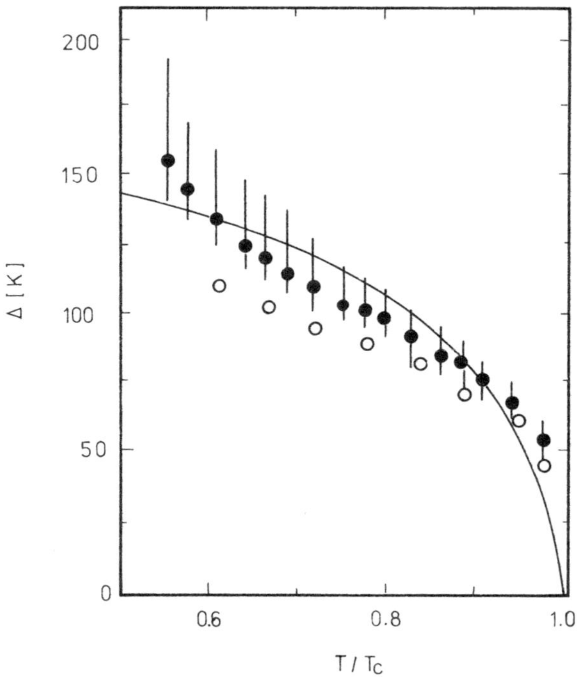

Fig.15 Superconducting energy gap *vs.* reduced temperature (obtained from BRT dependence $\kappa_{phs}/\kappa_{phn}$, where *s* denotes superconducting state and *n* — normal one): • — Eu-Ba-Cu-O, ○ — Y-Ba-Cu-O [70].

for weak electron-phonon interaction in BCS theory. Also Morelli *et al.* [44] estimated some characteristic parameters of superconductors (basing on thermal conductivity and specific heat data) in the frames of BCS theory. However, it seems that there are still not enough experimental data to be able to calculate or even estimate values of κ_e and κ_{ph} in superconducting state of high-temperature superconductors in the frames of BCS model (taking into considerations effects of strong electron-phonon coupling, if necessary). Open remains the question if BCS theory can be applied to the high-temperature superconductors. It seems that the answer is negative.

Change of the temperature dependence of κ by the superconducting transition and the presence of the maximum on the curves κ(T) was in many works (*e.g.*: [15,36,44,64]), as it was noted above, being connected with the phenomenon of electron-phonon interaction, just as it was done for metallic superconductors. However, similar course of the κ(T) curves for samples differing wide in the value of thermal conductivity coefficient and resistivity as well as in the critical temperature indicates necessity of taking into consideration other mechanisms responsible for for the mentioned effects including other possible mechanism of phonon relaxation. For example, singularities analogical to those observed in the course of κ(*T*), C(*T*) and $l_{ph}(T)$ (linear and cubic term in low temperature specific heat, monotonic increase of the mean free path *l* with temperature lowering) are also characteristic in amorphous dielectric (*cf.*,*e.g.*, [8]). Decreasing of the maximum for samples of $YBa_2Cu_3O_{7-x}$

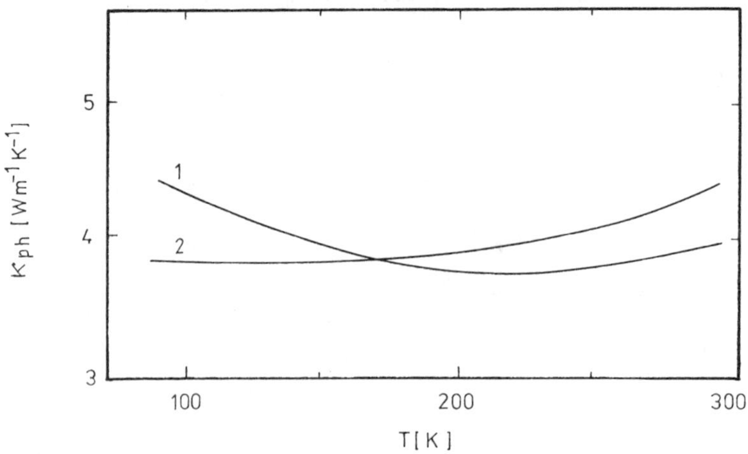

Fig. 16 Temperature dependence of phonon thermal conductivity ($\kappa_{ph} = \kappa_{exp} - \kappa_e$) for samples of different contribution of electron components to the total thermal conductivity: 1 — $YBa_2Cu_3O_{6.81}$, $\kappa_e \cong 20\%$; 2 — $YBa_2Cu_3O_{6.54}$, $\kappa_e \cong 1\%$ [39].

with T_c below 90 K (*e.g.*, for $x = 0.46$ and $T_c \cong 60$ K [39]) and very weak effect or total lack of it near T_c for La-Sr-CuO samples, in which the magnitude of electron component is similar to that of Y-Ba-CuO, also calls for searching of other mechanisms. In the last case, however, one can explain the lack of effect thus: diminishing with temperature den-density of phonons effectively compensates the reduction of the amount of phonon-carrier scattering, what may cause washing out and disappearance of the $\kappa(T)$ peak [36]. Therefore for explanation of the role played by the electron-phonon interaction in thermal conductivity of superconducting ceramics samples further investigations are necessary.

Obviously one cannot univocally state that the electron-phonon interaction is strong and, basing solely on that assumption, interpret the increase of $\kappa(T)$ below T_c. Comparisons of the value of the phonon component of thermal conductivity κ_{ph} obtained from the relation $\kappa_{ph} = \kappa_{exp} - \kappa_e$ for samples with various oxygen content (below $x = 0.1$), show that the values as well as characteristics $\kappa_{ph}(T)$ are similar for comparable samples [39] (Fig. 16). This means that they do not depend on the value of the electron component, which differs between investigated samples within almost two orders of magnitude. Identical results were obtained for several samples Gd-Ba-CuO with various magnitude of electron component [39]. These results may evidence the weakness of the electron-phonon interaction [39].

Independently, similar conclusion can be drawn from the results of electrical resistivity investigations in high temperatures. In this case the main, interesting effect (obviously besides the temperature region where if falls down to zero) is the linear increase of resistivity with temperature in the majority of samples. This property is particularly interesting within the context of the above considerations of eventual mechanism of κ increase. The problem of interpretation of the linear resistivity dependence with assumption of strong electron-phonon coupling follows the fact that linearity extends to much higher temperatures

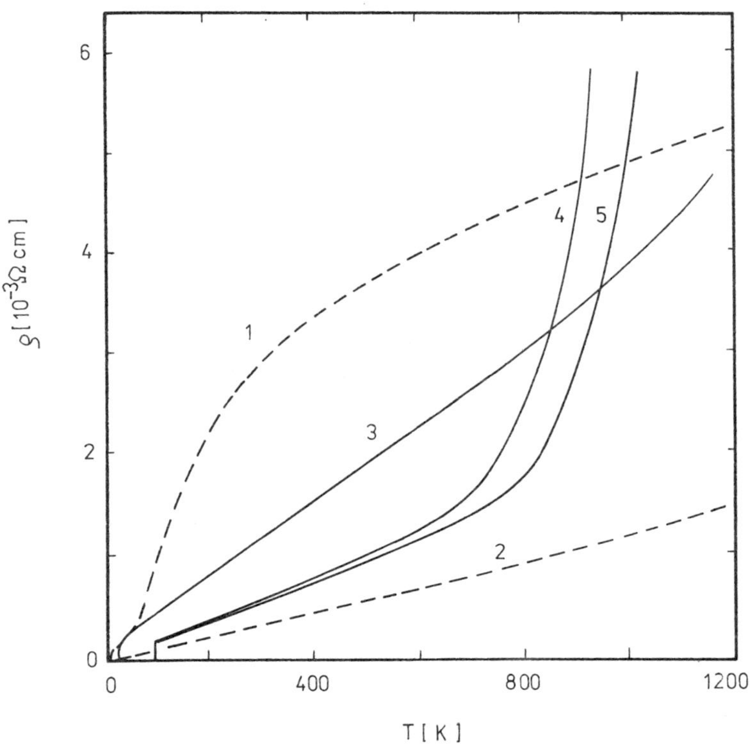

Fig. 17 Comparison of temperature dependence of electrical resistivity of high-temperature superconductors with other metals [75]: 1 — V_3Si (×40), 2 — Cu (×175), 3 — La-Sr-CuO, 4 — Y-Ba-CuO (polycrystal), 5 — Y-Ba-CuO (monocrystal).

than one may suspect (for La-Sr-CuO, Y-Ba-CuO and Y-Ba-CuO single crystal - above 1000 K [74,75], see Fig. 17). Increase of resistivity in temperatures above 600 K for polycrystalline Y-Ba-CuO is interpreted as a result of oxygen desorption.

As it is known (*e.g.*, [76]), when electron mean free path l_e diminishes with respect to interatomic distances in metal, the characteristccs of resistivity as function of temperature exhibits the effect of "saturation", *i.e.* the rate of change of increase of resistance with temperature is reduced in some range of temperature. In case of La-Sr-CuO the lack of saturation was stated — up to do 4000 $\mu\Omega$cm and for monocrystalline Y-Ba-CuO - up to do 1000 $\mu\Omega$cm [75]. Estimating the mean free path l_e for given carrier density n, from the relation of Boltzmann type

$$\rho = \hbar\ (3\pi^2)^{1/3}/n^{2/3}\ e^2 l_e \qquad (10)$$

and comparing with Cu-O bond length (about 2 Å) one finds that in order to substantiate the lack of saturation one must adopt for electron density a number less than about $10^{22}cm^{-3}$ (for $n = 10^{22}cm^{-3}$, $l_e = 1$ Å). Linear characteristics of ρ extending to such high temperatures calls for a mean free path of the order of 10 Å [74,75] what corresponds to electron density of $\sim10^{20}cm^{-3}$. If this is the case, one could hardly imagine how such small number of carriers can so effectively scatter phonons as it happens by thermal conductivity. Thus measurements of electrical resistivity also substantiate notion about small part of the electrons - phonons mutual interaction.

If one apply the relation of Allen *et al.* [77] for the resistivity of metal limited by phonon scattering

$$d\rho/dT = 2\pi k_B \,/\, \hbar e^2 (m/n)_{eff} \lambda_{tr} \ , \qquad (11)$$

where: λ_{tr} — electron-phonon coupling constant, m_{eff} — effective mass connected with increase of carrier mass in high temperatures, one finds that for the observed in experiments (for La-Sr-CuO and Y-Ba-CuO) value $d\rho/dT \cong 3 - 4\ \mu\Omega cm\ K^{-1}$ the diminishing of carrier density n causes simultaneous diminishing of the electron-phonon interaction strength. Estimated value of the electron-phonon coupling [75] amounts for Y-Ba-CuO: $\lambda_{tr} \leq 0.3$ and for La-Sr-CuO: $\lambda_{tr} \leq 0.1$. This is one of the smallest known magnitudes of coupling, similar to the value of λ_{tr} for copper ($\lambda_{tr} > 1$ corresponds to strong coupling). Obviously the above considerations do not constitute univocal proof but they are only indication of the possibility of presence of weak electron-phonon coupling in high-temperature superconductors, what may be the evidence of non-phonon mechanism of superconductivity. It follows that further investigations of electrical resistivity in high temperatures are necessary, particularly in samples of high degree of disorder (some of them indicated traces of "saturation" effect [78]).

If one assume that the scattering of phonons on electrons is weak, one cannot connect the difference between thermal conductivity in normal and in superconducting state with the reduction of phonon component above T_c resulting from the electron-phonon interaction. There is also possibility that the mechanism of electron-phonon scattering in not the only one responsible for thermal conductivity behaviour below superconducting transition. It seems that presented here discussion indicates this possibility clear.

In [79] Jeżowski suggested adopting for explanation of the discovered effects the convective mechanism of heat transport. This phenomenon is known from the theory of classical superconductors, as one of the mechanisms responsible for increase of κ below T_c . The nature of the effect [80,81] is as follows: arising of a gradient of temperature ΔT in the sample causes that besides normal heat stream also normal electric current j_n appears, evoked by the same excitations that take part in energy transport responsible for thermal conductivity of superconductor. In isotropic case this current is fully compensated by superconducting current j_s, so that $j_n + j_s = 0$. However, contrary to superconducting component, entropy flux is connected with the movement of normal component and although the net charge transport is zero, the above mechanism leads to appearance of additional component of thermal conductivity — — κ_c. As follows from theoretical estimation for classical superconductors [10,14] the ratio κ_c/κ (where κ is the total thermal conductivity) should be of the order of $k_B T_c/\epsilon_F$ (where ϵ_F is Fermi energy). It turns out that for high-temperature superconductors κ_c/κ (where $\kappa_c = \kappa_{max} - \kappa_n$ and κ_n is the thermal conductivity coefficient in normal state) in the region of maximum is of the order of $\sim 10^{-2}$- 10^{-1} and in new superconductors $k_B T_c/\varepsilon_f$ is of the same order of magnitude [82]. Thus we see that the mechanism of convective heat conductivity may be considerable contribution to the observed increase of κ below T_c . In classical superconductors additional component of thermal conductivity below superconductive transition have not been observed, what is not surprising because κ_c/κ is for them a quantity of the order of $\sim 10^{-4}$ and the convective heat transport is therefore negligible in comparison to the total

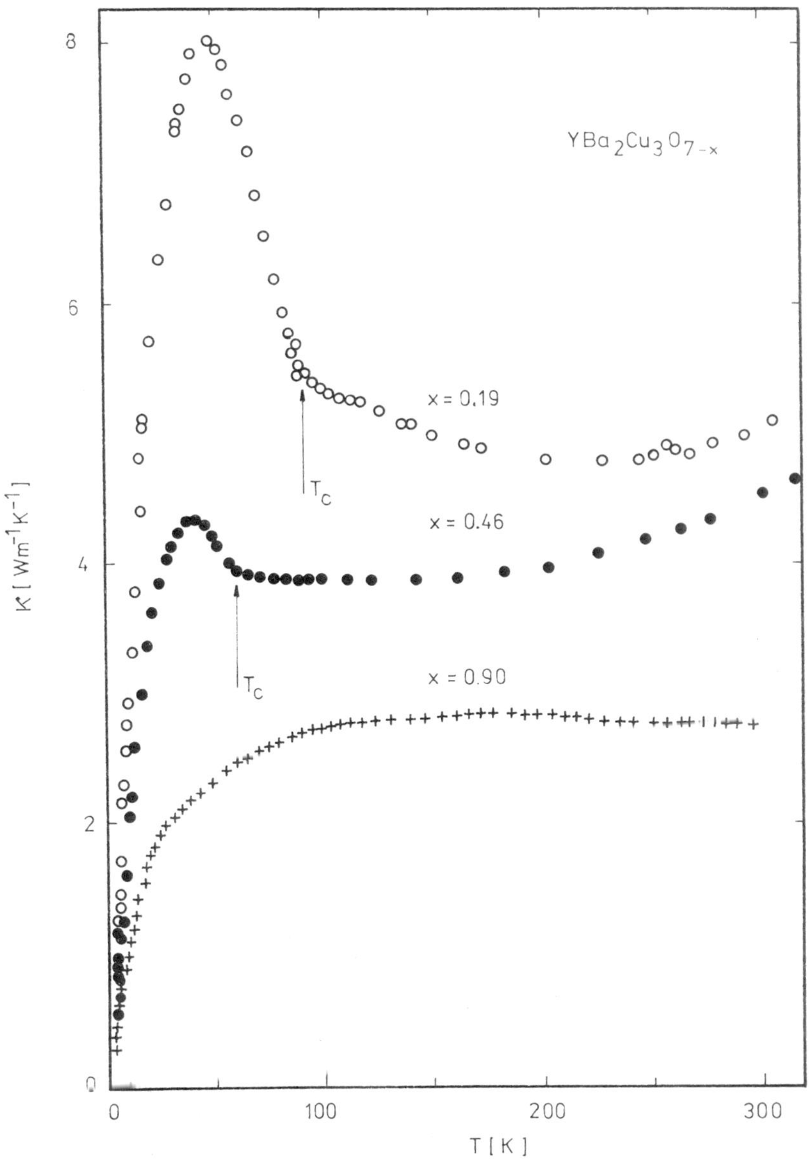

Fig.18 Thermal conductivity of $YBa_2Cu_3O_{7-x}$ samples with various oxygen content [41].

thermal conductivity. One can suppose that on the same reason the effect is not observed in La Sr-CuO samples, and the observed one for Y-Ba-CuO samples having $T_c \cong 60$ K is much smaller than for the samples with $T_c \cong 90$ K. The above does not exclude possibility of contributions of other mechanisms, *e.g.*, mechanism connected with oxygen atoms movement resulting from decomposition of orthorhombic phases, or the mechanism connected with appearance of libron oscillations at T_c [83,84]. In case of correlation between these oscillations they may constitute means for additional heat transfer, in the other one they may be source of additional scattering, leading to decrease of thermal conductivity.

2.4. Influence of Oxygen Content on the Value and Temperature Dependence of Thermal Conductivity

Just in the first publications devoted to new superconductors it was pointed out, how important is T_c dependence on oxygen content. One therefore might suppose that investigations on kinetic characteristics of $YBa_2Cu_3O_{7-x}$ compounds with varying x would give particularly interesting results. However, the majority of authors presently limit their interest to measurements on samples with small x or do not report the content of oxygen in their samples at all. Besides, it applies not only to investigations of kinetic properties, but also to other fields of research.

One of the most interesting results was discovery of change of thermal conductivity value and temperature dependence of κ with change of x change of oxygen content). For the first time these dependencies were reported in works of Jeżowski *et al.* [39] and Salce *et al.* [21], later appeared papers by Zavaritskii *et al.* [85,86] and Efimov *et al.* [24]. In the Fig.18 thermal conductivity of $YBa_2Cu_3O_{7-x}$ in the temperature range 4 - 300 K for three compositions: x = 0.19, 0.46 and 0.90. With the decrease of oxygen concentration the value of thermal conductivity also decreases and the maximum value lowers and shifts towards low temperatures with gradual flattening. For non-superconducting samples one observes monotonic decrease of thermal conductivity with lowering of temperature (no maximum of κ). This indicates change of character of phonon scattering in the whole volume of the sample, or vanishing of additional component of thermal conductivity resulting from structure changes taking place when oxygen content decreases. In the first case appearance of oxygen vacancies in Cu—O chains can strongly scatter short-wavelength phonons characteristic for the region $T > \Theta/4$. In the second additional component of thermal conductivity would be connected just with the presence of Cu—O chains. Appearance of breaks in the chains — along with diminishing amount of oxygen in the material would cause decrease of thermal conductivity.

The vanishing of maximum of the characteristics $\kappa(T)$ on transition from superconducting to non-superconducting material (from conductor to semiconductor) have not to support the hypothesis for the decisive role played by phonon-electron scattering in thermal resistance, as it has been stated in [85,86]. It is sure that the increase of κ below T_c is the intrinsic property of samples of high critical temperature, being therefore connected with onset of superconductivity or having with superconductivity common cause. Responsible for the increase of κ may be other mechanism, not necessary connected with electron-phonon interaction.

Particularly interesting is the fact that thermal conductivity of $YBa_2Cu_3O_6$ samples turned out to be considerably higher (+80% to +150% at ~200 K) than for samples with $6 < x \leqslant 7$ [20,21,86] — see the Fig.19. This fact was connected with the lack of electron component κ_e , as confirmation of strong electron—phonon interaction. This explanation does not seem true, since the samples with $x < 0.4$ also exhibit semiconducting character of the electrical resistivity temperature dependence. It is hard to believe that very small amount (below few percent) of the orthorhombic phase might have so decisive influence on the value of thermal conductivity. The sound velocity in the samples with $x = 1$ varies from these with $x = 0$ only minutely (9%) [87], likely as specific heat does [35,88]. In that situation one may suppose that the basic reason for different κ characteristic for samples with $x = 1$ may be lack

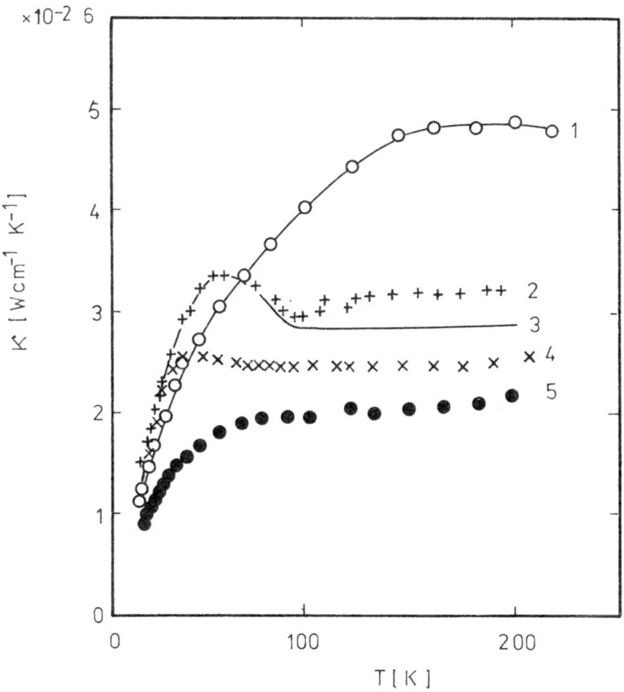

Fig. 19 *a* Temperature dependence of thermal conductivity of $YBa_2Cu_3O_{7-x}$ sample with various oxygen content due to different thermal treatment: 1 — x = 1, 2 — x = 0, 3 — x = 0.23, 4 — x = 0.30, 5 — x = 0.69 [85,86].

of oxygen vacancies. It should be, however, pointed out that the work [79] reported strong decrease of thermal conductivity in samples of composition very close to $YBa_2Cu_3O_7$. This may suggest the possibility of additional dependence on oxygen amount, the greatest value of thermal conductivity coefficient should correspond to x = 0.1 — 0.2.

Basing on comparisons of the values of thermal conductivity of samples $YBa_2Cu_3O_6$ and $YBa_2Cu_3O_7$ [86] the relaxation time for phonons scattered on electrons ($\tau_{ph}{}^e$) was estimated, exploiting the relation $\kappa \sim Cv^2\tau$. This needed assumptions that the different mechanisms of phonon scattering are independent (of each other) {*cf.* equation (1)} and that in the sample $YBa_2Cu_3O_6$ there is lack of scattering on electrons and oxygen vacancies. The value $\tau_{ph}{}^e \sim 10^{-12}$s for T = 100 K was obtained. This value is comparable with the relaxation times $\tau_{ph}{}^e$ in classical superconductors, basing on which the authors [86] conclude the strong electron-phonon coupling. It is worth noting that the above value differs significantly from the smaller value $\tau_{ph}{}^e \cong 10^{-15}$s obtained in [44].

The authors of the work [89] turned attention to the dependence of the electron component of thermal conductivity on the amount of oxygen in the sample. Its diminishing causes decrease of κ_e the total thermal conductivity as well as the maximum height. Besides, the difference of κ_e of about order of magnitude does not causes so significant change of height of thermal conductivity maximum as it should be expected from the mechanism of phonon-electron scattering, what was pointed out earlier.

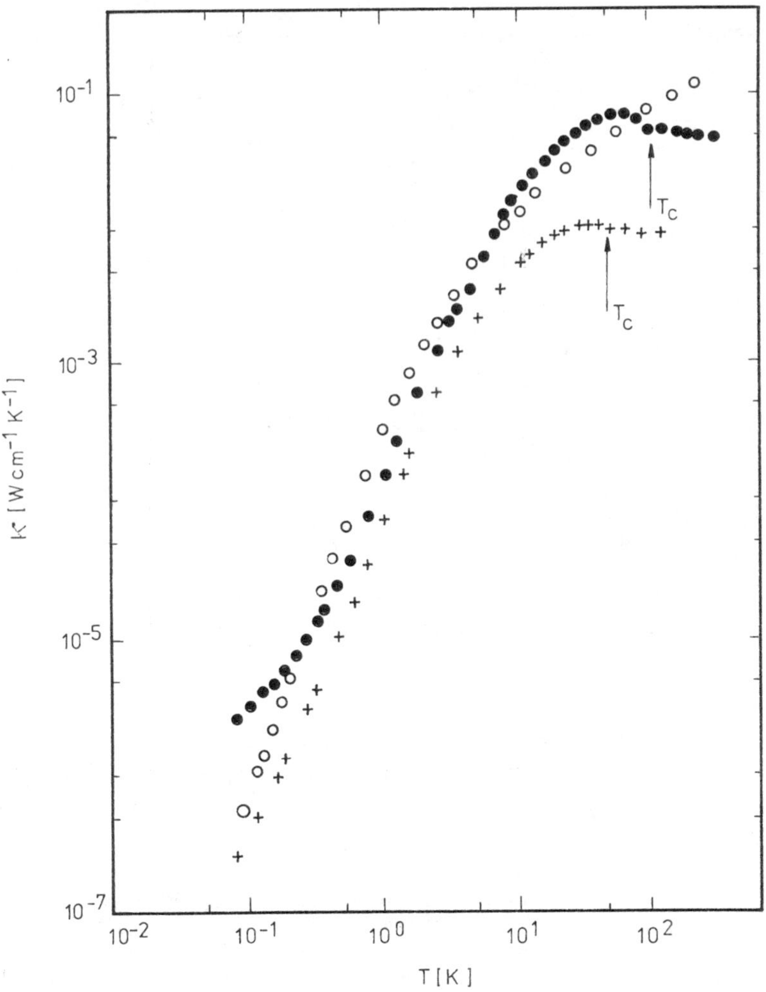

Fig. 19 *b* Temperature dependence of thermal conductivity of various $YBa_2Cu_3O_{7-x}$ samples with different oxygen content: + — $x = 0.45$, • — $x = 0.10$, o — $x = 0.90$ [21].

The measurements of thermal conductivity in the family of compounds RE-Ba-CuO (RE ≡ Gd [40,44], Eu [44,70], Dy [44], Sm [40], Er [44] did not give fundamentally different results. In the Fig.20 the temperature dependence of κ for Y-, Gd-, Sm—Ba-CuO in temperature range 4.2 - 300 K was depicted. The measurements of $\kappa(T)$ of the compound Tl-Ca-Ba-CuO, having T_c = 119 K [42] (Fig.21), besides the shift of the beginning of increase (derivative jump) of κ and the temperature of the appearance of maximum, have given similar dependence of κ as for Y-Ba-CuO.

It seems that, in order to obtain univocal conclusions on the nature of maximum and $\kappa(T)$ characteristics, measurements of κ and ρ in monocrystalline samples in normal and superconducting state, the more that the results can have fundamental meaning for explanation of many properties of high-temperature superconductors.

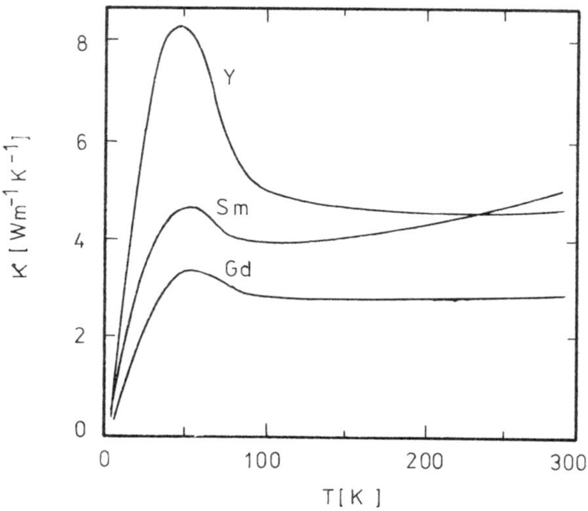

Fig.20 Temperature dependence of thermal conductivity of RE-Ba-CuO samples, where RE ≡ Y, Sm, Gd [40].

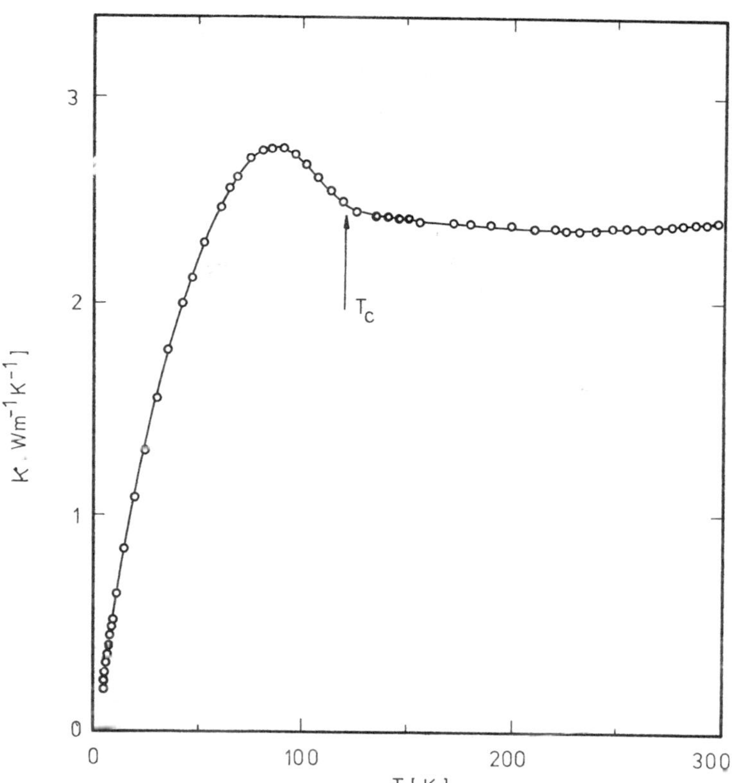

Fig.21 Thermal conductivity of $Tl_2Ca_2Ba_2Cu_3O_x$ sample [42].

2.5. Anomalies of Thermal Conductivity Above the Critical Temperature

From the very moment of discovery of high-temperature superconductivity in ceramic Y-Ba-CuO with oxygen deficiency many methods of physical investigations were applied to study their properties. Some of them allow to discover effects connected probably with order-disorder type transitions in the system of oxygen vacancies and instabilities of the crystal structure in temperatures much above T_c and, as some suppose, near that temperature. It seemed that thermal conductivity, sensitive to oxygen content of samples (as we have known from the preceeding paragraph) should constitute a method for investigation of oxygen vacancies ordering and eventual, connected with the last, structural phase transitions. As one can suppose, the grater κ the more ordered is oxygen distribution with the same mean stoichiometry of various samples. Thermal conductivity investigations conducted with particular aim to search for anomalies corresponding to structural transitions have indeed given interesting results.

It should be pointed out here that the first work that formulated hypothesis about a phase transiton in high temperature in new supercon-

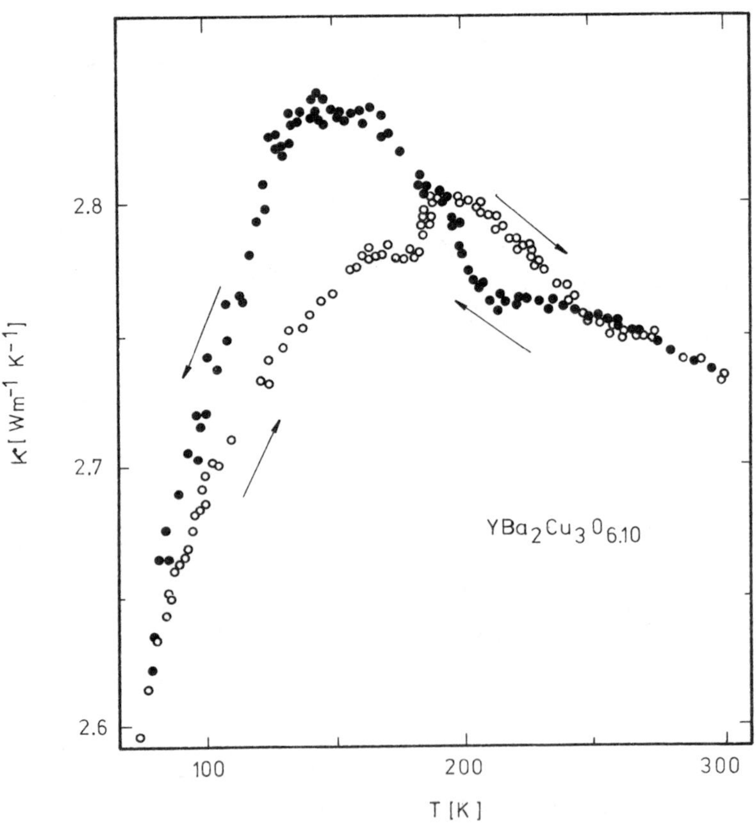

Fig.22 *a* Hysteresis of thermal conductivity of $YBa_2Cu_3O_{6.10}$ sample [41]. • — measurement on cooling, o — on heating.

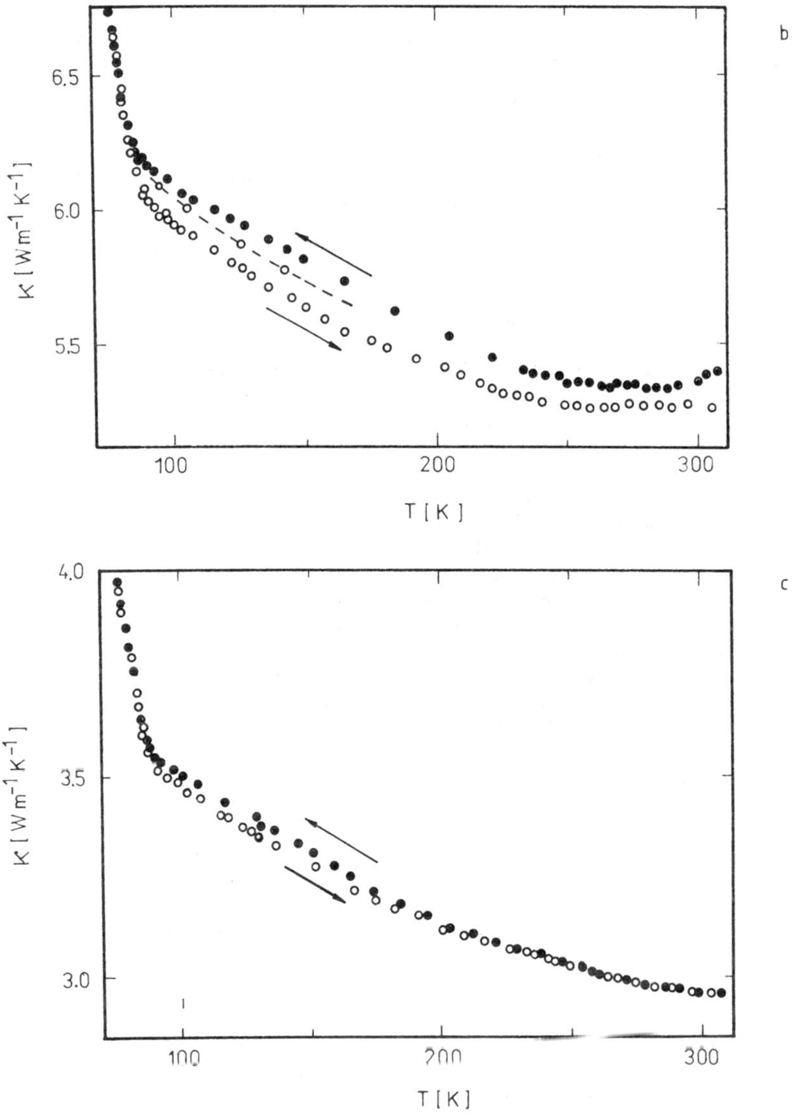

Fig.22 *b, c* Hysteresis of thermal conductivity of $YBa_2Cu_3O_{7-x}$ samples [79]. ● — measurement on cooling, o — on heating. (*b*) - $YBa_2Cu_3O_{6.90}$; (*c*) - $YBa_2Cu_3O_{6.95}$. The broken line in part (*b*) corresponds to "intermediate" courses.

ductors was the work on their thermal conductivity. Jezowski *et al.* [15] observed appearance of a small anomaly in the course of $\kappa(T)$ and $\rho(T)$ in temperature range 250 - 270 K, which, as we now think, was connected with the process of ordering of oxygen vacancies. As it latter turned out, for the evidencing of anomalies with help of thermal conductivity investigations very precise measurements of $\kappa(T)$ are necessary. Besides, it was confirmed that the discovered effects are closely connected with the oxygen content in the samples and, as the amount of oxygen increases the quantity and temperature range of their appearance diminishes. It should be noted that the measurements of other physical quantities disclosing anomalies were not conducted as depending on oxygen content. It prevents their comparison and drawing conclusions on the base of it.

Among the quantities showing anomalies, elastic properties shall be mentioned since they are much more sensitive to processes connected with atoms movement and phase transitions, much more than, *e.g.*, thermal properties. Interesting is the fact that it was thermal conductivity of superconductors that allowed to observe hysteresis effects, similar as observed in elastic properties investigations. It should be stressed that hysteresis of thermal conductivity of solids has not been observed up to now.

The measurements of $\kappa(T)$ made by Jeżowski *et al.* [41,79] showed hysteresis in samples of Y-Ba-CuO of various oxygen content. Besides the hysteresis, other anomalies were observed in the temperatures 80 - - 300 K, appearing particularly in non-superconducting sample of tetragonal structure. In the Figs.22 a,b,c the results of measurements obtained both during heating and cooling on samples of non-stoichiometric composition with respect to oxygen content: a) — $YBa_2Cu_3O_{6.1}$ b) — $YBa_2Cu_3O_{6.90}$, c) — $YBa_2Cu_3O_{6.95}$. Below the temperature $T_2 \cong 90$ K (depending on the sample $80 \leq T_2 < 90$ K; for superconducting samples this temperature is close to T_c) thermal conductivity does not depend on the way of approach to the particular temperature and $\kappa(T)$, both for heating and for cooling is identical. Above this (T_2) temperature the distinct hysteresis of $\kappa(T)$ dependence is observed in all investigated samples, both orthorhombic and tetragonal. The maximum difference between κ values amounts: a) — 4% (T = 135 K), b) — 3% (T = 150 K), c) — 1,5% (T = 130 K). In case of samples *a* and *c* closing of the hysteresis loop is observed at the temperature T_1 amounting, respectively, ~240 and ~210 K. On the other hand, for the sample *b* estimated value of T_1 is ~400 K. Above T_1 the value of thermal conductivity again is not depending of the way of approaching particular temperature (heating or cooling). The important fact is that the values of thermal conductivity coefficient in the region of hysteresis appearance are stable and reproducible. The temperatures T_1 and T_2 limit from above and from below, respectively, the region of existence of hysteresis. The course of $\kappa(T)$ in the region of hysteresis do not depend on the speed of cooling or heating; also no changes connected with elapse of time were found (samples were stored in vacuum). The reason for this hysteresis is still not clear.

Considering generally the character of $\kappa(T)$ dependence one can formulate hypothesis that, on cooling, at temperature T_1 begins a transition from phase A to phase B. This transition extends up to temperature T_2, below which "normal" temperature dependence for the phase B is observed, down to the lowest temperatures. If this effect is connected to a phase transition, besides its origin, unclear is also why the process extends to such wide range of temperatures. It may, *e.g.*, mean that this first-order phase transition has diffusive character and is connected with ordering of oxygen vacancies. The lack of anomalies (not counting the small anomaly at T_1) in the course of electrical conductivity of the investigated samples, and, first of all, small contribution of electron component to the total thermal conductivity seems to indicate that the hysteresis is not linked with the electron component of thermal conductivity. It is to be noted here that small anomalies in electrical resistivity at T_1 were observed only sporadically, *e.g.*, [15,90,91]. Thus discovered $\kappa(T)$ dependence peculiarities are connected with changes of phonon component and can be results of lattice instability, which is confirmed in other investigations.

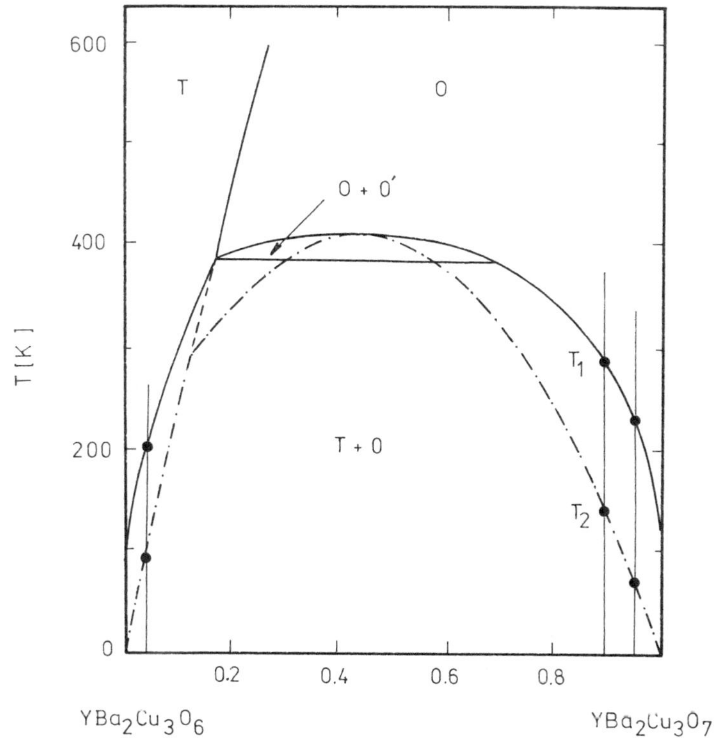

Fig.23 Equilibrium phase diagram of $YBa_2Cu_3O_{7-x}$ system ($0 < x < 1$). T — disordered tetragonal phase, O, O' — ordered orthorhombic phases of different oxygen index. Vertical lines are described in text. [112,113].

As it was mentioned above, particularly interesting turned out to be the measurements of elastic properties of Y-Ba-CuO [72,90,92-103]. In many of cited works, *e.g.*, [94,97,100,102] effects of hysteresis were observed, occurring mostly above T_c, but in different regions of temperature by various authors. Many of these investigations indicate, that lattice stiffening increase below the superconducting transition temperature. Among others, Ledbetter *et al.* [102] stated that the observed hysteresis of sound velocity suggested a wide phase transition. Ewert *et al.* [99] see the reason for hysteresis in porosity of the material, while Kurtz *et al.* [103] suppose that temperature range of the effects observed in investigations of elastic properties speak strongly (besides other features) for the presence of ferroelectric phase. Some of the measurements of specific heat in orthorhombic Y-Ba-CuO show, besides known jump at T_c, also anomaly in the range 200 - 240 K [90, 104,105] connected, according to those authors, with ordering of oxygen in Cu—O chains. In dilatometric measurements of $YBa_2Cu_3O_{7-x}$ small changes of the linear thermal expansion were found near T_c [106] and near 160 and 240 K [107]. Small hysteresis of thermal expansion by cooling and heating was observed by Kirschner *et al.* [108] in temperature range 273 — 100 K. They tried to explain this phenomenon as the result of thermomechanical corrosion processes. Zhang Yuheng *et al.* [91] discovered in the sample of tetragonal structure anomalies in thermal expansion, electrical resistivity and specific heat. The anomaly at about 103 K they ascribe to a partial phase transition from tetragonal to orthorhombic structure. The difference in temperatures marking the beginning and end of hysteresis, obtained by measurements of various physical quantities, can be explained by small difference in the energy of phases A and B in the region of phase transition. The measurement itself, being a disturbance to physical system, *e.g.*, in acoustic investigations, can lead to a change of system state, for example from a metastable phase A to stable phase B. The mechanism of transition (ordering of oxygen vacancies) and its diffusive character may indicate the low energy barrier between A and B states what could be trated as argument for the presented interpretation. The maximum T_1 and minimum T_2 would determine temperatures of spinodal decomposition.

The presence of irregularities in the course of temperature dependence of lattice constants at 90 and 240 K was found in neutron diffractography [109]. Electron microscopy investigations [110] as well as X-ray high-resolution diffraction [111] in the Y-Ba-CuO compound allow to discover local changes of $a-b/b+a$ (a and b being the lattice parameters) in temperatures above T_c . These few examples indicate clearly that the effect observed in thermal conductivity may also originate from a change of structural type, and therefore result from phonon spectrum reconstruction. A support for the supposition about the existence of the structural phase transition near 240 K, displaying character of the spinodal decomposition, can be found in papers on lattice parameters [112] or Raman spectra [113] in these compounds.

Because of the character of the observed changes of temperature range (at the high-temperature side) where the hysteresis begins, authors of [41,79] base the interpretation on the phase diagram of the system $YBa_2Cu_3O_{7-x}$ (Fig.23). It was proposed theoretically by Khachaturyan *et al.* [114,115], who, starting from Nernst Principle, forecasted instability of the crystal structure of non-stoichiometric Y-Ba-CuO. Vertical lines in the Fig.23 correspond to thermodynamical process (with assumption of constant oxygen content in the sample), during which anomalies of thermal conductivity appear at temperatures similar to these observed in experiments [41,79].

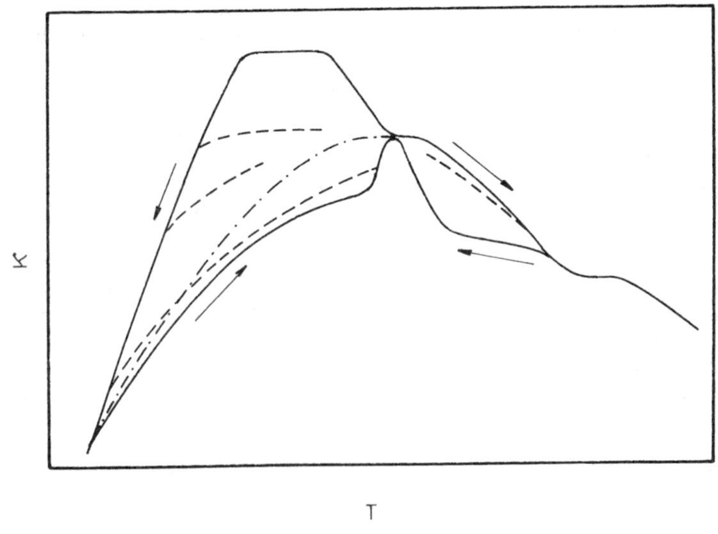

Fig.24 Schematic presentation of hysteresis of thermal conductivity of $YBa_2Cu_3O_{6.10}$ [41].

During measurements of thermal conductivity of superconducting samples (on cooling) after crossing temperature T_1 one finds in the region where two or more orthorhombic phases coexist having the same structure but different oxygen content. Close to temperature T_2 decay of the orthorhombic phases to tetragonal and orthorhombic ones due to spinodal decomposition takes place. The amount of oxygen in one of the orthorhombic phases increases while in the other decreases tending towards construction of the phase $YBa_2Cu_3O_6$. Due to structural changes at T_2 the measurement of thermal conductivity on heating gives different value of the thermal conductivity coefficient. Return to original state occurs again in temperature T_1 . In case of tetragonal sample of $YBa_2Cu_3O_{6.1}$, besides hysteresis, anomalies on the curves κ(T) were observed: one may link these anomalies with discovery of direct decomposition of tetragonal phase (below T_1) into mixture of two phases, orthorhombic and tetragonal, which process ends at T_2 . The observed increase of κ (240 — - 180 K on cooling) would be in that case caused by oxygen atoms migration — jump from *a* axis onto *b* axis in the Y-Ba-CuO elementary cell and appearance of Cu—O chains. Different temperatures T_1 , being starting points for structural changes and resulting from different amounts of oxygen, agree with the presented phase diagram.

Still two effects deserve to be mentioned, reported in works [41, 79]. The first is connected with occurrence of "intermediate" curves (*cf.* Figs.22 b and 24) in the region of the hysteresis loop; they may indicate the presence of metastable states, in samples of both tetra-

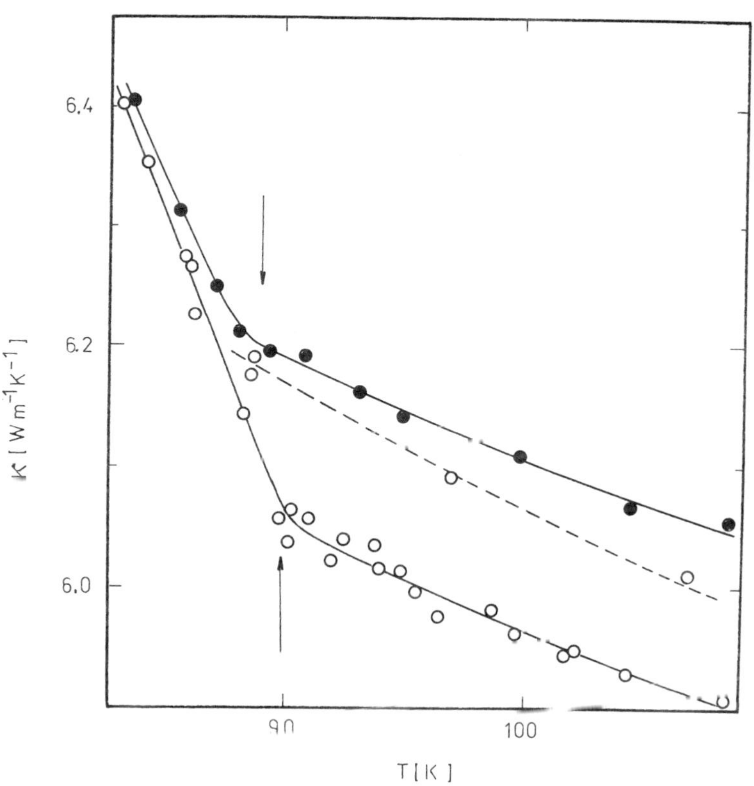

Fig.25 *a* Temperature dependence of thermal conductivity of $YBa_2Cu_3O_{6.90}$ sample in the vicinity of T_c [79].

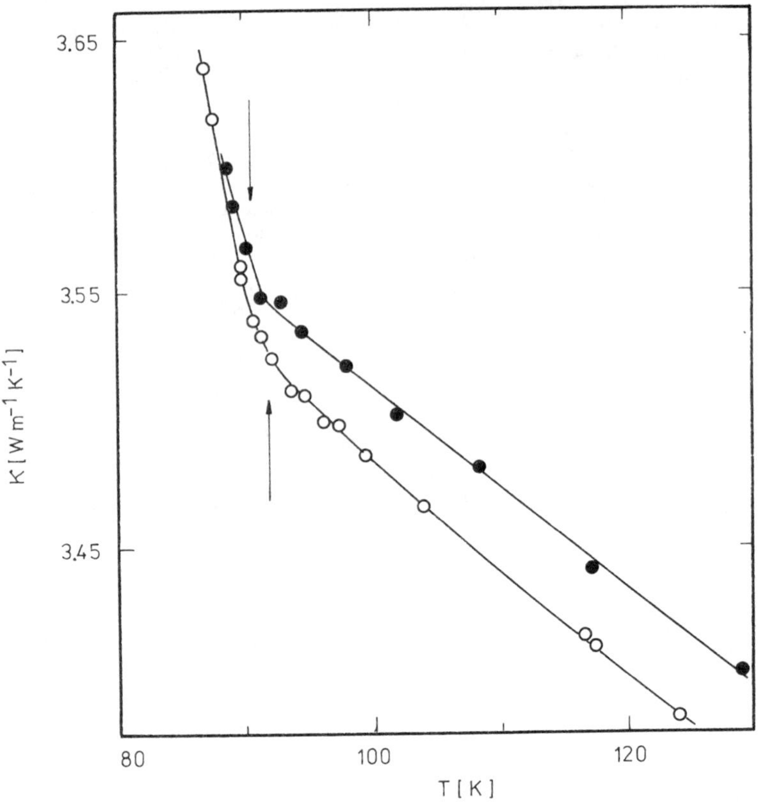

Fig.25 *b* Temperature dependence of thermal conductivity of $YBa_2Cu_3O_{6.95}$ sample in the vicinity of T_c [79].

gonal [41] and orthorhombic [79] structure. The second effect consists in a shift of the derivative jump of $\kappa(T)$ near T_c towards higher temperatures (Fig.25 - shift indicated by arrows). This is the result of processes occurring in the sample with elapse of time, when it finds below the superconducting transition temperature ($80 < T < 90$ K). This shift may mean some small change (increase) of T_c . Raising the sample temperature above T_1 causes return of the point of considered jump of derivative of $\kappa(T)$ back to the initial place. This phenomenon could help explaining T_c shifts observed by thermal cycling of Y-Ba-CuO sample during electrical resistivity measurements in, *e.g.*, [116].

Closing the discussion of anomalies in $YBa_2Cu_3O_{7-x}$ we want yet call attention to the changes of temperature dependence of κ following from changing amount of oxygen in the investigated samples. Basing on analysis of the hitherto obtained results [69] and on comparisons with other discussed above literature data one can present the course of $\kappa(T)$ as it is depicted in the Fig.26. Decrease of thermal conductivity for $T > T_{max}$ is connected with domination of the scattering of phonons on defects. The characteristic for these processes dependence is $\kappa \propto T^{-1}$ and similar to it dependence is indeed observed in experiments for samples with $x < 0.1$ (Fig.26 c). On the other hand, the observed in several works increase

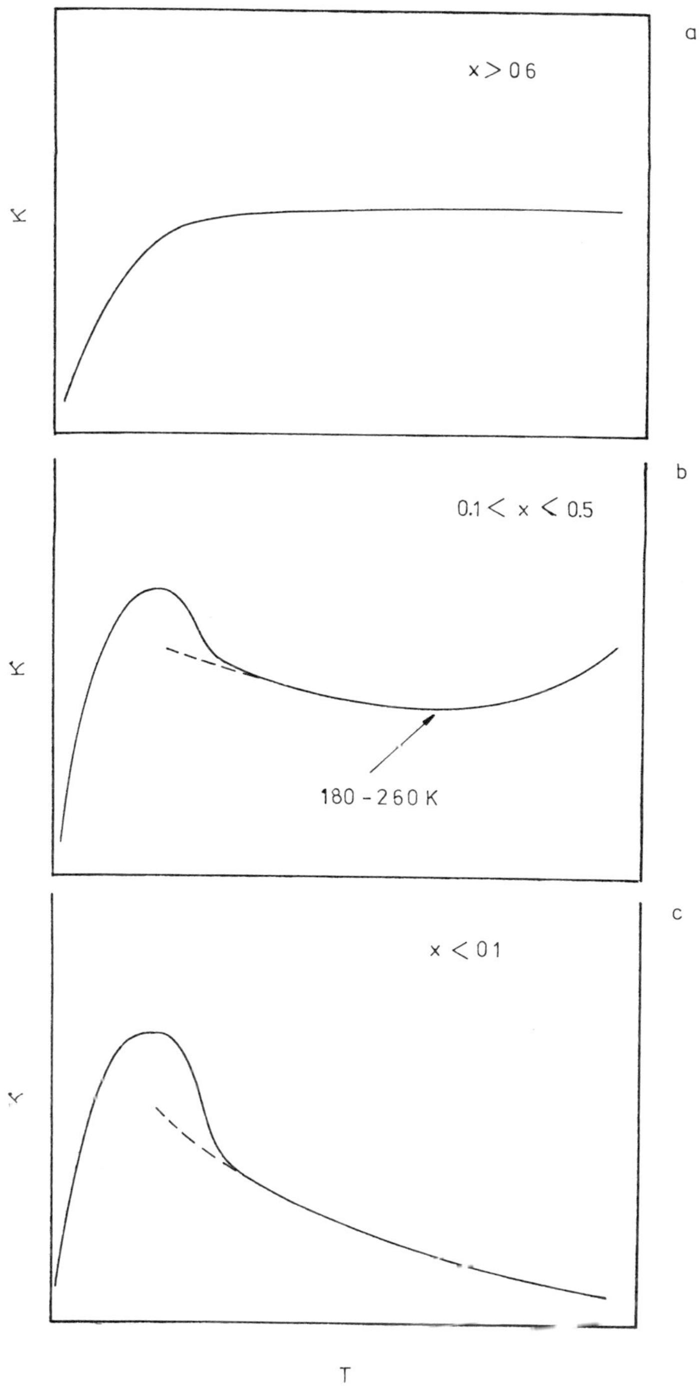

Fig.26 Schematic presentation of thermal conductivity temperature dependence of $YBa_2Cu_3O_{7-x}$ *versus* oxygen index x [69].

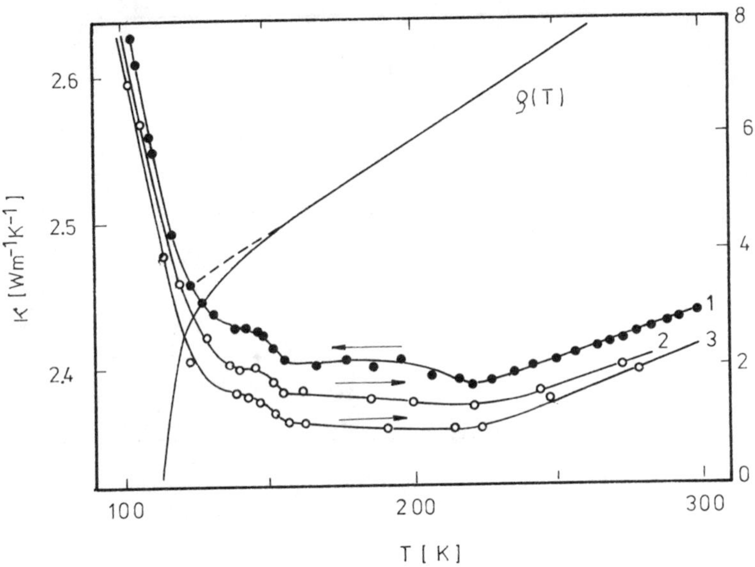

Fig.27 Peculiarities of $\kappa(T)$ dependence of $Tl_2Ca_2Ba_2Cu_3O_x$ [42]. 1 — first cooling; 2 — first heating after keeping below T_c for 5 hours; 3 — heating after keeping below T_c for 11 hours. Right-hand scale: resistivity.

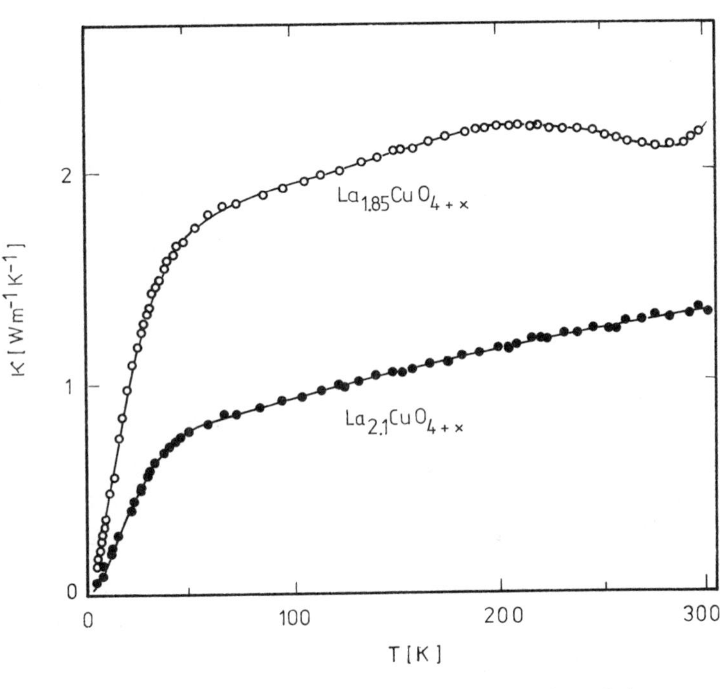

Fig.28 Thermal conductivity of samples of: $La_{2.1}CuO_{4+x}$ (nonsuperconducting) and $La_{1.85}CuO_{4+x}$ (superconducting) [43].

of κ, beginning at various temperatures (180 - 260 K) — Fig.26b, may result from the processes of ordering of oxygen vacancies.

Hysteresis was also found in $Tl_2Ca_2Ba_2Cu_3O_x$ compound — Fig.27 [42]. It follows henceforth that it is the property of the materials with high T_c, connected with their structure and, indirectly, possibly also with superconductivity. However it should be remembered that analogous hysteresis was also observed in tetragonal, non-superconducting Y-Ba-CuO. In the case of Tl-Ca-Ba-CuO sample, besides hysteresis, also small and abrupt increase of thermal conductivity at the temperature, where the characteristics of the electrical resistivity (~155 K) declines from linearity, calls attention. This effect may be the result of a structural change being the reason for resistivity fall down. D.c. measurements of magnetic susceptibility did not show at that temperature any changes that might evidence the existence of additional high-temperature phase. Thermal conductivity of compounds based on La in high temperature range has been measured only for La_2CuO_4 with oxygen surplus: superconducting $La_{1.85}CuO_{4+x}$ and non-superconducting $La_{2.1}CuO_{4+x}$ [43]. The hysteresis effect has not been found in these samples, however in the superconducting one a reversible (with respect to heating and cooling) anomaly has been found in the course of $\kappa(T)$ in the range 200 -290 K (Fig. 28). This anomaly can be connected with a structural phase transition — from high temperature tetragonal phase to low temperature — orthorhombic.

Thus thermal conductivity measurements of high-temperature superconductors proved to be particularly sensitive, allowing discovery of even small structure changes, in many cases impossible to reveal by other experimental methods. Of course, further systematic investigations are necessary for explanation (confirmation or rejection) of several hypotheses, which have been collected and presented in this review.

REFERENCES

[1] J.L.Olsen, *Proc.Phys.Soc.* A **65**, (1952) 518.

[2] N.V.Zavaritskii, *Zh.Eksp.Teor.Fiz.* **33**, (1957) 1085; **34**, (1958) 1116.

[3] K.Mendelsohn and J.Olsen, *Proc.Phys.Soc.* A **63**, (1950) 2.

[4] G.Renald and C.Reynolds, *Phys.Rev.* B **2**, (1970) 3533.

[5] S.Laredo, *Proc.Roy.Soc.* A **229**, (1955) 473.

[6] J.Hulm, *Proc.Roy.Soc.* A **204**, (1950) 98.

[7] J.M.Ziman, *Electrons and Phonons*, Oxford Univ.Press (Oxford 1960).

[8] R. Berman, *Thermal Conduction in Solids*, Clarendon Press (Oxford 1976).

[9] *Thermal conductivity*, (Ed. R.P.Tye), Academic Press, (London 1969).

[10] B.T.Geilikman and V.Z.Kresin, *Kineticheskie i nestatsionarnye yavleniya v sverkhprovodnikakh*, Nauka, (Moskva 1972).

[11] G.Rickayzen, *Theory of Superconductivity*, Interscience Publishers, John Wiley & Sons, (New York 1965).

[12] J.Bardeen, G.Rickayzen, and L.Tewordt, *Phys.Rev.* **113**, (1959) 982.

[13] B.T.Geilikman and V.Z.Kresin, *Zh.Eksp.Teor.Fiz.* **36**, (1959) 959. B.T.Geilikman and V.Z.Kresin, *Dokl.Akad.Nauk SSSR*, **123**, (1958) 259.

[14] B.T.Geilikman, *Zh.Eksp.Teor.Fiz.* **34**, (1958) 1042

[15] A.Jeżowski, J.Mucha, K.Rogacki, R.Horyń, Z.Bukowski, M.Horobiowski, J.Rafałowicz, J.Stępień-Damm, C.Sułkowski, E.Trojnar, A.J.Zaleski, and J.Klamut, *Phys.Lett.* A **122**, (1987) 431.

[16] K.Rogacki, C.Sułkowski, A.J.Zaleski, Z.Bukowski, P.Greń, R.Horyń, E.Trojnar, and J.Klamut, *phys.stat.sol.* (b) **146**, (1988) K103.

[17] J.E.Graebner, L.F.Schneemeyer, R.J.Cava, J.W.Waszczak, and E.A.Rietman, in: *High-Temperature Superconductors Fall 1987*, (Eds. M.B.Brodsky, R.C.Dynes, K.Kitazawa, and H.L.Tuller), Proc. Mat.Res.Soc.Symp. vol.**99**, Mat.Res.Soc. (Pittsburgh 1988) p.745.

[18] J.J.Freeman, T.A.Friedmann, D.M.Ginsberg, J.Chen, and A.Zangvil, *Phys.Rev.*B **36**, (1987) 8786.
[19] W.Gottwick, R.Held, G.Sparn, F.Steglich, H.Rietschel, D.Ewert, B.Renker, W.Bauhofer, S.von Molnar, M.Wilhelm, and H.E.Hoenig, *Europhys.Lett.* **4** (1987) 1183.
[20] G.Sparn, W.Schiebeling, M.Lang, R.Held, U.Gottwick, F.Steglich, and H.Rietschel, *Physica* C **153-155** (1988) 1010.
[21] B.Salce, R.Calemczuk, C.Ayache, E.Bonjour, J.Y.Henry, M.Raki, L.Forro, M.Couach, A.F.Khoder, B.Barbara, P.Burlet, M.J.M.Jurgens, and J.Rossat-Mignod, *Physica* C **153-155** (1988) 1014.
[22] A.Bernasconi, E.Felder, F.Hulliger, H.R.Ott, Z.Fisk, F.Greuter, and C.Schueler, *Physica* C **153-155** (1988) 1034.
[23*a*] J.L.Cohn, S.D.Peacor, and C.Uher, *Phys.Rev.*B **38** (1988) 2892;
[23*b*] C.Uher and J.L.Cohn, *J.Phys.C: Solid State Phys.* **21** (1988) L957.
[24] V.B.Efimov, A.A.Levchenko, L.P.Mezhov-Deglin, R.K.Nikolaev, and N.S.Sidorov, *Fiz.Nizk.Temp.* **14** (1988) 1301;
– *Materialy I° Vsesoyuznogo soveshchaniya po vysokotemperaturnoĭ sverkhprovodimosti*, vol.3 (Khar'kov 1988) p.201.
[25] M.P.Zaitlin and A.C.Anderson, *Phys.Rev.*B **12**, (1975) 4475.
[26] A.Migliori, Ting Chen, B.Alavi, and G.Grüner, *Solid State Commun.* **63** (1987) 827.
[27] R.Beyers, G.Lim, E.M.Engler, R.J.Savoy, T.M.Shaw, T.R.Dinger, W.J.Gallagher, and R.L.Sandstrom, *Appl.Phys.Lett.* **50** (1987) 1918.
[28] A.Missenard: *Conductivité thermique des solides, liquides, gaz et de leurs melanges*, Éditions Eyrolles (Paris 1965).
[29] N.E.Philips, R.A.Fisher, S.E.Lacy, C.Morcenat, J.A.Olsen, W.K.Ham, A.M.Stacy, J.E.Gordon, and M.L.Tan, *Physica* B **148** (1987) 360.
[30] M.Ishikawa, T.Takabatake, and Y.Nakazawa, *Physica* B **148** (1987) 332.
[31] S.J.Collocott, G.K.White, S.X.Dou, and R.K.Williams, *Phys.Rev.*B **36** (1987) 5684.
[32] S.Vieira, S.Bourgeal, R.Villar, A.Aguilo, M.A.Ramos, and C.Moure, *Physica* C **153-154** (1988) 1007.
[33] S.D.Peacor and C.Uher, *Phys.Rev.* B **39** (1989) 11559.
[34] S.von Molnar, A.Torressen, D.Kaiser, F.Holtzberg, and T.Penney, *Phys.Rev.*B **37** (1988) 3762.
[35] D.Eckert, A.Bezinge, A.Junod, and J.Muller, *Physica* C **153-155** (1988) 1036.
[36] C.Uher and A.B.Kaiser, *Phys.Rev.*B **36** (1987) 5680.
A.B.Kaiser and C.Uher, *Aust.J.Phys.* **41** (1988) 597.
[37] M.Núñez Regueiro, D.Castello, M.A.Izbizky, D.Esparza, and C.D'Ovidio, *Phys.Rev.*B **36** (1987) 8813.
[38]. M.A.Izbizky, M.Núñez Regueiro, P.Esquinazi, C.Duran, and C.Fainstein, *Phys.Lett.*A **129** (1988) 71.
[39] A.Jeżowski, K.Rogacki, R.Horyń, and J.Klamut, *Physica* C **153-155** (1988) 1347.
[40] A.Jeżowski, A.J.Zaleski, M.Ciszek, J.Mucha, J.Olejniczak, E.Trojnar, and J.Klamut, *Helv.Phys.Acta* **61** (1988) 438.
[41] A.Jeżowski, J.Klamut, R.Horyń, and K.Rogacki, *Supercond.Sci.& Technol.* **1** (1989) 296.
[42] A.Jeżowski, A.J.Zaleski, H.Misiorek, E.P.Khlybov, and V.V.Evdokimova, *Phys.Lett.* A, *in print.*
[43] J.Mucha and A.Jeżowski, *to be published.*
[44] J.Heremans, D.T.Morelli, G.W.Smith, and S.C.Strite III, *Phys.Rev.*B **37** (1988) 1604.
– J.Heremans, D.T.Morelli, G.W.Smith, and S.C.Strite III, *Research Publication* GMR-6049 (1987).

[45] B.A.Merisov, G.Ya.Khadzhaí, M.A.Obolenskií, and O.A.Gavrenko, *Materialy I° Vsesoyuznogo soveshchaniya po vysokotemperaturnoí sverkhprovodimosti*, vol.3 (Khar'kov 1988) p.146.
[46] *Topics in current physics*, vol. 24: Amorphous solids, low temperature properties (Ed. W.A.Philips) Springer Vg. (Berlin 1981).
[47] *Topics in Applied Physics*, vol.46: Glassy Metals I. Ionic structure, electronic transport, and crystalization (Eds. H.-J.Güntherodt and H.Beck) Springer Vg. (Berlin 1981).
[48] B.Golding, N.O.Birge, W.H.Haemmerle, R.J.Cava, and E.Rietman, *Phys.Rev.*B **36** (1987) 5606.
[49] A.K.Raychaudhuri and S.Hunklinger, *Z.Phys.*B **57** (1984) 113.
[50] P.Esquinazi, J.Luzuriaga, C.Duran, D.Esparza, and C.D'Ovidio, *Phys.Rev.*B **36** (1987) 2316.
[51] J.J.Freeman and A.C.Anderson, *Phys.Rev.*B **34** (1986) 5684.
[52] M.Núñez Regueiro, P.Esquinazi, M.A.Izbizky, C.Duran, D.Castello, J.Luzuriaga, G.Nieva, and D'Ovidio, *Physica* C **153-155**, 1016 (1987) 1016.
[53] M.Núñez Regueiro, M.A. Izbizky, and P.Esquinazi, *Solid State Commun.* **67** (1988) 401.
[54] K.Kawazaki, *Progr.Theor.Phys.* **29** (1963) 801.
[55] G.A.Slack, *Phys.Rev.* **126** (1962) 427.
[56] S.Simizu, S.A.Friedberg, E.A.Hayri, and M.Greenblatt, *Phys.Rev.*B **36** (1987) 7129.
[57] A.P.Ramirez, L.F.Schneemeyer, J.V.Waszczak, *Phys.Rev.*B **36** (1987) 7145.
[58] M.E.Reeves, D.S.Citrin, B.G.Pazol, T.A.Friedman, and D.M.Ginsberg, *Phys.Rev.*B **36** (1987) 6915.
[59] B.Barbara, I.Felner, E.Bonjour, and R.Calemczuk, *Physica* C **153-155** (1988) 914.
[60] K.Kadawaki, F.E.Kayzel, and J.J.M.Franse, *Physica* C **153-155** (1988) 1028.
[61] B.D.Dunlap, M.Slaski, D.H.Hinks, L.Soderholm, M.Beno, K.Zhang, C.Segre, G.W.Grabtree, W.K.Kwok, S.K.Malik, I.K.Schuller, J.D.Jorgensen, and Z.Sungaila, *J.Magn.Magn.Mat.* **68** (1987) L139.
[62] K.Mori, K.Noto, M.Sasakawa, Y.Isikawa, K.Sato, N.Kobayashi, and Y.Muto, *Physica* C **153-155** (1988) 1515.
[63] A.Ježowski, A.J.Zaleski, M.Ciszek, J.Olejniczak, and J.Klamut, *Proc. of Conf.* INCONCRYO '88 (Calcutta 1988) *in press*.
[64] V.Bayot, F.Delannay, C.Dewitte, J-E.Erauw, X.Gonze, J-P.Issi, A.Jonas, N.Kinany-Alaoui, M.Labricht, J-P.Michenaud, J-P.Minet, and L.Piraux, *Solid State Commun.* **63** (1987) 983.
[65] D.T.Morelli, J.Heremans, and D.E.Swets, *Phys.Rev.*B **36** (1987) 3917.
[66] W.L.Kozhevnikov, A.T.Lonchakov, V.A.Fotev, I.M.Tsivilkovski, and S.M.Cheshnitski, *Fiz.Metall.Metalloved.* **64** (1987) 191.
[67] K.Bartkowski, R.Horyń, A.J.Zaleski, Z.Bukowski, M.Horobiowski, C.Marucha, J.Rafałowicz, K.Rogacki, A.Stępień-Damm, Cz.Sułkowski, E.Trojnar, and J.Klamut, *phys.stat.sol.* (a) **103** (1987) K37.
[68] B.A.Merisov, G.Ya.Khadzhaí, M.A.Obolenskií, and O.A.Gavrenko, *Fiz.Nizk.Temp.* **14** (1988) 643.
[69] A.Ježowski, *unpublished data*.
[70] M.Izbizky, M.Núñez Regueiro, P.Esquinazi, and C.Fainstein, *Phys.Rev.*B **38** (1988) 9220.
[71] M.V.Newitt, S.W.Grabtree, and T.E.Klippert, *Phys.Rev.*B **36** (1987) 2307.
[72] T.Lægreid, K.Fossheim, and F.Vassenden, *Physica* C **153-155** (1988) 1096.
[73] M.Lang, T.Lechner, S.Riegel, F.Steglich, G.Weber, T.J.Kim, B.Lüthi, B.Wolf, H.Rietschel, and M.Wilhelm, *Z.Phys.*B **69** (1987) 459.
[74] M.Gurvitch and A.T.Fiory, *Phys.Rev.Lett.* **59** (1987) 1337.
[75] M.Gurvitch, A.T.Fiory, L.S.Schneemeyer, R.T.Cava, G.P.Espinosa, and J.V.Waszczak, *Physica* C **153-155** (1988) 1369.

[76] M.Gurvitch, *Phys.Rev.*B **24** (1981) 7404.
[77] P.B.Allen, T.P.Beaulac, F.S.Khan, W.H.Butler, F.J.Pinski, and J.C.Swihart, *Phys.Rev.*B **34** (1986) 4331.
[78] A.Jeżowski and K. Rogacki, *unpublished data.*
[79] A.Jeżowski, *Solid State Commun.*, **79** (1989) 419.
[80] V.L.Ginzburg, *Zh.Eksp.Teor.Fiz.* **21** (1951) 979.
[81] C.Gorter, *Can.J.Phys.* **34** (1956) 1334.
[82] V.L.Ginzburg, *unpublished data.*
[83] J.Hanuza, J.Klamut, R.Horyń, B.Jeżowska-Trzebiatowska, *Mod.Phys.Lett.*B **2** (1988) 667.
[84] J.Klamut, T.Głowiak, J.Hanuza, Z.Henkie, R.Horyń, B.Jeżowska--Trzebiatowska, R.Kubiak, K.Łukaszewicz, J.Stepień-Damm, and A.Zygmunt, *Physica* C **153-155** (1988) 288.
[85] N.V.Zawaritsky, A.V.Samoilov, and A.A.Yurgens, *Pis'ma v Zh.E.T.F.* **48** (1988) 221.
[86] A.V.Samoilov, A.A.Yurgens, and N.V. Zawaritsky, in: *Progress in High Temperature Superconductivity,* vol.**11**, "High T_c from Russia" World Scientific Publ.Co. (Singapore 1989).
[87] M.Suzuki, Y.Okuda, I.Iwasa, and A.J.Ikushima, *Jap.J.Appl.Phys.* **27** (1988) L308.
[88] A.Junod, *Physica* C **153-154** (1988) 1078.
[89] A.Jeżowski, J.Mucha, A.J.Zaleski, M.Ciszek, J.Olejniczak, K.Rogacki, C.Sułkowski, M.Wołcyrz, and J.Klamut, *Phys.Lett.*A **127** (1988) 225.
[90] R.Calemczuk, E.Bonjour, J.Y.Henry, L.Forro, C.Ayache, M.J.M.Jurgens, J.Rossat-Mignot, B.Barbara, P.Burlet, M.Couach, A.F.Khoder, and B.Salce, *Physica* C **153-155** 960 (1988) 960.
[91] Zhang Yuheng, Chen Zhaoija, Ruan Yaozhong, Wang Yun, Mao Xianlei, Yang Hongshun, Li Liping, Chen Lin, Cheng Tingzhu, Meng Guangyao, Peng Dingkun, and Hu Junbao, *Physica* C **153-155** (1988) 978.
[92] T.Lægreid and K.Fossheim, *Europhys.Lett.* **6** (1988) 81.
[93] W.Yening, S.Huimin, Z.Jinsong, X.Ziran, G.Min, N.Zhongmin, and Z.Zhifang, *J.Phys.C: Solid State Phys.* **20** (1987) L665.
[94] G.Cannelli, R.Cantelli, F.Cordero, G.A.Costa, M.Ferretti, and G.L.Olcese, *Phys.Rev.*B **36** (1987) 8907. *Physica* C **153-155** (1988) 298.
[95] H.Yusheng, Z.Baiwen, L.Sihan, X.Jiong, L.Yongming and Ch.Haoming, *J. Phys.F: Met.Phys.* **17** (1987) L243.
[96] Y.Horie, Y.Terashi, H.Fukuda, T.Fukami, and S.Mase, *Solid State Commun.* **64** (1987) 501.
[97] V.Müller, K.de Groot, D.Maurer, Ch.Roth, K.H.Rieder, H.Eickenbusch, and R.Schöllhorn, *Japan.J.Appl.Phys.* **26** (1987) 2139.
[98]. P.Lemmens, F.Stellmach, S.Ewert, S.Guo, J.Wynants, G.Arlt, A.Comberg, H.Passing, and G.Marbach, *Physica* C **153-155** (1988) 294.
[99] S.Ewert, S.Guo, P.Lemmens, F.Stellmach, J.Wynants, G.Arlt, D.Bonnenberg, H.Kliem, A.Comberg, and H.Passing, *Solid State Commun.* **64** (1987) 1153.
[100] G.Mamsurova, K.S.Pigalkii, V.P.Sakun, L.G.Shcherbakova, I.E.Graboi, and A.R.Kaul, *Phys.Lett.*A **131** (1988) 538.
[101] T.Datta, H.M.Ledbetter, C.E.Violet, C.Almasan, and J.Estrada, *Phys.Rev.*B **37** (1988) 7502.
[102] H.M.Ledbetter and S.A.Kim, *Phys.Rev.*B **38** (1988) 11857.
[103] S.K.Kurtz, L.E.Cross, N.Setter, D.Knight, A.Bhalla, W.W.Cao, and W.N.Lawless, *Mater.Lett.* **6** (1988) 317.
[104] T.Lægreid, K.Fossheim, E.Sandvold, and S.Julsrud, *Nature* **330** (1987) 637.
[105] T.Lægreid, K.Fossheim, O.Trætteberg, E.Sandvold, and S.Julsrud, *Physica* C **153-155** (1988) 1026.
[106] G.K.White, S.J.Collocott, R.Driver, R.B.Roberts, and A.M.Stewart, *J.Phys.C: Solid State Phys.* **21** (1988) L631.

[107] M.A.Obolenskii, V.P.Popov, K.B.Chashka, and V.A.Storodub, *Fiz. Nizk. Temp.* **14** (1988) 863.
[108] T.Kirschner, T.Trager, J.Mátrai, T.Porjesz, and J.György, *Physica*, C **153-154** (1988) 1419.
[109] M.François, A.Junod, K.Yvon, A.W.Hewat, J.J.Capponi, P.Strobel, M.Marezio, and P.Fischer, *Solid State Commun.* **66** (1988) 1117. *Physica* C **153-155** (1988) 962.
[110] M.Sarikaya and E.A.Stern, *Phys. Rev.* B **37** (1988) 9373.
[111] H.You, J.D.Axe, X.B.Kan, S.C.Moss, J.L.Liu, and D.J.Lam, *Phys. Rev.* B **37** (1988) 2301.
[112] U.Murek, K.Keulerz, and J.Rohler, *Physica* C **153-155** (1988) 270; R.Srinivasen, K.S.Giriajan, V.Ganesan, V.Radhakrishnan, and G.V.Subba Rao, *Phys. Rev.* B **38** (1988)
[113] Ming-sheng Zang, Chen Qiang, Sun Dahun, Ji Rong-fu, Qin Zheng-hao, Yu Zheng, and J.F.Scott, *Solid State Commun.* **65** (1988) 487; L.A.Rebane, T.A.Fimberg, E.M.Feter, G.E.Blumberg, and E.R.Joon, *Solid State Commun.* **65** (1988) 1535; J.Hanuza, J.Klamut, A.J.Zaleski, R.Capeletti, M.Ciszek, J.Olejniczak, and J.Stępień-Damm, *Mod. Phys. Lett.* B (1989), *in press*.
[114] A.G.Khachaturyan, S.B.Semenovskaya, and J.W.Morris *Jr.*, *Phys. Rev.* B **37** (1988) 2243.
[115] A.G.Khachaturyan and J.W.Morris Jr., Phys.Rev.Lett. **61** (1988) 215.
[116] R.N.Bhargava, S.P.Herko, and W.N.Osborne, *Phys. Rev. Lett.* **59** (1987) 1468.

A NEW THEORETICAL APPROACH TO ELECTRON PAIRING AND SUPERCONDUCTIVITY IN HIGH Tc SUPERCONDUCTORS

A. Nakamura
Japan Atomic Energy Research Institute, Naka-gun, Ibaraki, Japan

INTRODUCTION

Recently, the present author proposed a new theoretical approach to the electron pairing and superconductivity in high T_c superconductors, i.e., Multi Valence Resonance-Condensation Model(MVRC Model) and demonstrated its practical applicability to various high T_c oxide systems such as strongly anisotropic cuprates and almost isotropic bismuthates, A-15 type intermetallic Nb_3Ge, etc., suggesting the Model to be a possible novel and universal origin of superconductivity[1,2].

In the present article, the author gives a detailed review of this new theoretical approach to high T_c superconductivity, together with the subsequently made several progresses on the analyses based on the Model and also on the Model itself.

Basic picture of high T_c(as well as low T_c) superconductivity that the MVRC Model suggests appears to be remarkably simple, transparent and universal, so that we can almost immediately grasp the essence of superconductivity as an instinctive visual image. This is briefly summarized as follows: Multi Valence Resonance Condensation(MVRC) state of superconductivity is the Bose condensate of the microscopic quantum chemical resonances inside the crystal molecule of the responsible chemical species Cu(Bi,Pb,Nb, etc.) as shown in Fig. 1. This state is realized through the self-organization process of the electron system itself as the fundamental and universal chemical bonding force(valence fluctuation⟶ valence ordering⟶ resonance-condensation), self-driven by the symmetry of the respective electron orbitals.

Two modes of these resonance hybrids shown in Fig. 1 are in principle possible; the one and the other which have either higher or lower average valence than the intermediate valence state Ⓣ(Cu^{2+}, Bi^{4+}) are designated as oxidized and reduced type resonance hybrids(ORH and RRH), respectively.

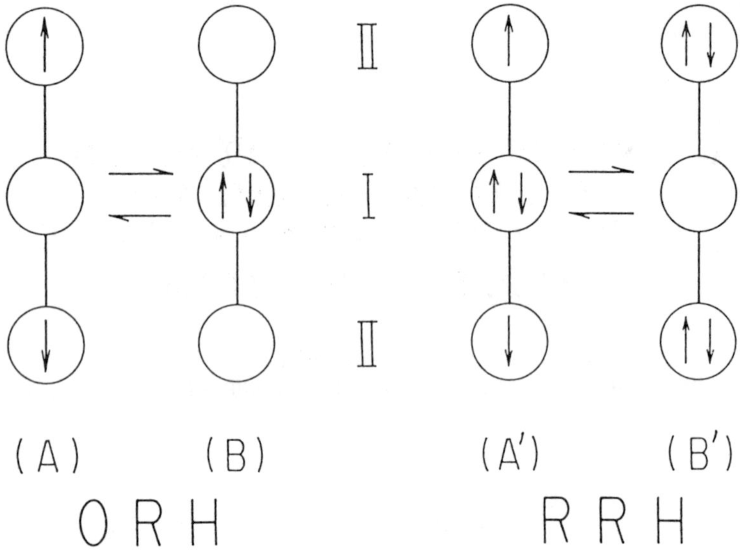

Fig. 1. Oxidized and Reduced type Resonance Hybrids(ORH and RRH)

Of course, in oxides, the Cu-Cu(Bi-Bi) bond is not a direct one, but an indirect one via oxide ion(O^{2-}), whereas in Nb_3Ge the Nb-Nb bond is a direct one. In the former systems, depending on the detailed nature of the hybrid MO(molecular orbital) bond between various cations and oxygen, hole (or electron) in the ORH(or RRH) type conductor would exhibit versatile feature from the cations' d(s)- to oxygen 2p orbitals. But, the essential point here seems to be the fact that such hybrid MO bond exists there and indeed charge transfer process; $\underline{Cu}^{3+}+O^{2-} \rightleftharpoons Cu^{2+}+\underline{O}^{-}$ occurs between them[3,4]. Then, the formal valence representation[5,6] is certainly a well defined convenient description of various superconductors with various kinds of multi valent type cations such as $Cu^{1,2,3+}$, $Pb^{2,3,4+}$, $Bi^{3,4,5+}$, etc.

As is apparent in Fig. 1, these microscopic resonances inside the crystal molecule are attained by the simultaneous symmetrical paired motion of two electrons(spins) to the opposite perpendicular directions. So, if some chemical and structural conditions are satisfied and indeed these microscopic quantum chemical resonances exist there inside the crystal molecule, these can propagate and extend macroscopically throughout the whole crystal almost spontaneously and in the perfectly coherent manner, giving rise to the MVRC state of superconductivity, that is, the macroscopically coherent quantum chemically resonating state with real space electron pair in close correspondance to the so-called Cooper pair in k(momentum)-space in the BCS theory[7].

Validity of the Model can be checked by the experimental findings of various superconductors using the following two-fold criteria:

(a). Multi valence state, i.e., consecutive three different valence states of various cations should have the spin degeneracy designated in Fig.1 : $Cu^{1+}(Bi^{3+})$=(↑↓), $Cu^{2+}(Bi^{4+})$=(↑), $Cu^{3+}(Bi^{5+})$=( ). For this to be realized, for other than s electron system(l=0), symmetry reduction around the cations is necessary enough to lift the degeneracy of the electron orbitals and then to provide the splitted single HLMO(highest lying molecular orbital) bond.

(b). Under the above criterion (a), the actual geometrical configuration of these two modes of resonance hybrids with different average cation valence depends crucially on the actual chemistry and crystal chemistry of the systems. Therefore, it is almost immediately judged at which compositions for a given system with a given structure the onset of superconductivity is optimized. And this theoretically predicted optimum composition is expected to coincide with the experimentally observed composition with the maximum transition temperature T_c of the system.

This was done detailedly in Ref.[1,2] for almost all the oxide superconductors reported so far and also for A-15 type intermetallic $Nb_3Ge(Ga)$, and is more concisely summarized in this paper. The results appear to be quite convincing and even decisive, demonstrating the practical applicability of the present model by filling up the hitherto existing frustrating deep gap between the theory and experiments, in other words, physics and chemistry of superconductivity. See Table 1 for brevity.

Also in the subsequent sections, we will see that the present model is also capable of giving a fair account on the normal state behaviour, nature and mechanism of superconductive transition, isotope effects, etc. of these superconductors, substantiating the basic validity of the whole scenario of superconductivity as the above mentioned self-organization process of the electron system itself elucidated by the present model.

Thus, the MVRC Model is expected to be useful as a guiding chemical and structural principle both in searching new superconductors and constructing new microscopic theory of superconductivity. Since quantum chemistry is now quite a versatile and common technique in chemical bond theory, this can be fully utilized for both of these purposes. During the course of such experimental and theoretical attempts, the author believes that the completely revised new vast scope of superconductivity research will emerge and evolve in the future before us.

DESCRIPTION AND APPLICATION OF THE MVRC MODEL

In this section, the author gives detailed description and application of the MVRC Model, taking $YBa_2Cu_3O_{7-x}$ as a paradigm and therefrom extending the discussion to various other oxide and non-oxide systems including also conventional low T_c metal, hydride and organic superconductors, etc. Main results obtained here are that (1) The MVRC state does indeed provide a possible unified picture of the superconducting state of all these systems, and (2) The derived expression of the transition temperature T_c can also give a fair account of experimentally found versatile isotope effects of these various superconductors in a systematic and consistent manner.

MVRC State of Superconductivity

90K class superconductor $YBa_2Cu_3O_7$(x=0)[8] with tri-layered perovskite structure[9] and the average Cu valence of 2.33+ offers a best and clearest example of the present model. It is immediately understood that (b-c) plane of this system shown below in Fig. 2 is nothing but the interconnected one dimensional array(ABAB···) of ORH in Fig. 1 to the right and left horizontal directions. We designate this tri-layered type quasi-one dimensionality(1D') in the following.

Thus, the microscopic intra-molecular resonance(A⇄B) inside the crystal molecule(ORH in Fig. 1), provided now with the intermolecular resonance, is considered to manifest itself as a macroscopic Multi Valence Resonance-Condensation state of superconductivity(MVRC state: ABAB···⇄BABA···), where every electron moves as a pair, i.e. two electrons of the opposite spins(↑ and ↓) moving to the opposite directions(±v). This gives the real space electron pair, as well as k-space electron pair(the Cooper pair in the BCS theory[7]), which constitutes the basic unit and therefore the essence of supperconductivity.

We may regard the MVRC state as the Bose condensate of two bosons(A,B) strongly interacting in real space. In Fig. 2, at first sight, to realize this macroscopic condensation, two coupled motions of two electrons(=total of four electrons = two pairs; the intra- and inter-molecular ones shown as solid- and dashed lines, respectively) appear to be necessary. But, if we recognize the facts that the original microscopic resonance ORH in Fig. 1 is the intra-molecular one involving only the three cations in the perpendicular direction(not the six cations of A+B) and therefore also in the MVRC state the electrons are completely delocalized in the coherent manner so as for these two configurations(ABAB··· and BABA···) to be perfectly equivalent and indistinguishable, it is easily understood that only one pair with its charge 2e and its mass 2m is indeed the basic unit of the MVRC state.

The other important aspect of this Cu-O bond related to the aforementioned criterion (a) is its non-cubic symmetry around Cu: Square planer and pyramidical configurations of oxygen around Cu in the middle chain and the

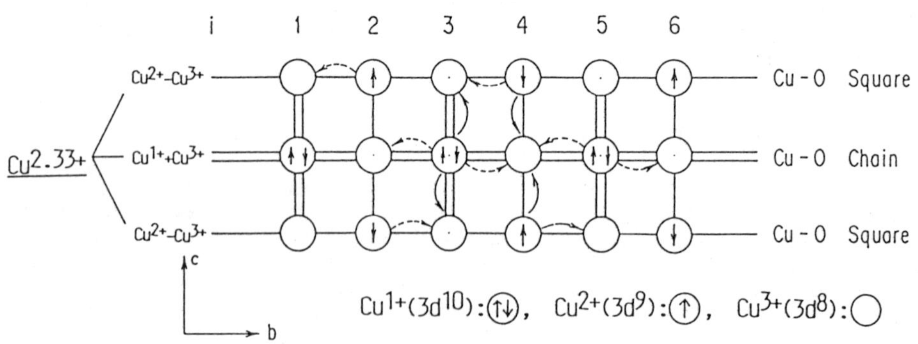

Fig. 2. Tri-layered type 1D' superconductivity in $YBa_2Cu_3O_7$: ORH.

upper- and lower 2D(two-dimensional) square planes respectively, are essential to lift the degeneracy of 3d orbitals(l=2) of Cu(Hund rule), firstly in the free atom(ion) state and secondly in the cubic octahedral symmetry, so that this multi valence state of Cu occurs inside a splitted single Cu-O orbital bond with the designated spin degeneracy shown in Figs. 1 and 2. From molecular orbital(MO) theory, this is to be the anti-bonding σ^* MO bond between Cu $3d_{x^2-y^2}$ and oxygen 2p orbitals[5,10]. This appears to be the definite reason why in d(or more properly other than s) electron systems, superconductivity is found only in anisotropic 1-, 2-, 3D systems, but not in isotropic 3D systems, whereas in s electron systems isotropic 3D superconductivity is possible as in $Ba(K)BiO_3$[11,12], which are briefly and clearly seen in Table 1.

Thus, in $YBa_2Cu_3O_7$ both quantum chemical- and crystal chemical conditions posed by criteria (a) and (b) are almost ideally realized, and so it seems quite reasonable to suppose that it is this full resonance through and between the middle Cu-O chain and the upper- and lower 2D Cu-O square planes that makes this system so an unprecedented high T_c superconductor with T_c= 92K. This interpretation is entirely consistent with experimental facts that so far reported any kinds of chemical substitutions of Cu, Ba and Y by any other elements[13], radiations[14,15] and nonstoichiometry to both $\pm x$ directions[16], which inevitably disturb the valence state and/or the structure of the (b-c) plane shown in Fig. 2, deteriorate the superconductive property of this system, exhibiting the remarkable uniqueness of this particular chemical structure. Whereas, only trivalent rare earth elements Re^{3+}, even the magnetic ions, can substitute Y well, giving a series of $ReBa_2Cu_3O_7$ with the comparable T_c[17]. This is rather natural because these Re^{3+} isovalent with Y and located well outside of the (b-c) plane do neither alter the valence state of Cu nor interfere much with the structure of the (b-c) plane.

A flat pancake-shaped structure of the electron pair[18](the coherent length $\xi \simeq 3.8$- 7.0 and 30-35Å for parallel and perpendicular to the c-axis directions, respectively[19]), the orientational anisotropy of the upper critical field[20] more properly described by 3D anisotropic model than 2D thin film model indeed reveals the original quantum chemical resonating nature of the crystal molecule along the c-axis, which is liberated along the b-axis through their interconnection via intervenuing oxygens. Apparently, this resonance along the c-axis, i.e. 3D anisotropy can by no means be neglected.

Above discussion also implies that only if we can further increase the strength of both the intra- and inter-molecular resonance in Fig. 2 simultaneously, we might be able to have a superconductor with the much higher transition temperature T_c which exhibits 3D anisotropy, possibility of which yet remains to be seen in the future research. It is surprising that release of microscopic quantum chemical resonance in organic molecules such as benzene and naphthalene in the macroscopic level once dreamed of by Little[21] is realized in this rather poorly metallic inorganic oxide in so an unprecedented and different manner. To this point, we will return later more deliberately when we discuss the nature and mechanism of superconductive transition.

It is now well recognized that $YBa_2Cu_3O_{7-x}$ exhibits the second superconductive plateau of T_c=55-60K around $0.25 \leq x \leq 0.50$[22] with the quasi 2D super-

Table 1

Brief Summary of the Present Analysis on Various Superconductors

Dimension	Systems (Electron Conf.)	T_c(max) (K)	Cation orbital	Cation Valence (ORH or RRH)	Valence State (↑↓) (↑) ○
1D Chain	$\underline{Nb}_3Ge(4d^45s^1)$	23.7	d_{z^2} σ	2.25+(ORH)	$Nb^{1,2,3+}(4d^{4,3,2})$
1D' Chain	$YBa_2\underline{Cu}_3O_7(3d^{10}4s^1)$	92	$3d_{x^2-y^2}$ σ*	2.33+(")	$Cu^{1,2,3+}(3d^{10,9,8})$
2D Square Plane	$La_{1.75}Sr_{0.25}\underline{Cu}O_4$	37	"	2.25+(")	" (")
	Tl(Bi)-$\underline{Cu}$-O Systems	130	"	" (")	" (")
	$Nd_{1.85}Ce_{0.15}\underline{Cu}O_{3.95}$	20	"	1.75+(RRH)	" (")
2- to 3D' Network	$Li_{0.8}\underline{Ti}_{2.2}O_4(3d^24p^2)$	12.2	" σ	3.25+(ORH)	$Ti^{2,3,4+}(3d^{2,1,0})$
	$K_{0.75}\underline{Mo}O_3(4d^55s^1)$#	4.3	$4d_{x^2-y^2}$ σ	5.25+(")	$Mo^{4,5,6+}(4d^{2,1,0})$
	$K_{0.75}\underline{W}O_3(5d^46s^2)$#	1.5	5d " "	" (")	$W^{4,5,6+}(5d^{2,1,0})$
	$K_{0.25}\underline{Re}O_3(5d^56s^2)$#	3.6	5d " "	5.75+(RRH)	$Re^{5,6,7+}$(")
	$\underline{Ag}_7O_8NO_3(4d^{10}5s^1)$#	1.0	4d " σ*	2.25+(ORH)	$Ag^{1,2,3+}(4d^{10,9,8})$
	$Ba\underline{Pb}_{3/4}Bi_{1/4}O_{3.935}(6s^26p^2)$	13	6s σ	3.25+(")	$Pb^{2,3,4+}(6s^{2,1,0})$
3D Network	$Ba_{3/4}K_{1/4}\underline{Bi}O_3(6s^26p^3)$	30	"	4.25+ (")	$Bi^{3,4,5+}$(")

Underlined cationic species denote those responsible for their superconductivity.

#: For these systems, theoretically most optimized compositions(cation valence) shown in this table are not realized experimentally.

conductivity[20,23] similar to that of single layered $La(Sr)_2CuO_4$[24,25] and multi layered Tl(Bi)-Cu systems[26,27]. Further 2D extension of trilayer structure in Fig. 2 composed of the middle Cu-O chain(I) and the upper- and lower Cu-O square planes(Ⅱ) to the upper and lower directions as IⅡIⅡI ··· is considered to provide the required 2D Cu-O square plane((a-b) plane) responsible for their 2D superconductivity. This is the isotropic 2D array of A and B of ORH in Fig. 1 with overall optimum average Cu valence of +2.25 and shown in Fig. 3 below. We also know in Table 1 that as its symmetrical counterpart, 2D superconductivity of RRH(electron type) of Fig. 1 is indeed realized in recently found $Nd_{2-x}Ce_xCuO_{4-y}$[28,29] with the optimum Cu valence of +1.75. 3D network system is obtained simply by combining 2D square planes shown in Fig. 3 as (a-b), (b-c) and (a-c) planes in 3D manner. Therefore, also in 3D systems as in 2D systems, the superconductivity is most optimized at the average cation valence of ±0.25 from the intermediate valence(↑).

Further reduction of $YBa_2Cu_3O_{7-x}$ to x=1.0 leads to the formation of the magnetic insulator(or semiconductor) $YBa_2Cu^{1+}Cu^{2+}{}_2O_6$, which is of particular interest from the present point of view. Above shown valence state of this system is just that of A' of RRH in Fig. 1. Then, it seems quite reasonable to expect that this system could be a superconductor by the quantum chemical resonance with B' only if oxide ions(O^{2-}) exist and interconnect the Cu ions inside the chain. Unfortunately, these oxide ions are completely missing in this system[5], so microscopic resonance in its crystal molecule, if ever it exists there, can not at all extend over the whole crystal via intermolecular resonance. Then, this A'-B' pair may exist as static two multi valence

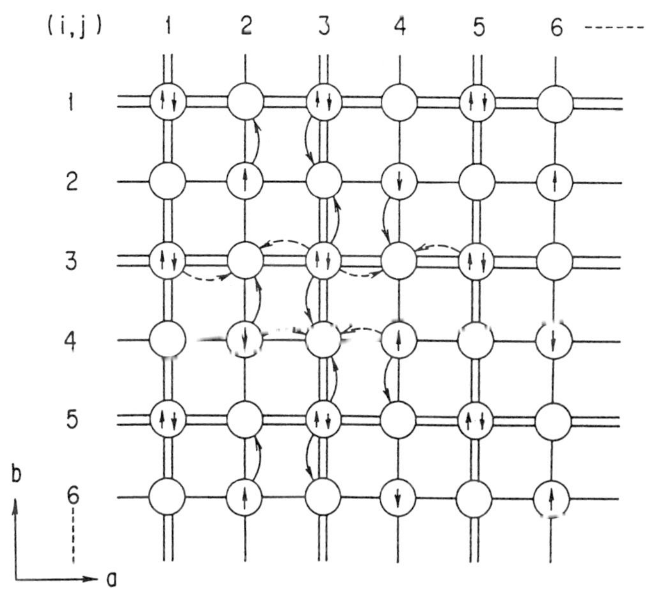

Fig. 3. 2D square plane responsible for 2D superconductivity in layered cuprates, etc. (ORH).

states of Cu. In any event, the electron system in $YBa_2Cu_3O_6$ and the analogous La_2CuO_4 systems is sufficiently localized so as not to exhibit the itinerant character, and this seems to be the main reason of the antiferromagnetic ordering of these systems[30,31].

In Table 1, the following several comments on other oxide superconductors than cuprates would be adequate: In d electron systems including 3d-, 4d- and 5d electron systems, isotropic 3D superconductivity does not indeed exist in accordance with the aforementioned criterion (a). In spinel type lithium titanate system, the onset of the maximum T_c for the single phase material is reported for the chemical composition of $Li_{0.8}Ti_{2.2}O_4$[32]. Detailed chemical formula of this composition is considered to correspond to $(Li^{1+}_{0.8}Ti^{3+}_{0.1}Ti^{4+}_{0.1})_t[Ti^{3+}_{1.50}Ti^{4+}_{0.50}]_oO_4$, where $(\ )_t$ and $[\ ]_o$ designate the tetrahedral and octahedral sites, respectively. The average valence state of $Ti^{3.25+}$ on the octahedral site is just the expected one from the 3D network of the 2D square planes of ORH in Fig. 3. So the quasi 3D superconductivity in the slightly distorted spinel structure is indeed realized to its maximum at this composition[1]. As the tetrahedral sites in the spinel lattice do not form the directly connected 3D network, the above overall valence distribution between both these sites appears to be the most favourable one energetically.

Also for several Mo-, W- and Re bronze type oxides with perovskite related structure[33,34] and silver clathrate salts[35], criterion (a) applies quite well without exception, and the superconductivity is found only in the anisotropic quasi 3D systems, but not in the isotropic cubic systems, e.g. metallic cubic bronze. In all these systems, as indicated in Table 1, the theoretically optimum superconductive composition is not realized experimentally probably due to some crystallographic reasons[1]. This in turn means that if we could build in their optimum superconductive conditions in some other crystal chemical forms, we would be able to have the higher T_c superconductors involving these cations. We also know in Table 1 that in one other system than the above mentioned $Nd(Sr,Ce)_2CuO_{4-y}$, i.e. K doped ReO_3, RRH (electron) type superconductivity is realized, suggesting a lot of future experimental scope of finding the RRH type symmetrical counterparts of the ORH(hole) type superconductors already found and listed there.

As for s electron systems(l=0) which have only two spin degeneracy from the beginning, anisotropy and/or symmetry reduction around the cations are not necessary. Indeed in $Ba(K)BiO_3$[11,12], isotropic 3D superconductivity is found on the octahedral(B-site) (Bi-O) 3D network plane in the cubic perovskite structure. Reported onset of maximum T_c around K content of 25 atomic % on the A-site[12], i.e. $Ba^{2+}_{0.75}K^{1+}_{0.25}Bi^{4+}_{0.75}Bi^{5+}_{0.25}O_3$ again agrees well with the theoretically expected average cation valence state of $Bi^{4.25+}$ for 3D network of 2D planes of ORH shown in Fig. 3.

Fig. 4 shows the schematic of the onset of the maximum T_c around x=0.25 in somewhat complicated solid solution system $BaPb_{1-x}Bi_xO_{3-y}$ with slightly distorted perovskite structure[36,37]: One fourth(=0.25) of the B-site is occupied by Bi^{5+}(rather than Bi^{4+}) and these Bi^{5+} are arranged in the perfectly ordered manner as shown there. This gives the 2D (Pb-O) square planes(j= 2,4,6···) responsible for the quasi 2D superconductivity. In this plumbate superconductor(6s electron system), anisotropy is not essential and rather

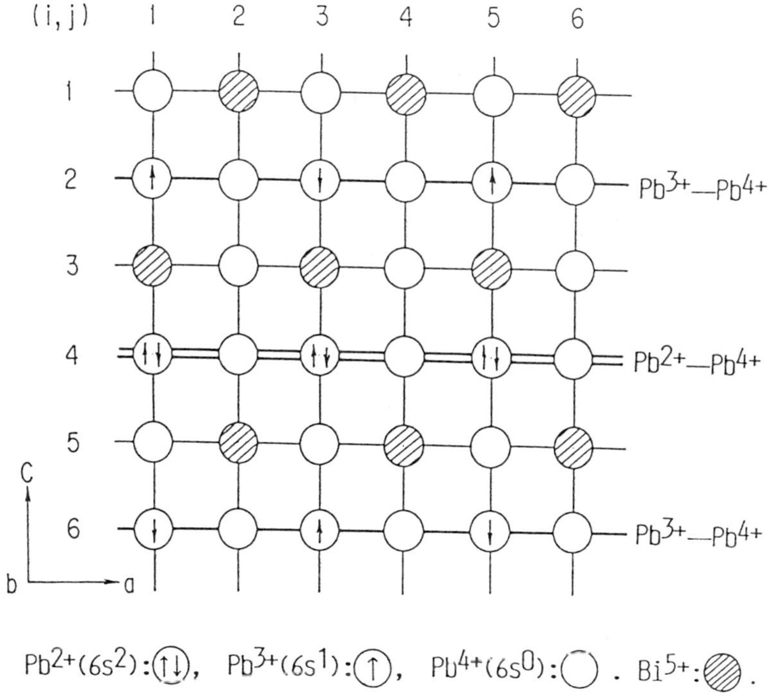

Fig. 4. 2D' superconductivity in $BaPb_{0.75}Bi_{0.25}O_{3.9375}$: ORH.

brought about by this extrinsic cause of Bi existence in the same B-sites. These ORH type 2D (a-b) square planes are separated by the intervenuing (Bi^{5+}+Pb^{4+}) (a-b) planes(j=1,3,5···). Then, simple valence counting shows that oxygen deficiency of the system should be y=0.0625 and the exact chemical formula of this system at x=0.25 is $BaPb_{0.75}Bi_{0.25}O_{2.9375}$. This ordered Bi on the B-site also gives the often reported slight distortion with tetragonal symmetry(a=b≠c) from the ideal cubic structure. Since the perfect Bi order does not usually occur, the tetragonal distortion and therefore the 2D superconductivity in this system is condidered to be only the level of limited quasi 2 dimensionality(2D') almost close to 3D isotropy as a bulk property. Since this Bi concentration(x=0.25) appears to be too small for the Bi-Bi direct connection(via oxide ion) to extend as the 2 to 3D networks which is the prerequisite for the onset of MVRC state in the molecular level in Fig. 1, the above interpretation seems to be the only possibility for the superconductivity of this system.

Application of the present model to non-oxide systems appears also promising. The result is schematically depicted in Fig. 5 for A-15 type

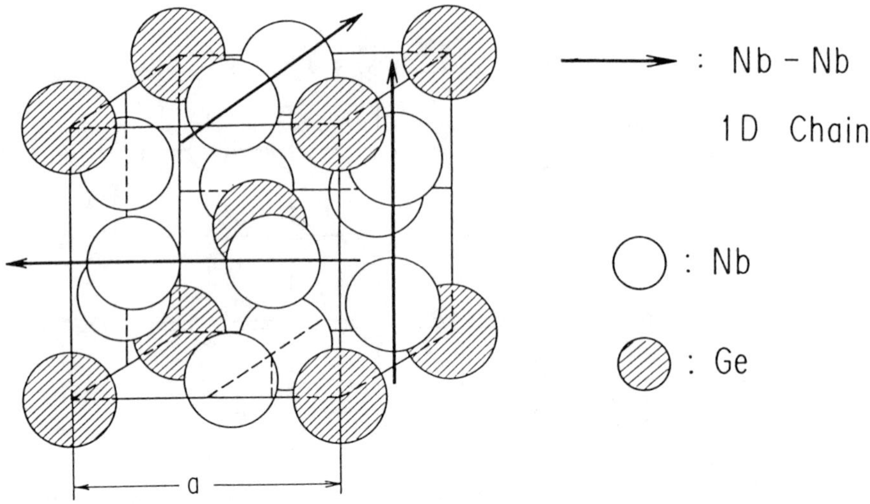

High Tc of Nb_3Ge

1D Extension of ORH :

$[Nb^{2.25+}(4d^{2.75})]_3^{6.75+}$ Ge $(4S^2\,4P^2\,\underline{4P^4\,4d^{2.75}})^{6.75-}$

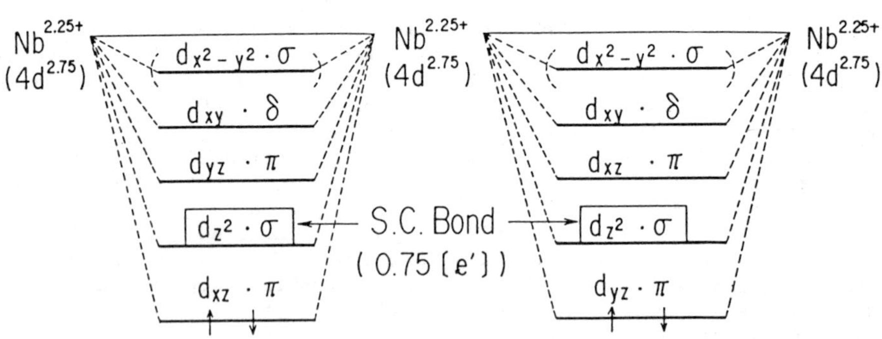

Fig. 5. Crystal structure and high T_c superconductivity of A-15 Nb_3Ge.

intermetallic Nb_3Ge with true 1D Nb-Nb chain. One dimensionality around Nb in this system is already enough to lift the degeneracy of 4d orbitals of $Nb(4d^4 5s^1)$: Since the chemical bonding force between the direct d-d orbitals[10] is expected to be in the order shown in the lower part of Fig. 5, first the lowest lying d_{xz}(or d_{yz}) π MO bond is considered to be filled by two electrons and unites Nb firmly to z direction in 1D manner. Here, due to the non-cubic configuration of Ge around Nb, the role of d_{xz}- and d_{yz} MO bonds exchanges in every Nb-Nb connection alternately as shown there. So for the superconductive bond, we should go to the second lowest lying d_{z^2} σ MO bond for the first time in the present analysis. 1D extension of the completely intermingled resonating state A(A')$\rightleftharpoons$B(B') of ORH(RRH) in Fig. 1 to the upper and lower directions gives the average electron concentration per cation of 0.75(1.25) for ORH(RRH) in this superconductive bond. In other words, also in 1D case the optimum average valence of cation(Nb) for the onset of superconductivity is just ±0.25 from the intermediate $Nb^{2+}(4d^3)$ valence state for ORH(RRH). If we choose the ORH(hole) type as the more plausible one and denote the remaining electrons from Nb to Ge, the perfectly isoelectronic configuration is obtained for Nb and Ge, as shown in the middle part of Fig. 5. This fact is possibly related to the highest T_c of 23.7K of Nb_3Ge[38] among so far reported all the metallic and intermetallic superconductors, implying the plausibility of the present interpretation. Straightforward extension of the present model to other various A-15 type and also another type intermetallic systems will lead to a new model-based systematics of these various types of intermetallic superconductors.

Results of the above mentioned extensive analyses are briefly summarized in Table 1 in the order of increasing dimensionality of the systems and demonstrate the almost universal applicability of the MVRC Model to variously structured various types of superconductors with at least nine kinds of different cationic species responsible for their superconductivity.

Expression of Transition Temperature T_c in MVRC Model

So far, only focussing on the MVRC state of superconductivity to get the quick visual image of high T_c superconductivity, the present author showed that the proposed MVRC state can indeed provide a possible unified picture of the superconductive state of up to now reported almost all the oxide superconductors and also A-15 type intermetallic Nb_3Ge. Before proceeding further, it would be worth while summarizing here concisely the main conclusions obtained above as follows: 'Superconductivity is rather a versatile (and even a common) phenomenon occasionally found among the systems with the multi valent type cations and in addition with the splitted single HLMO bond in which the microscopic quantum chemical resonance shown in Fig. 1 can be realized macroscopically in the coherent manner with the designated spin degeneracy. Its onset is most optimized at chemical compositions of the systems which give the average cation valence of ±0.33 and ±0.25 from the intermediate valence state Ⓘ for tri-layered type quasi 1D' systems and for other 1-, 2-, 3D systems, respectively. Here, + and - correspond to the ORH and RRH type superconductivity, respectively.'

A global image of high T_c superconductivity emerging from here is a very instinctive one; the macroscopically coherent quantum chemically resonating chemical bond with real space electron pair which is realized according to the most fundamental rule of the chemical bond theory in terms of quantum chemistry, that is, symmetry of the electron orbitals of the responsible chemical species. This novel and almost universal picture of (high T_c) superconductivity seems to be most clearly illustrated above notably for $YBa_2Cu_3O_7$ and Nb_3Ge, the most representative high T_c oxide and non-oxide superconductors, respectively, so that this appears to be also valid basically for other various types of superconductors not mentioned above such as conventional low T_c metals, organic superconductors (and heavy fermions, etc.).

Although for such a generalization of the present model we should carefully await detailed analyses on these various and specific superconducting systems as was done for oxide superconductors in the last section, the following experimental facts might be listed here briefly as giving qualitative supports to the present point of view:

(a) Even in pure metal systems, similar to above mentioned A-15 type intermetallics, the relatively high T_c superconductors are often found in the multi valent type elements such as Nb, Pb, Bi, Sn, Hg, Ir, Tc, V, Ti, and so on[39].

(b) As for so-called organic superconductor, the completely intermingled quantum chemically resonating state of A(A') and B(B') of ORH(RRH) in Fig. 1 does not at all rely on whether its original normal state had been the multi valence state of cationic- or anionic one, or their various hybrid versions, or even whether it had been really the distinctive multi valence state or not. So, it appears to exist almost no reason to deny the extension of the present model to so far reported such organic type superconductors[40], in which their superconductivity seems to be owned basically by the partially filled single pπ HLMO bond of the chalcogen-chalcogen(S,Se,Te) 1- to 2D network planes, essentially a charge transfer complex type normal state conductor quite similar to the above mentioned inorganic superconductors.

Thus, now that we proceed to discuss nature and mechanism of superconductive transition and derive the expression of the transition temperature T_c, from the foregoing discussions, we can naturally conceive the following quite simple and universal basic scenario equally relevant to any kind of phase transition in general that phase transition proceeds to the direction in which the more stable chemical bond(and crystal structure) is formed through the self-organization process of the electron system itself as the fundamental and universal chemical bonding force according to the general quantum chemical and thermodynamic rules with decreasing temperature, competing against the thermal agitation of the system.

More specifically in the present case, as was delineated in detail in Ref.[1], this would be; disordered multi valence state at high temperature (valence fluctuation)→A(A')-B(B') pair type binary ordered state at intermediate temperature(valence ordering)→MVRC state of superconductivity at low temperature(resonance-condensation). This is explained more concisely in the following taking oxide superconductors as the examples.

The systems may approach to these particular superconductive chemical compositions(valence states of the cations) shown in Table 1 by chemical doping of aliovalent oxides and/or oxygen nonstoichiometry, etc., either from the insulating side(La_2CuO_4, $YBa_2Cu_3O_6$, $Li_{1.3}Ti_{1.7}O_4$, $BaBiO_3$) or from the metallic side($BaPbO_3$, ReO_3). Thereby, their initial stable electronic states, more or less firmly attached to e.g. magnetism(La_2CuO_4, $YBa_2Cu_3O_6$), lattice distortion($Li_{1.3}Ti_{1.7}O_4$, $BaBiO_3$), or strongly hybridized metallic covalent bond($BaPbO_3$, ReO_3), would be significantly destabilized, driven to the half way just in the middle of the so-called metal/insulator transition region. In this sense, the electron system in the normal state of these high T_c superconductors may be regarded to be metastable or even unstable, and therefore it begins to move collectively and cooperatively suddenly from now on, seeking for an alternate stable chemical bond in the self-organizing manner making full use of its intrinsic nature as the chemical bonding force.

First, providing the splitted single HLMO bond by the symmetry reduction of the systems if necessary, the electron system exerts the dynamic valence exchange(valence disproportionation[5]): A(A')$\rightarrow$B(B') or vice versa, to generate the multi valence state, for example, as shown in Fig. 2 for $YBa_2Cu_3O_7$. This is because the resultant three(multi) valence state can stabilize the system both energetically and entropically much more than the extreme two kinds of two valence states A(A') or B(B'): Referring to Fig. 2, this can give rise to the strong electron(⇅)-hole(○) attractive interaction both inside and between the chains(planes); we denote here the former intrachain(plane) and the latter interchain(plane) electron-hole attractive interaction energies as E_a(exciton) and E_a(CDW), respectively. Moreover, 1:1 concentration ratio of Cu^{1+}(electron) and Cu^{3+}(hole) in the middle chain and of Cu^{2+} and Cu^{3+} in the upper and lower 2D square planes in Fig. 2 is what is optimum both for the maximum gain in the configurational entropy in high temperature disordered state and for the maximum critical temperature T_c' for the ordering of this A-B binary alloy if there is attractive interaction between them. These combined effects of both energy and entropy terms definitely stabilize the multi valence state of Cu(cation) in the normal state both at high and low temperatures.

At high enough temperature where the thermal agitation of the system is quite strong(the main part of which is thermal vibration of the crystal lattice(phonon) and the configurational disorder), these multi valent Cu ions are more or less randomly distributed between the Cu sites in the chain and square planes as the thermally fluctuating disordered multi valence state with lots of unsymmetrical electron(spin) configurations in Fig. 2 in the small polaron type localized hopping conduction regime. Then, with decreasing temperature, using the above introduced E_a(exciton) and E_a(CDW) type electron-hole attractive interaction as the ordering energy of the process, this high temperature disordered multi valence state would gradually start to order, competing against the thermal agitation of the system.

When the temperature of the system is further lowered and the thermal agitation of the system becomes sufficiently weak, the ordering of the system reaches a critical level around which the net resonance(-condensation) stabilization energy of the electron system $J=E_S-E_N$ becomes almost zero to negative($J\leq0$), the electron system would drop almost spontaneously into the macroscopically coherent most stable MVRC state of superconductivity, simultaneously changing the nature of the chemical bond drastically from basically the electrostatically bound excitonic and CDW type one to the perfectly uniform strong resonance bond.

This defines the quite unique nature of this superconductive transition as a kind of sequential double second order phase transition in which the high temperature disordered multi valence state once transforms to a classical A-B type ordered alloy just shown in Fig. 2 and thereupon again undergoes a novel transition which sublimates its order to the MVRC state of superconductivity. Contrary to the rather paradoxical phonon mediated BCS picture of superconductivity proposed for conventional low T_c metal superconductors[7], basic character of this whole process of high T_c superconductivity described above in the present model appears to be much more straightforward and transparent: High T_c superconductivity is literally the decoupling process from 'phonon'; we see here a finest example of the self-organization process of the electron system both the origin and the driving force of which owned by itself as the fundamental and universal chemical bonding force, always trying to escape and decouple from this almost universal energy destabilizer and the electron scatterer of the system through a self-motivated highly correlated collective motion, firstly rather imperfectly in its own electronically driven ordered state with the metallic like normal state conductivity, and then finally almost perfectly in the self-organized Bose condensate state with remarkable zero resistivity attained via macroscopically coherent paired symmetrical motion of two electrons as the real space electron pair($M=2m$, $V=(+v)+(-v)=0$).

From the above qualitative but sufficiently detailed arguments on the nature and mechanism of superconductive transition, it almost directly follows that the superconductive transition temperature T_c of the system may be given by the following expression;

$$kT_c = 1.14|E_N|\exp(-|E_N|/|E_S-E_N|) = 1.14|E_N|\exp(-1/\lambda) \qquad (1)$$

where k is the Boltzmann's constant, and E_N and E_S designate the energy density of the system in the normal and superconductive state at OK, respectively, and are given by,

$$E_N = E_a(\text{exciton})+E_a(\text{CDW})+E_N(\text{phonon}),\quad E_S = E_S(\text{MVRC})+E_S(\text{phonon}),$$

$$J = E_S-E_N,\quad \lambda = J/E_N,\quad \text{all at OK.} \qquad (2)$$

Here, the so-called coupling constant λ is defined by the ratio of the resonance-condensation energy J and the normal state energy E_N as J/E_N and gives a measure of the energy gain J obtained by the electron pair formation(the superconductive state) relative to the randomly moving unpaired electronic state(the normal state).

The completely dissociated multi valence state is chosen as the standard state of energy($E=0$), so E_N at the perfect ordered state at OK and E_S at the perfect MVRC state at OK, both are expected to take generally the large negative values, since the major terms in them, E_N(exciton), E_N(CDW) and E_S(MVRC) are essentially the chemical bonding energy in the normal and superconductive state, respectively. Whereas, $E_N(\text{phonon}) \simeq E_S(\text{phonon}) = hw_D$ at OK(the zero point oscillation in the Debye model) only takes the relatively small positive value.

Preexponential term of Eq.(1), $T_c'=1.14|E_N|$ is the critical temperature for the ordering of 1:1 type alloy on the 2D plane with the coordination number z'=4 as shown in Fig. 2. In the present case, this ordered state should further transform to the MVRC state. Then, by multiplying T_c' by this relative stability factor $\exp(-|E_N|/|J|)=\exp(-1/\lambda)$, Eq.(1) is simply obtained. Eq.(1) gives the reasonable limiting cases that when $\lambda=J/E_N \longrightarrow 0$ or ∞, $T_c \longrightarrow 0$ or T_c'. These limiting cases mean that 1) when $J=0(\lambda=0)$, i.e. no resonance occurs due to the so-called compression energy[41], T_c should be zero, and (2) on the other hand, when the superconductive state is much more stable than the normal state and therefore $\lambda=J/E_N \longrightarrow \infty$, T_c should approach to T_c' infinitely, because once the odering begins, simultaneously the resonance-condensation starts to evolve.

Eq.(1) says that in principle there exists no upper limit of T_c theoretically, in accordance with our instinctive understanding in Figs. 1 and 2 that T_c of the system might be risen infinitely in principle only if the strength of the microscopic resonance, i.e. the chemical bonding force is further increased. This in turn suggests the following two basic requisites for the onset of high T_c superconductivity: To get the higher T_c, (a) Hybridization(covalency) of the HLMO bond in Fig. 1 should be increased, and in close connection with this, (b) energy difference between two different multi valence state A(A') and B(B') of ORH(RRH) in Fig. 1 should be minimized possibly to zero. This means that when the A(A')-B(B') energy difference becomes larger, the resonance(-condensation) energy J becomes smaller due to the above mentioned compression energy[41,42], and finally exceeded by the latter(no resonance). Although quantitative argument of these two basic factors is not possible until the reliable quantum chemical calculations are available, to what extent these two conditions (a) and (b) are attained in the actual crystals of these various superconductors is considered to determine basically the magnitude of the transition temperature T_c of the systems.

Analyses of Isotope Effects of Superconductors

In the following, the author describes the first application of Eq.(1) for the isotope effects analyses of various superconductors including oxide superconductors, conventional low T_c metals and palladium hydrides, etc. The aim of the analyses is two-fold; firstly to demonstrate the validity of Eq.(1) as the expression of T_c almost universally applicable to various types of superconductors, and secondly to quantify at least phenomenologically the above mentioned qualitative discussions of the magnitude of T_c of these superconductors.

First, oxygen isotope effects of several oxide superconductors reported in literatures were analyzed[1,2] on the basic assumptions that $w_D \propto M^{-1/2}$ and $\Delta E_{N,S}=h\Delta w_D$(i.e. E_N(phonon)$=E_S$(phonon) and also the effect of $O^{16} \rightarrow O^{18}$ exchange on $E_{N,S}$(phonon), $\Delta E_{N,S}$(phonon)$=\Delta E_{N,S}$ is the same for both the normal and superconductive states, keeping other major terms of $E_{N,S}$ in Eq.(2) constant unaffected by this exchange, as are usually done). The results are listed in Table 2 giving a fair explanation on the experimental observations, but before discussing these details, the process of the analysis is briefly illustrated below for the sake of clarity, taking $YBa_2Cu_3O_7$ as an example.

Table 2

Results of Oxygen Isotope Effect Analysis of Several Oxide Superconductors

Systems	T_c(K) Exp. (Calc.)	ΔT_c(K) ($O^{16\to18}$)	$\frac{T'_c}{T_c}$	Θ_D(K) ($\Delta E_{n,s}(p)$)	E_s	E_n (meV)	$J=E_s-E_n$	λ ($=J/E_n$)
Bi-Cu-O (n=2+3)	110 (112.5)	-0.4	0.99636	418 (-2.06)	-3162.4	-2695.4	-467.0	0.1773
$YBa_2Cu_3O_7$	92 (91.6)	-0.5	0.99457	383 (-1.88)	-1796.7	-1515.5	-281.2	0.1856
$La(Sr)_2CuO_4$	37 (36.0)	-1.0	0.97297	376 (-1.85)	-286.4	-233.9	-52.5	0.2245
$BaPb(Bi)O_3$	12 (11.8)	-0.6	0.95000	196 (-0.97)	-75.14	-60.74	-14.4	0.2371
$Ba(K)BiO_3$	29.4 (29.6)	-1.4	0.95238	196[a)]	-47.65	-34.95	-12.7	0.3634
	" (29.7)	"	"	376[b)]	-138.5	-110.2	-28.3	0.2569
	" (29.4)	"	"	451[c)]	-190.1	-153.8	-36.3	0.2360

a): The same value as in $BaPb(Bi)O_3$ is assumed. b): The same value as in $La(Sr)_2CuO_4$ is assumed. c): The calculated value so as to give λ=0.2360. Exp.T_c, ΔT_c, Θ_D: Literature values.

In Eq.(1), we can choose any arbitrary set of E_N and E_S only to fulfill the condition of e.g. T_c=92K for $YBa_2Cu_3O_7$. To fix these energies E_N and E_S uniquely, we need at least one other relevant information on these quantities. Since even a rough theoretical estimate is available now on any of E_N, E_S and J, for this purpose, we utilize here the experimental data of the oxygen isotope effect of this system shown in Table 2, neglecting the relatively large experimental uncertainty of this value[42,43]. Adopting the value of $E_{N,S}$(phonon)=383K=33meV[45], the basic assumption of the isotope effect: $E_{S,N}(\text{phonon})=hw_D \propto M^{-1/2}$ gives the value of $\Delta E_{N,S}$=-1.8873meV for this system. This results in the expected maximum isotope effect of $\Delta T_c=T_c'-T_c$= -5.44K and T_c'/T_c=0.943 in the conventional BCS theory, which is by far stronger than the experimentally observed quite small maximum isotope effect in this system listed in Table 2. Here, T_c' denotes the T_c of the O^{18}-substituted sample.

On the other hand, from Eq.(1), we obtain the following equation for T_c'/T_c:

$$\frac{T_c'}{T_c} = \frac{|E_N'|}{|E_N|} \exp(-\frac{|\Delta E_{N,S}|}{|E_S-E_N|}) = \frac{|E_N|+|h\Delta w_D|}{|E_N|} \exp(-\frac{|h\Delta w_D|}{|J|}) \qquad (3)$$

for due to the aforementioned basic assumption, J'=J holds, because both E_N and E_S are stabilized by the same amount $\Delta E_{N,S}=h\Delta w_D$ by this isotope exchange. Insertions of above mentioned numerical values of $\Delta E_{N,S}$ and T_c'/T_c to Eq.(3) gives for this system;

$$\frac{T_c'}{T_c} = 0.994565 = \frac{|E_N|+1.8873}{|E_N|} \exp(-\frac{1.8873}{|J|}) \qquad (3')$$

Since the preexponential term of Eq.(3')(and (3)) is always greater than 1, the exponential term should be smaller than the experimental T_c'/T_c=0.994565, and this sets the condition that $|J|=|E_S-E_N|<325.28$meV.

The only one unambiguous unique sets of E_S and E_N satisfying both Eqs. (1) and (3) are listed in Table 2 for several oxide superconductors with other various relevant parameters. Experimentally it is found in Table 2 that higher the T_c of the system, larger the Debye temperature Θ_D of the system and smaller the isotope effect. The present analyses give a physically reasonable quantitative account on this clear experimental trend that higher the T_c of the system, larger the chemical bonding forces $|E_S|$, $|E_N|$ and therefore the resonance-condensation energy $|J|$, whereas the coupling constant $\lambda(=J/E_N)$ itself decreases with T_c, suggesting particularly the importance of the chemical stability of the normal state for high T_c superconductivity.

Indeed, the relatively large absolute values of $E_N(\approx E_a(\text{exciton})+E_a(\text{CDW}))$ and $E_S(\approx E_S(\text{MVRC}))$ especially in high T_c systems such as multi-layered Bi Cu-O system(the mixture sample of the Cu layer number n=2+3) and $YBa_2Cu_3O_7$ in Table 2 appear to be interpretable only if we regard them as the chemical bond energies at OK in the respective normal and superconductive states of the Cu-O containing whole system. This seems to be definitely so because the Debye temperature $k\Theta_D(=hw_D)$ of these systems employed here is nothing but the zero-point oscillation of the system seen from the oxygen side in

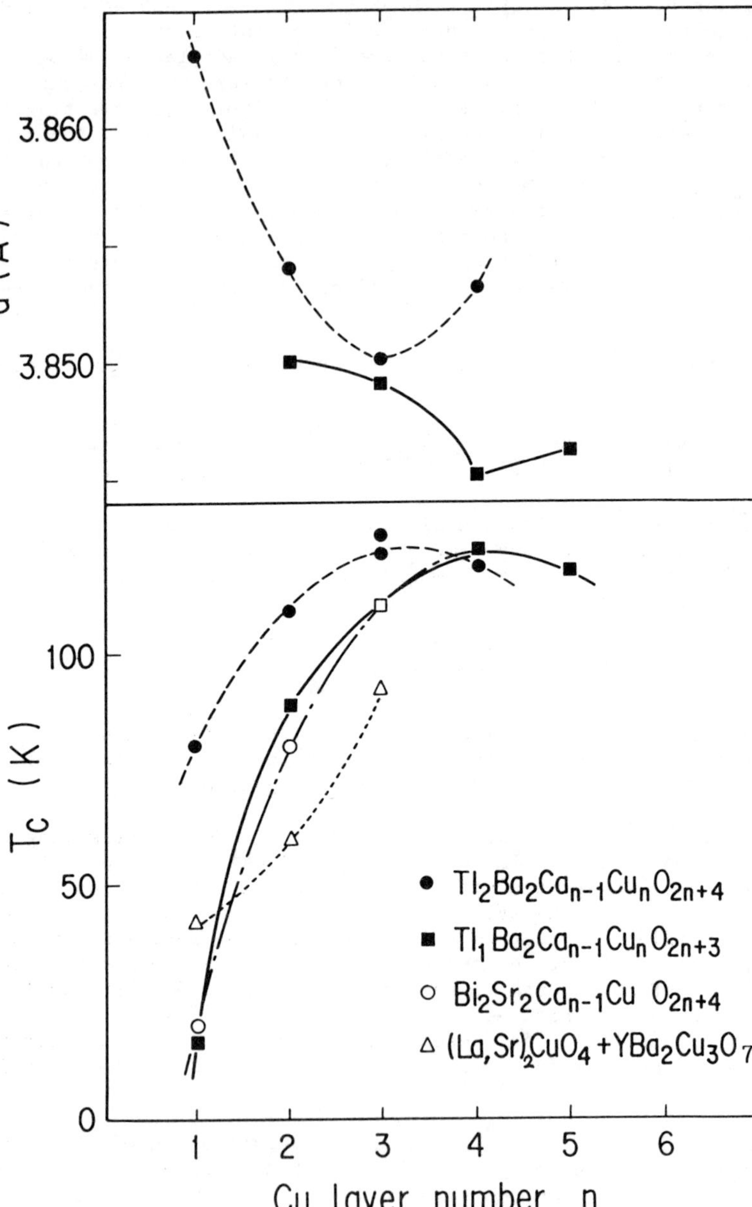

Fig. 6. Intra-planar lattice constant a(Å) and T_c vs. Cu layer number n in layered cuprates (Refs: a(Å);[46], T_c;[47]).

the potential well of the whole chemical bond energy of the system.

Fig. 6 shows the intra-planar lattice constant a(Å) and T_c vs. Cu layer number n relationships in several ORH(hole) type cuprate superconductors. In the lower part of this figure for T_c, $La(Sr)_2CuO_4$(n=1) and $YBa_2Cu_3O_{7-x}$ systems are also included regarding $YBa_2Cu_3O_{6.50-6.75}$ as n=2 and hypothetically $YBa_2Cu_3O_7$ as n=3. The fact that both for single Tl and double Tl layered Tl-Ba(Ca)-Cu-O systems, the systems with the highest T_c(n=4 for the former and n=3 for the latter) have the shortest a(Å) seems to reinforce the above obtained conclusion from the isotope effects analysis that the chemical bonding force is the major driving force of this superconductive transition. Then, it appears also reasonable to speculate that the tremendous rise of T_c of these Tl and Bi based multi-layered cuprates going from n=1 to n=3~4 is mostly due to the increase of the E_a(exciton) and E_a(CDW) type normal state chemical bond energies concomitant with the lattice contraction of the systems with n.

Now, quite a straightforward interpretation of the isotope effect of these superconductors in general appears to be possible based on the results shown in Table 2: As a general rule, higher the T_c of the system, smaller the isotope effect of the system, and this is a mere reflection of the obvious physical reality of the system that the system with the stronger chemical bond and therefore as a result with the higher Debye temperature, has the smaller relative weight of the phonon energy term in the total energy of the system, i.e. $hw_D/(|E_a(\text{exciton})+E_a(\text{CDW})|)$, and therefore naturally the relative retardation effect of the phonon to its superconductive transition is smaller.

The above conclusion on the isotope effect can be written concisely in the follwing form: Putting $|E_N|=nhw_D$ and so by definition $|J|=n\lambda hw_D$ regarding n as a numerical parameter, from Eq.(3) under a good first approximation that $h\Delta w_D/|J|<<1$,

$$\frac{T_c' M'^{1/2}}{T_c M^{1/2}} = 1 - \frac{\{(M'/M)^{1/2}-1\}\{1-(n+1)\lambda\}}{n\lambda} \qquad (4)$$

As is seen in Table 2, for relatively low T_c systems it is generally true that the experimental isotope effect is rather strong approaching to that of the BCS limit, and from the present analysis $|E_N|$ becomes only several times as large as hw_D(n=1~4) and also λ~0.25. In this case, the second term in the rhs of Eq.(4) is negligible compared to the first term, i.e. 1(especially when $\lambda=1/(n+1)$(e.g. n=3 and λ=0.25), this is exactly equal to zero). So, we get the following conventional isotope effect relationship occasionally found in the low T_c metal superconductors in quite a transparent way;

$$T_c' M'^{1/2} = T_c M^{1/2} = \text{constant} \qquad (5)$$

When n becomes larger, the isotope effect naturally becomes weaker, and in the extreme limit that the T_c of the system becomes sufficiently high and therefore n>>1(nλ>>1), $T_c=T_c'$(no isotope effect).

So far, we analysed the isotope effect assuming $\Delta E_N=\Delta E_S=\Delta E_{N,S}$(phonon) by this isotope exchange of the system. But, more generally speaking, Since the superconductive state is more stable than the normal state at OK in these superconductors, i.e. $J=E_S-E_N<0$, the potential well of E_S would be generally steeper and deeper than that of E_N. Then, it would be natural to expect that the Debye frequency w_D of the system would rise with the superconductive transition and therefore $|\Delta w_D|$ by the isotope exchange in the superconductive state would also be larger than that in the normal state.

Reverse isotope effects in H⟶D substituted Pd hydride[48] and organic salt[49] are understandable in this context, where the tremendous mass difference between H and D can magnify such a rather minor effect. On the assumption that $\Delta E_S(\text{phonon})=(1+\lambda)\Delta E_N(\text{phonon})(\lambda=\text{constant})$, this reverse isotope effect was analysed for PdH(D). Its numerical result is listed in Table 3 [2] and its potential well description in Fig. 7. In Table 3, Θ_N in the normal state in the respective systems are the Einstein temperature of hydrogen in PdH and that of deuterium in PdD, and $\Theta_D(\text{Pd})$ shown below Table 3 is the Debye temperature of Pd in PdH(D), all reported in Ref.[48]. So, we can here directly calculate $\Delta E_N(\text{phonon})(\text{PdH}\rightarrow\text{PdD})$ in the above assumption $\Delta E_S(\text{phonon}) =(1+\lambda)\Delta E_N(\text{phonon})(\lambda=\text{constant})$ as $\Delta E_N(\text{phonon})=k\Theta_N(\text{H})-k\Theta_N(\text{D})=-20.6\text{meV}$. Then, from Eq.(1), we can easily determine all the relevant energy parameters such as $\Theta_S(\text{H,D})$, $E_{N,S}$, J, λ, etc. listed in Table 3 by the following relation;

$$\frac{T_c(\text{PdD})}{T_c(\text{PdH})} = \frac{11.05}{8.8} = 1.25568 = \frac{|E_N(\text{PdH})|+20.6}{|E_N(\text{PdH})|} \qquad (6)$$

for owing to the latter assumption of $\lambda=\text{constant}$, the exponential term of Eq.(1); $\exp(-1/\lambda)$ is the same for the both systems and drops automatically in Eq.(6).

As is clearly seen in Fig. 7, for both PdH and PdD systems the depth of the potential well $E^0_{N,S}$ is the same both in the normal and superconductive states. But, due to the steeper and deeper potential well of E_S than that of E_N, the superconductive state is more stabilized than the normal state by this isotope exchange and this gives rise to the reverse isotope

Table 3 Analysis of reverse isotope effect in PdH(D) systems

System	T(K)	Θ_N (K) (E_N(phonon))	Θ_S (K) (E_S(phonon))	E_N	E_S (meV)	J	$\lambda=J/E_N$
PdH	8.8	720 (62.0)	871 (75.0)	-80.86	-97.71	-16.85	0.20831
PdD	11.05	480 (41.4)	580 (50.0)	-101.54	-122.69	-21.15	0.20831

For both PdH and PdD : $E_N^0(\text{well}) = E_N - E_N(\text{phonon}) = -142.9$ meV
$\Theta_D(\text{Pd}) = 270$ K, $E_S^0(\text{well}) = E_S - E_S(\text{phonon}) = -172.7$ meV
$\Delta E^0_{N\rightarrow S}(\text{well}) = -29.8$ meV

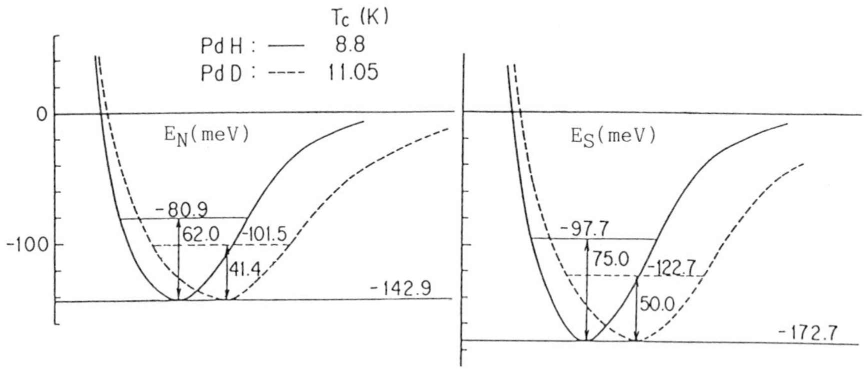

Fig. 7 Potential well description of the reverse isotope effect in PdH(D) systems.

effect of the PdH(D) systems.

Summarizing the discussions of the isotope effects of various superconductors, it seems enough correct to mention that these superconductors are in general in the weak coupling limit of $0<\lambda<0.25$ irrespective their magnitude of T_c, and are more so, higher the T_c of the system, although the meaning of the coupling constant λ is completely different from the conventional one. Because of the relatively large experimental uncertainty in determining the isotope effects of these systems and also of the rather simplified and idealized assumptions made here for their analysis, the energy parameters $E_{N,S}$ and J of these superconductors listed in Tables 2 and 3 should only be the rough estimates and need to be elaborated by other relevant experimental data and/or the reliable quantum chemical calculations. Yet, it appears that Eq.(1) can give a fair account on the versatile features of the isotope effects of these various superconductors in a consistent and systematic manner.

DISCUSSIONS

As was shown in the foregoing, the present model seems to have succeeded to some extent in providing a possible unified physical picture of (high T_c) superconductivity, that is, the macroscopically realized MVRC state of the microscopic quantum chemical resonance shown in Fig. 1 through the self-organization process(valence fluctuation→valence ordering→resonance-condensation) of the electron system as the fundamental and universal chemical bonding force. The key ingredients of this whole story of superconductivity appear to be 'Symmetry' in a word: According to the symmetry of the electron orbitals, anisotropy(or symmetry reduction) of the crystal system appears to be essential in other than s(l=0) electron systems for the whole process to occur inside the splitted single HLMO bond with the designated spin degeneracy shown in Fig. 1. As for the symmetry between electron and hole, this, by constituting itself the basic structure of the resonance hybrids as A(A') -B(B') pair in Fig. 1, firstly provides the driving force of the ordering process in the normal state(the electron-hole attractive interaction energy: E_a(exciton)+E_a(CDW)), and finally enables the electron system to evolve this ordering energy totally in the superconductive state, manifesting itself as the resonance-condensation energy E_S(MVRC). This further provides the so-called 'hole'(ORH)- and 'electron'(RRH)-type superconductors as their respective symmetrical counterparts. So, lastly, it seems enough reasonable to speculate in Fig. 1 that the simultaneous symmetrical paired motion of two electrons as the real space electron pair would be a possible origin of the zero resistivity and the perpetual current of superconductivity, which enables the electron system to escape and decouple nearly perfectly from the almost universal energy destabilizer and electron scatterer of the system, i.e. phonon.

One major advantage of the present model seems to exist in its unmatchable practical applicability to interprete, reproduce and predict the most optimum conditions(chemical composition, valence state, crystal chemistry, etc. of the system) for the onset of superconductivity of various systems with one to three dimensionality. For this purpose, so far proposed variety of mechanisms such as plasmon, exciton, CDW, SDW, (bi)polaron, magnetic, etc. mechanisms including also the most representative phonon-mediated BCS mechanism appear to be so much helpless and powerless, that one may rather inclined to wonder and speculate whether something more essential and crucial might be missing in those various models. To the author those do not at all appear to be the real origin of superconductivity, but do appear to be more like a type of triggers which drive the system to the superconductive state, or more like a type of inhibitors which retard the onset of superconductivity. Indeed, various triggers are possible so far as the origin is there, but the reverse is not true.

As is apparent from the foregoing discussions, the excitonic type E_a(exciton) and the charge density wave type E_a(CDW) defined in the present model as the intra- and inter-planar electron-hole attractive interaction energy seems definitely to be the trigger or more than that, for this promotes the ordering of the electron system which is regarded as a precursor process of the coming superconductivity. Note here also the following facts :(1) electron-electron coulombic repulsive interaction usually postulated as the major object to be overcome by other some mechanism(s) which supplies the electron-electron attractive interaction energy to form the Cooper type electron pair is here totally switched over by this electron-hole attractive interaction of the electron system itself, and (2) notably moreover, due to the quantum mechanical wave nature of the electron-hole interaction both inside and between the planes, these basically electrostatical interaction in the normal state is considered to change over themselves to the quantum chemical resonance-condensation energy(delocalization energy) of the electron system E_S(MVRC) in the superconductive state, after the ordering of the electron system is attained to some critical point.

As for the spin-spin(magnetic) interaction, though one major theoretical attempts are certainly devoted to this direction, it seems somewhat unnatural for the author to seek the origin of superconductivity to such a magnetic interaction, whatever the type of it, in view of the obvious fact that in the superconductive state such a static magnetic correlation is totally absent and also the magnetic field always acts to destabilize and decompose the superconductivity of any kind of systems. This is in line with our more general point of view that magnetism disturbs and/or inhibits superconductivity, and indeed experimentally superconductivity is usually found in systems with its responsible chemical species essential non-magnetic in nature, which appears also true for high T_c systems. If we take the following experimental fact properly into account that even in cuprate systems for which the relevance of such a model is suggested, superconductivity does not manifest itself until the magnetic order of the system sufficiently slow down as mentioned previously, it seems more appropriate to regard such a magnetic interaction as the retarder(and inhibitor) of superconductivity.

As for phonon, from the present point of view which defines superconductivity literally as the decoupling process from phonon, this appears to be an almost universal retarder and disturber of superconductivity because of its intrinsic nature as the thermal agitator and the electron scatterer of the system. Since the present model can give a fair explanation on the versatile isotope effects of these superconductors, one major motive in support of the phonon-mediated BCS picture of superconductivity(and also its subsequent several extended versions) appears to be lost now. Instead, we may argue that superconductivity is possible only in case the above E_a(exciton)+E_a(CDW) type attractive interaction overcomes this phonon energy term E_N(phonon) at 0K in the normal state and the electron system can order by itself competing against this disturbance from phonon, that is, in Eq.(2) E_N=E_a(exciton)+E_a(CDW)+E_N(phonon)<0, preventing the whole superconductive system from decomposing to the thermally agitated completely random state. Note that Eq.(1) can also give this another correct limiting case of $T_c \rightarrow 0K$ when $|E_N| \rightarrow 0$.

From the above argument of phonon energy term, it is worth while here to give a thought on the following question: Is it practically meaningful to consider the above mentioned limiting case of $T_c \rightarrow 0K$ when $|E_N| \rightarrow 0$? Such a situation seems to be indeed possible and is given an explanation referring to Fig. 5 on Nb_3Ge for the sake of clarity; the phonon energy term E_N(phonon)=hw_D at 0K in Eq.(2) should be the zero point oscillation of the whole system in the potential well of the total chemical bond energy of the whole system. This total chemical bond energy $\mathring{E}_N$(total) in turn is the gross sum of the lowest lying stable $d_{xz}(d_{yz})$ π bond energy, the second lowest lying superconductive d_{z^2} σ bond energy in Fig. 5, and further the ionic(+covalent) bond energy between $3Nb^{2.25+}$ and $Ge^{6.75-}$. The second order derivative of this total chemical bond energy E_N(total) in the potential well description as shown in Fig. 7 for PdH(D) with respect to the reaction coordinate r_i is the force constant $k(\simeq k_{ij}=\partial^2 E_N(\text{total})/\partial r_i \partial r_j)$, which appears in the expression $w_D=(k/m)^{1/2}$. Therefore, it is quite conceivable that occasionally the chemical bond energy of the superconductive d_{z^2} σ bond $\mathring{E}_N$=E_a(exciton)+E_a(CDW), only a part of $\mathring{E}_N$(total), may not be large enough to overcome E_N(phonon)=hw_D and bring the superconductive system to E_N=$\mathring{E}_N$+E_N(phonon)<0, so in that event, such a system would not exhibit superconductivity.

In principle, through the detailed quantum chemical calculations, the theoretical estimates are possible for these various chemical bond energies and therefore also for the detailed structure of the phonon spectrum of the system including the anharmonic term so far neglected in the present analysis. Such a theoretical approach would enable us to elucidate more quantitatively the basic various properties of superconductivity; the fascinating quantum chemical (resonance-condensation) process developing in real space just before us. Importance of such an approach can never be overrated.

As for Eq.(1), it seems to the author that the basic validity of this equation as the expression of superconductive transition temperature T_c is verified to a considerable extent by the consistent analysis of versatile isotope effects of these superconductors. As its support, the author adds that Eq.(1) gives the reasonable limiting cases of $T_c \to 0K$ when $|J| \to 0$ and also $|E_N| \to 0$, and $T_c \to T_c'$ when $\lambda = J/E_N \to \infty$, as mentioned formerly.

Obviously, the present model is only a tiny first step toward the theoretical understanding of high T_c superconductivity, and much should be done in the future both experimentally and theoretically to understand and clarify various properties of this fascinating new version of chemical bond, i.e. the superconductive bond, at its microscopic level in Fig. 1 and at its macroscopic level in Figs. 2, 3, 4 and 5, etc. as its crystal scale 1-, 2- and 3D extensions. As for the theoretical works, these are e.g. to give a clear-cut explanation on the most fundamental properties of superconductivity such as zero resistivity, perpetual current and Meissner effect, etc. consistent with the more general phenomenological description of superconductivity not dependent on the microscopic interpretation of its origin, and also for doing this, to interrelate the basic energy parameters E_N, E_S and J in Eqs. (1) and (2) to other various basic physical parameters which characterize the superconductive(and normal) state such as penetration depth $\lambda(T)$, coherent length $\xi(T)$, energy gap $\Delta(T)$, the specific heat $c_p(T)$ and its gap at T_c $\Delta c_p(T_c)$ and so on as was done in the BCS theory[7].

As for the experimental works, making full use of the present model as a guiding chemical and structural principle, there appears to exist almost immense experimental scope of finding new superconductors, for example, those with the other responsible chemical species and with higher T_c whether inorganic or organic type ones, the RRH(electron) type symmetrical counterparts of already found ORH(hole) type superconductors, etc.. Though qualitative at present, it would be appropriate to summarize shortly here what the present model suggests to such experimental efforts:

(1) To bring the system to the optimum valence state of the responsible chemical species, that is, ±0.33 for tri-layered type 1D' system and ±0.25 for other 1-, 2- and 3D systems from the intermediate valence state Ⓣ.

(2) To check the system with other than s(l=0) electron systems whether it has enough anisotropic crystal structure to provide a splitted single HLMO bond for the responsible chemical species.

(3) To make the system have the stronger chemical bond(E_N) with the more rigid crystal lattice(Θ_D).

Eq.(1) says that theoretically in principle there exists no upper limit of T_c. But, of course, this does not at all mean that there exists no physically realistic upper bound of T_c, for the basic energy parameters E_N, E_s and $J=E_S-E_N$ in Eq.(1) naturally have their own physically reasonable upper bounds which can be realized actually in the solid state crystal system. In this context, whether room temperature- and even the higher temperature superconductors are indeed possible or not, is a particularly interesting major subject of the present and future challenge, which would bring about a tremendous impact on science and technology of superconductivity.

CONCLUSION

In this article, a new theoretical approach for the electron pairing and superconductivity in high T_c superconductors; Multi Valence Resonance-Condensation Model(MVRC Model) proposed by the present author[1,2] has been described extensively and in detail, mainly focussing on high T_c(as well as low T_c) oxide superconductors and A-15 type intermetallic Nb_3Ge system.

Starting from the microscopic quantum chemical resonance hybrids inside the three cations crystal molecule shown in Fig. 1 as ORH and RRH, the Model shows that MVRC state of superconductivity is the Bose condensate of these microscopic resonances of the various responsible chemical species such as Cu, Bi, Pb, Nb, etc.(Table 1), which is realized macroscopically in the perfectly coherent manner through the synergetical- and/or the self-organization process of the electron system itself as the fundamental and universal chemical bonding force, self-driven by the symmetry of the electron orbitals of the respective responsible chemical species, that is, valence fluctuation (disordered multi valence state)→valence ordering(ordered A(A')-B(B') type binary state)→resonance-condensation(MVRC state of superconductivity).

This was demonstrated to provide a possible unified picture of superconductive state of all these systems for both the so-called 'hole'(ORH)- and 'electron'(RRH) type superconductors in the perfectly symmetrical manner, as was briefly summarized in Table 1, and was further suggested shortly to provide that of other types of superconductors such as conventional low T_c metals and organic superconductors, etc..

Expression of transition temperature T_c, Eq.(1) derived on this basis was shown to give a fair account on the so far reported oxygen isotope effects of high T_c oxide superconductors(Table 2), the $T_cM^{1/2}$=constant relationship in low T_c systems, and also the reverse isotope effect in PdH(D) system (Table 3) in a consistent and systematic manner.

Most importantly, the present model reveals the inherent deep connection of superconductivity with the most fundamental and universal nature of matter, i.e. chemical bond. Thus, by elucidating a new simple and beautiful symmetrical structure of this novel reality of our world, high T_c superconductivity, MVRC Model is expected to be a chemical and structural principle in both searching new superconductors and constructing a new microscopic theory of superconductivity.

Acknowledgements

The present author wishes to express his hearty thanks to Dr. H. Ihara of Electrotechnical Laboratory for kindly permitting him to reproduce Fig. 3 of Ref.[47] as the lower part of Fig. 6 of this article with slight changes mentioned in the text.

REFERENCES

[1] A. Nakamura, submitted to Jap. J. Appl. Phys., March 1989.
[2] A. Nakamura, to be published in 'Proc. of Tsukuba Seminar on High T_c Superconductivity', May 31 - June 2, 1989, Tsukuba, Japan.
[3] S. Hufner, Z. Phys. B61, (1985) 135.
[4] N. Nucker, J. Fink, J.C. Fuggle, P.J. Durham and W.M. Temmerman, Physica 153-6, (1988) 119.
[5] A.W. Sleight, Science 242, (1988) 1519.
[6] J.A. Wilson, J. Phys. C21, (1988) 2067.
[7] J. Bardeen, L.N. Cooper and J.R. Schrieffer, Phys. Rev. 108, (1957) 1175.
[8] M.K. Wu, J.R. Ashburn, C.J. Torg, P.H. Hor, R.L. Gao, Z.J. Huang, Y.Q. Wang and C.W. Chu, Phys. Rev. Lett. 58, (1987) 908.
[9] T. Siegrist, S. Sunshine, D.W. Murphy, R.J. Cava and S.M. Zahurak, Phys. Rev. B35, (1987) 7137.
[10] T. Yonezawa, T. Nagata, H. Kato, A. Imamura and K. Morokuma, Ryōshi Kagaku Nyūmon (Introduction to Quantum Chemistry), Kagaku Dōjin, Tokyo, (1983) chapter 11.
[11] R.J. Cava, B. Batlogg, J.J. Krajewski, R. Farrow, L.W. Rupp, Jr, A.E. White, K. Short, W.F. Peck and T. Kometani, Nature 332, (1988) 814.
[12] D.G. Hinks, B. Dabrowski, J.D. Jorgensen, A.W. Mitchell, D.R. Richards, Shiyou Pei and Donglu Shi, Nature 333, (1988) 836.
[13] O. Inoue, Y. Takahashi, H. Hirano, S. Adachi, T. Wada, S. Kawashima and T. Mihara, J. Ceram. Soc. Jpn. 96, (1988) 512.
[14] T. Kato, M. Watanabe, Y. Kazumata, H. Naramoto, T. Iwata, Y. Ikeda, H. Maekawa and T. Nakamura, Jap. J. Appl. Phys. 27, (1988) L2097.
[15] A. Hofmann, H. Kronmuller, N. Moser, R. Reisser, P. Schule and F. Dworshak, Physica C156, (1988) 528.
[16] C.N.R. Rao, Progress in High Temperature Superconductivity, Vol. 7(Ed. C.N.R. Rao), World Scientific, Singapole, (1988) p. 1.
[17] J.M. Tarascon, W.R. Mckinnon, L.H. Greene, G.H. Greene, G.W. Hull and E.M. Vogel, Phys. Rev. B36, (1987) 226.
[18] W.A. Little, Science 242, (1988) 1390.
[19] T.K. Worthington, W.J. Gallagher and T.R. Dinger, Phys. Rev. Lett. 59, (1987) 1160. A. Yamagishi, H. Fuke, K. Sugiyama, M. Date, T. Tagima, M. Hikita, T. Ishii, A. Katsui, Y. Hidaka, T. Iwata and S. Tsurumi, Physica C153-5, (1988) 1459.
[20] M.J. Naughton, R.C. Yu, P.K. Davies, J.E. Fisher, R.V. Chamberlin, Z.Z. Wang, T.W. Jing, N.P. Ong and P.M. Chaikin, Phys. Rev. B38, (1988) 9280.
[21] A.W. Little, Phys. Rev. 134, (1964) A1416.
[22] R.J. Cava, B. Batlogg, A.P. Ramirez, D. Weder, C.H. Chen, E.A. Rietman and S.M. Zahurak, High Temperature Superconductors (Eds. M.B. Brodsky et al.), MRS, Pittsburgh, (1988) p. 19.
[23] K. Kanoda, T. Kawagoe, M. Hasumi, T. Takahashi, S. Kagoshima and T. Mizoguchi, J. Phys. Soc. Jpn. 57, (1988) 1554.
[24] J.G. Bednortz and K.A. Muller, Z. Phys. B64, (1986) 189. J.G. Bednortz, M. Takashige and K.A. Muller, Europhys. Lett. 3. (1987) 379.
[25] S. Uchida, H. Takagi, K. Kitazawa and S. Tanaka, Jap. J. Appl. Phys. 26, (1987) L1. K. Kishio, K. Kitazawa, S. Kanbe, I. Yasuda, N. Sugii, H. Takagi, S. Uchida, K. Fueki and S. Tanaka, Chem. Lett., (1987) 429.
[26] Z.Z. Sheng and A.M. Hermann, Nature 332, (1988) 55.

[27] H. Maeda, Y. Tanaka, Y. Fukutomi and T. Asano, Jap. J. Appl. Phys. 27, (1988) L209.
[28] Y. Tokura, H. Takagi and S. Uchida, Nature 337, (1989) 345.
[29] Y. Tokura, Kotai Butsuri (Solid St. Phys.) 24, (1989) 559.
[30] J.H. Brewer, E.J. Ansaldo, J.F. Carolan, A.C.D. Chaklader, W.N. Hardy, D.R. Harshman, M.E. Hayden, M. Ishikawa, N. Kaplan, R. Keitel, J. Kempton, R.F. Kiefl, W.J. Kossler, S.R. Kreitzman, A. Kulpa, Y. Kuno, G.M, Luke, H. Miyatake, K. Nagamine, Y. Nakazawa, N. Nishida, K. Nishiyama, S. Ohkuma, T.M. Riseman, G. Roehmer, P. Schleger, D. Shimada, C.E. Stronach, T. Takabatake, Y.J. Uemura, Y. Watanabe, D. L. Williams, T. Yamazaki and B. Yang, Phys. Rev. Lett. 60, (1988) 1073.
[31] A. Khurana, Physics Today September, (1987) 17.
[32] D.C. Johnston, H. Prakash and W.H. Zachariasen, Mater. Res. Bull. 8, (1973) 777.
[33] A.R. Sweedler, C.J. Raub and B.T. Matthias, Phys. Lett. 15, (1965) 108.
[34] A.W. Sleight, T.A. Bither and P.E. Bierstedt, Solid St. Comm. 7, (1969) 299.
[35] M.B. Robin, K. Andres, T.H. Geballe, N.A. Kuebler and D.B. Mcwhan, Phys. Rev. Lett. 17, (1966) 917.
[36] S. Uchida, S. Tagima, K. Kitazawa and S. Tanaka, Kotai Butsuri (Solid St. Phys.) 20, (1985) 955.
[37] M. Suzuki and T. Murakami, Solid St. Comm. 53, (1985) 691.
[38] J.R. Gavaler, Appl. Phys. Lett. 23, (1973) 480.
[39] S.V. Vonsovsky, Yu.A. Izyumov and E.Z. Kurmaev, Superconductivity of Transition Metals, Springer Verlag, Berlin, (1982) chapter 1.
[40] G. Saito, Bull. Ceram. Soc. Jpn. 7, (1987) 574. K. Bechgaad and D. Jerome, Betsusatsu Science 84, (1987) 105.
[41] R. Mcweeny, Coulson's Valence, Oxford Univ. Press, UK, (1979) p. 247.
[42] S. Mizushima and S. Nagakura, Ryōshi Kagaku (Quantum Chemistry), Iwanami Shoten, Tokyo, (1968) p. 125.
[43] K.J. Leary, H.C. zur Loye, S.W. Keller, T.A. Faltens, W.K. Ham, J.N. Michaels and A.M. Stacy, Phys. Rev. Lett. 59, (1987) 1236.
[44] H. Katayama, H. Yoshida, T. Hirooka, A.J. Mascarenhas, Y. Okabe, T. Takahashi, T. Sasaki, A. Ochiai, T. Suzuki, J.I. Pankove, T. Ciszek and S.K. Deb, Jap. J. Appl. Phys. 26, (1987) L2085.
[45] M.B. Maple, Y. Dalichaouch, J.M. Ferreira, P.R. Hake, S.E. Lambert, B.W. Lee, J.J. Neumeier, M.S. Torikachvili, K.N. Yang, H. Zhou, Z. Fisk, M.W. Mcelfresh and J.L. Smith, Novel Superconductivity (Eds. S. A. Wolf and V.Z. Kresin), Plenum Press, N.Y., (1987) 839.
[46] Y. Syono, M. Kikuchi, S. Nakajima, T. Suzuki, T. Oku, K. Hiraga, N. Kobayashi, H. Iwasaki and Y. Muto, IMR Preprint, No. 1014, May 1, 1989.
[47] H. Ihara, Parity 11, (1988) 47.
[48 H. Hemmes, A. Driessen, R. Griessen and M. Gupta, Phys. Rev. B39, (1989) 4110.
[49] K. Oshima, H. Urayama, H. Yamochi and G. Saito, J. Phys. Soc. Jpn. 57, (1988) 730.

HIGH RESOLUTION POLARIZED X-RAY ABSORPTION SPECTROSCOPY: SYMMETRY OF THE HOLE STATES IN HIGH-T_c SUPERCONDUCTORS

A. Marcelli
INFN-Laboratori Nazionali de Frascati, Frascati, Italy

INTRODUCTION

Oxygen deficient-perovskite materials with expected non integer formal valence state of the Cu ions higher than two, have first attracted the scientific interest of the group of Caen university [1-3] because some of these ceramic materials are very good conductors at room temperature. Following the discovery of the high temperature superconductivity in this class of materials [4,5] an impressive number of scientific investigation have been carried out on the electronic structure of these systems to understand the microscopic mechanism of high temperature superconductivity.

Principal aim of this short review is to underline, in the dark jungle of the experimental techniques, the role of the x-ray absorption spectroscopy (XAS) as one powerful technique to investigate high temperature superconductors materials. Recent results obtained by polarized x-ray absorption spectroscopy with synchrotron radiation and polarized x-ray absorption calculation in the frame of the multiple scattering theory on high temperature superconductors are reported. They show that XAS can be used to investigate not only the amounts of Cu^{1+}, Cu^{2+} or Cu^{3+} configurations in the ground state and their relative changes with the oxygen content, moreover the detection of the symmetry of the hole state can be obtained giving unique information on the electronic states near the Fermi energy.

Core level spectroscopies provide a direct tool to investigate the electronic structure and the charge carriers in these materials. [6] In particular x-ray photoemission spectroscopy (XPS) [7-12] is essentially determined by the initial density of occupied states.
Resonant XPS experiments at the Cu 3p threshold which shown the presence of strong electron correlation around the copper atoms demonstrate the existence of a two-hole bound state, [10,13-17] while Cu 2p XPS shown the close similarity between high temperature superconductors and charge transfer type insulators [18] and strongly suggest superconductivity mechanism via oxygen 2p holes. [7,19-21]

Core-level x-ray absorption spectroscopy (i.e. at copper K-edge, [22-27] at copper L_3-edge, [28-31] and at oxygen K-edge [20]) probes the different unoccupied electronic states of angular moment $\ell'=\ell\pm1$ at selected atomic sites, moreover polarized x-ray absorption spectroscopy spectra probe the orientation of the unoccupied orbitals. Thus polarized XAS spectroscopy of a single crystal is a suitable probe to determine experimentally the symmetry of the molecular orbitals where the electronic holes induced by doping in a deficient-perovskite are formed.

In the classical quantum theory the absorption cross section is given by many-body excitations of the N-electron system. Following the interaction with the photon beam with energy ω, the system is excited from the initial state i at energy E_i to a final state f at energy E_f. In the one-electron approximation the x-ray absorption is described by single-particle processes. The N-electron system is separated into two parts: the single electron in the core level which is excited into an unoccupied level and the N-1 passive electrons which do not participate directly in the electronic transition. The one-electron transition takes place in a static potential determined by a single configuration of the N-1 passive electrons.

The absorption coefficient $\mu_c(E)$ for the transitions from the core level $c(n,\ell,J)$ with energy E_c and wavefunction φ_{cM}, in the one-electron approximation, can be expressed in atomic units as

$$\mu_c(E) = 4\pi^2 \alpha F_c(E)/(\Omega/\nu) \qquad E>E_F \tag{1}$$

where $\alpha^{-1}=137.036$ is the inverse fine-structure constant, Ω is the volume of the primitive cell, ν is the number of contributing atoms in the primitive cell and $F_c(E)$ is the spectral distribution of oscillator strength [32]

$$F_c(E)= (\omega/3) \Sigma_{\mathbf{k},j} \Sigma_{M=-J} r^2_{cM,\mathbf{k}j}\, \delta(E - E_{\mathbf{k}j}) \tag{2}$$

$$\mathbf{r}_{cM,\mathbf{k}j} = \int \varphi_{cM}(r)\, \mathbf{r}\, \Psi_{\mathbf{k}j}(r)\, d^3r = < \varphi_{cM} \mid \mathbf{r} \mid \Psi_{\mathbf{k}j}> \tag{3}$$

$\Psi_{\mathbf{k}j}$ and $E_{\mathbf{k}j}$ are the wavefunction and energy of the j-th conduction band at reduced vector **k**, ω is the photon energy $\omega = E - E_c$ and the integration is carried out over the volume of the primitive cell. Since the dipole transitions dominate the process of photoabsorption, an electron from a core level having angular momentum ℓ is excited into the $\ell\pm1$ final states. Neglecting the spin-orbit coupling for the band states, $F_c(E)$ can be written as [33,34]

$$F_c(E) = (\omega/3)\, (2J+1)/2(2\ell+1)\, \{ \ell/(2\ell-1) f_{c,\ell-1}(E) + (\ell+1)/(2\ell+1)\, f_{c,\ell+1}(E)\} \tag{4}$$

where the partial strength $f_{c,\ell}(E)$ can be factorized into

$$f_{c,\ell}(E) = \rho_{c,\ell}\, N_\ell(E) \tag{5}$$

$N_\ell(E)$ is the angular projected density of states defined as

$$N_\ell(E) = 2\, \Sigma_{\mathbf{k},j} \Sigma_m |<Y_{\ell m} \mid \Psi_{\mathbf{k}j} >|^2\, \delta(E - E_{\mathbf{k}j}) \tag{6}$$

where the energy band states are labelled by the reduced wave vector **k** and the band index j. The effective matrix element $\rho_{c,\ell}(E)$ is given by

$$\rho_{c,\ell}(E) = < \varphi_c \mid \mathbf{r} \mid \phi_\ell(E) >^2 / < \phi_\ell^2(E) > \tag{7}$$

where the wave function $\phi_\ell(E,r)$ is a solution of the radial Schrödinger equation inside the muffin-tin (MT) sphere of radius S (r<S), and outside the MT sphere (r≥S), is given by

$$\phi_\ell(E,r) = [\cos \delta_\ell(E)]\, J_\ell\,[(E-V_o)^{1/2}\, r] - [\sin \delta_\ell(E)]\, n_\ell\,[(E-V_o)^{1/2}\, r] \tag{8}$$

where $\delta_\ell(E)$ is the ℓ-th phase shift of the muffin tin potential, V_o is the muffin tin zero of the potential and J_ℓ and n_ℓ are spherical Bessel functions.

The spectra of crystals are therefore interpreted in terms of the product of the partial (of selected orbital momentum $\ell \pm 1$) density of states and of the matrix element. Müller *et al.* [33,34] have shown that the x-ray absorption near edge structure (XANES) spectra can be understood as the product of an atomic-like term and a solid state term. The factorization of the partial oscillator strength into the solid state term $\chi_\ell(E)$ and atomic term $f^{at}{}_{c,\ell}(E)$ is given by

$$f_{c,\ell}(E) = f^{at}{}_{c,\ell}(E) \cdot \chi_\ell(E). \tag{9}$$

The atomic term is obtained by considering a single muffin tin potential confined in the Wigner-Seitz (WS) sphere of radius S_{WS} and the spectrum for the transition from the core level in the single atom is

$$f^{at}{}_{c,\ell}(E) = (2\ell+1)\, N^{FE}(E) < \varphi_c \mid \mathbf{r} \mid \phi_\ell(E,r) >^2 \tag{10}$$

where the integration is over the Wigner-Seitz (WS) sphere and $N^{FE}(E) = (E - V_o)^2/2\pi$ is the free electron density of states. The solid state term is

$$\chi_\ell(E) = N_\ell(E) / N^{at}{}_{c,\ell}(E) \tag{11}$$

where

$$N^{at}{}_{c,\ell}(E) - 1/\Omega\, (2\ell+1) N^{FE}(E) < \phi_\ell^2(E,r) > \tag{12}$$

is the projected density of states for the single-sphere problem.

This approach leads to the following understanding of the x-ray absorption spectra: i) the overall magnitude and shape of a particular spectrum is determined by the corresponding atomic transition rate $f^{at}{}_{c,\ell}(E)$ and ii) the fine structure of the spectrum is determined by the solid state factor $\chi_\ell(E)$ which is proportional to the density of band states with $\ell \pm 1$ orbital characters. The ℓ-1 term is often ignored because usually it exhibits a much smaller amplitude than the ℓ+1 term. However in some cases, like in the first six electron volt above threshold in the Si L_3-edge, the ℓ+1 density of states is negligible and the ℓ-1 term is the most important.

In some cases the atomic part exhibits strong resonances in a small energy range therefore the contribution of the solid state factor $\chi_{\ell}(E)$ can be neglected. In these cases the spectra of solids can be interpreted in terms of atomic transitions (see for examples the $L_{2,3}$ edges of Ni and Cu in insulating systems). [35-37] In other cases the atomic part exhibits a smooth structureless spectrum over a large energy range and therefore the spectral features can be assigned to the solid state factor χ_{ℓ} (E).

In the case of metals, the core hole in the static final state potential is fully screened by valence electrons close to the Fermi level. Most of the XANES spectra, in this case, can be interpreted in the frame of the von Barth and Grossmann final state rule, [38,39] which states that the wave function of the excited photoelectron is determined by the final state potential with the core hole and the relaxed N-1 electrons. Therefore in most cases the one-electron transitions are assumed to take place in the potential of the fully relaxed configuration of the N-1 passive electrons in the presence of the core hole. The fully relaxed configuration is defined as the final state many-body configuration with the lowest energy. The final state rule breaks down for transitions to nearly filled bands like in the case of $2p \rightarrow \varepsilon d$ transition in palladium or in nickel metal.

Many-body effects in XANES can arise because of the presence of different many-body final state configurations of the N-1 passive electrons. Configuration interaction of many-body final states can appear because of core hole perturbation in the final state and/or configuration interaction in the ground state like in valence fluctuating materials [40] or correlated systems such as NiO, CeO_2, PrO_2 and CuO. The electronic structure of these insulating correlated oxides has been object of large interest in these last years. [41,42] In these compounds a strong hybridization between a localized metal orbital and the delocalized 2p oxygen states coexist with a large electronic d-d or f-f Hubbard repulsion. The cuprate high-T_c superconductors belong to this class of materials. The specific electronic structure of these systems is characterized by localized states on the metal sites with a large electronic correlation. The ground state of the system can be described by mixing of localized Cu $3d^9$ and Cu $3d^{10}\underline{L}$ configurations where $\underline{L}$ indicates a hole in the oxygen derived valence band. In the stoichiometric insulating systems the two ionic configurations are separated by the energy ΔE. The mixing between the localized electronic configurations is due to a large oxygen-metal hybridization T ($T \sim \Delta E$).
This type of electronic structure is possible, when some orbitals have an atomic-like localized character, the energy separation between the delocalized ligand band and the localized metal orbital is negligible and there is a large metal oxygen covalency. This seems to be the case of high temperature superconducting perovskites where the Cu 3d states are atomic-like and both the metal 3d and the oxygen 2p band cross the Fermi level. [6] Moreover since the first experimental investigation by core-level spectroscopies [7,8,28,29] the d-d Coulomb repulsion and the Cu-O hybridization energies were found to be about 6-8 eV and 2-3 eV respectively.

POLARIZED X-RAY ABSORPTION SPECTROSCOPY

Polarized x-ray absorption spectra show a strong dichroism for anisotropic sites. The strong dichroism of XANES spectra of single crystals was observed by several groups and in different materials. [43-57] The dichroism for electronic transitions is well known in the classical quantum theory of the absorption cross section. By the interaction with the photon beam the system is excited from the initial state i at energy E_i to the final state f at energy E_f and in the dipole approximation the total absorption cross section is given by

$$\sigma(\omega) \sim \omega \ \Sigma_f | M_{if} |^2 \ \delta(E_i - E_f + \omega) \quad (13)$$

where the sum is extended over all the possible final states f and M_{if} is the matrix element involving the initial Ψ_i and final Ψ_f^* many-body radial wave functions

$$M_{if} = \int \Psi_f^*(r_1, r_2,...., r_n) \ \Sigma_n (\mathbf{r}_n \cdot \mathbf{E}) \ \Psi_i(r_1, r_2,...., r_n) \ dr \quad (14)$$

where **E** is the unitary polarization vector of the electric field and the $\mathbf{r}_n$ is the vector describing the position of the n-th electron. Therefore, in anisotropic systems the matrix element is dependent on the direction of the polarization vector via the terms $(\mathbf{r}_n \cdot \mathbf{E})$.

In the multiple scattering description of the absorption cross section in real space the polarization effects are determined by combination of terms like $|(\mathbf{r}_i \cdot \mathbf{E})|^2$ where $\mathbf{r}_i$ are the positions of neighboring atoms respect with the central absorbing atom. Just one term of this type remains in the single scattering approximation which describe the polarized extended x-ray absorption spectra (EXAFS) [54] in which oscillations originating at a particular scattered atom can appear or disappear depending on whether the electric field is directed towards that atom or not. This property has been widely used in surface EXAFS studies to determine separately the interatomic distances parallel and normal to the surface plane. [55,56]

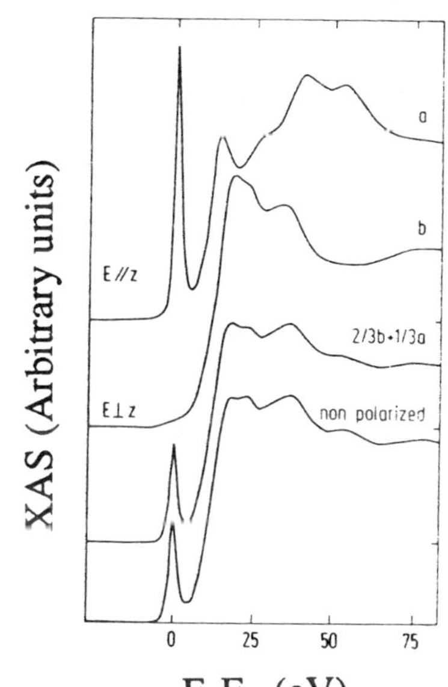

Fig. 1
Polarized experimental XANES spectra of V_2O_5 dehydrated sol-gel. Curve a: the electric field **E** is parallel to the normal **z** of the V_2O_5 layers, i.e. along the axis of the square pyramids, curve b: the electric field **E** is parallel to the V_2O_5 layers; the lowest curve is the unpolarized spectrum and above it the the weighted spectrum from the two polarized spectra.

The multiple scattering contributions due to a set of atoms in a particular direction or on a plane can be selected in the XANES spectra of anisotropic clusters using the polarization dependence of the absorption cross section Therefore, the orientation and angular distribution of neighboring atoms can be determined by changing the relative position of the incident beam.

Fig. 1 reports the vanadium K-edge XANES spectra of a spontaneously dehydrated sol-gel sample of V_2O_5. This sample is formed of oriented sheets separated by about 10 Å of water molecules. Each sheet consists of square plane bipyramid with the base on the **xy** plane and a short double bond of 1.58 Å on one side and a long bond with the oxygen of a water molecule on the other side, in the **z** direction. The experimental spectra for the electric field **E** oriented along the **z** direction (**E**||**z**) and parallel to sheet planes (**E**⊥**z**) are shown. These spectra are a good example of angular resolved XANES because the dichroism is very large and it appears both in the threshold region for the molecular like bound excitations, giving the white line, and in the continuum region. The white line is completely suppressed for the polarization vector oriented on the plane, showing that it is produced by the short double V-O bond. [52]

The interpretation of polarized XANES spectra in terms of multiple scattering allows the determination of the role of a selected group of atoms. This is illustrated in Fig. 2 where the angular resolved XANES calculations are reported for two type of VO_6 clusters. The calculation clearly shows that the white line appears only with the electric field along the V-O double bond. The main features of the experimental spectrum (**E**||**z**) are well reproduced by theoretical calculations for both clusters with and without (upper curve) vanadium atom on the **xy** plane.

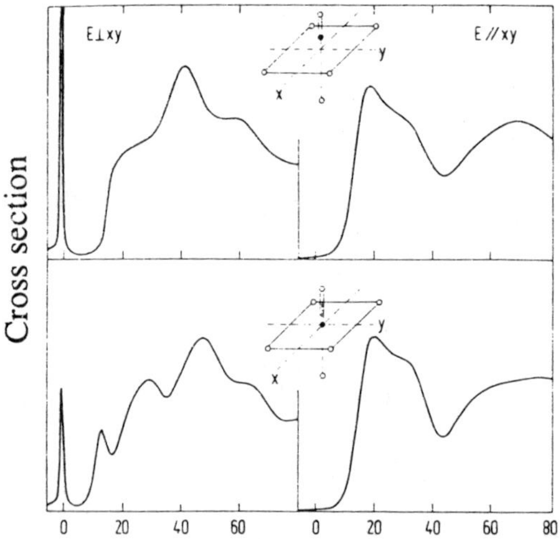

Fig. 2
The theoretical angular resolved multiple scattering calculation for a square bipyramid with a short double V-O bond (d=1.58Å) and with a long V-O bond (d=2.6Å) along the **z** axis due to the oxygen of a water molecule and four oxygens on the **xy** plane at 2.0 Å. Polarized spectra with the V out of plane by 0.5 Å (upper curves) and for a similar cluster with the same distances but the with the V atom in the plane (lower curves).

The interpretation of angular resolved XANES is very useful for understanding and interpreting the spectra of unpolarized spectra. Some features in unpolarized spectra can be assigned to multiple scattering with a selected number of atoms, like for the structures at about 25 eV and that at 60 eV in the unpolarized spectrum in Fig. 1 which can be associated with multiple scattering in the plane and along the normal respectively. Starting with simple models a qualitative agreement with experimental results can be reached which is improved as soon as more complicated clusters are used.

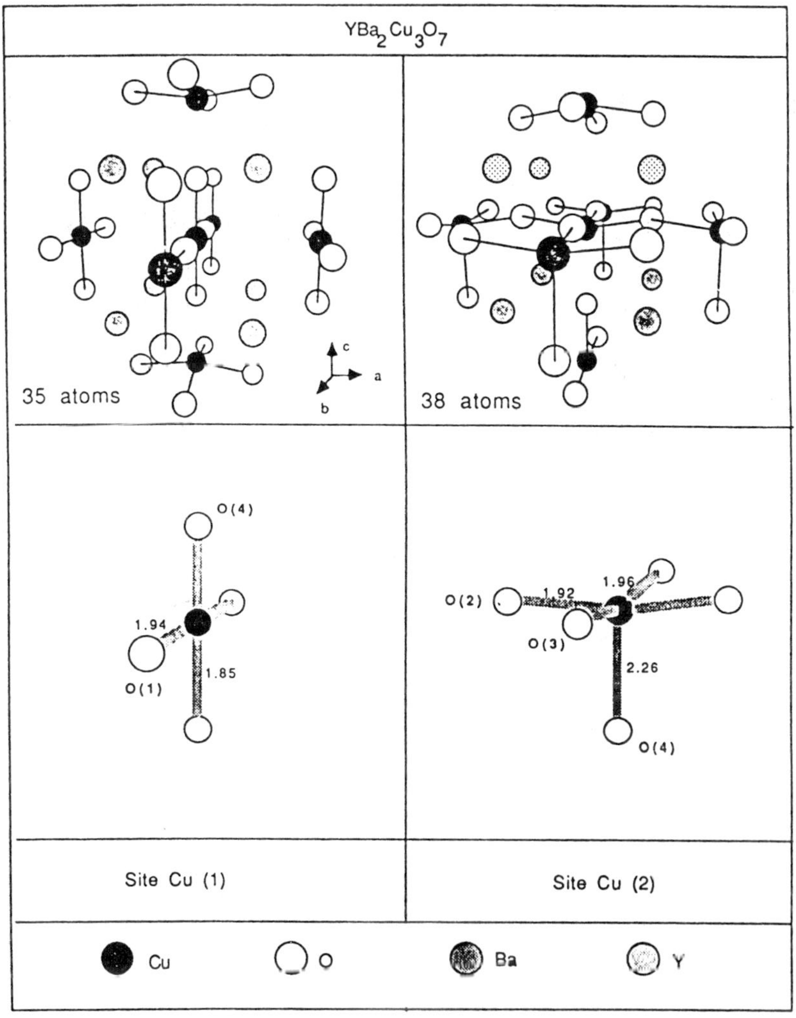

Fig. 3
Structure of the clusters of neighbor atoms surrounding the Cu(1) and Cu(2) sites in $YBa_2Cu_3O_7$.

The Angular Resolved XANES at the Cu K-edge

Due to the local character of this experimental probe, angular resolved XANES are used to determine the orientation of biological complexes and of chemisorbed molecules on surfaces. [49,57,58] Recently was applied to the study of the Cu K-edge XANES to determine the structural changes associate with the loss of oxygen going from the high temperature superconductor $YBa_2Cu_3O_7$ to the antiferromagnetic insulator $YBa_2Cu_3O_6$. [26,27,59-63]

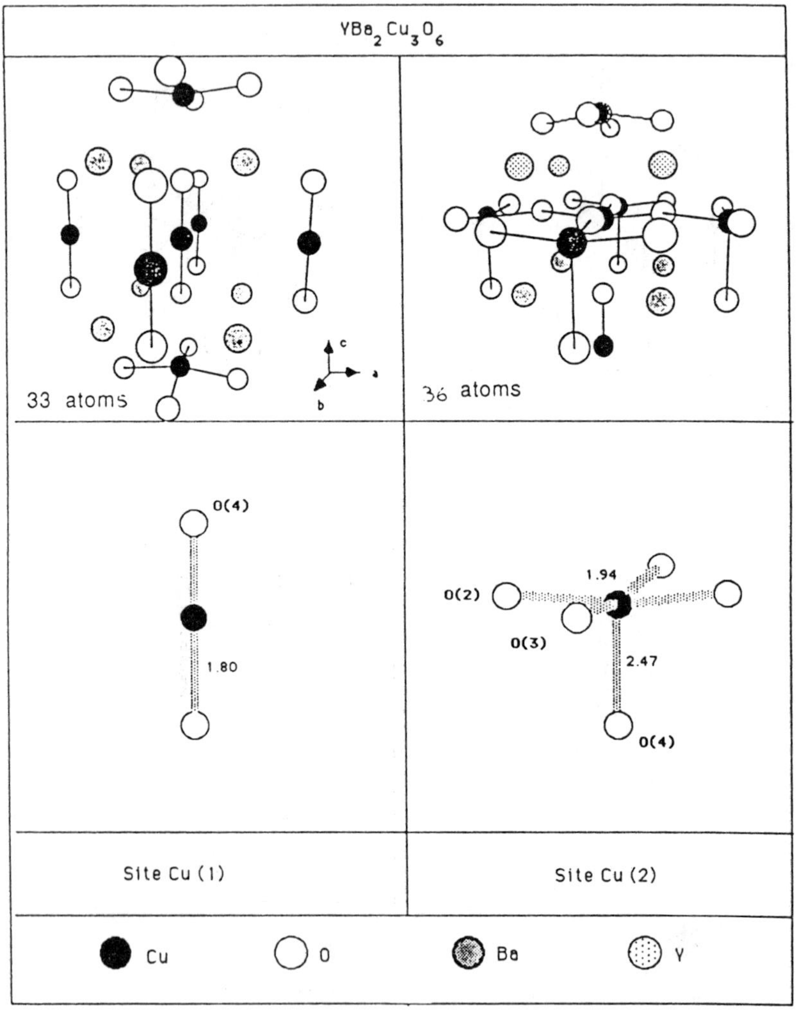

Fig. 4.
Structure of the clusters of neighbor atoms surroundings the Cu(1) and Cu(2) sites in $YBa_2Cu_3O_6$.

XANES calculation in the frame of multiple scattering (MS) theory has been performed to interpret a large variety of materials, from the transition metal oxides, to biological molecules andmolecular complexes and recently to the study of the distortions of Cu site. [64]

In the case of copper oxides the orthorhombic crystalline structure of $YBa_2Cu_3O_7$ changes to the tetragonal structure of $YBa_2Cu_3O_6$ (see for example Cava, *et al.*, [65]). There are two types of Cu sites, named Cu(1) and Cu(2), in the crystalline cell. Therefore two clusters of about 35 atoms, relevant for the Cu K-edge XANES, within a sphere of 5Å radius around each of the two central Cu ions for $YBa_2Cu_3O_7$ ($YBa_2Cu_3O_6$), are shown in Fig.3 (Fig.4). [66]

The Cu K-edge XANES probes the local structure of a cluster of atoms of finite size around the absorbing Cu ion. In a recent XANES calculation made on $YBa_2Cu_3O_7$ using a multiple scattering approach, to get an an agreement with the experimental data a large clusters of about 35 atoms within a sphere of 5Å radius was necessary. [67] In agreement with the experimental data on the copper K-edge, multiple scattering calculation shows that changes in the XANES spectra induced by oxygen vacancies are the elongation of the Cu(2)-O(4) distance (between the Cu(2) in the CuO_2 plane and its apical oxygen) which induces the shift of the main peak in the polarized XANES spectra with **E**||**c** toward lower energy. The formation of the oxygen vacancies induces changes also in the **E**$\perp$**c** spectra for the Cu(1) ion in the linear chains. [66]

Starting from the oxygen rich phase $YBa_2Cu_3O_7$ the structural changes observed in the Cu coordination spheres by multiple scattering calculation are:
i) the Cu ion in site Cu(1) changes from a square plane configuration with 4 oxygens in the **bc** plane (Fig. 3) to a linear coordination (Fig. 4) in $YBa_2Cu_3O_6$;
ii) the Cu ion in site Cu(2) has a fivefold square pyramid configuration where the Cu(2)-apical oxygen O(4) distance changes from 2.26 Å to 2.47 Å in $YBa_2Cu_3O_6$.
The axial distances of Cu(1)-O(4) and Cu(2)-O(4), measured by neutron diffraction experiments [65], are shown in Table 1. Moreover the different Cu(2)-O(2) and Cu(2)-O(3) nearest neighbor distances (Fig. 3) become the same in tetragonal $YBa_2Cu_3O_6$ (Fig. 4).

Table 1
Axial distances from neutron diffraction data [65]

oxygen stoichiometry 7-x	0.0	0.2	0.5	1.0
Cu(1)-O(4) distance (Å)	1.84	1.84	1.79	1.79
Cu(2)-O(4) distance (Å)	2.30	2.32	2.43	2.47

The **E**$\perp$**c** polarized Cu K-edge XANES of single crystals of $YBa_2Cu_3O_{7-x}$ has been measured for two different oxygen concentration (x=0.15 and x=0.8) by Tolentino *et al.* [62] The main difference between the two spectra is the presence of two peaks at 1.6 and 8.6 eV (the zero in energy has been chosen at the first point of inflection of the absorption spectrum of metallic Cu, E_0=8976 eV). These peaks are associated to the formation of linear chains Cu(1)O(4) along the **c** axis, and to the change of formal valence of the Cu(1) from +2 to +1

(i.e. from the electronic configuration mainly $|3d^9\underline{L}\rangle$ to mainly $|3d^{10}\rangle$). Moreover only a red-shift in energy of about 1 eV has been found in unpolarized XANES [62] going from x=0.05 to x= 0.15 in agreement with findings of several groups. [59,61,63,68].
This calculation shows that changes of the XANES spectra going from $YBa_2Cu_3O_7$ to $YBa_2Cu_3O_6$ are due to structural changes in the Cu coordination shells based on changes in the local structure induced by depletion of oxygen.

<u>The **E||c** polarized XANES spectra.</u> The main changes observed [68] in the polarized **E||c** XANES spectrum going from $YBa_2Cu_3O_7$ to $YBa_2Cu_3O_6$ are: i) a red-shift of the energy of the absorption edge (features at 7.3 eV and at 14.6 eV) and ii) a lowering of the intensity of the main absorption peak (at 17.1 eV), both measured relative to the Cu metal K-edge for $YBa_2Cu_3O_7$.
The crystallographic data show that in the direction of the **c** axis, the main structural changes are: i) the contraction of the axial distance Cu(1)-O(4) and ii) the elongation of Cu(2)-O(4) distance, as shown in Fig. 3.

It is well established that the bond length variations induce an energy shift of the XANES peaks according with the rule $(E_c-E_o)d^2$ = constant (in which E_o is a reference energy and d is the interatomic distance). [69] This rule is expected to be valid for small bond length variation $\Delta d/d$. In order to test this rule, XANES have been calculated using as reference energy the intersphere constant V_o, which is the value of the potential in the interstitial volume between the muffin tin spheres, for both linear and square plane coordinated copper ions by varying the Cu-O distance in the range $1.85<d<3.5$ Å. Fig. 5. shows that these calculations give an important deviation from the linear trend for d >2.6 Å. Therefore a linear red-shift versus decreasing $1/d^2$ (increasing bond distance) in the range of Cu-O distances $1.8<d<2.5$ Å is expected.

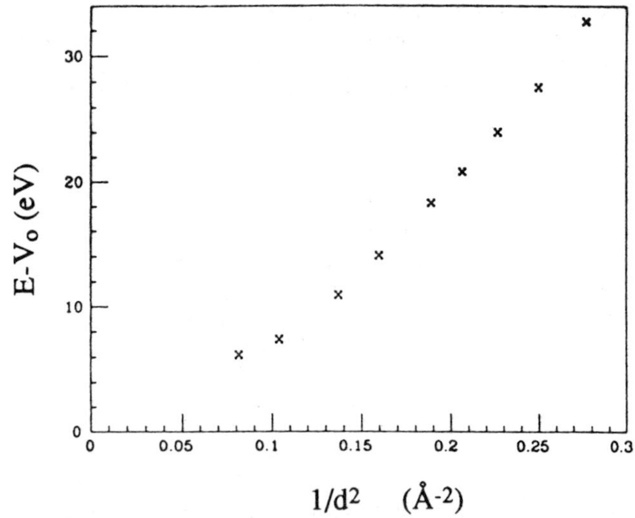

Fig. 5.
Energy of the strongest peak in Cu K edge XANES due to the multiple scattering resonance determined by the first shell of oxygens for a CuO_2 linear cluster as a function of $1/d^2$ where d is the Cu-O distance.

The calculated **E**||c XANES for the two Cu(1)O(4)$_2$O(1)$_2$ and Cu(2)O(2)$_2$O(3)$_2$O(4) clusters for $YBa_2Cu_3O_7$ are compared in Fig. 6 with Cu(1)O(4)$_2$ and Cu(2)O(2)$_2$O(3)$_2$O(4) for $YBa_2Cu_3O_6$. The variation observed in the spectrum of Cu(2)O(2)$_2$O(3)$_2$O(4) going from $YBa_2Cu_3O_7$ to $YBa_2Cu_3O_6$ in panel b gives an energy red shift induced by the elongation of the Cu(2)O(4) distance by 0.21 Å.

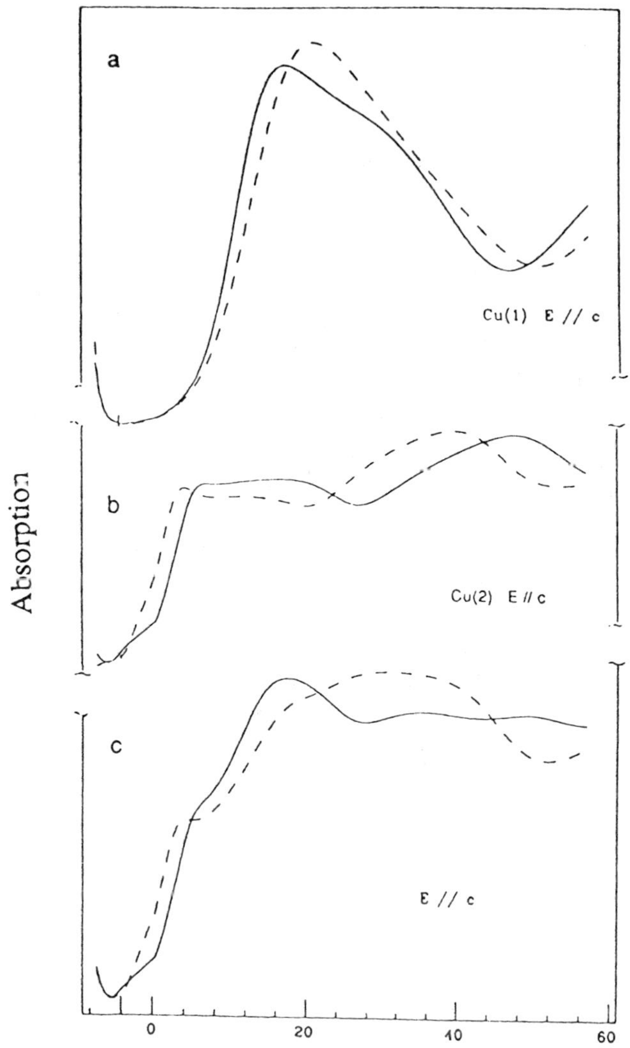

Fig. 6.

Calculated polarized Cu K-edge XANES with the electric field **E**||c axis for one oxygen shell clusters. Solid line: $YBa_2Cu_3O_7$, dashed line $YBa_2Cu_3O_6$. *Panel a)* Cu(1) site: Cu(1)O(4)$_2$O(1)$_2$ cluster of $YBa_2Cu_3O_7$ and Cu(1)O(4)$_2$ cluster of $YBa_2Cu_3O_6$, *panel b)* Cu(2) site: Cu(2)O(2)$_2$O(3)$_2$O(4) cluster for $YBa_2Cu_3O_7$ and for $YBa_2Cu_3O_6$, *panel c)* weighted sum of the two Cu sites in the unit cell.

The small contraction of the Cu(1)O(4) distance and the change of the overall shape of the **E||c** spectrum of Cu(1) site (panel a) due to the change from square plane to linear coordination, induces a blue shift.

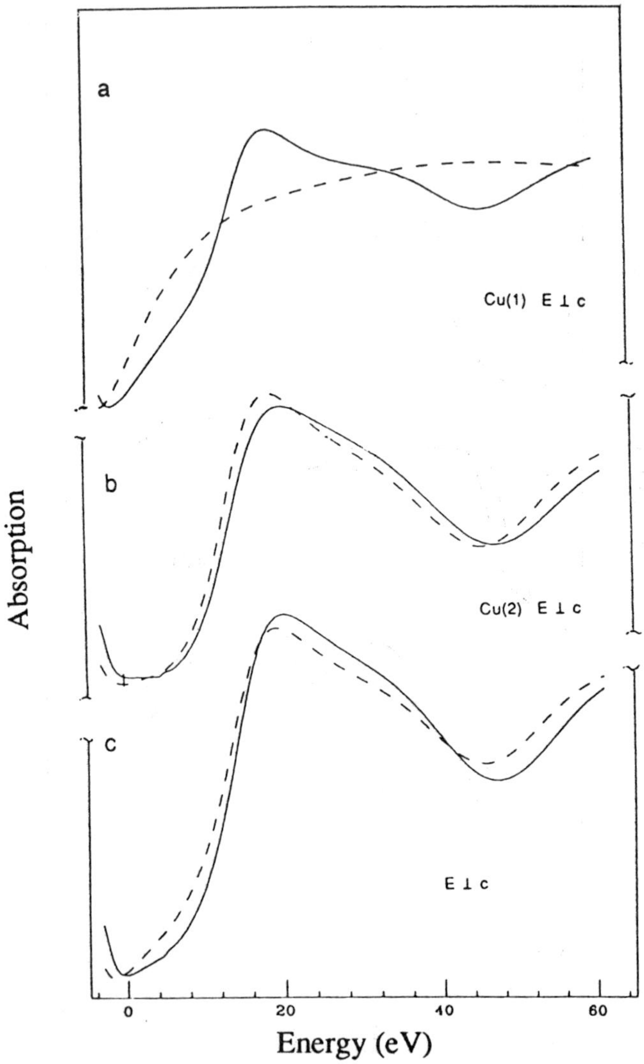

Fig. 7

Calculated polarized Cu K-edge XANES with the electric field **E⊥c** axis, for one oxygen shell cluster. Solid line $YBa_2Cu_3O_7$, dashed line $YBa_2Cu_3O_6$. From top to bottom as in Fig. 6, a) Cu(1) site, b) Cu(2) site, c) weighted sum of the two Cu sites in the unit cell.

The experimental polarized spectra are determined by the weighted sum 1:2 of the spectra of Cu(1) and Cu(2) site. The sum of the two spectra is shown in Fig. 6, panel c. The weighted sum of the two sites, is in good agreement with the experiment of Heald, *et al.*, [68] moreover the calculated spectra predict the lowering of intensity of the main absorption peak (about 9% from theory, versus 5% from experiment). The calculated energy interval between the shoulder on the rising edge and the main peak in calculation is about 12 eV, and it is consistent with experimental. The missing peak at 14.6 eV is assigned to a contribution of further shells.

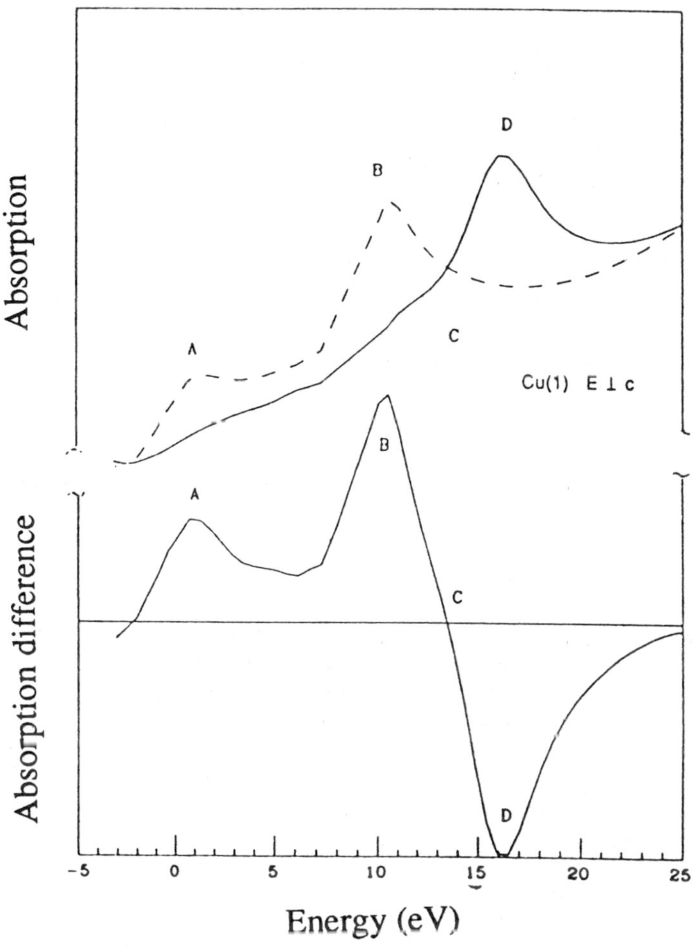

Fig. 8.
Upper panel: calculated polarized Cu K-edge XANES of Cu(1) site for two shell clusters (including Ba ions) with the electric field **E**⊥**c** axis. Solid line $YBa_2Cu_3O_7$, dashed line $YBa_2Cu_3O_6$. Lower panel: difference between the spectra of $YBa_2Cu_3O_6$ and $YBa_2Cu_3O_7$ in the upper part of the figure.

<u>The **E⊥c** polarized XANES spectra</u>. The experimental data of Tolentino, *et al.*, [62] show the formation of two peaks named A and B at 1.6 and 8.6 eV respectively and the lowering of the main peak D at 16.5 eV going from $YBa_2Cu_3O_7$ to $YBa_2Cu_3O_6$. The main structural changes in the directions parallel to the **ab** plane are the missing of O(1) ions in the Cu(1) site of $YBa_2Cu_3O_6$. Looking at both Fig. 3 and 4 the Cu(2) clusters of the two crystalline structures are very similar up to the 3th shell, therefore in the **E||ab** polarized Cu XANES spectra the most prominent differences are expected to arise only from the site Cu(1).

Fig. 7 shows the results of the XANES calculation of the same clusters as in Fig. 6 but for the **E⊥c** polarization. The spectra for the Cu(2) sites, on panel b, exhibits only small differences between the two compounds that can be assigned to the orthorhombic to tetragonal transition, and to the effect of changing Cu(2)O(4) distance. The XANES calculation for the Cu(1) cluster in panel a shows a large change determined by the fact that there are no atoms in the **ab** plane in the Cu(1) site of $YBa_2Cu_3O_6$, therefore the spectrum **E⊥c** of Cu(1) in $YBa_2Cu_3O_6$ is determined mainly by the atomic muffin tin potential.
The spectrum c in the bottom of Fig. 7 is obtained summing the absorption of the two sites, and shows only the main absorption peak similar to experimental finding by Tolentino, *et al* ., [62] without the features A and B.

In Fig. 8 the calculation of the polarized **E⊥c** XANES for the Cu(1) site for a larger cluster including the second shell of 8 Ba ions shows two new features A and B in the spectra of $YBa_2Cu_3O_6$, at energy ~ 1.5 and ~10 eV. The difference spectrum between the calculated spectra, shown in Fig. 6, reproduce well the experimental features named A, C and D in the difference spectrum of the experimental results [62] while a difference of about 1.5 eV for the B feature remains.

DETERMINATION OF THE STATES INDUCED BY DOPING AT THE Cu L_3 EDGE

The Case of Strontium Doped La_2CuO_4

In the many body description La_2CuO_4 is expected to be a mixture of $3d^9$ and $3d^{10}\underline{L}$ configuration, like CuO and other divalent copper compounds. [70] The doping by oxygen should induce new states $3d^8$ and $3d^9\underline{L}$, if the hole is created at the Cu or at the oxygen site respectively and $3d^{10}\underline{L}^2$ states if both holes are in the oxygen site. Moreover, a distortion of the structure can induce a variation of the mixing between the $3d^{10}\underline{L}$ and $3d^9$ configuration. Therefore the ground state of $La_{2-x}Sr_xCuO_4$ is expected to be a mixture of the $3d^9$ and $3d^{10}\underline{L}$ configurations for the local clusters $(CuO_4)^{6-}$ and of the $3d^9\underline{L}$, $3d^{10}\underline{L}^2$ and $3d^8$ configurations for the local clusters $(CuO_4)^{5-}$ induced in strontium doped sites. The weight of these configurations depends on their relative energy position and hybridization.

The analysis of both the Cu L_3-edge x-ray absorption and the photoemission spectra from the same core level of the divalent compound La_2CuO_4, allows the determination of the ground state parameters such as the energies to create a d hole (ε_d) and an oxygen hole (ε_L), as function of T, the hybridization energy in the ground state. This approach can be used also to determine the energy parameters of formal trivalent Cu compound such as $NaCuO_2$ which are described by the mixing of the two-hole configurations $3d^9\underline{L}$, $\underline{L}^2$ and $3d^8$ in the ground state, and to discuss the important role of the two holes Coulomb interaction U_{cL} between the core hole c and the ligand hole $\underline{L}$ in the final state and the interaction U_{dL} between the d hole and the ligand hole $\underline{L}$.

The ground state energies for the two-hole configurations of the strontium induced states in the superconducting compound La-Sr-Cu-O can be made with the same calculation. It shows that the two hole d-L and c-L Coulomb interaction play an important role both in the energy position of the initial and the final states and that the final states are determined by the $3d^8$ contribution in the initial state, also with an hybridization energy of T=2.2 eV, are too weak (~1%) to be detected experimentally. [71]

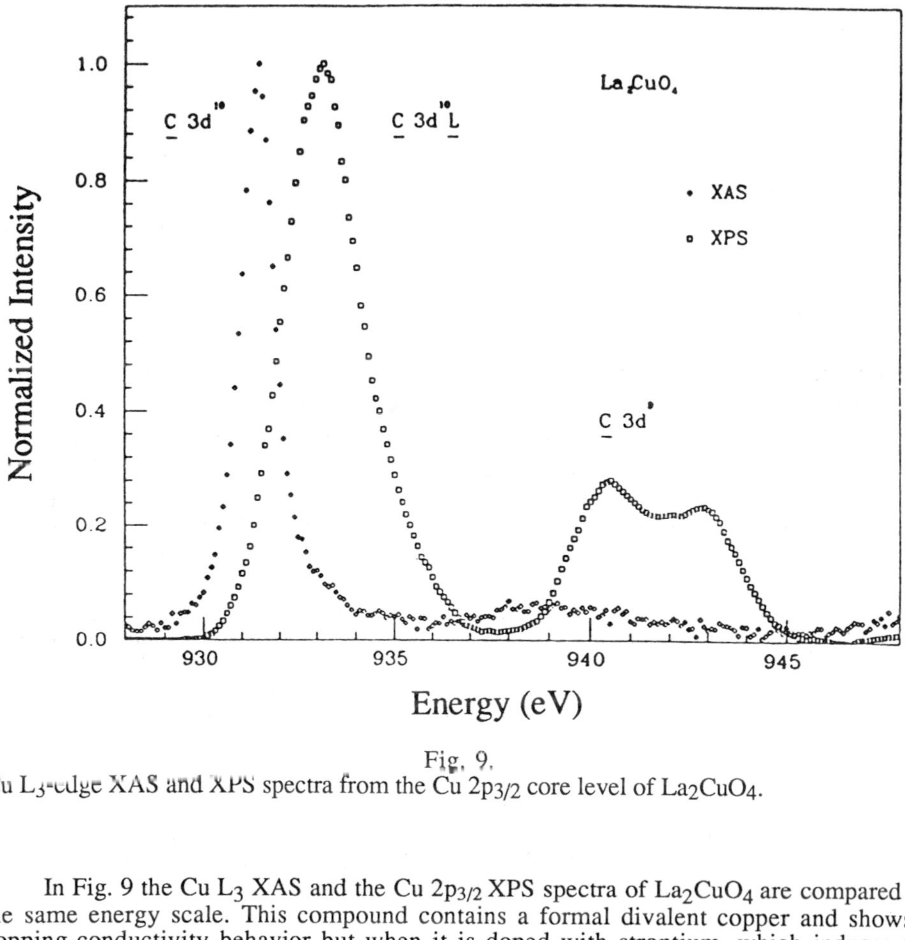

Fig. 9.
Cu L_3-edge XAS and XPS spectra from the Cu $2p_{3/2}$ core level of La_2CuO_4.

In Fig. 9 the Cu L_3 XAS and the Cu $2p_{3/2}$ XPS spectra of La_2CuO_4 are compared on the same energy scale. This compound contains a formal divalent copper and shows a hopping-conductivity behavior but when it is doped with strontium, which induces the increase of the Cu formal valence state, it shows the superconductivity behavior.
The one hole ground state wave function for the divalent copper site is a mixing of the two pure configurations: $|3d^9>$ with one hole on the copper site and $|3d^{10}\underline{L}>$ with one hole on the oxygen site

$$\psi_g=a_g\,|3d^9>+b_g\,|3d^{10}\underline{L}> \qquad (15)$$

The $|3d^{10}\underline{L}>$ is the first excited state of the ionic configuration obtained by a charge transfer excitation giving one hole in a 2p orbital. Thus the XAS white line is due to the $\underline{c}3d^{10}$ final state while the two XPS peaks are due to the hybridization of the two configurations $\underline{c}3d^9$ and $\underline{c}3d^{10}\underline{L}$.
In the frame of this simple orbital model it is possible give an estimation of the final state and ground state energies as a function of the parameters Δ, ε_L, ε_d, Q, U_{cL} and T, where $\Delta=\varepsilon_L-\varepsilon_d$ is the energy to transfer a hole from the copper site to the oxygen site ($\underline{d}$->$\underline{L}$) , ε_L and ε_d are the energies for creating one hole in the oxygen 2p orbital and in the Cu 3d orbital, Q is the $\underline{d}$-$\underline{c}$ hole interaction, T is the hybridization integral between the two configurations ($T=<3d^9|H|3d^{10}\underline{L}>$) and U_{cL} is the pair hole interaction between c and $\underline{L}$ orbital.

Starting from the energy separation between the two XPS peaks: E_2^{XPS} which is the energy of the $\underline{2p}3d^{10}\underline{L}$ main line and E_1^{XPS} which is the energy of the $\underline{2p}3d^9$ satellite line (experimental value $E_2^{XPS}-E_1^{XPS}= 8.5$ eV) and their intensity ratio ($I_2/I_1=0.43$) it is possible the determination of both $Q-U_{cL}$ and Δ as a function of the hybridization T.
ε_L is given by the expression of the energy separation between the XAS main line and XPS main line, which is strongly U_{cL} dependent :

$$E_1^{XPS}-E^{XAS}= \varepsilon_L+U_{cL}+[(Q-U_{cL})-\Delta]/2-\{[(Q-U_{cL})-\Delta]^2+4T^2\}^{1/2}/2 \qquad (16)$$

The introduction in the calculation of the parameter U_{cL}, which determines simply a renormalization of the Q term, reduces the calculated ε_L value and the measured energy separation ($E_1^{XPS}-E^{XAS}=1.8$ eV) becomes mainly a final state effect. The values of ε_L, ε_d and Q, calculated for several values of U_{cL} clearly show that the energy of the ground state, with respect to the energy of the pure copper $3d^{10}$ (Cu^+) oxygen $2p^6$ (O^{2-}) ionic configurations, decreases by increasing the c-L repulsion in the final state. A characteristic set of data obtained by this approach is $\varepsilon_L=1.3$ eV, $\varepsilon_d=0.3$ eV and Q=9.5 eV for $U_{cL}=1.3$ eV and T=2.2 eV).

Formally trivalent copper $NaCuO_2$ compound extends this approach to the case of two holes configurations in the ground state in the XAS final state and to the three holes in the XPS final state. The cluster appropriate for $NaCuO_2$ is a square plane CuO_4 where the diamagnetic ground state is considered to be a singlet pair of two holes given by the linear combination of the $3d^9\underline{L}$, $3d^{10}\underline{L}^2$ and $3d^8$ configurations. The final states are formed by the hybridization of $\underline{c}3d^9\underline{L}$, $\underline{c}3d^{10}\underline{L}^2$ and $\underline{c}3d^8$ for XPS and of $\underline{c}3d^9$, $\underline{c}3d^{10}\underline{L}$ for XAS.

A diagram of the energy levels probed by X-ray photoemission and absorption is shown in Fig. 10 for the trivalent case. The experimental XPS [72] and XAS [73] spectra are shown in Fig. 11, where the white line in the absorption spectrum is due mainly to the $\underline{c}3d^{10}\underline{L}$ and the two XPS peaks come principally from $\underline{c}3d^{10}\underline{L}^2$ and $\underline{c}3d^9\underline{L}$ final states.

The input to extract the energy values of ε_L, ε_d and Q as a function of the pair-hol interactions (U_{dd}, U_{cL}, U_{dL} and U_{LL}) and T, are again the experimental value of the energ separation of the photoemission peaks $E_2^{XPS}-E_1^{XPS}=9.2$ eV, their intensity ratio $I_2/I_1=0.24$, th energy separation between XAS and XPS: $E_1^{XPS}-E_4^{XAS}=1.8$ eV (see Fig. 10) and the effectiv parameters U_{dd}, U_{cL}, U_{dL} and U_{LL}. With $U_{dd}=8$ eV, chosen in agreement with the experimenta valence band photoemission spectra, as in the divalent case, the introduction of hole interaction involving ligand holes reduces in this case both the single-hole energy values and the ground stat energy. In Fig. 12 are reported the values Δ, ε_L, ε_d, Q as a function of T, obtained with $U_{cL}=U_{dL}=1.5$ eV and $U_{LL}=0.5$ eV. The energy parameters ε_L and ε_d are lowered respect to the calculation without pair hole interactions U_{cL}, U_{dL} and U_{LL}, giving the values (at T=2.2 eV) of $\varepsilon_L=0.3$ eV, $\varepsilon_d=0.2$ eV and $\Delta=0.1$ eV.

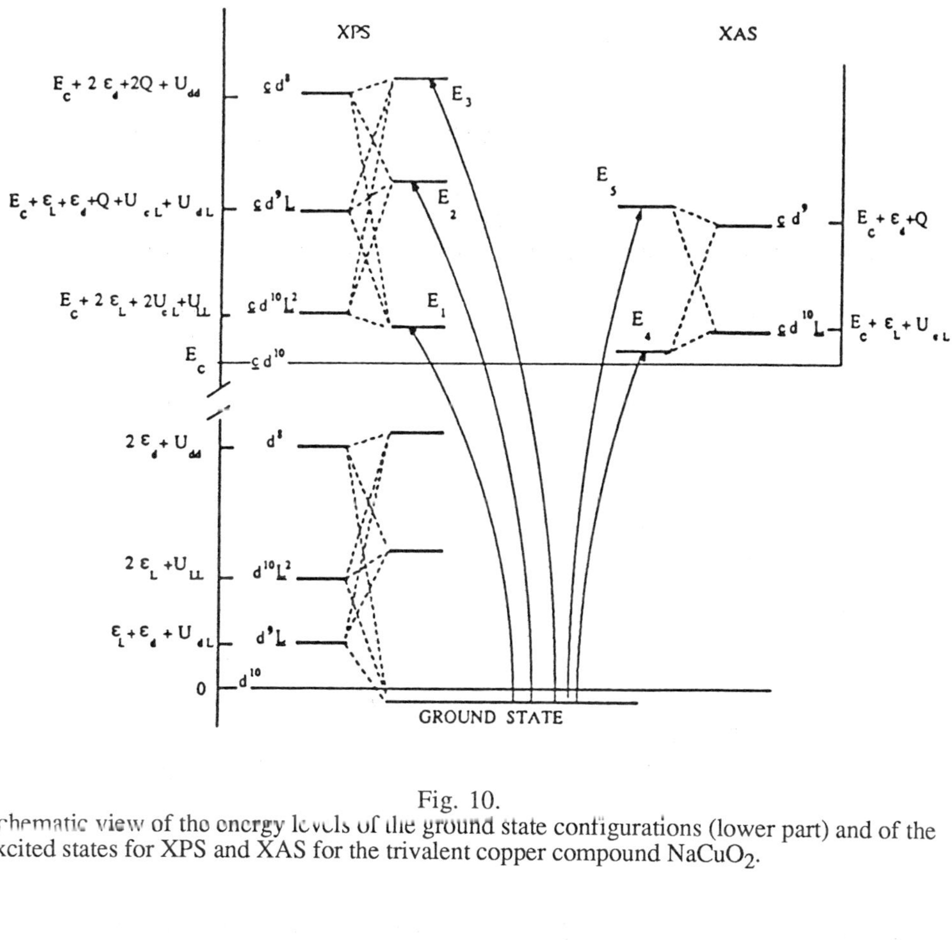

Fig. 10.
Schematic view of the energy levels of the ground state configurations (lower part) and of the excited states for XPS and XAS for the trivalent copper compound $NaCuO_2$.

The important result for the trivalent compound is that the cost to create one hole on the ligand orbital is reduced as well as the charge transfer energy Δ. Using these parameters is possible calculate the intensity for the transition at higher energy: about 2% for the XAS and less for the third XPS feature. The XAS and XPS transitions at higher energy named E_5^{XAS} and E_3^{XPS}, due mainly to the $3d^8$ contribution, in fact are not detected experimentally.

The energies of the ground state and of the two excited levels obtained from the hybridization of the ionic configurations $d^{10}\underline{L}^2$ and d^8 are plotted in Fig. 13 as function of the hybridization energy, while Fig. 14 shows the weight of the ionic configurations $3d^9\underline{L}$ (A),

$3d^{10}\underline{L}^2$ (B), $3d^8$ (C) in the ground state of $NaCuO_2$. The contribution of the $3d^9\underline{L}$ and $3d^{10}\underline{L}^2$ is nearly the same while the weight of the $3d^8$ is only about 4%.

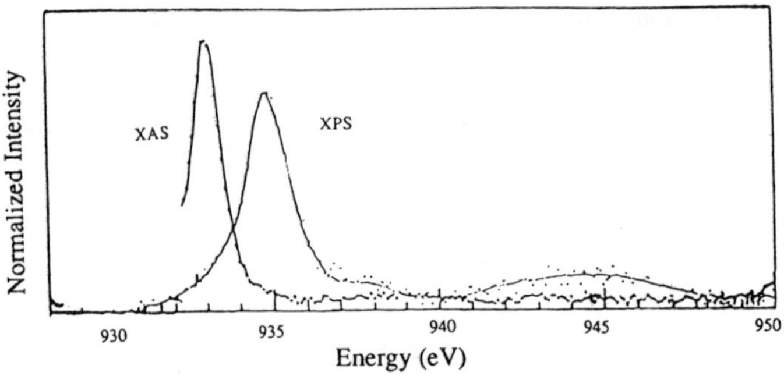

Fig. 11.
L3-edge x-ray absorption and x-ray photoemission spectra of $NaCuO_2$ compound.

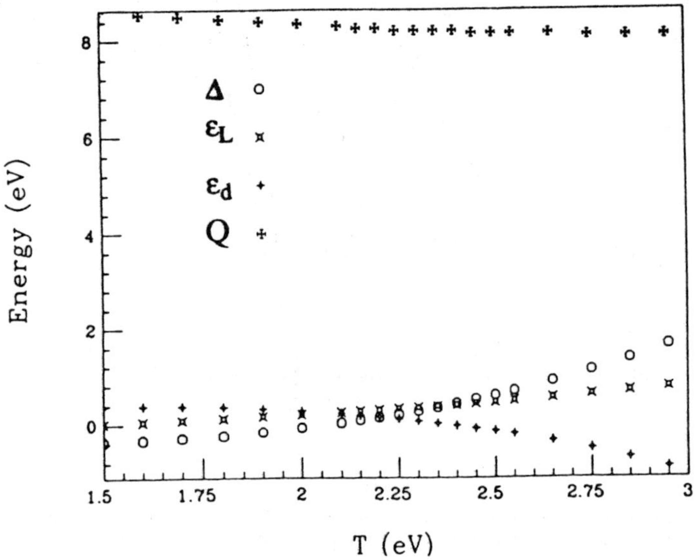

Fig. 12.
Variation of the energy parameters Q, Δ, ε_L, ε_d as function of the hybridization energy T in the case of formally trivalent $NaCuO_2$.

For doped La-Sr-Cu-O superconductor with formally mixed copper valence the ground state is a superposition of the nominally divalent Cu with one-hole states $3d^9$ and $3d^{10}\underline{L}$, and nominally trivalent Cu with two-hole states $3d^8$, $3d^9\underline{L}$, and $3d^{10}\underline{L}^2$ states. Although extra holes move around and are not localized in actual systems the ground state is constructed by hybridization of the one-hole and two-hole configurations. However, neglecting at first approximation the mixing between divalent and trivalent configuration, by subtracting the La_2CuO_4 spectra to the superconductor ones a simulation of the spectra of a "trivalent" compound could be obtained. In Fig.15 the differences between the L_3 core level XAS and $2p_{3/2}$ XPS spectra of the superconducting compound $La_{1.15}CuO_4$ and La_2CuO_4 are plotted. Similar experimental results are obtained for the differences of XPS spectra of Y-Ba-Cu-O system. [72]

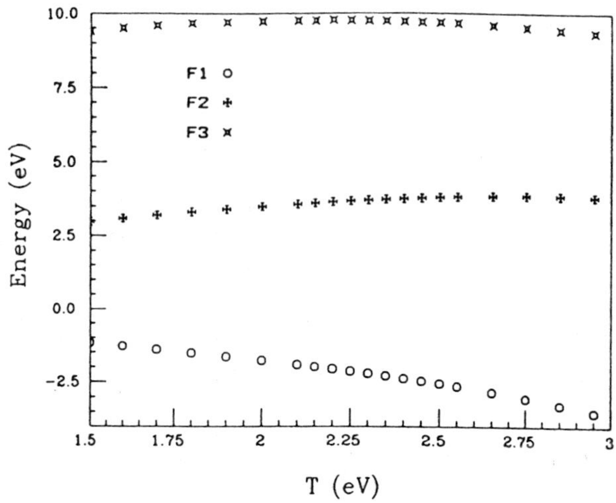

Fig. 13.
Energies of the ground state (F1) and of the two excited levels F2 and F3 obtained from the hybridization of the ionic configuration $d^{10}\underline{L}^2$ and d^8 as a function of the hybridization energy T for the formally trivalent compound $NaCuO_2$.

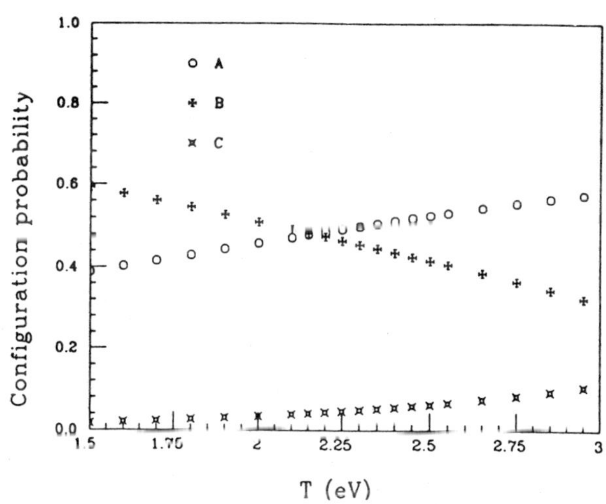

Fig. 14.
Weight of the ionic configurations $3d^9\underline{L}$(A), $3d^{10}\underline{L}^2$(B) and $3d^8$(C) in the ground state of $NaCuO_2$.

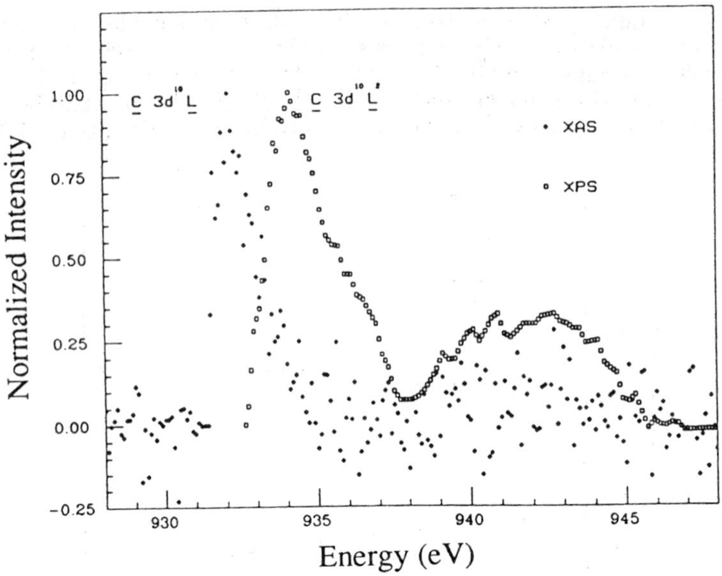

Fig. 15.
Difference between the L_3 XAS and Cu $2p_{3/2}$ XPS spectra of $La_{1.85}Sr_{0.15}CuO_4$ and La_2CuO_4.

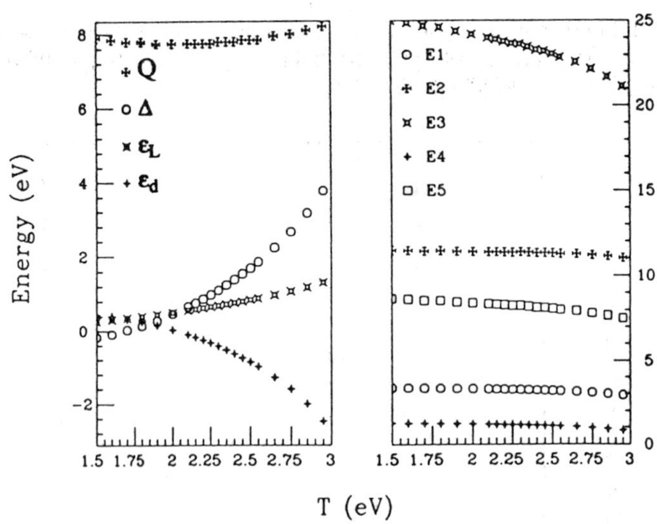

Fig. 16.
Left panel: variation of the energy parameters Q, Δ, ε_L and ε_d as a function of the hybridization energy T for the two hole states induced by doping. These curves were found by considering the difference of the XPS and XAS spectra shown in Fig. 15 between the superconducting compound $La_{1.85}Sr_{0.15}CuO_4$ and the divalent one La_2CuO_4. *Right panel*: XAS and XPS final state energies. The E3 and E5 transition not observed in the experiment are obtained using as imput the experimental E1, E2 and E4 values and the parameters in left side panel.

Similar analysis of $NaCuO_2$ can be carried out on these difference spectra using the experimental values of E_2^{XPS}-E_1^{XPS}=8.1 eV, I_2/I_1=0.3 and E_1^{XPS}-E_4^{XAS}=2.1 eV, which gives the curves reported in Fig.16. The ε_L, ε_d, Δ and Q curves are obtained for parameters values of $U_{cL}=U_{dL}$=1.3 eV and U_{LL}=0.8 eV (U_{dd}=8 eV) with T=2.2 eV obtaining now the values of ε_L=0.5 eV, ε_d=0 eV and Δ=0.5 eV.
The energy values of the three hybridized states for the two hole states are reported in Fig. 17. The configuration F1 is the ground state. The excited configuration F3 at the highest energy is mainly due to the $3d^8$ configuration. Its energy is calculated to be at about 9 eV above the energy of the pure copper $3d^{10}$ (Cu^+) oxygen $2p^6$ (O^{2-}) ionic configurations. The contribution of the $3d^8$ configuration to the ground state (Fig. 18) is of the order of 6%.

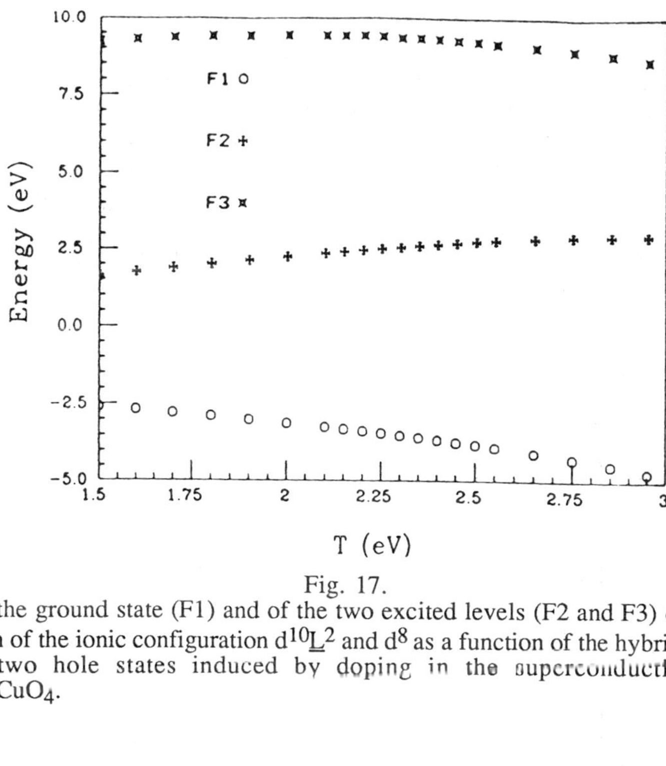

Fig. 17.
Energies of the ground state (F1) and of the two excited levels (F2 and F3) obtained by the hybridization of the ionic configuration $d^{10}\underline{L}^2$ and d^8 as a function of the hybridization energy T for the two hole states induced by doping in the superconducting compound $La_{1.85}Sr_{0.15}CuO_4$.

CORRELATION BETWEEN T_c AND THE ENERGY SPLITTING OF POLARIZED Cu 2p->3d TRANSITION

Sr doping in La_2CuO_4 system or oxygen doping in $YBa_2Cu_3O_{6.5+x}$ show the formation of $3d^9\underline{L}$ states at the copper L_3 X-ray absorption spectra. These states are due to many body configuration with itinerant holes in the oxygen valence band coupled with localized 3d electrons on the Cu sites which are close to the Fermi level in the superconducting phase. A key point in a deeper investigation of the electronic structure is the determination of the changes of unoccupied states induced by oxygen doping in the antiferromagnetic insulating background.

High resolution polarized Cu L_3 x-ray absorption spectra of single crystals of three classes of high temperature superconductors $La_{2-x}Sr_xCuO_4$ (x=0.04 and x=0.1) $Bi_2CaSr_2Cu_2O_8$ and $YBa_2Cu_3O_{6-\delta}$ have been measured in order to determine via the different direction of the polarization vector where extra holes exist in a high temperature oxide superconductors.

Polarized spectra of single crystals with several incidence angles between the electric field of the photon beam **E** and the sample surface normal **z** can be simply recorded, using synchrotron radiation linearly polarized in the horizontal plane. [74]
For the case of $Bi_2Sr_2Ca_1Cu_2O_8$ and $YBa_2Cu_3O_{7-\delta}$ the **z** axis is parallel to the **c** axis while for $La_{2-x}Sr_xCuO_{4-y}$ the **z** axis is parallel to the **b** axis. The full polarized XAS spectra (**E**$\perp$**z**) (**E** parallel to the **xy** plane of the CuO_2 layers) and (**E**||**z**) can be obtained by extrapolation of the spectra taken at four different incidence angles. The study of the polarization dependence of the "white line" in the Cu L_3-edge XAS due to $2p_{3/2}$->3d transition in the antiferromagnetic insulating La_2CuO_4 and superconducting crystals of $La_{2-x}Sr_xCuO_4$ (x=0.04 and x=0.1), $Bi_2CaSr_2Cu_2O_8$ and $YBa_2Cu_3O_{7-\delta}$ (δ=4.5 and δ=0.2) shows the presence of the electronic transition polarized in the **z** direction which probe the components of the orbitals of the d holes in the **z** direction. [75] The investigation of the energy splitting $\Delta\varepsilon$ between the in-plane and out-plane 2p->3d transition shows that the splitting for each class of superconductors is correlated with the critical temperature of the superconducting crystals.

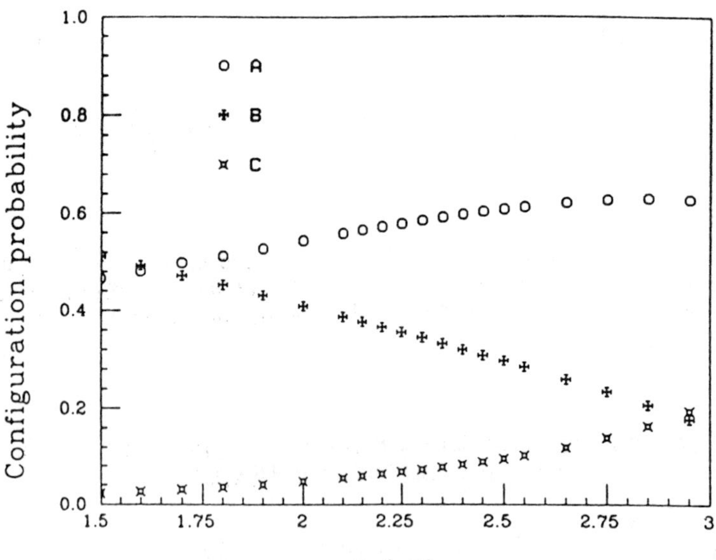

Fig. 18.
Weight of the ionic $3d^9\underline{L}$ (A), $3d^{10}\underline{L}^2$ (B) and $3d^8$ (C) configurations in the ground state of superconducting compound $La_{1.85}Sr_{0.15}CuO_4$.

While it is well established that the holes induced by doping are mainly in the oxygen derived valence band i.e. ligand holes giving the $3d^9\underline{L}$ configuration, [28,30,71,76-78] the determination of the symmetry of the two holes states found by photoinduced IR [79] and by many other methods is still an open problem.

As is well known, the 5 Cu 3d orbitals are splitted by the crystalline field with D_{4h} symmetry into four energy levels: the b_1 (x^2-y^2) orbital which corresponds to the highest level, the a_1 $(3z^2-r^2)$ orbital, the b_2 (xy) orbital and e (d_{xz},d_{yz}) orbitals. The ligand hole $\underline{L}$ wave functions consist of linear combinations of oxygen 2p orbitals with the above symmetries.
The electronic configuration of the Cu ions in the antiferromagnetic insulating compounds is expected to be the Cu $3d^9$ configuration, where the d hole is in the Cu $3d_{x^2-y^2}$ orbital of b_1 molecular symmetry in the CuO_4 square plane cluster. This is expected for a perfect square plane but it is well known that a simple system of regular CuO_4 square plane does not exhibit high temperature superconductivity above 20 K.
The tilting of the elongated octahedron of the $La_{2-x}Sr_xCuO_4$ and the distortions of the CuO_5 square pyramids of $Bi_2CaSr_2Cu_2O_8$ and $YBa_2Cu_3O_{7-\delta}$ induces a mixture of $3d_{3z^2-r^2}$ symmetry (a_1 molecular symmetry) or of the $3d_{xz}$ (e molecular symmetry) with the $3d_{3z^2-r^2}$ in the ground state. [80,81]
There are several experimental indications that the holes are not present only in the upper valence band σ bond formed by the $3d_{x^2-y^2}$ and the $O(2p_x)$ or $O(2p_y)$ in the CuO_2 layers. The holes induced by doping in $YBa_2Cu_3O_{7-\delta}$ (T_c=80K system) are present in the apical oxygen O(4) of Cu(2) which is also in the linear Cu(1)O chains.[74,78].
Evidence for d-d excitations in $YBa_2Cu_3O_7$ has been also reported, [82] and the study of the polarization dependence of the "white line" in the Cu L_3-edge XAS of $Bi_2CaSr_2Cu_2O_8$ has also indicated the presence of 10%-20% of Cu 3d holes oriented in the **z** direction. [75,83]

Several theoretical models for high T_c superconductivity such as the three band model [84,85] the spin polaron pairing mechanism [86] the d-d excitation model, [87-90] the Kanamori model [91] and the interlayer pairing [92-94] require the presence of electronic states with two holes in two different orbitals.

Fig. 19 shows the white line due to Cu 2p->3d transitions of the polarized, **E**$\parallel$**z** and **E**$\perp$**z**, Cu L_3 XAS of the insulating La_2CuO_{4-y} and of superconducting $La_{1.96}Sr_{0.04}CuO_4$ and $La_{1.9}Sr_{0.1}CuO_4$. The curves in Fig. 19 are normalized to the white line maximum to show the energy shift. The white line maximum for the **E**$\parallel$**z** polarization is always at lower energy than for the **E**$\perp$**z** polarization. The energy separation $\Delta\varepsilon$ between the maxima of the two polarized white lines can be obtained by curve fitting avoiding the artifact of the variation of the lineshape with the incidence angle. The energy separation $\Delta\varepsilon$ decreases from 230 ± 50 meV to 80 ± 50 meV going from La_2CuO_{4-y} to $La_{1.9}Sr_{0.1}CuO_4$. The intensity of the **E**$\parallel$**z** white line is about 25% of the intensity of the **E**$\perp$**z** white line with an error in the measure of the relative intensity of about 15-20%. The presence of a small component of the white line for **E**$\parallel$**z** polarization indicates that there are Cu d holes having a component of their orbital in the **z** direction d_z (d_z is introduced here to indicate one of the three possible 3d orbital: $3d_{3z^2-r^2}$, d_{xz} or d_{yz} that this experiment cannot separate).

Fig. 20. reports the polarized Cu L_3-edge XAS of $YBa_2Cu_3O_{6.5}$ (T_c=60K), b) $YBa_2Cu_3O_{6.8}$ (T_c=82.5K) in a larger energy range including the white line and the shoulder at about 1.5 eV above it. The energy separation $\Delta \varepsilon$ between the white lines in the two polarizations decreases from 200 meV to 120 meV going from T_c=60K to T_c=82.5K sample.

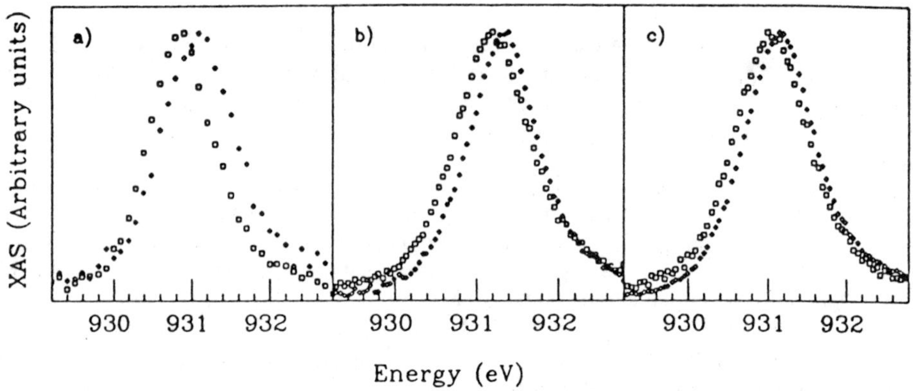

Fig. 19
Polarized L_3-edge XAS spectra measured with electrical field parallel (**E**||**z**) (□) and perpendicular (**E**⊥**z**) (◇) to the sample **z** axis, for: a) La_2CuO_{4-y} , b) $La_{1.96}Sr_{0.04}CuO_4$, and c) $La_{1.9}Sr_{0.1}CuO_4$

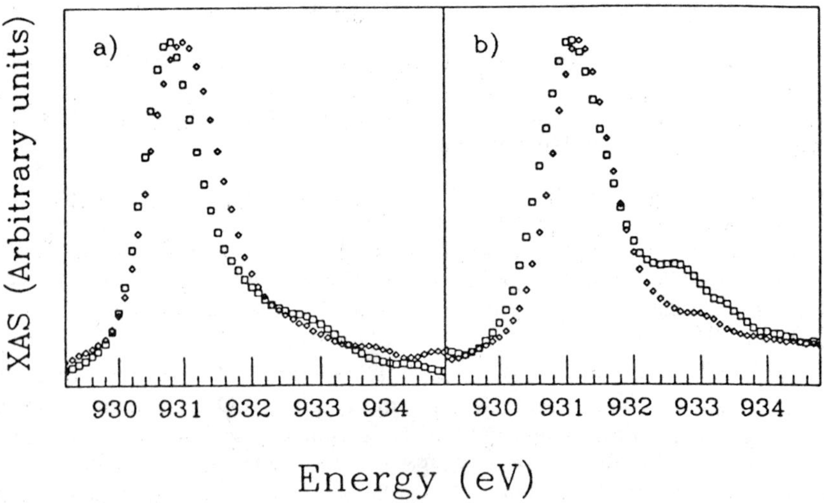

Fig. 20
Polarized L_3-edge XAS spectra measured with electrical field parallel (**E**||**c**) (□) and perpendicular (**E**⊥**c**) (◇) to the sample **c** axis, for: a) $YBa_2Cu_3O_{6.5}$ (T_c=60K), b) $YBa_2Cu_3O_{6.8}$ (T_c=82.5K).

In Fig. 21 the white lines of polarized Cu L_3-edge XAS of a highly oriented film $Bi_2Sr_2Ca_1Cu_2O_y$, (T_c=61K) with incidence angle ϕ=10° and 75° are shown. [83] The spectra of a CuO polycristalline sample of the same size have been recorded in the same run in order to keep under control the energy calibration and any other possible effects on the data collection. The CuO spectra do not change by changing the angle of incidence while the white line of $Bi_2Sr_2Ca_1Cu_2O_y$ film shows a large anisotropy. This data show that the white line has two components separated by an energy splitting $\Delta\varepsilon$ of about 400 eV. The peak B at higher energy observed for the polarization direction **E**⊥**c** is interpreted as an evidence for the presence of d holes with $3d_{3z^2-r^2}$ simmetry.

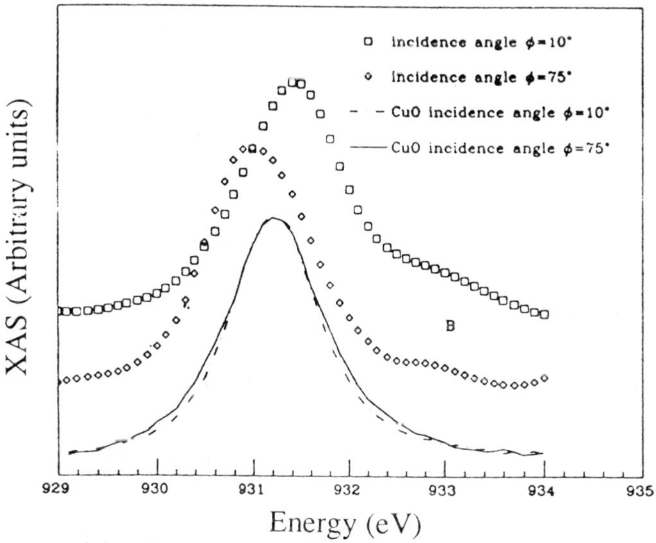

Fig. 21.
Polarized Cu L_3-edge XAS spectra for $Bi_2Sr_2Ca_1Cu_2O_y$ oriented film (T_c=61K) for two extreme incidence angles ϕ=10°, i.e. with the electric field nearly parallel to the CuO_2 planes (**E**⊥**c**)(□), and for ϕ=75°, with the electric field nearly normal to the CuO_2 planes (**E**||**c**)(◇).

Fig. 22 reports the energy difference $\Delta\varepsilon$ between the maximum of the in-plane (the **xy** plane of CuO_2 layers) **E**⊥**z** polarized white line and the out-of-plane **E**||**z** polarized line for the three classes of compounds as function of the critical temperature of each sample.
In the series of $La_{2-x}Sr_xCuO_4$ the values of $\Delta\varepsilon$ reaches its maximum for the insulating x=0 system. These results show that the doping induces an important change in the d unoccupied states by reducing the energy separation between the unoccupied states with components along **z** and that with components along the **xy** plane.

Bianconi *et al.* [30] in order to discuss the problem of the symmetry of the $3d^9\underline{L}$ states in the doped $La_{2-x}Sr_xCuO_4$, have shown that the $3d^9\underline{L}$ states induced by doping gives an high energy tail of the white line of unpolarized La_2CuO_4. The difference between the **E**⊥**c** spectra of $La_{2-x}Sr_xCuO_4$ for x=0.04 minus x=0.10, recorded under the same experimental conditions and with the same instrumental broadening, is reported in Fig. 23. While no detectable variation are found in the difference between the **E**||**c** spectra, the spectrum of the difference is similar to that of the unpolarized spectra, [30] therefore the states induced by doping in the low T_C regime (T_C=12K) are mainly $\underline{d}_x\underline{L}$. From the results of De Sanctis *et al.* [71] the observed shoulder can be associated with the formation of singlet states $\underline{d}(b_1)\underline{L}(b_1)$ with U_{dL}=1.3 eV, therefore both $La_{2-x}Sr_xCu_2O_4$ (x=0.1) and $Bi_2Sr_2CaCu_2O_8$ [83] crystals show the presence of a resolved shoulder $\underline{2p}3d^{10}\underline{L}$ in the range 1-2 eV above the white line for the $E \perp z$ polarization indicating the presence of $\underline{d}(b_1)\underline{L}(b_1)$ states.

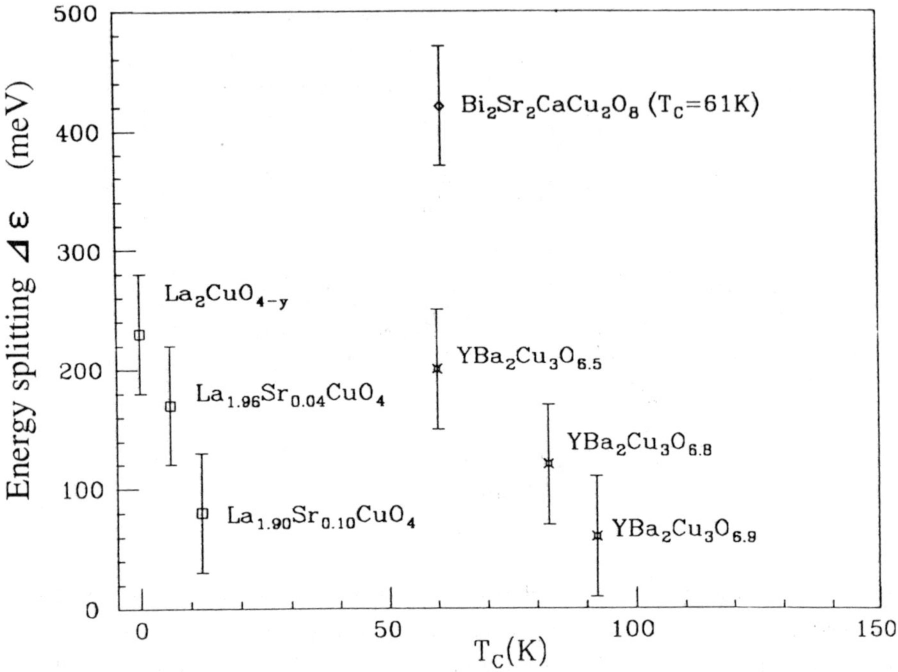

Fig. 22.
Energy splitting between the peaks of polarized white lines in the L_3-edge XAS (**E**⊥**c** peak minus **E**||**c** peak) as function of critical temperature (T_c) for the following classes of superconductors: $La_{2-x}Sr_xCu_2O_4$ (□), $YBa_2Cu_3O_{7-\delta}$ (+), and $Bi_2Sr_2CaCu_2O_8$ (◇).

The triplet states $\underline{d}(a_1)\underline{L}(b_1)$ are not detected by the Cu L_3 XAS spectroscopy because the split-off of the shoulder from the energy region of the white line is determined by the Coulomb repulsion $U_{cL} \sim U_{dL}$ which for the $\underline{d}(a_1)\underline{L}(b_1)$ is expected to be much smaller than 1.3 eV for $\underline{d}(b_1)\underline{L}(b_1)$. [87-90] Because of the small value of U_{cL}, the final states determined by $\underline{d}(a_1)\underline{L}(b_1)$ configuration are expected to be unresolved from the **E**||**z** white line. The presence of the white line component **E**||**z** indicates the possible presence of the $\underline{d}(a_1)\underline{L}(b_1)$ or

$\underline{d}(e)\underline{L}(b_1)$ configurations with $U_{cL} \sim U_{dL}$ smaller than 1 eV. In fact the $\underline{L}(b_1)$ was detected by oxygen K-edge absorption of $Bi_2Sr_2CaCu_2O_8$ by Fink *et al.* [77] and the holes $\underline{d}_z$ with Cu 3d orbital components in the **z** direction ($\underline{d}_z$ indicates one of the 3d orbital: $3\underline{d}_{3z^2-r^2}$ (a_1) or $\underline{d}_{xz}$, $\underline{d}_{yz}$ (e)) are detected by the polarization dependence of the L_3 XAS white line.

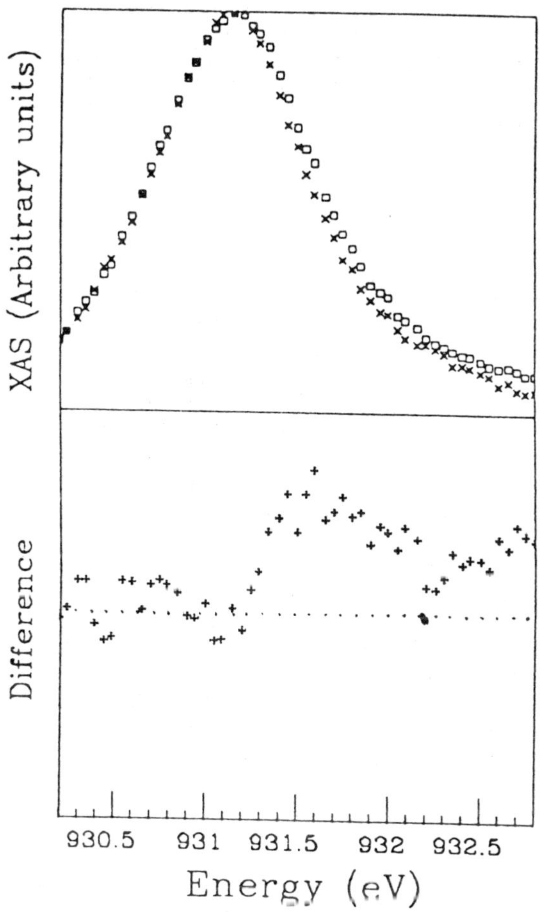

Fig. 23.
Comparison between the white lines of L_3-edge XAS spectra measured with an angle of 80° between the photon electric field **E** and the **z** axis of $La_{2-x}Sr_xCu_2O_4$ x=0.04 (+) and x=0.1 (□). In the lower panel the difference between the two spectra is plotted to show the spectrum of strontium induced transitions with a peak at about 931.7 eV. The dotted line indicates the zero of the difference.

The presence of the holes out of the CuO_2 plane is found in the high T_c range of the $YBa_2Cu_3O_{7-\delta}$ system. Moreover the shoulder at 1.5 eV above the white line in Fig. 20 is much weaker in the spectrum of the T_c=60K sample than in the spectrum of the T_c=82.5K sample. [74] The $\mathbf{E}\perp\mathbf{z}$ spectrum of the T_c=60K sample shows an asymmetric high energy tail of the white line indicating the presence of a $3d^9\underline{L}$ configuration with $\underline{d}$ unoccupied orbital $\underline{d}_x\underline{L}$ in the CuO_2 plane. Most of the weight of the 1.5 eV shoulder observed in the $\mathbf{E}\parallel\mathbf{z}$ spectra of the T_c~80K is formed with the transition from the low T_c range (50-60K) to the high T_c range (80-90K). Taking into account that the presence of the hole in the apical O(4) is detected by the analysis of the x-ray absorption at the Cu(1) ions, these results are in agreement with the theoretical prediction of Fujimori, [95] which shows that the ligand hole is stabilized in the apical oxygen by decreasing the Cu(2)O(4) distance. Therefore for 50-60K superconductivity both $\underline{d}_x\underline{L}$ and $\underline{d}_z\underline{L}$ configuration have comparable probability while in the 80-90K superconductivity the holes are found to be mostly out of the CuO_2 plane.

SUMMARY

Joint analysis of XPS and XAS spectra of the same core level is a simple experimental approach to obtain the energies of the two hole configurations in a formally trivalent compound and for the states induced by doping in superconductor systems like $La_{1.85}Sr_{0.15}CuO_4$. Without neglect the limitations of this simple approach which does not consider the multiplet splitting, the bandwidth, and takes in consideration only one type of orbital symmetry for the d and L holes it shows that in the copper-oxygen compounds the pair-hole interaction in the final state U_{cL}, strongly affects the energy difference between the x-ray absorption (XAS) and the x-ray photoemission (XPS) main lines, and the pair-hole interactions U_{dL} and U_{LL} both in the ground state and in the final state are important to calculate the energies of the two holes ground state.

There is evidence from polarized L_3 XAS spectroscopy of a correlation between the critical temperature of a superconductors and the energy splitting of the $\mathbf{E}\parallel\mathbf{z}$ and $\mathbf{E}\perp\mathbf{z}$ polarized "white lines" in agreement with the predictions of the d-d excitation model. [87-90]. The expected formation of singlet states $\underline{d}(b_1)\underline{L}(b_1)$ states have been found but $\underline{d}_z$ holes are also present. In particular in the $YBa_2Cu_3O_7$ system for the higher critical temperature there is evidence of oxygen induced ligand hole out of the CuO_2 planes. The intensity of the $\underline{d}(b_1)\underline{L}(b_1)$ is expected to decrease with the decrease of the ratio between $d_{x^2-y^2}$ and d_z states over the U_{cL} interaction. On the contrary the configurations $\underline{d}(a_1)\underline{L}(b_1)$ [87-90] or $\underline{d}(b_1)\underline{L}(a_1)$, where the ligand holes are out of the CuO_2 plane [92,93] become more important with increasing T_c.

Acknowledgements - I gratefully acknowledge Prof. A. Bianconi who involved me in the study of high-temperature superconductivity and for his encouraging support in all these years. A special thank to A. Kotani for many stimulating discussion and to A.M. Flank, A. Fontaine and P. Lagarde for their invaluable support. I have also benefited from the close collaboration with many colleagues, especially P. Castrucci, A. Clozza, M. De Santis, S. Della Longa, M. De Simone, A. Di Cicco, A. Fabrizi, A. Gargano and M. Pompa.

REFERENCES

[1] C. Michel and B. Raveau, J. Solid State Chem. 43, (1982) 73.
[2] C. Michel and B. Raveau, Rev. Cim. Mineral., 21, (1984) 407.
[3] C. Michel, L. Er-Rakho and B. Raveau, Mater. Res. Bull. 20, (1985) 667.
[4] J.C. Bednorz and K.A. Müller, Z. Phys. B - Condensed Matter 64, (1986) 189.

[5] J.C. Bednorz, M. Takashige and K.A. Müller, Europhys. Lett. 3, (1987) 379.
[6] K. Okada and A. Kotani, J. Phys. Soc. Japan 58, (1989) 1095.
[7] A. Fujimori, E. Takayama-Muromachi, Y. Uchida and B. Okai, Phys. Rev. B 35, (1987) 8814.
[8] A. Fujimori, E. Takayama-Muromachi and Y. Uchida, Solid State Commun. 63, (1987) 857.
[9] A. Bianconi, A. Congiu Castellano, M. De Santis, P. Delogu, A. Gargano and R. Giorgi, Solid State Commun. 63, (1987) 1135.
[10] Z. Shen, J.W. Allen, J.J. Yeh, J.-S. Kang, W. Ellis, W. Spicer, I. Lindau, M.B. Maple, Y.D. Dalichaouch, M.S. Torikachvili, J.S. Sun and F. Geballe, Phys. Rev. B 36, (1987) 8414.
[11] P. Steiner, S. Hufner, V. Kinsinger, I. Sander, B. Siegwart, H. Schmitt, R. Schulz, S. Junk, G. Schwitzgebel, A. Gold, C. Politis, H.P. Muller, R. Hoppe, S. Kemmeler-Sack and C. Kunz, Z Phys. B - Condensed Matter 69, (1988) 449.
[12] G. Gourieux, G. Krill, M. Maurer, M.F. Ravet, A. Menny, H. Tolentino and A. Fontaine, Phys. Rev. B 37, (1988) 7516.
[13] R. Kurtz, R.L. Stochbauer, D. Mueller, A. Shih, L.E Toth, M. Osofsky and S.A. Wolf, Phys. Rev. B 35, (1987) 8818.
[14] M. Onellion, Y. Chang, D.W. Niles, R. Joynt, G. Margaritondo, N.G. Stoffel and J.M. Tarascon, Phys. Rev. B 36, (1987) 819.
[15] J.A. Yarmoff, D.R. Clark, W. Drube, U.O. Karlsson, A. Taleb-Ibrahimi and F.J. Himpsel, Phys. Rev. B 36, (1987) 3967.
[16] N.G. Stoffel, J.M. Tarascon,Y. Chang, M. Onellion, D.W. Niles and G. Margaritondo, Phys. Rev. B 36, (1987) 3986.
[17] T. Takahashi, F. Maeda, H. Arai, H. Katayama-Yoshida, Y. Okabe, T. Suzuki, S. Hosoya, A. Fujimori, T. Shidara, T. Koide, T. Miyahara, M. Onoda, S. Shamoto and M. Sato, Phys. Rev. B 36, (1987) 5686.
[18] J. Zaanen, G.A. Sawatzky and J.W. Allen, Phys. Rev. Lett. 55, (1985) 418.
[19] P.Steiner, J. Albers, V. Kinsinger, I. Sander, B. Siegwart, S. Hufner and C. Politis, Z. Physik B - Condensed Matter 66, (1987) 275.
[20] N. Nucker, J. Fink, B. Renker, D. Ewert, C. Politis, P.J.W. Weijs and J.C. Fuggle, Z. Phys. B - Condensed Matter 67, (1987) 9.
[21] T. Takahashi, F. Maeda, H. Katayama-Yoshida, Y. Okabe, T. Suzuki, A. Fujimori, S. Hosoya, S. Shamoto and M. Sato, Phys. Rev. B 37, (1988) 9788.
[22] J.M. Tranquada, S.M. Heald, A. Moodenbaugh and M. Suenaga, Phys. Rev. B 35, (1987) 7187.
[23] E.E. Alp, G.K. Shenoy, D.G. Hinks, D.W. Capone II, L. Soderholm, H.B. Schuttler, J. Guo, D.E. Ellis, P.A. Montano and M. Ramanathan, Phys. Rev. B 35, (1987) 7199.
[24] Y.B. Boyce, F. Bridges, T. Claeson, T.H. Geballe, C.W. Chu and J.M. Tarascon, Phys. Rev. B 35, (1987) 7203
[25] H. Oyanagi, H. Ihara, T. Matsubara, T. Matsushita, M. Hirabayashi, M.Tokumoto, K. Murata, N. Terada, K. Senzaki, T. Yao, H. Iwasaki and Y. Kimura, Jpn. Journ. Appl. Phys. 26, (1987) L1233 and ibidem 26, (1987) L638.
[26] A. Bianconi, A. Congiu Castellano, M. De Santis, C. Politis, A. Marcelli, S. Mobilio and A. Savoia, Z. Physik B - Condensed Matter 67, (1987) 307.
[27] F. Baudelet, G. Collin, E. Dartyge, A. Fontaine, J.P. Kappler, G. Krill, J.P. Itié, J. Jegoudez, M. Maurer, Ph. Monod, A. Revcolevschi, H. Tolentino, G. Tourillon and M. Verdaguer, Z. Physik B - Condensed Matter 69, (1987) 141.
[28] A. Bianconi, A. Congiu Castellano, M. De Santis, P. Rudolf, P. Lagarde, A.M. Flank and A. Marcelli, Solid State Commun. 63, (1987) 1009.
[29] A. Bianconi, A. Clozza, A. Congiu Castellano, S. Della Longa, M. De Santis, A. Di Cicco, K. Garg, P. Delogu, A. Gargano, R. Giorgi, P. Lagarde, A. M. Flank and A. Marcelli, Int. J. Modern Phys. B 1, (1987) 853.
[30] A. Bianconi, J. Budnick, A.M. Flank, A. Fontaine, P. Lagarde, A. Marcelli,

H. Tolentino, B. Chamberland, C. Michel, B. Raveau and G. Demazeau, Phys. Lett. A 127, (1988) 285.
[31] A. Bianconi, M. De Santis, A. Clozza, A. Congiu Castellano, A.M. Flank, A. Fontaine, P. Lagarde, J. Budnick and A. Gargano, Core Level Spectroscopy in Condensed Systems, (Eds.J. Kanamori and A. Kotani), Springer Verlag, Berlin, (1988), pp. 146.
[32] U. Fano and J.W. Cooper, Rev. Mod. Phys. 40, (1968) 441.
[33] J.E. Müller, O. Jepsen and J.W. Wilkins, Solid State Commun. 42, (1982) 365.
[34] J.E. Müller and J.W. Wilkins, Phys. Rev. B 29, (1984) 4331.
[35] J. Fink, Th. Müller-Heinzerling, B. Scheerer, W. Speier, F.U. Hillebrecht, J.C. Fuggle, J. Zaanen and G.A. Sawatzki, Phys. Rev. B 32, (1985) 4899.
[36] J. Zaanen, G.A. Sawatzki, J. Fink, W. Speier and J.C. Fuggle, Phys. Rev. B 32, (1985) 4905.
[37] G.A. Sawatzki, Core Level Spectroscopy in Condensed Systems, (Eds.J. Kanamori and A. Kotani), Springer Verlag, Berlin, (1988), pp.99 and reference therein.
[38] U. von Barth and G. Grossmann, Solid State Commun. 32, (1979) 645.
[39] U. von Barth and G. Grossmann, Phys. Rev. B 25, (1982) 5150.
[40] A. Kotani, T. Jo and J. C. Parlebas, Adv. Phys. 37, (1988) 37.
[41] A. Bianconi, A. Kotani, K. Okada, R. Giorgi, A. Gargano, A. Marcelli and T. Miyahara, Phys. Rev. B 38, (1988) 3433.
[42] A. Bianconi, T. Miyahara, A. Kotani, Y. Kitajima, T. Yokoyama, H. Kuroda, M. Funabashi, H. Arai and T. Ohta, Phys. Rev. B 39, (1989) 3380.
[43] M. Piacentini, EXAFS and Near Edge Structure, (Eds. A. Bianconi, L. Incoccia and S. Stipcich), Springer Verlag, Berlin, (1983), pp.193.
[44] D.H. Templeton and L.K. Templeton, Acta Crystallogr. A38, (1982) 62.
[45] J.E. Penner-Hahn, R.A. Scott, K.O. Hodgson, S. Doniach, S.R. Desjardins and E.I. Solomon, Chem. Phys. Lett. 88, (1982) 595.
[46] T.A. Smith, J.E. Penner-Hahn, M.A. Berding, S. Doniach and K.O. Hodgson, J. Am. Chem. Soc. 107, (1985) 5945.
[47] N. Kosugi, T. Yokoyama, K. Asakura and H. Kuroda, Chem. Phys. 91, (1984) 249.
[48] F.W. Kutzler, R.A. Scott, J.M. Berg, K.O. Hodgson, S. Doniach, S.P. Cramer and C.H. Chang, J. Am. Chem. Soc. 103, (1981) 6083.
[49] A. Bianconi, A. Congiu Castellano, P.J. Durham, S.S. Hasnain and S. Phillips, Nature 318, (1985) 685.
[50] R.A. Scott, J. Penner-Hahn, S. Doniach, H.C. Freeman and K.O. Hodgson, J.Am. Chem. Soc. 104, (1982) 5364.
[51] S. Stizza, M. Benfatto, I. Davoli, G. Mancini, A. Marcelli, A. Bianconi, M. Tomellini and J. Garcia, J. Physique 46, C8, (1985) 255
[52] S. Stizza, M. Benfatto, J. Garcia and A. Bianconi, J. Physique 47, C8, (1986) 691.
[53] J.E. Penner-Hahn, M. Benfatto, B. Hedman, T. Takahashi, S. Doniach, J.T. Groves and K.O. Hodgson, Inorg. Chem. 25, (1986) 2255.
[54] P.A. Lee, Phys. Rev. B 13, (1976) 5261.
[55] S.M. Heald and E.A. Stern, Phys. Rev. B 17, (1978) 4069.
[56] J. Stöhr, X-ray Absorption: Principles and Techniques of EXAFS, SEXAFS and XANES, (Eds. R. Prinz and D. Konigsberger), Wiley, New York, (1986).
[57] J. Stöhr and R. Jaeger, Phys. Rev. B 26, (1982) 4111.
[58] J. Stöhr, J.L. Gland, E.B. Kollin, R.J. Koestner, A.L. Johnson, E.L. Muetterties and F. Sette, Phys. Rev. Lett. 53, (1984) 2161.
[59] T. Iwazumi, I. Nakai, M. Izumi, H. Oyanagi, H. Sawada, H. Ikeda, Y. Saito, Y. Abe, K. Takita and R. Yoshizaki, Solid State Commun. 65, (1988) 213.
[60] H. Oyanagi, H. Ihara, T. Matsubara, T. Matsushita, M. Hirabayashi, M. Tokumoto, K. Murata, N. Terada, K. Senzaki, T. Yao, H. Iwasaki and Y. Kimura, Jpn. Journ. Appl. Phys. 26, (1987) L1233.

[61] J.M. Tranquada, S.M. Heald, A.R. Moodenbaugh and Y. Xu, Phys. Rev. B 38, (1988) 8893.
[62] H. Tolentino, E. Dartyge, A. Fontaine, G. Tourillon, T. Gourieux, G. Krill, M. Maurer and M.F. Ravet, High T_c Superconductors: Electronic Structure, (Eds. A. Bianconi and A. Marcelli), Pergamon Press, Oxford, (1989).
[63] E.E Alp, S.M. Mini, M. Ramanathan, B.W. Veal, L. Soderholm, G.L. Goodman, B. Dabrowski, G.K. Shenoy, A. Bommanavar and O.B. Hyun, (1989) to be published
[64] G. Onori, A. Santucci, A. Scafati, M. Belli, S. Della Longa, A. Bianconi and L. Palladino, Chem. Phys. Lett. 149, (1988) 289.
[65] R.J. Cava, B. Batlogg, S.A. Sunshine, T. Siegrist, R.M. Fleming, K. Rabe, L.F. Schneemeyer, D.W. Murphy, R.B. van Dover, P.K.Gallagher, S.H. Glarum, S. Nakahara, R.C. Farrow, J.J. Krajewski, S. M. Zahurak, J.V. Waszczak, J.H. Marshall, P. Marsh, L.W. Rupp, Jr. W.F. Peck and E.A. Rietman, Physica C 153-154, (1988) 560.
[66] S. Della Longa, M. De Simone, C. Li, M. Pompa and A. Bianconi, High T_c Superconductors: Electronic Structure, (Eds. A. Bianconi and A. Marcelli), Pergamon Press, Oxford, (1989).
[67] K.B. Garg, A. Bianconi, S. Della Longa, A. Clozza, and M. De Santis and A. Marcelli, Phys. Rev. B 38, (1987) 244.
[68] S.M. Heald, J.M. Tranquada, A.R. Moodenbaugh and Y. Xu, Phys. Rev. B 38, (1988) 761.
[69] A. Bianconi, M. Dell' Ariccia, A. Gargano and C.R. Natoli, EXAFS and Near Edge Structure, (Eds. A. Bianconi, L. Incoccia and.S. Stipcich), Springer Verlag, Berlin, (1983), pp.57.
[70] G. van der Laan, C. Westra, C. Haas and G.A. Sawatzky, Phys. Rev. B 23, (1981) 4369.
[71] M. De Santis, A. Bianconi, A. Clozza, P. Castrucci, A. Di Cicco, M. De Simone, A M. Flank, P. Lagarde, J. Budnick, P. Delogu, A. Gargano, R. Giorgi and T.D. Makris, High T_c Superconductors: Electronic Structure, (Eds. A. Bianconi and A. Marcelli), Pergamon Press, Oxford, (1989).
[72] P. Steiner, S. Hufner, A. Jungmann, V. Kinsinger and I. Sander, High T_c Superconductors: Electronic Structure, (Eds. A. Bianconi and A. Marcelli), Pergamon Press, Oxford, (1989).
[73] A. Bianconi, A.M. Flank, A. Fontaine, P. Lagarde, A. Marcelli, M. Verdaguer, J. Jegoudez, A. Revcolevschi and G. Demazeau, Physica C 153-155 (1988) 117.
[74] A. Bianconi, M. De Santis, A. Di Cicco, A.M. Flank, A. Fontaine, P. Lagarde, H. Katayama-Yoshida, A. Kotani and A. Marcelli, Phys. Rev. B 38, (1988) 7196.
[75] A. Bianconi, P. Castrucci, A. Fabrizi,A.M. Flank, P. Lagarde, S. Della Longa, A. Marcelli, Y. Endoh, H. Katayama-Yoshida and Z.X. Zhao, High T_c Superconductors: Electronic Structure, (Eds. A. Bianconi and A. Marcelli), Pergamon Press, Oxford, (1989).
[76] G. Kaindl, D.D. Sarma, O. Strebel, C.T. Simmons, U. Neukirch, R. Hoppe and H.P. Muller, Physica C 153-155, (1988) 139.
[77] J. Fink, N. Nucker, H. Rombeg and S. Nakai, High T_c Superconductors: Electronic Structure, (Eds. A. Bianconi and A. Marcelli), Pergamon Press, Oxford, (1989).
[78] A. Bianconi, M. De Santis, A.M. Flank, A. Fontaine, P. Lagarde, A. Marcelli, H. Katayama-Yoshida and A. Kotani, Physica C 153-155, (1988) 1760.
[79] C. Taliani, R. Zamboni, G. Ruani, A.J. Pal, F.C. Matacotta, Z. Vardeny and X. Wea, High T_c Superconductors: Electronic Structure, (Eds. A. Bianconi and A. Marcelli), Pergamon Press, Oxford, (1989).
[80] M. Bacci, Chemical Physisc 104, (1986) 191.
[81] M. Bacci, Japanese Journal of Applied Physics 27, (1988) L1699.
[82] H.P. Geserich, G. Scheiber, J. Geerk, H.C. Li, G. Linker, W. Assmus and

W. Weber, Europhysics Letters 6, (1988) 277.
[83] A. Bianconi, M. De Santis, A. Di Cicco, A. Fabrizi, A.M. Flank, P. Lagarde, A. Marcelli, C. Politis, H. Katayama-Yoshida, A. Kotani and Z.X. Zhao, Modern Physics Letters B 2, (1988) 1313.
[84] C. Castellani, C. Di Castro and M. Grilli, Physica C 153-155, (1988) 1659.
[85] C. Castellani, C. Di Castro and M. Grilli, Proceedings of the Conference Towards the theoretical understanding of high T_c superconductors - Trieste, (Eds. S.Lundqvist, E.Tosatti, M.P. Tosi and Yu Lu), Int. Journal of Modern Physics 1, (1988) 659.
[86] H. Kamimura, Proceedings of the Conference Towards the theoretical understanding of high T_c superconductors - Trieste, (Eds. S.Lundqvist, E.Tosatti, M.P. Tosi and Yu Lu), Int. Journal of Modern Physics 1, (1988) 699.
[87] W. Weber, Z. Phys. B - Condensed Matter 70, (1988) 323.
[88] W. Weber, Advances in Solid State Physics 28, (1988) 141.
[89] M. Jarrell, H.R. Krishnamurthy and D.L. Cox, Phys. Rev. B 38, (1988) 4584.
[90] A.L. Shelankov, X. Zotos and W. Weber, Physica C 153-155, (1988) 1307.
[91] T. Nishino, M. Kikuchi and J. Kanamori, Solid State Commun. 68, (1988) 455.
[92] J. Askenazi and C.G. Kuper, Physica C 153-155, (1988) 1315.
[93] J. Askenazi and C.G. Kuper, High T_c Superconductors: Electronic Structure, (Eds. A. Bianconi and A. Marcelli), Pergamon Press, Oxford, (1989) and references therein.
[94] S. Kurihara, Physica C 153-155, (1988) 1247.
[95] A. Fujimori, High T_c Superconductors: Electronic Structure, (Eds. A. Bianconi and A. Marcelli), Pergamon Press, Oxford, (1989).

PHASE DIAGRAMS IN THE Y-Ba-Cu-O SYSTEM AND CRYSTAL GROWTH OF YBCO SUPERCONDUCTORS

Milos Nevriva
Czechoslovak Academy of Sciences, Prague, Czechoslovakia

1. INTRODUCTION

The recent discovery of high-temperature superconductivity led, among others, to an enormous research work on sintered polycrystalline samples. The low density, as well as the presence of weak link between grains, required the development of other methods for the preparation of ceramic superconductors; most of those seem to be on the preparation of single crystals. The study of physical phenomena on single crystals is also very important to achieve a theoretical understanding of high temperature superconductivity, to optimize material properties for various applications and hopefully to get higher values of the superconducting transition temperature than the current maximum, being about 150 K. Successful material processing, however, presupposes a good knowledge of phase equilibria in a given system. Such information is conveniently collected in phase diagrams, which are also extremely valuable for an optimization of crystal growth experiments. In oxide systems exist about 5000 binary systems. The phase diagrams for only some tenths of these are known. The number of ternary systems tremendously increase up to 300000 and the data of known phase diagrams are relatively much less favourable. Therefore, the increased interest is devoted to mathematical synthesis of phase diagrams because their direct determination is time consuming and experimentally exacting. The combination of methods of phenomenological and statistical thermodynamics, together with experimental data, can substantially minimize elaboration demand for the determination of phase diagrams.

[1] Currently visiting scientist in ICTP, P.O. BOX 586, 34100 Trieste, Italy, from March 17, 1989 to March 17, 1990.

2. PHASE DIAGRAM DETERMINATION IN THE Y-Ba-Cu-O SYSTEM.

The determination of phase diagrams requires a great amount of experimental work connected with preparation of equilibrium samples and various analyses. The data obtained might not be quite reliable due to changes of high-temperature state during quenching, an insufficient rate of stable equilibrium attainment (mainly at the dynamic DTA and TG methods), a chemical contact between a sample and a crucible material, etc. All the mentioned problems have occurred during experiments focused to determine the phase relations in Y-Ba-Cu-O system and that, in fact, also explains contradictory results in the literature. In this respect, no systematic study of all the phase compatible relations in this system has been reported.

2.1. Experimental Methods.

The experimental techniques for the phase equilibria determination are conveniently divided into static and dynamic methods, according to the temperature regime applied [1, 2, 3]. A more objective classification seems to be as follows:

(i) Direct methods of observation of the phases creation and/or disappearance include all kind of microscopy and X-ray or electron diffraction analysis (metalography, microprobe, EMP, SEM, STEM, etc), thermogravimetry, differential pressure analysis and/or EGA.

(ii) Indirect methods, where the changes of the state of a system are indicated by the detection of a thermophysical property such as in various methods of thermal analysis (heating and cooling curves determination, DTA, thermodilatometry, thermomechanical measurements, emanation thermal analysis, electrometry and magnetometry, thermosonic measurements, and other nonconforming methods [4, 5, 6], EMR, EPR, Mössbauer spectroscopy, etc.

From the point of view of the determination of entire phase diagrams, it is important to emphasize the method of investigation; that is, if the material is studied at the actual temperature, or the high-temperature structure is frozen-in to laboratory temperature by quenching. The latter method does not guarantee the preservation of the high-temperature state, so that we cannot be sure of the actual phase diagram.

To eliminate the crystallization of secondary phases from the melt and the contamination of samples by the crucible material, a simple method called "soaking method" can be used [7]. During the experiment the liquid phase is soaked among grains of the substrate material until the interface between the substrate and the solid residue becomes almost free of liquid. Since the pellets are easily separable from the substrates, no substantial contamination of the solid residue by the substrate material occurs. The experimental arrangement of this method can be seen in fig. 1. The studied sample placed on a supporting pellet, pressed from a suitable material, is placed inside a horizontal tube furnace and heated at a chosen temperature and atmosphere until the liquid phase is soaked into the supporting pellet, and the solid residue is then obtained and used for the following analyses.

The interpretation of the data obtained depends on the character of the experimental study, i.e., whether we want to construct the equilibrium phase diagram by extrapolation to static ("zero heating rate") conditions or, whether we remain satisfied with the partly non-equilibrium phase boundary determined under steady heating rate, often satisfactory enough for practical purposes of a real system . In order to obtain the reliable experimental data for the construction of phase diagrams, it is suitable to employ both direct and indirect methods and both experimental manners at dynamic and static conditions.

2.2 Thermodynamics of Phase Equilibria

Despite a great simplification, the most useful equation is that derived independently by Le Chatelier [8] and Schröder [9] in the form:

$$\ln\, x_A = \Delta H_m^A (1/T_A^{melt} - 1/T)/R \tag{1}$$

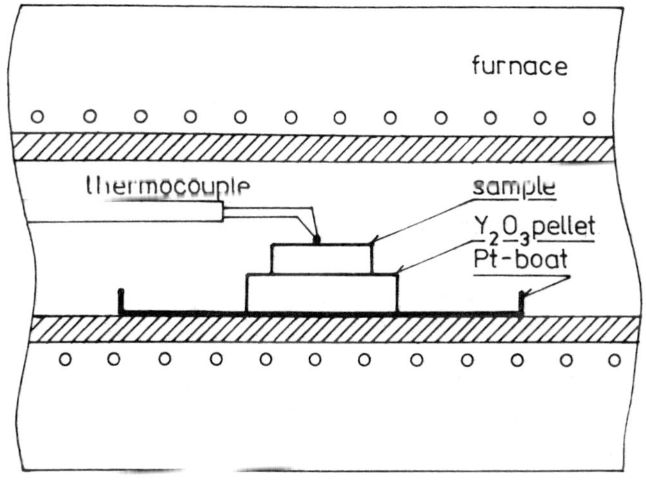

Fig.1. Scheme of arrangement for soaking experiments.

describing solubility of pure component A in an ideal solution, where $x_A, \Delta H_m^A, T_A^{melt}$, and T are the molar fraction, the enthalpy change of melting, the temperature of melting and the actual temperature of the solution, respectively. By substitution of the concentration by the activity in equation (1) it enables one to express the system non-ideality. According to Lewis [10] the activity is given as a multiple between concentration and the activity coefficient, γ, i.e., $a_i = x_i \gamma_i$. Considering the formation of a binary solution from the pure components A and B, the integral change of Gibbs energy is given by

$$\Delta G_{mix} = RT(x_A \, ln \, x_A + x_B \, ln \, x_B) + \Delta G^{EX} \tag{2}$$

In practice we can distinguish four basic types of real solutions:

(i) ideal solutions (then $\gamma = 1; \Delta G^{EX} = 0$).

(ii) regular solutions [11] (then $\gamma =$ const; $\Delta G^{EX} = \Omega \, x_A \, x_B$).

(iii) quasi-regular solutions [12] (then $\gamma = f(x); \Delta G^{EX} = \Omega \, x_A \, x_B \, (1 - k x_A)$ where k and Ω are interaction parameters).

(iv) coordination solutions [13, 14] (then $\gamma = f(x, T); \Delta G^{EX} = \Omega \, x_A \, x_B (1 - kT/\Omega)$.

Three approaches are given in the contemporary literature for the consequent thermodynamic analysis of phase boundaries:

(i) Joint solution of two differential equations, describing the equilibrium between two coexisting phases, usually written through the classical form of chemical potential μ_i as follows [15]:

$$\mu_A = G_A - x_A (\partial G_A / \partial x_A) \tag{3}$$

$$\mu_B = G_B + (1 - x_A)(\partial G_B / \partial x_B) \tag{4}$$

(ii) Common tangent to the Gibbs energy dependence of coexisting phases called as geometric method [12, 16, 17] (see fig.2) expressed by the system of equations (5).

$$\frac{G_m^\alpha - G_m^l}{x^\alpha - x^l} = \frac{dG_m^\alpha}{dx} = \frac{dG_m^l}{dx} \tag{5}$$

(iii) Calculation of the equilibrium melt composition from the soaking experiments using the equations derived for the $YO_{1.5}$-BaO-CuO ternary system [7] from the lever rule in the following forms:

$$X_c^L = (X_c^O \, M_R - X_c^R M_O \, \omega_R)/(M_R - M_O \, \omega_R) \tag{6}$$

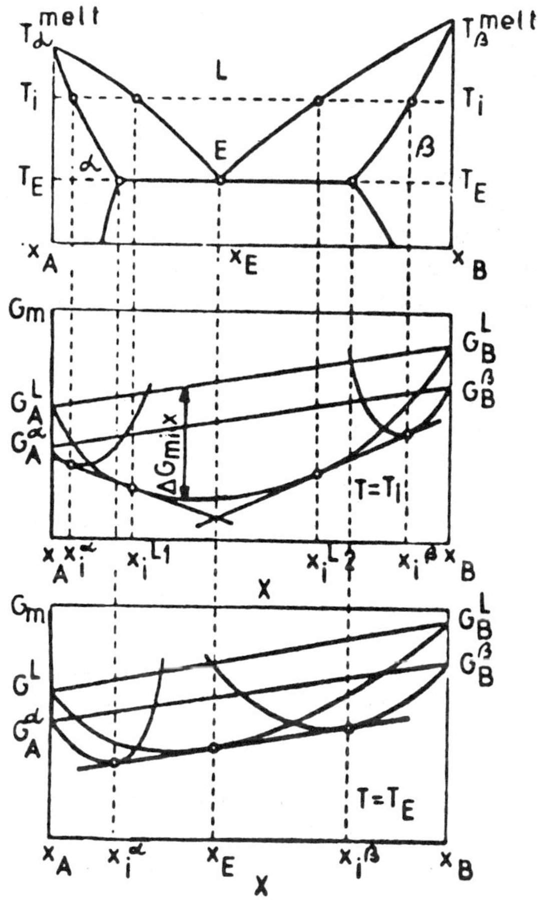

Fig. 2. Correlation between a phase diagram of eutectic type and dependence of molar Gibbs energy, G_m, on composition x, where subscripts A, B, E, l, α and β mean pure components A and B, eutectic composition, l is liquid, α and β solid phases, respectively.

$$X_y^L = (X_y^O\, M_R - X_y^R.M_O\; \omega_R/(M_R - M_O\; \omega_R) \tag{7}$$

where the following abbreviations are used $y = YO_{1.5}$, and $c = CuO$. Equations (6) and (7) allow to determine the composition of equilibrium melt (X_c^L , and X_y^L) from the composition of the original sample (X_c^O , and X_y^O), the composition of the solid residue (X_c^R, X_y^R), their "molecular weights", and the weight ratio of the solid residue ($\omega_R = w_R/w_O$).

2.3 Binary Phase Diagrams

The phase diagram of the Cu-CuO system was first published by Roberts and Smith [18]. They found two eutectics, the first one of the composition of 95.7 wt% Cu and 4.3 wt% O at a temperature of 1060°C and the second one of the composition of 86.2 wt% Cu and 13.8 wt% O at a temperature of 1083°C. The isobaric phase diagram of the Cu_2O-CuO system at 1 atm pressure of oxygen was constructed by Vogel and Pôcher [19] on the basis of thermal and microscopic data and dissociation-pressure/temperature data of a number of previous authors. CuO and Cu_2O were indicated as phases of fixed composition with negligible depatures from stoichiometry. This oxides form an eutectic at about 13 at% oxygen corresponding to 68 mol% Cu_2O and 32 mol% CuO approximately, and melting at 1075°C, which is in a good agreement with earlier determination by Foote and Smith [20] and Roberts and Smith [18], giving 1070°C and 1080°C, respectively. Vogel and Pôcher [19] indicated that, when CuO is heated under the oxygen pressure of 1 atm, melting occurs at about 1110°C with loss of oxygen and formation of an oxygen deficient liquid phase. With further rise of temperature, the loss of oxygen is more pregressive and at about 1400°C a saturated solution of oxygen in liquid copper is formed. The dissociation curves of CuO at oxygen pressure of 0.21, 0.33, 0.50, and 1.0 atm determined by Gadalla et al are shown in [21] . At the first three pressures dissociation to Cu_2O occurs in solid state at 1026, 1050, and 1072°C respectively, with no mutual solid solubility between oxides. At 1.0 atm CuO melts with loss of oxygen at 1096°C.

The preliminary shape of the CuO-BaO phase diagram was recently reported by Roth et al [22] Phase diagram of the CuO-$BaCuO_2$ system determined on the basis of X-ray and DTA measurements by Nevřiva et al [23] in oxygen atmosphere is shown in fig. 3. There is one eutectic point at 72 mol% CuO and 28 mol% BaO and at the temperature of 925°C. The eutectic composition about 72 mol% CuO was also reported by Ballestrino et al [24] in this system, although the eutectic was about 870°C. Below the eutectic temperature there only CuO and $BaCuO_2$ solid phases coexist. Nevertheless, more precise measurements made recently in this system showed the composition of the eutectic to be shifted to higher content of CuO (about 75 mol%) and also a limited solubility of BaO (about 4 mol%) in CuO [7]. The congruent melting of $BaCuO_2$ phase is in good agreement with earlier reports by Midgeon et al [25] and Roth et al [22] giving the temperature of melting between 1010-1020°C. Additional binary phases else reported in this system are as follows: Ba_2CuO_3 [22, 26-29], Ba_3CuO_4 [26, 29], $Ba_3Cu_5O_8$ [29] $BaCu_2O_2$ [30], Ba_4CuO_5 [31], and Ba_6CuO_7 [31]. Discrepancy of the reported data existing in the literature is probably due to the use of various starting materials as well as various crucible materials used during the fabrication of samples and a consequent analysis (DTA).

The preliminary phase diagram of the CuO-$YO_{1.5}$ binary system was published by Roth et al [22] and the phase diagram of the CuO - $YCuO_{2.5}$ system constructed on the basis of the X-ray and DTA measurement under oxygen atmosphere by Nevřiva et al [23] is given in fig. 4.

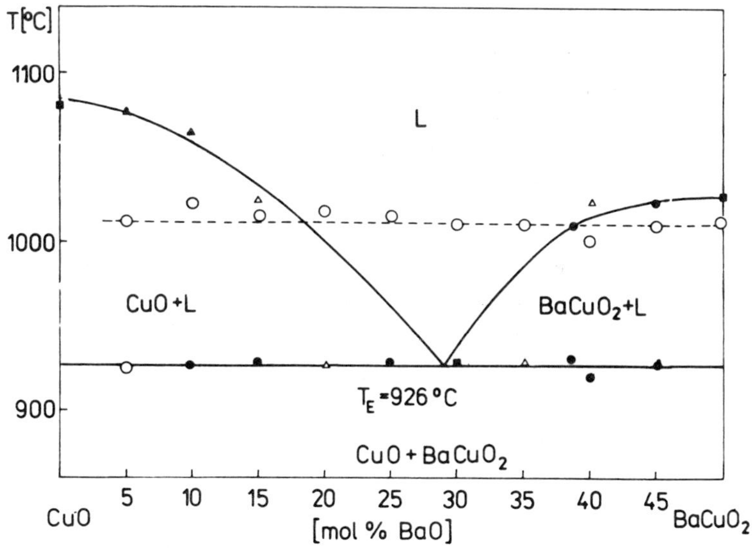

Fig. 3. Phase diagram of the CuO-BaCuO$_2$ binary system determined by Nevřiva et al [23] in O$_2$. Measuring points represent enthalpy change of DTA endothermic effects: (○) smallest; (●) 25% (△) 50%; (▲) 75%; (■) largest.

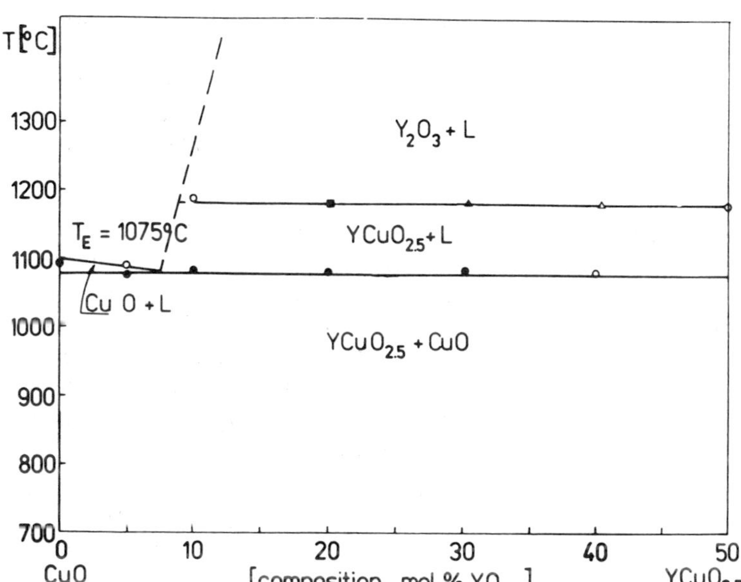

Fig. 4. Phase diagram of the CuO-YCuO$_{2.5}$ binary system determined by Nevřiva et al [23] in O$_2$. Measuring points represent enthalpy change of DTA endothermic effects: (○) smallest; (●) 25%, (△) 50%; (▲) 75%; (■) largest.

There are two invariant points, one eutectic of the composition of about 7.5 mol% $YO_{1.5}$ and the temperature of 1075°C and one peritectic with the temperature of 1180°C. The $YCuO_{2.5}$ phase incongruent melting has been already determined by Lambert [32]. Two other phases were also reported in this system: the $YCuO_2$ [33], and $YCuO_3$ [34], however, both of them were not detected in the current studies of the Y-Ba-Cu-O phase equilibria.

The $YO_{1.5}$-BaO phase diagram was reported by Lopato et al [35]. Two compound Y_2BaO_4 and $Y_4Ba_3O_9$ were found to be stable up to 1400°C and 2200°C, respectively. Two other phases $Y_2Ba_2O_5$ and $Y_2Ba_2O_7$ reported by Kwestroo [36] were not confirmed and are suspected to be carbonates [29, 37].

2.4 Pseudobinary and Ternary Phase Diagrams

Phase equilibria in the quaternary system Y-Ba-Cu-O are commonly studied by dynamic methods under conditions which do not guarantee the constant content of oxygen in the system. So that it cannot be classified as a closed one in thermodynamical sense. In fact the system is closed only for fixed components Y,Ba,Cu (or $YO_{1.5}$, BaO, CuO) but open for oxygen (free component). Therefore the above mentioned system should be designed as Y-Ba-Cu-(O) to express the diverse role of the oxygen (with the respect to the fixed components) in the equilibrium conditions. It follows from the general principles that the complexity of the equilibrium phase relationships in the quaternary system obtained at chosen constant P_{O2} is equal to that of the ternary system, if we disregard a variable oxygen content. The equilibrium phase relations at given P_{O2} then can be represented in a triangle diagram with fixed components at the corners, if the barycentric coordinates are chosen to be independent on the real oxygen content.

The determination of temperature-composition cuts through this ternary system, particularly those including YBCO superconducting phase were in the centre of interest. The $YCuO_{2.5}$-$BaCuO_2$ tie line is one of the most intensively studied cut [22, 26, 28, 38, 39] through Y-Ba-Cu-O system. The $YCuO_{2.5}$-$BaCuO_2$ pseudobinary phase diagram constructed from the results of X-ray phase analysis and DTA by Nevřiva et al [38, 39] is given in fig. 5. The respective phase coexisting at lower temperature region are as follows: $YCuO_{2.5}$ + Y_2BaCuO_5 + CuO for 0-33.3 mol% $BaCuO_2$, YBCO + Y_2BaCuO_5 + CuO for 33.3-66.6 mol% $BaCuO_2$ and YBCO + $BaCuO_2$ for 66.6-100 mol% $BaCuO_2$. Peritectic decomposition of the YBCO phase at about 1000°C lead to coexistence of Y_2BaCuO_5 + CuO + $BaCuO_2$ and after melting of the last one at 1025°C the Y_2BaCuO_5 + CuO + L coexist. Above 1075°C the situation becomes more complicated due to the presence of liquid phase. Y_2O_3 together with four other unknown phases are present there. Three of these were later indicated as Cu_2O, $BaCuO_2$. and substituted $BaPt_4O_6$ phase [40] arising due to the melt reaction with Pt crucible.

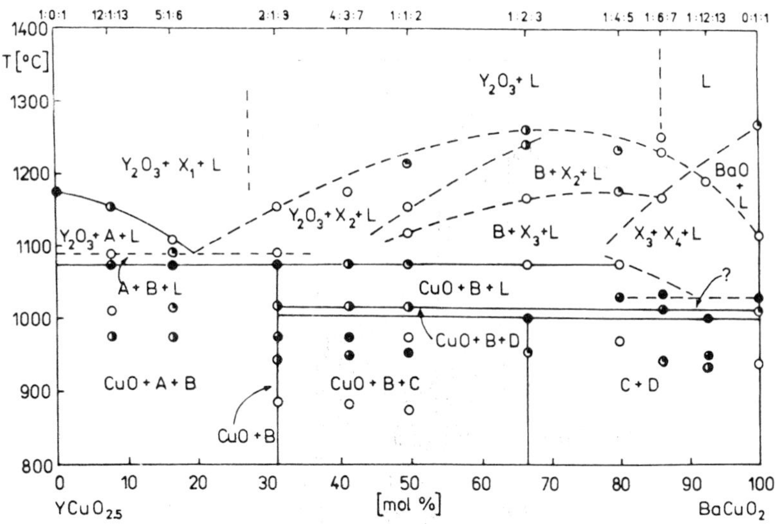

Fig. 5. Phase diagram of the $YCuO_{2.5}$-BaCuO pseudobinary system by Nevřiva et al [38, 39] (A-$YCuO_{2.5}$; B-$YCuO_{2.5}$; B-Y_2 $BaCuO_5$; C-$YBa_2Cu_3O_{7-\gamma}$; D-$BaCuO_2$).

From the crystal growth point of view more interesting diagram which has already been reported by several authors [41-44]. The most reliable results on this system are given by Oka et al [44] which focused in detail the measurement upon the area surrounded by CuO-$BaCuO_2$-YBCO triangle suggested to be favourable for growing the YBCO crystals in many reports. Their measurements were carried out along four radial directions starting from YBCO as the common end member, towards CuO, $Ba_3Cu_{12}O_{15}$, $Ba_3Cu_7O_{10}$ and $Ba_3Cu_5O_8$, respectively, as the opposite-end members. The results are stereographically illustrated in fig. 6.

Phase relationships in the CuO-$BaCuO_2$-YBCO system at 950°C in air and at 975°C in oxygen were determined by Nevřiva et al [7] using the procedure based on the separation of the melt from an equilibrated sample and the subsequent analysis of the solid residue. The resulting cuts of ternary phase diagrams with approximating liquidus curves and boundaries of four primary crystallization fields which were estimated are shown in Figs. 7 and 8. The abbreviations of coexisting phases are used as follows: T=CuO, D=$BaCuO_2$, G=Y_2BaCuO_5 and S=$YBa_2Cu_3O_{7-\gamma}$. At 975°C under the oxygen atmosphere (see fig. 12) three three-phase regions (S+D+L, S+G+L, and T+G+L), four two-phase regions (D+L, S+L, G+L,and T+L), and the region of homogeneous melt (L) occur. This is in agreement with our previous work [34].

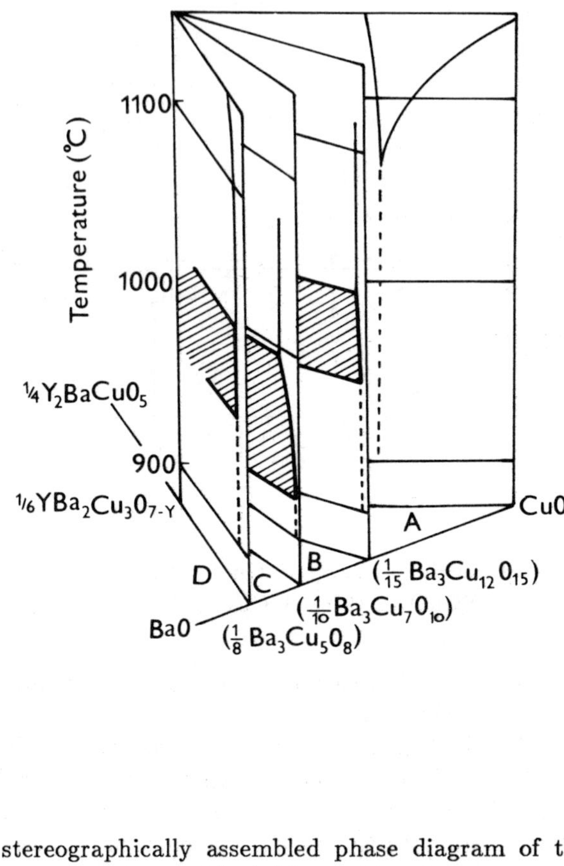

Fig. 6. A stereographically assembled phase diagram of the CuO-BaCuO$_2$ - YBa$_2$Cu$_3$O$_{7-\gamma}$ system by Oka et al [44] (the region of coexistence YBCO with liquid phase is hatched).

The boundary of homogeneous melt regions is drawn in fig. 7 by solid lines for 975°C and for 950°C by dashed-dotted lines. Dotted lines represent projections of the estimated boundaries of primary crystallization fields -the intersect at composition 73.8 mol% CuO and 1.0 mol% YO$_{1.5}$ represents the estimated ternary eutectic point (E). With respect to crystal growth of YBCO the important question arises how the melt equilibria depend on the P_{O2}. The phase diagram constructed from the results obtained under air at 950°C (fig. 8) shows the fields of the same phases as at 975°C in O$_2$, however, the compositions of the equilibrium melts (positions of the liquidus curves) are different.

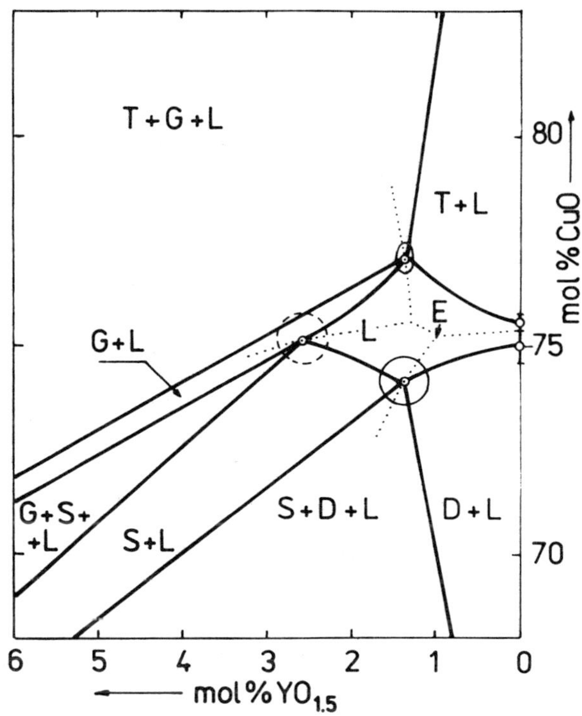

Fig. 7. Isothermal cut of the CuO-$BaCuO_2$-$YBa_2Cu_3O_{7-\gamma}$ ternary system at 950°C in air by Novřiva ot al [7].

The ternary eutectic point (E) was estimated to be at 75.2 mol% CuO and 1.0 mol% $YO_{1.5}$. Uncertainties of the individual melt composition have been expressed by the solid and dashed ellipses the axes of which correspond to the calculated deviations (see Figs. 7 and 8).

Aselage and Keefer [45] determined phase diagram in CuO-rich corner of $YO_{1.5}$-BaO-CuO ternary system in air. They used DTA and direct visual observations with optical pyrometer telescope and/or hot stage optical microscope. They identified eleven invariant points (e_1-890°C; e_2-920°C; p_1-940°C; p_2-975°C; e_3, p_3-1000°C; m_1, m_2-1015°C; e_4-1110°C; m_3-1122°C and m_4-1270°C).

Incongruent melting of YBCO (m_1) results in L+Y_2BaCuO_5. The latter also undergoes incongruent melting producing L+Y_2O_3. As far as their data were insufficient to locate the liquid compositions accurately only a schematic ternary phase diagram of the mentioned system is drawn in fig. 9.

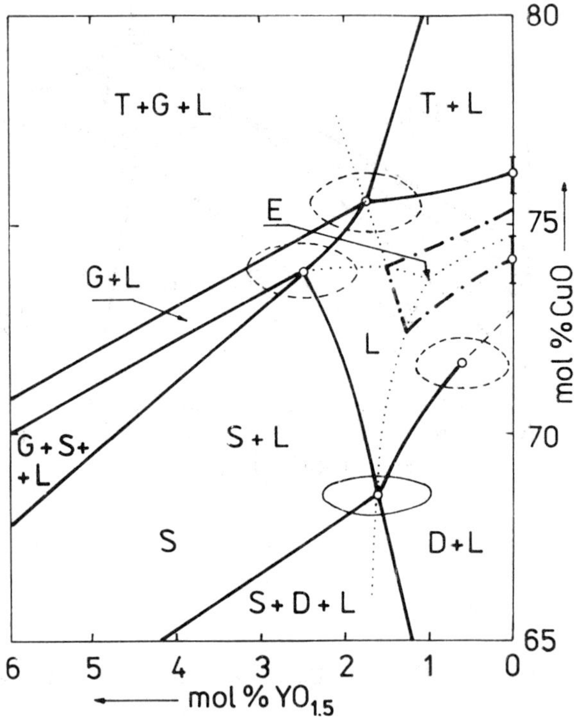

Fig. 8. Isothermal cut of the CuO-$BaCuO_2$-$YBa_2Cu_3O_{7-\gamma}$ ternary system at 975°C in oxygen by Nevřiva et al [7].

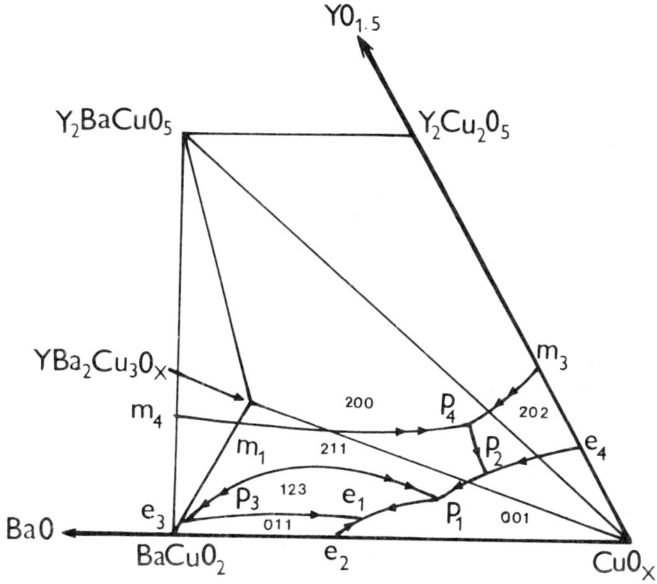

Fig. 9. Schematic ternary phase diagram by Aselage et al [45]. Eutectics are denoted as "e", peritectics as "p", and melting points as "m". The primary crystallization fields are labelled as follows: 200-Y_2O_3; 202-$Y_2Cu_2O_5$; 001-CuO; 011-$BaCuO_2$; 211-Y_2BaCuO_5; 123-$YBa_2Cu_3O_x$.

3. CRYSTAL GROWTH OF $YBa_2Cu_3O_{7-\gamma}$ SUPERCONDUCTORS

Of the various methods for the single crystals growth of YBCO only two have been found to achieve crystals of a sufficiently good quality. The first is solid state reaction which leads to small crystallites and the second is crystal growth from high temperature solutions using nonstoichiometric melts. The latter is currently the most promising for obtaining high quality bulk crystals of YBCO and/or other related phases. Full understanding of crystal growth processing is, however, necessary.

3.1 Fundamental Principles

Principle of single crystal growth from high-temperature solution can be seen in fig. 10, where a typical eutectic phase diagram is given. Liquidus curve (drawn by full line in fig. 10) represents thermodynamic boundary between solvent and solvent + solute regions.

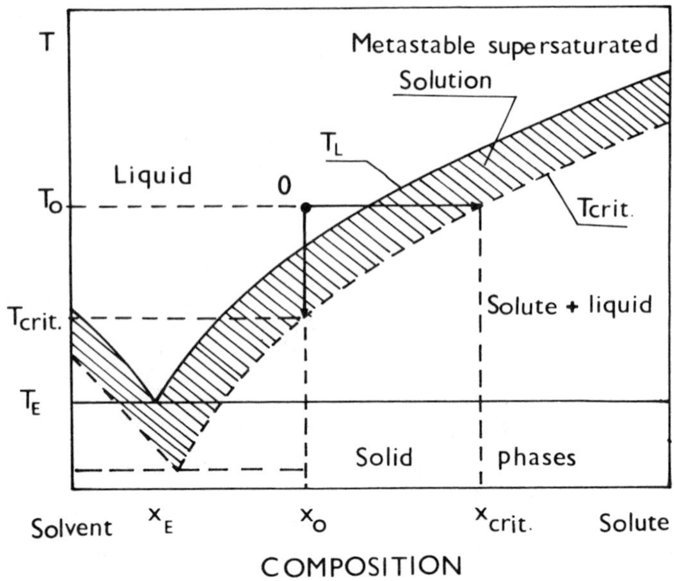

Fig. 10. Hypothetic pseudobinary phase diagram of solvent-solute system.

Between both the liquidus curve and that drawn by dashed line there exists the metastable region of undercooled solution, called also Ostwald-Miers region, as a consequence of the necessity to form a nucleus of the critical size before the precipitation of a crystal. The width of metastable region significant for growth of large crystals depends above all on the properties of the melt and the experimental arrangement. There exist three fundamental growing modes, corresponding to various techniques producing supersaturation:

(i) Solvent evaporation at a constant temperature during the growth. It results in homogeneous concentration of dopants or impurities; homogeneous density of equilibrium defects; no variation of the composition in the case of solid solutions and possibility to grow materials with a specified valence state. Evaporation rate can be controlled by the temperature, gas flow removing the solvent vapours, free surface area of the melt, etc. The main disadvantage arises from the solvent vapours being very corrosive and attacking the insulation ceramic materials as well as heating elements.

(ii) Slow cooling with spontaneous uncontrolled nucleation requiring a small vertical temperature gradient keeping the bottom of the crucible at lower temperature than the upper part. A continuous changing of the temperature during the growth can lead due to the differences in the solubility of the individual components to inhomogenities and concentration gradients in the grown crystals. The dimensions of the crystals can be enlarged by limiting the number of spontaneously nucleated crystals, using e.g. temperature oscillations during the decreasing temperature [46]. Another method restricting the number of nucleation centers is to spot cool the crucible bottom by an air jet or more simply by a ceramic or platinum rod [47]. Nevertheless, due to the temperature gradient, where the top of the melt is hotter than the bottom, the natural convection is minimized and the tendency to a layering of the solution appears. In this case a stirring technique, e.g. Accelerated Rotation Crucible Technique (ACRT) [48] should be applied.

ACRT represents an important method to preserve the experimental conditions such as homogeneity of the melt and a sufficient material transport to the surface of the grown crystals during the crystallization process. During acceleration and decceleration of crucible following three main flow mechanism occur: spiral shearing distortion flow, Ekman-layer flow and Couette flow [49, 50]. The most effective technique of nucleation control is top seeded growth. In contrast to previous method the top seeded growth on Pt wire or holder requires an inverse temperature gradient, i.e. the top of the solution is colder than the bottom [51, 52]. Homogenization of the solution is assured by natural convection as well as by the rotation of the seed. A high rotation rate required for a good homogenization reduce the width of Ostwald-Miers region so that spontaneous nucleation can occur. Therefore an optimum rotation rate has to be obtained and may vary from several tens to hundreds rotation per minute. The special case of this technique is arrangement described by Bennet [53] and Tolksdorf [54, 55]. However, the technique requires sophisticated and very expensive platinum crucibles and also furnace which allows the complicated manipulation.

The most extended method of the growth of thin films from high temperature solutions, liquid phase epitaxy [56], is similar to the top-seeded growth. Nevertheless, it exhibits some peculiarities as follows: instead of the seed a substrate with the same eventually closed related structure and compatible lattice parameters is used; growth is carried out isothermally on a substrate dipped into a supersaturated melt, the temperature of which is equilibrated before dipping; the residual melt is removed from the surface of grown films by centrifuging or tilting the substrate with film before pulling out [57].

(iii) Transport techniques like transport in a temperature gradient, travelling solvent zone technique, electrolytic growth in high temperature solution, etc. Obvious advantage of these techniques is isothermal growth. In the case of transport in a temperature gradient a nutrient is held at higher temperature near the bottom of the crucible and the seed below the liquid level at the lower temperature. Mass transport from the dissolved nutrient to the seed is assured either by the natural convection and diffusion through the boundary layer or by stirring. A real application meets difficulties because of the necessity to dissolve the nutrient crystals just before the growing process starts. The problem can be avoided e.g. by cooling the melt down the liquidus temperature after the soak period placing the preheated amount of a nutrient into a perforated platinum tube and immersing in the melt.

3.2. Experimental Arrangement

An important problem in the growth of YBCO crystals is the choice of suitable crucible materials. Platinum crucibles, usually used for the growth of oxide materials, are heavily attacked by BaO-CuO based melts. This corrosion can be decreased using e.g. Au crucibles as reported by Kaiser [58]. Several other crucible materials have ever been suggested and approved like MgO [59], ZrO_2 [60], or ThO_2 [61]. The most widely used crucible material is alumina as high purity and density as possible in order to avoid Al contamination of the crystals. Al enters the YBCO structure particularly during the growth at higher temperature and slower cooling rate. Al ions have significant effect on reducing T_c and increasing tetragonality [62].

The general procedure for growth of YBCO crystals consists of mixing starting materials, compacting them into crucible, heating to a maximum temperature for several hours and then cooling at a suitable rate to room temperature. Three techniques which have the advantage of separation of grown crystals from the melt have been successfully tested as follows:

(i) Cavity method [63] based on a formation of cavities in the melt most probably due to the oxygen evolution at higher temperature than 1000°C. The crystals grown in these cavities are separated from the melt by its contraction during the solidification and cooling to room temperature. They must be recovered mechanically by breaking all the mass together with the crucible. Therefore it is advantageous to use ceramic crucibles like alumina for this method.

(ii) Melt transport method [58] utilizing wetting and creeping of the liquid solutions on metals like gold and platinum so that quasi free crystals can be obtained. Moreover the lower homogenization temperature in comparison to that used at the case of previous method of about 1000°C is sufficient. The melt is spontaneously transported over the crucible walls, therefore another external crucible must be used. Single crystals are particularly grown on outer surface of the internal crucible perpendicularly to its bottom and could be easily taken out. In the case of platinum material the internal crucible are often damaged due to the formation of Pt containing phases. Using Au crucible in this method the contamination is not so pronounced. However lower homogenization temperature should be used because of the lower melting point of Au.

(iii) Crystal growth from homogeneous melt [64] represents probably the best way for the growth of large crystals and to separate them at the same time. It consists of achieving the proper concentration and temperature conditions for complete dissolution of solid phases and to remove the melt from the crystals by decanting or pouring the liquid through a sieve or by tilting the crucible. These conditions, however, can be achieved only through a systematic study of phase relationships in the system and optimization of growth conditions.

3.3. Solvents

The choice of a suitable solvent for the crystal growth from high-temperature solutions is a complex problem involving both the physical and the chemical properties.

(i) Physical properties —obviously the most important is the melting temperature, in addition the viscosity, density and volatility are important, particularly in lead compounds.

(ii) Chemical properties —there are two important relations solvent-crystal: no component of the solvent enters the grown crystals and the other in which at least one component of the solvent is simultaneously a componenent of the the grown crystal. According to acid-base concept [65, 66] based on the equilibrium reaction

$$Base = Acid + X^{m-} \tag{8}$$

the solvents used for the growth of crystals belong to two groups: basic compounds supplying anions into solution and acidic compounds, forming complexes by binding of anions from the solution, e.g. the oxides form the following series in terms of decreasing basicity: alkaline oxides $>$ BaO $>$ PbO $>$ Bi_2O_3 and in terms decreasing acidity oxides: B_2O_3 $>$ SiO_2 $>$ V_2O_5 $>$ WO_3 $>$ MoO_3 $>$ P_2O_5. The oxygen ion concentration in a solvent is controlled by the ratio of the present basic and acidic components those are present and their relative basicity or acidity. It was found that a successful growth may require an excess of the acidic oxide, as in the garnet systems. Relatively high acidity causes formation of large complex anion of the type MO_x where x varies from 3 to 6 depending on the ionic radius of M and the temperature. They can, impede the growth process and unfavourably influence habit of the arising crystals. If the basic oxides are relatively concentrated formation of complexes of small coordination numbers may be expected. Then due to a sufficient growing rate, the number of arising nuclei is restricted [67-68].

(iii) Solubility of the crystallizing phase —is given by the slope of liquidus curve and has basic importance for the crystal growth. If it is to mild then a small change of the temperature causes large change in the melt composition resulting in a variation of the composition of the grown crystal, particularly in multicomponent system. If it is to sharp then a wide temperature range has to be used for the growth of large crystals. The liquidus (saturation temperature) can be determined using the equation (1) if the respective data are accessible or could be obtained experimentally using the optical methods [69], growing methods [70], DTA, or solubility models [71] etc.

Many various solvents [72] than BaO-CuO were tried for growth of YBCO crystals. However, such factors as limited dissolution, formation of more stable phases with Y- or Ba- ions, the oxidation state of CuO etc excluded their use. Perhaps an addition of small percentage of B_2O_3 [72], Cu_2O [73] or In_2O_3 [74] to BaO-CuO based melts may be useful to obtain lower eutectic temperature and consequently larger temperature range for stable growth of YBCO crystals.

3.4. Growth Conditions

The upper limit of the temperature range for crystal growth is given by the incongruent melting point of YBCO in air and oxygen being between 1000°C and 1050°C. The temperature of ternary eutectic determined at 905 ± 30°C then indicates the lower limit for growth of YBCO in the given ternary system. Hence it is very important to optimize growing conditions with the melt having the liquidus temperature just below the temperature of peritectic melting of YBCO, i.e. with temperature range as large as possible. For growth of large crystals very slow growth rates are usually required. Nevertheless long-time experiments especially in alumina crucibles are not favourable so that the rate from 0.3 to 2°C per hour represent necessary compromise between maximum stable growth rate and minimum substitution of copper in chains. Temperature stability during growth is not as critical as in growth of many other oxide materials such as garnets, etc. However temperature control of better than ± 0.2°C and temperature programming from 0.1 to 10°C are needed. The simple muffle or tube furnace with Kanthal or Crusillite heating element are often sufficient.

The melt equilibria in this system were discussed in detail elsewhere. Nevertheless, there are other data on liquid equilibria concerning the crystal growth in the literature. Takekawa and Iyi [75] also suggested the primary crystallization field of YBCO (in oxygen atmosphere) as a triangle surrounded by primary crystallization fields of CuO, $BaCuO_2$ and $YBaCuO_5$. Schneemayer et al [61] prepared the best crystals from composition $YBa_4Cu_{10}O_x$ in air with starting temperature 1000°C. Kaiser et al [58] reported the region of extensive melting at about 72 mol% CuO in the temperature range of 950-1000°C in air. Scheel and Licci [72] pointed out the oxygen partial pressure as important parameter for crystal growth of the YBCO phase successfully occurred from starting composition of $Y_{4.2}Ba_{29.3}Cu_{66.5}O_x$. In their further work Licci et al [64] reported a pseudobinary phase diagram of YBCO - $Ba_{.28}Cu_{.72}O_x$ system evaluated from DTA measurement in air with an eutectic at 835 ± 10°C, composition $Y_{.2}Ba_{28.1}Cu_{71.7}O_x$ and a peritectic reaction at 935°C. Fischer et al [76] determined isotherms of liquidus surfaces for 1000, 1050,and 1100°C in air and found that the composition of the melt coexisting with CuO and YBCO is near to 76 mol% CuO and 24 mol% BaO. The survey of the melt compositions used for the growth of YBCO crystals was reported by Scheel and Licci [77].

From our systematic study of this system [80-81], the region of optimum compositions for crystal growth is located in the field of the primary crystallization of YBCO close to the boundary between YBCO + L and YBCO + $BaCuO_2$ regions and depends on the soaking temperature and growth rate (see the region indicated by dashed line in fig. 11; points marked by ● correspond to the melt composition under-studied). Our results exhibit a shift of the optimum melt composition to the higher content of CuO similar to that of binary eutectic of CuO-BaO system compared with data reviewed by Scheel and Licci [77]. It is probably due to occurrence of disturbances as a consequence of the actual experimental conditions. These could be as follows:

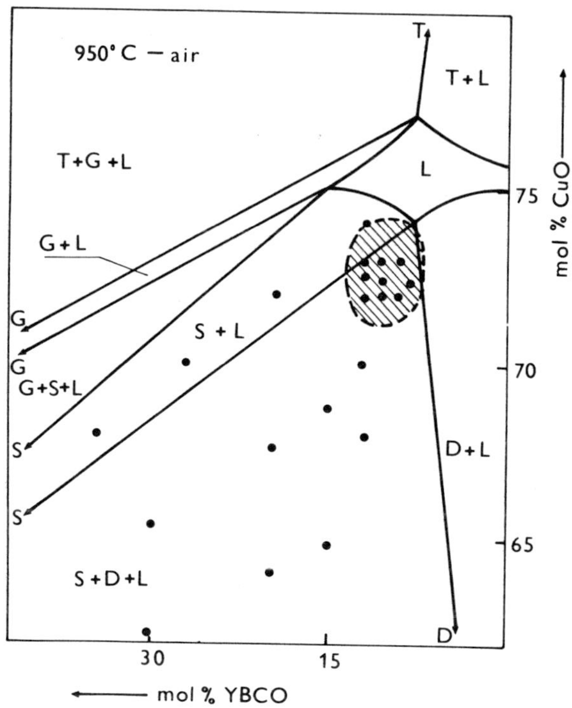

Fig. 11. Graphically represented composition of melts that used during our systematic study [80-81].

(i) The reactions between the crucible materials and the melts. Contamination products can effect uncontrolled changes of the melt composition and consequently its saturation temperature.

(ii)The reaction of BaO component with CO_2 always present in air resulting in formation of $BaCO_3$ and hence to changes of the melt composition and its saturation temperature. These complications were pointed out in the recent work of Leeuw et al [29] where presence of carbonates due to use of $BaCO_3$ as the starting material was suggested to be the reason for discrepancies in the phase relations determined in air at 950°C [22,26,28]. Besides this explanation there is another one following from thermodynamic data [78]. It was found that $BaCO_3$ can be formed at 950°C if the following inequality [7] is valid:

$$log\, a_{BaO} + log\, P_{CO2} >= -3.0 \qquad (9)$$

(a_{BaO} is activity of BaO in oxide melt, P_{CO2} = partial pressure of CO_2 in gaseous atmosphere). It means that $BaCO_3$ is stable if e.g. a = 0.25 and p > 0.4 kPa, so that BaCO3 can be formed under conditions usual for the crystal growth carried out in air.

(iii) The equilibrium interaction between the melt and oxygen affecting the total oxygen content and the phase relationships in the system.

The above discussed facts explain the mutual inconsistency of results in the literature. It can be concluded that the conditions for YBCO crystal growth should be optimized individually in the framework of the ambient atmosphere, crucible material, temperature regime and melt composition.

3.5. YBCO Crystals

The shape of the grown YBCO crystals are square plates usually extremely thin with dimensions up to 10 mm in size and thickness from .005 to .5 mm. It was suggested [79] that the various habits of the grown crystals are related to the growth conditions —cubes and thick square plates correspond to stable growth, whereas dendrites and thin plates indicate morphological instability. Whereas the crystal size in cavities reaches only a few millimetres, occasionally large plates up to 8mm can be found on the surface of the melt [72,81] (see fig. 12).

Tetragonal crystals formed initially sometimes transform *in situ* to the superconducting phase depending on the oxygen pressure and the cooling rate. Therefore to obtain high T_c superconducting phase in some cases post-annealing procedure, below 600°C is necessary. The tetragonal-orthorhombic transformation leads to twinning and also to formation of single-domain fragments thereby affecting critical current densities. Superconducting critical current measured on as grown single crystal $YBa_2Cu_3O_{7-\gamma}$ with dimension 3x2.5x.055mm [81], exhibiting liquid nitrogen superconductivity, reached the value 250 A/cm^2 at liquid nitrogen temperature in earth magnetic field and displayed weak decrease with magnetic field up to 150 Öe [82]. The polycrystalline $YBa_2Cu_3O_{7-\gamma}$ samples lose substantially their critical current already in 10 Öe magnetic field. The magnetic field dependence of the single crystals is on the other hand very weak. There remains observable critical current even in 1000 Öe perpendicular field. Since the absolute value of critical current density remains low, it seems that the crystal studied still contains great number of internal Josephson junctions located on twin boundaries. Nevertheless, these junctions are well protected from external magnetic field. It is probably due to the fact that the crystal is dense and displays no pores forming internal loops sensitive to magnetic field as found in polycrystalline ceramics.

Fig. 12. YBCO crystal grown on the top of the melt by Nevřiva [81].

Acknowledgements

The critical current measurements and systematic study of melt compositions for growth of YBCO crystals were made during our stay on Second Experimental Workshop on High Temperature Superconductors by International Centre for Theoretical Physics in Trieste. We thank to the organizers for providing the necessary apparatus.

REFERENCES

[1.] R.E. Newnham and R. Roy, Treats on Solid. State Chem. (Eds. N.B. Hannay),Plenum Press, N.Y.,(1973) p.437.

[2.] F.N. Rhines, Application of Phase Diagrams in Metallurgy and Ceramics, Vol.1 (NBS SP-496), Geithersburg, (1977) p. 147.

[3.] J.B. MacChesney and P.E. Rosenberg, Phase Diagrams in Material Science and Technology, Vol.1 (Eds.A.M. Alper), Academic Press, N.Y.(1970), p.113.

[4.] R. Fükiger and J.L. Jorda, ref. 2, p.375.

[5.] F.P. Glasser, ref. 2, p.407.

[6.] Phase diagrams in Material Science and Technology, Vol. 1-5(Eds. A.M. Alper), Academic Press N.Y.(1970)

[7.] M. Nevřiva, P. Holba, S. Durčok, D. Zcmanová, E. Pollert and A. Triska, Physica C 157, (1989) 334.

[8.] H. Le Chatelier, Compt. Rend. 100, (1885) 50.

[9.] J. Schröder, Z. Physik. Chem. 11, (1893) 449.

[10.] G.N. Lewis, Z. Physik. Chem. 61, (1908) 129.

[11.] J.H. Hildebrand, J. Amer. Soc. 51, (19290 66.

[12.] P. Holba, Silikáty 20, (1976) 193.

[13.] C.H.P. Lupis and J.P. Elliot, Acta Met. 15, (1967) 265.

[14.] J.C. Matthieu, F. Durand and E. Bonnier, J. Chim. Phys. 11, (1965) 1289, 1297; 12, (1967) 261.

[15.] J. Šesták, Thermophysical Properties of Solids, Elsevier, Amsterdam, (1984)

[16.] A.C. Van Rijn van Alkemade, Z. Physik. Chem. 11, (1893) 289.

[17.] H.W.B. Roozeboom, Z. Physik. Chem 30, (1899) 385, 413.

[18.] F.H. Smith and H.S. Roberts, J. Amer. Chem. Soc 42, (1920) 2582; 43, (1921) 1061.

[19.] R.V. Vogel and W. Pôcher, Z. Metallkunde 21, (1929) 333.

[20.] H.W. Foote and K.J. Smith, J. Amer. Chem. Soc. 30, (1908) 1345.

[21.] A.M.M. Gadalla, W.J. Ford and J. White, Trans. Brit. Ceram. Soc. 62, (1963) 57.

[22.] R.S. Roth, K.L. Davis and J.R. Dennis, Ceramic Superconductors, Advan. Ceram. Mater. 2, (1987) 303

[23.] M. Nevřiva, E. Pollert, L. Matejková and A. Tříska, J. Crystal Growth 91, (1988) 434.

[24.] G. Ballestrino, S. Barbanera and P. Paroli, J. Crystal Growth 85, (1987) 585.

[25.] H. Midgeon, F Jeannot and M. Zanne, Rev. Chim. Mineral 13,(1976) 440.

[26.] K.G. Frase and D.R. Clarke, ref.22, p.295.

[27.] W.K. Wong-Ng, K.L. Davis and R.S. Roth, J. Amer. Cer. Soc.71, (1988) C 64.

[28.] G.Wang, S.J. Hwu, S.N. Song, J.B. Keterson, L.D. Marks,K.R. Poeppelmeier and T.O. Mason, ref.22, 313.

[29.] D.M. De Leeuw, C.A.H.A. Mutsaers, C. Langreis, R.C.A. Smoorenberg and P.J. Rommers, Physica C 152,(1988) 39.

[30.] C. Teske and H. Muller-Buschbaum, Z. Anorg. Allg.Chem. 371, (1969) 325.

[31.] R.S. Roth, C.S. Rawn, F. Beech, J.D. Whitler and J.O. Anderson, NBS Geithesburg, preprint No. 6146.

[32.] E. Lambert, JCPDS Report No. 33-511 of Mineral Petrogr. Inst, University of Heidelberg, (1981).

[33.] T. Tshiogaro, N. Ishizawa, N.Mizutani and M. Kato, J. Sol. St. Chem. 49, (1983) 232.

[34.] M.Arjomand and D.S. Machin, J. chem. Soc. (1975) 1061.

[35.] L.M. Lopato, I.J. Majster and A.V. Švčenko, Izv. Acad. Nauk USSR, Neorg. Mat 8,(1972) 861.

[36.] W. Kwestroo, H.A.M. van Hal and C. Langreis, Mat. Res. Bull. 9,(1974) 1631.

[37.] G.A. Costa, M. Ferreti, F.A. Feaceschi and G.L. Olcese, Thermochim. Acta 133,(1988) 377.

[38.] M. Nevřiva, E. Pollert, J. Sestak, L. Matějková and A. Tříska, Phisica C 153-155,(1988) 377.

[39.] M. Nevřiva, E. Pollert, J. Šesták, L. Matějková and A. Tříska, Thermochimica Acta 130,(1988) 263.

[40.] S.J. Schneider and C.L. Daniel, J. Amer. Cer. Soc. 52, (1969) 518.

[41.] R.A. Laudise, L.F. Schneemeyer and R.L. Barn, J. Crystal Growth 85,(1987) 569.

[42.] R.W. Mc Callum, R.N. Shelton, M.A. Nock, J.D.Swenson, M.A. Demenko, K.A. Gschneider, E.D. Gibson and A.R. Moodenbaugh (Ames Lab.), Microstructure and Superconductivity in High Tc Materials, preprint prepared for publictation.

[43.] Z. Jishan, J. Xiaoping, H. Jianguo, J. Ming, Y. Yang, Z. Zhia, G. Yanlong, T. Yajun, Q. Guiwen and H. Zhuangqi, Supercond. Sci. Tech. 1,(1988) 107.

[44.] K. Oka, K. Nakane, M. Ito, M. Saito and H. Unoki, Jap. J. Appl. Phys. 27,(1988) L 1065.

[45.] T. Aselage and K. Keefer, J. Mater. Res. 3,(1988)1279.

[46.] H.J. Scheel and D. Elwell, J. Crystal Growth 12, (1972) 153.

[47.] S. Bornmann, E. Glauche, P. Görnert, R. Hergt and C. Becker, Kristall und Technik 9, (1974) 895.

[48.] H.J. Scheel and E.O. Schulz-Du Bois, J. Crystal Growth 8, (1971) 304.

[49.] H.J. Scheel, J. Crystal Growth 13/14, (1972) 560.

[50.] E.O. Sshulz-Du Bois, J. Crystal Growth 12, (1972) 81.

[51.] S. Durcok and E. Pollert, J. Crystal Growth 82, (1987) 501.

[52.] M. Nevřiva, J. Crystal Growth 83, (1987) 543.

[53.] G.A. Bennet, J. Crystal Growth 3/4, (1968) 458.

[54.] W. Tolksdorf, J. Crystal Growth 3/4, (1968) 463.

[55.] W. Tolksdorf and F. Welz, Crystal Growth of Magnetic Garnets from High-Temperature Solutions in Crystals for Magnetic Applications, Springer Verlag Berlin, Heidelberg N.Y., (1978).

[56.] J.E. Davies and E.A.D. White and J.D.C. Wood, J. Crystal Growth 27, (1974) 227.

[57.] G. Bartels and G. Passing, J. Crystal Growth 44, (1978) 363.

[58.] D.L. Kaiser, F. Holtzberg, M.F. Chisholm and T.K. Worthington, J. Crystal Growth 85, (1987) 593.

[59.] K.L. Keester, R.M. Housley and D.B. Marshall, J. Crystal Growth 91, (1988) 295.

[60.] M.J.V. Menken and A.A. Meňovský, J. Crystal Growth 91, (1988) 264.

[61 .] L.F. Schneemeyer, J W. Waszczak, T. Siegriest, R.B van Dover, L.W. Rupp, B. Batlogg, R.J. Cave and D.W. Murphy, Nature 328, (1987) 601.

[62.] T. Siegriest, L.F. Schneemeyer, J.V. Waszczak, N.P. Singh, D.L. Opila, B. Batlogg, L.W. Rupp and D.W. Murphy, Phys. Rev. 336, (1987) 8385.

[63.] H. Katayama-Yoshida, I. Okabe, T. Takahashi, T. Sasaki, T. Hirooka, T. Suzuki T. Ciszek and K. Deb, Jap. J. Appl. Phys. 26, (1987) L 2007.

[64.] F. Licci, H.J. Scheel and T. Besagni, Physica C 153-155, (1988) 431.

[65.] H. Lux, Z. Elektrochem. Angew. Phys. Chem. 45, (1939) 303.

[66.] H. Flood and T Förland, Acta Chem.Scand. 1, (1947) 592.

[67.] B.M. Wanklyn, J. Crystal Growth 65, (1983) 533.

[68.] K. Fischer and E. Sinn, Cryst. Res. Technol. 21, (1986) 711.

[69.] H.G. Tang, B.J. Garrard and B.M. Wanklyn, J. Crystal Growth 58, (1982) 111.

[70.] S.H. Smith and D. Elwell, J. Mater. Sci 2, (1967) 297.

[71.] M. Nevřiva and K. Fischer, J. Crystal Growth 89, (1988) 571.

[72.] H.J. Scheel and F. Licci, J. Crystal Growth 85, (1987) 607.

[73.] S. Shamoto, Sol. State. Comm. 66, (1988) 1151.

[74.] A. Ono, H. Nozaki and Y. Ishizawa, Jap. J. Appl. Phys. 27, (1988) L 340.

[75.] S. Takekawa and N. Iyi, Jap. J. Appl. Phys. 26, (1987) L 851.

[76.] K. Fischer, R. Hergt and D. Linzen, Cryst. Res. Technol. 23, (1988)...

[77.] H.J. Scheel and F. Licci, MRS Bulletin October, (1988) 56.

[78.] I. Barin, O. Knäcke and O. Kubaschewski, Thermochemical Properties of Inorganic Substances, Suppl., Springer, Berlin, (1977)

[79.] H.J. Scheel, Physica C 153-155, (1988) 44.

[80.] M. Nevřiva, P. Holba, S. Durčok, D. Zemanová, E. Pollert and A. Tříska, Proc. ICF'5, Bombay, (1989) in print.

[81.] M. Nevřiva, Second Experimental Workshop on High Temperature Superconductors, Trieste, 1989 (unpublished results).

SUBJECT INDEX

A-15 superconductors 311, 321, 322, 335
Accelerated rotation crucible technique (ARCT) 385
Aerosols spray route 147
Amorphous film 92, 96, 100
Anisotropy of HTSC 59
Antiferromagnetic correlations 170
Antiferromagnetic exchange
 coupling 129
 interaction 205
Antiferromagnetic insulating phase 3, 8, 10, 11, 12, 14, 19, 116, 175, 346, 361
 stability of 14, 19, 23
Antiferromagnetic ordering 12, 22, 163, 175, 204, 280, 318
Antiferromagnetism 75
Atomic core level excitation 115
Auger Cu-O spectrum 133
Auger decay 130
 line shape 132, 133
 parameter 130
 spectrum 134, 147
Axial channeling 240

Band structure of triple layered system 19-22
BCS theory 2, 10, 268, 287, 312 332
 strong coupling 288, 292 293
 weak coupling 1-25, 287 290
 weak coupling limit 8
BCS type pairs 270
Bardeen-Frohlich electron-phonon interaction 178
Bidimensional critical fluctuations 9, 22
 effect on ordering 22
Bidimensional dispersion relation 5
Bidimensional reciprocal space 6
Binary phase diagrams (Y-based cuprate) 376-378
Bi-O planes 137
BIS 113, 114, 134-135
Bismuth based cuprate 217
 high Tc phase 217, 220, 225, 227, 229, 235
 low Tc phase 217, 220, 225, 227, 235
Bloch wavefunction 12, 13, 22
Bose condensate 311, 314 324
Bose fluid 211
Brillouin zone of single Cu-O layer 7
BSCCO system 39, 40
 phase diagram 41
BSCO system 40-49
 ED patterns 44, 45, 47
Buckled layers 167

Carrier density 286
Carrier-free tracer 162
Cationic nonstoichiometry 114
Ceramic superconductors 113
Ceramic technology 64, 76
Charge carriers 27
 concentration 109
 sign of 27
Charge compensation mechanisms 116
Charge fluctuations 116
 mechanism 202
 state 143
 transfer gap 115, 116
Charge transfer-type insulator 339
Chemical disorder 100
 doping effects 145

Cluster model 124
Cooper (type) pairing 128, 332
Coordination number 145, 153
Core hole 342
Core edge satellites 114
Core level electron energy loss 130
Core electron excitation 115, 122
Core level photoemission 118, 123
Core level spectroscopies 32, 114, 115, 121, 339-342
Correlation effects 114
  energy 125
Coster-Kronig Auger decay 116
Coupled phonon-plasmon mode 201
Critical current (density) 59, 64, 75, 76, 123, 147, 239, 390
  intergranular 59, 75, 82
  field independent 82
  intragranular 59, 75, 76, 147
Critical state model 75, 81
Crystal-field effect 60, 167
Crystal growth of YBCO 383-391
  fundamental principles 383-385
  experimental arrangement 386
  solvents 387
  growth conditions 388-390
Crystal growth from homogeneous melt 386
  melt transport method 386
Crystalline anisotropy 75
Crystalline to amorphous transformation 243
Crystallization fields 380
Cu NMR 159
Cu-O chains 116, 148, 159, 167, 292, 299, 315, 317, 347, 361
Cu-O hybridization 125, 126, 129
Cu-O planes 9, 52, 87, 88, 125, 129, 130, 143, 145, 147, 148, 167, 186, 315, 317, 365, 366
Cu-O square planar cluster 361
  Excitation of 116
Curie-Weiss law 60

DC (static) susceptibility 12, 30, 305
d-d excitation model 361
Debye frequency (pulsation) 2, 272, 330
Debye temperature 189, 265, 327, 329
Debye-Waller factor 145, 156, 178, 179, 181, 185, 186, 189
Decay channels 115
Demagnetization current 80
  intragranular 80
Depth profile 240
Depth resolution 248-249
Direct excitation 116
Direct transfer integral 5, 6, 13, 23
Doniach-Sunjie tail 125
DTA 87-91
Dynamic valence exchange 323

EDX 100
EELS 113, 114, 135-138
Effective BCS interaction 205
Effective mass 290
Effective transfer integral 1
EFG 153, 156, 167
Elastic recoil detection (ERD) 240, 258
Elastic scattering data analysis (ELSA) 250
Electron condensate 264
Electronegativity 143
Electron-electron interaction 114
Electron-phonon coupling constant 290, 324, 331
Electron-phonon interaction 292-295
Electron superconductor 5, 32, 37, 318
Electronic density of states 2, 3, 17, 143, 341
  splitting of 15
Electronic properties 113, 114
Electronic stopping power 240, 241-244
  stopping cross-section 243
Elementary excitations 264, 272
EMR 372
Endothermic peaks 87, 89-91
EPMA 220, 221, 250, 372
EPR 372
Escape depth 117
EXAFS 143, 343
Exchange coupling constant 204

Extrinsic surface effects 118

f-electron wavefunction 167
Fermi-Dirac occupation factor 11, 16
Fermi energy (level) 3, 6, 9, 20, 22, 113, 125, 132, 135 339, 342
Fermi liquid 200
chargeless 200
noninteracting 200
Fermi surface nesting 11
Ferromagnetic coupling 128
Fluctuating valence elements 116
Fluctuation effects 60
Fluorination 145
Flux creep 60, 67, 73-75
Flux lattice melting 148
Flux penetration 75
Flux (fluxoid) pinning 75, 81, 148

Gamma resonance 188
Grain aligned samples 153, 170
Grain orientation 97

Hall coefficient 218
Hall effect 230, 284
Heisenberg antiferromagnet 200
High excitation spectroscopies 113, 138
High energy spectroscopies 113
High frequency phonons 181
High spin configuration 160
Hilbert transform 130
Hole doping 116
Holons 211, 212
charged 211
Holon hamiltonian 212
Hopping conductivity 353
integral 200
HRTEM 39, 42, 99, 107, 108, 147
Hubbard band 115, 116
hamiltonian 23, 201
repulsion 126, 130, 342
Hybridization integral 126
Hyperfine interaction 170
structure 148, 160, 175, 178

Impurity phases 114, 147
Incommensurate modulation 39, 116
Inductive coupled plasma 219
Insitu cleaved crystals 117
Insulator-metal transition 116
Interface reaction 250
Intergrain regions 117
Intergrainular phases 123
Intergrowth 99
Interplanar interactions 145
coupling 170
pairing 361
Intra-atomic coulomb repulsive energy 12
Intra-atomic screening 134
Intraplanar interactions 145
Inverse photoemission spectra 135
Ion beam analysis 239
Ion beam deposition 217
Ion beam induced damage 243
mechanism of 243
Ion bombardment 117
Ion implantation 163
IR active mode 181
IR spectroscopy 147, 284
Isomer-shift 151, 159, 160, 167, 170, 175
Isotope effect 10, 325-331, 333
reverse 330
Isotope exchange 330

Josephson junctions 147
Josephson-like behaviour 67

Kanamori model 361
Kapton absorber 253
Kinematic factor 246, 240
Kinematic mass resolution 246
Kramer-Kronig relation 136
k-space pairing 314

Lattice parameters 28, 32, 47
Layered perovskite 100
Lead doping process 92
flow chart 93
Libron oscillations 291
Ligand holes 115, 116
Linear chain cluster model 162

Line defects 143
Liquid phase epitaxy 218, 385
Local screening 115
Lorentz number 265, 284
Low energy excitations 273
Lower critical field 75

Magnetic dipolar interaction 75
Magnetic excitations 5
fluctuations 170
ordering 163, 166, 167
168, 170, 177, 178
phase diagram 163
phase transition 279
Magnetic order parameter 205
Magnetic susceptibility 33, 35,
36, 163
Magnetic structure of Nd 123
174
of Y 123 177
Magnetic unit cell 175
Magnetization 75-83
hysteresis, 75, 81
hysteresis width 60, 75
Magnons 279
Many-body excitations 340, 342
Mean-field approximation 3, 6,
9, 12, 23
Mean square displacement (MSD)
179, 184
Metal-insulator transition 27, 30
Metallic doping sites 145
Microscopic quantum chemical
resonances 311, 315
Microstructure 145
Microtwin planes 160
Modulated perovskite 47
structure 107, 147
Modulation wavelength 47-49
Mossbauer f-factor 159, 188,
189
Mossbauer line-shape 178
Mossbauer-site intensity 151
Mossbauer spectroscopy 143, 145
Mott-Hubbard charge fluctuations
114
Mott-Hubbard hamiltonian 200
Mott insulator 1, 2, 200, 201
SR 143, 163
temperature dependence 167
Muffin-tin (MT) potential 341,
352
Muffin-tin (MT) sphere 341
Multi-layer deposition 88, 109, 239
Multiple layered lattice structure
12
band structure 12-14
paramagnetic susceptibility
14-19
Multiple scattering (MS) theory
347
Multiple-site occupancy 148
Multiple-split satellites 116
Multiple structure 114
Multi-valence resonance condensati
(MVRC) model 311
MVRC state of superconductivity
314, 335

Near surface stoichiometry 240
Neel state 204
Neel temperature 145, 163, 166
168, 170, 279
Neutron diffraction
scattering 145, 147, 166
170, 173, 179,
inelastic 179
NMR 143
Non Fermi-liquid theory 199
Non Rutherford back scattering
250-251
NQR 143, 163
Nuclear reaction analysis 240
Nuclear resonance broadening
(NRB) 258

One dimensional incommensurate
modulation 47, 52
Ordering
chemical 100
structural 106
Organic superconductors 322
Orthorhombic distortion 49, 145
Orthorhombicity 153, 179
Ostwald-Miers region 384, 385
O-T transformation 162, 300,
346
Oxygen core hole 127
Oxygen core level spectra 113
Oxygen deficient perovskite
339, 340
Oxygen hole mechanism 339
Oxygen holes 113
Oxygen loading 145, 153
Oxygen sublattice 249
Oxygen stoichiometry (content)
28, 39, 113, 143, 148, 240

Oxygen vacancy ordering 147
297

Pairing through antiferromagnetic
correlations 178
Paramagnetic phase 82, 167
Particle induced X-ray emission
(PIXE) 240, 251-255
Pauli paramagnetic susceptibility,
19, 20, 22
Peierls instability 179
Percolation 63, 72, 73, 200
Peroxitons 127
Phonon induced coupling constant
2
Phonon induced superconductivity
5, 23
Phonon instabilities 145
Phonon mean free path 272,
286
Phonon-phonon interaction 267
Phonon relaxation mechanism
287
Photo induced IR 361
Plasmon 136
Point defects 143
Polarised X-ray photoabsorption
130, 339, 340, 342-352, 360,
362
Principal axis system (PAS) 156
Proton induced gamma ray emission
(PIGE) 240
Pseudobinary and ternary phase
diagrams (Y-based cuprate)
376-378
Pyramidal cluster 128

Quadrupole doublet 153, 156
interaction 167, 169
moment 160
splitting 151, 159, 160
Quadrupolar effects 167
Quantum spin liquid phase 200
Quasi-octahedral chain site 160
Quasi-tetrahedral coordination
160

Raman active mode 181, 184
Raman scattering 143, 147,
178, 179, 181-184
Raman spectra 284, 300
RBS channeling 250
Real space pairing 311, 314, 322
Reciprocal lattice layers 161
Resonance scattering of phonons
273, 276
Resonant decay 116
Rietveld analysis 47
RKKY interaction 168
Rutherford back scattering (RBS)
spectrometry 240
RVB 199, 209

Seebeck coefficient 30, 32
temperature variation 30, 34
Segregation reaction 49
SEM 64, 104, 105, 273
Semiconducting antiferromagnet
145
Sequential deposition 106, 218
Shake-up features 116
Shake-up satellites 115, 122,
124, 125
Shielding current 81
SIMS 240, 258
Singlet pairs 2
Site intensity ratio 151
Softening of lattice vibrations
184
Sol-gel route
Specific heat 143, 178, 270,
272, 278, 299
Spectral distortion 123
Spin fluctuations 178
Spin glass 169
Spin liquid 200, 211, 213
Spinwave spectrum 213
Spinodal decomposition 300
Spinon 200, 203, 211
Spin-orbit doublet 121
scattering 60
Spin pairing 163
Spin polaron pairing mechanism
361
Spin-spin interaction 332
Spin state 145
Stacking faults 116
Strong coupling regime 129
superconductor 113
theory 270
Structural stability of deposited
films 102
Substrate-film reaction 239
Substrate temperature 100
Superconducting gap 113, 286
BCS value 113

Superconducting volume fraction
30, 52
Superstructures 117
Surface exsitu 113
Surface decomposition 114
degradation 117, 121, 133
morphology 113
reactivity 113, 117
scrapping 119
Symmetry of hole states 339

Ternary oxides 43
Tetragonal-orthorhombic change
147
Textured films 113, 123
TGA 148, 151, 250
Thermal conductivity 178,
263
anisotropic 272
anomalous 264
hysteresis 298
near Tc 282-292, 296-306
non-phononic component 273
phononic 267, 270
temperature dependence 292
Thermal expansion 299
linear 299
Thermally stable phase 99
Thermopower 27, 32, 35, 37
Thin films 88, 117, 126, 128
220, 221, 239, 240
EB evaporation 88
laser ablation 117, 129,
217, 239
molecular beam evaporation
239
sputtering 88, 94, 117,
217, 239
Thermal evaporation 239
Thomas-Fermi potential 243
Tight bonding approximation
13
Time of exposure 113
Training effect 81
Transition broadening 59, 63,
64, 73, 75
Transition (superconducting)
curve 59, 75
foot in 59, 64, 67
Triple cell (Tri-layered)
perovskite 145, 314
Tunnelling system (TS) 276
localised 277
resonance scattering by 278
Tunnelling states 278, 279
Twinning 117, 148
Tc enhancement 148
Twinning planes 147, 148, 159
153
Twin plane site 162
Two dimensional Wagner plot 131
Two hole resonance 128
Type-II superconductors 148

Ultra-light element detector (ULED)
220
Ultraviolet photoemission 128
Upper critical field 59, 60, 62-75
anisotropy of 61

Valence band emission 116
Van Hove singularities 3, 17
splitting 23
logarithmic singularities 6,
19, 20, 22
Vibrational density of states
(VDOS) 143, 145, 179, 181, 186
189,
Vibrational modes 179

Weak-links 63, 64, 67, 72, 80, 371
metallic character 67
effect of Ag doping 67
Weak superexchange 75
WHH theory 60
Wiedmann-Franz-Lorentz rule
265, 283, 284
Wigner-Seitz (WS) sphere 341

XANES 32, 341-352
anguler resolved 346-352
XAS 113, 114, 339, 353, 354, 355,
358, 359, 361-365
XFA 219
XMPA 64
XPS 113, 114, 147, 339, 353-358,
366
XRD 33, 40, 43, 48, 49, 50, 64,
92, 95, 96, 97, 100-104, 219,
228
intensity 89, 91

Zaanen-Sawatzky-Allen (ZSA) phase
diagram 115